Panzerkampwagen IV

The Spielberger German Armor and Military Vehicle Series

Panzerkampwagen IV

and its Variants 1935-1945
Book 2

Walter J. Spielberger • Hilary L. Doyle • Thomas L. Jentz

Schiffer Military History
Atglen, PA

Photo Credits:
Federal Archives / Military Archives (49), B.W.B. (4), Byrden, David (4), Chamberlain, Peter (1), Doyle, Hilary L. (3), Fransen, Peter A. (6), Photo Archives / Doerfler (1), Photo Archives / Vogg (1), Jentz, Thomas L. (7), Jentz, David (6), Kliment, charles K. (1), Lalak, Zbigniew (1), Meistmann, Kurt (1), Mathieu, Axel (1), Muench, Karl-Heinz (9), Munin Verlag (1), National Archives (7), Schneider, Oberstlt. Wolfgang (2), Schmidt, F. / Knittel, Hartmut (2), Spielberger, Walter J. (107), Tank Museum, Bovington UK (14), Turina, Peter (1), Yust, Charles H. (5).

The quality of the photos, some of which have been reduced, is attributable to their age and the conditions under which they were taken.

The drawings made by Hilary Louis Doyle in 1:24 scale were essentially reduced to approximately 1:35 scale.

Book translation by Dr. Edward Force

Book Design by Ian Robertson and Stephanie Daugherty.

Library of Congress Control Number: 2010941948

Printed in China.
ISBN: 978-0-7643-3756-7

This book was originally published in German under the title
BEGLEITWAGEN Panzerkampfwagen IV by Motorbuch Verlag

We are interested in hearing from authors with book ideas on related topics.

Published by Schiffer Publishing Ltd.
4880 Lower Valley Road
Atglen, PA 19310
Phone: (610) 593-1777
FAX: (610) 593-2002
E-mail: Info@schifferbooks.com.
Visit our web site at: www.schifferbooks.com
Please write for a free catalog.
This book may be purchased from the publisher.
Please include $5.00 postage.
Try your bookstore first.

In Europe, Schiffer books are distributed by:
Bushwood Books
6 Marksbury Avenue
Kew Gardens
Surrey TW9 4JF, England
Phone: 44 (0) 20 8392-8585
FAX: 44 (0) 20 8392-9876
E-mail: Info@bushwoodbooks.co.uk.
Visit our website at: www.bushwoodbooks.co.uk
Free postage in the UK. Europe: air mail at cost.
Try your bookstore first.

Contents

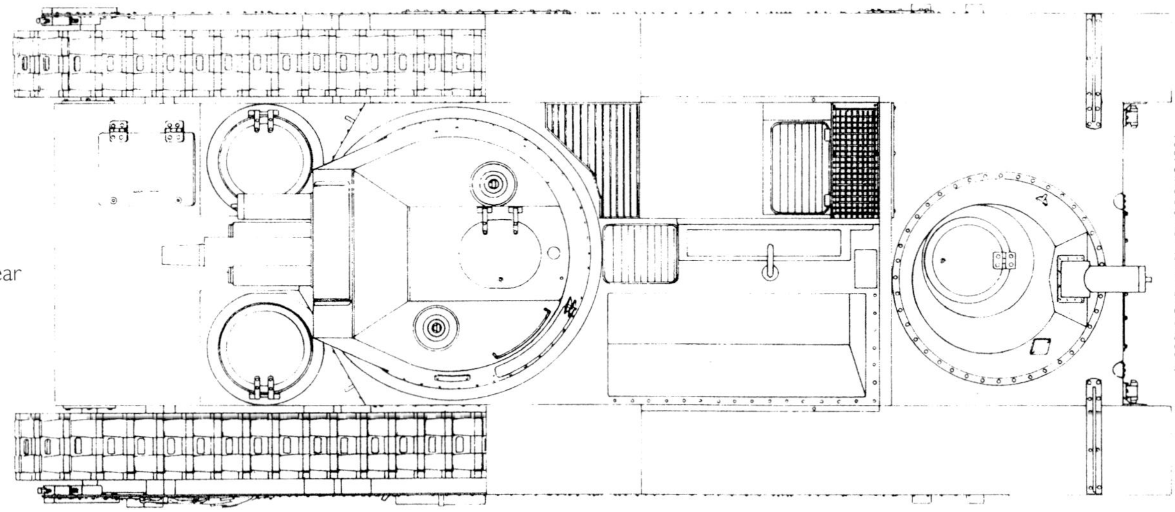

Krupp Large Tractor
with Modified Running Gear
(as of 1933).

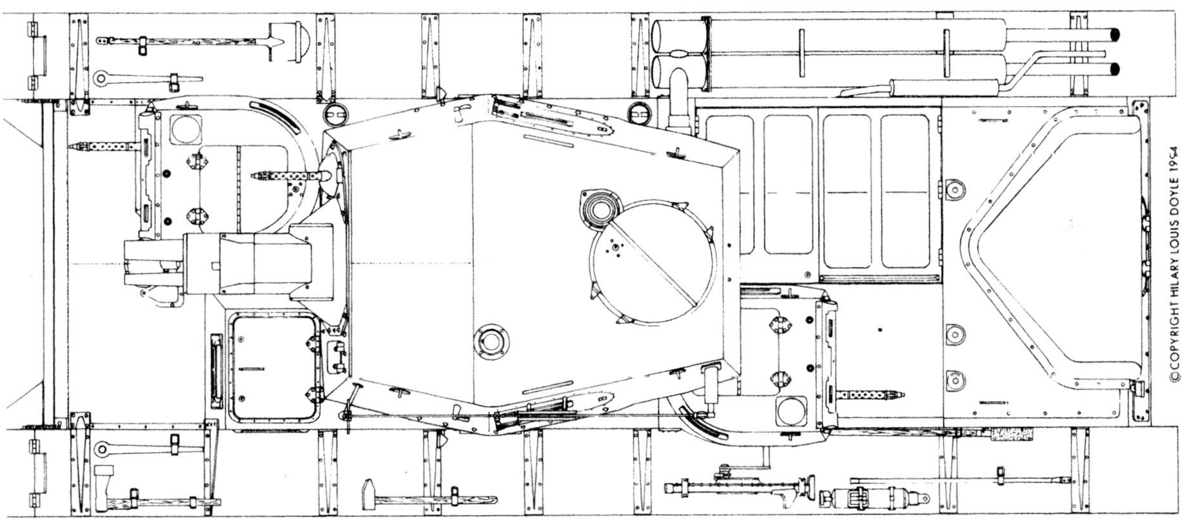

New Vehicle No. 3 to 5.

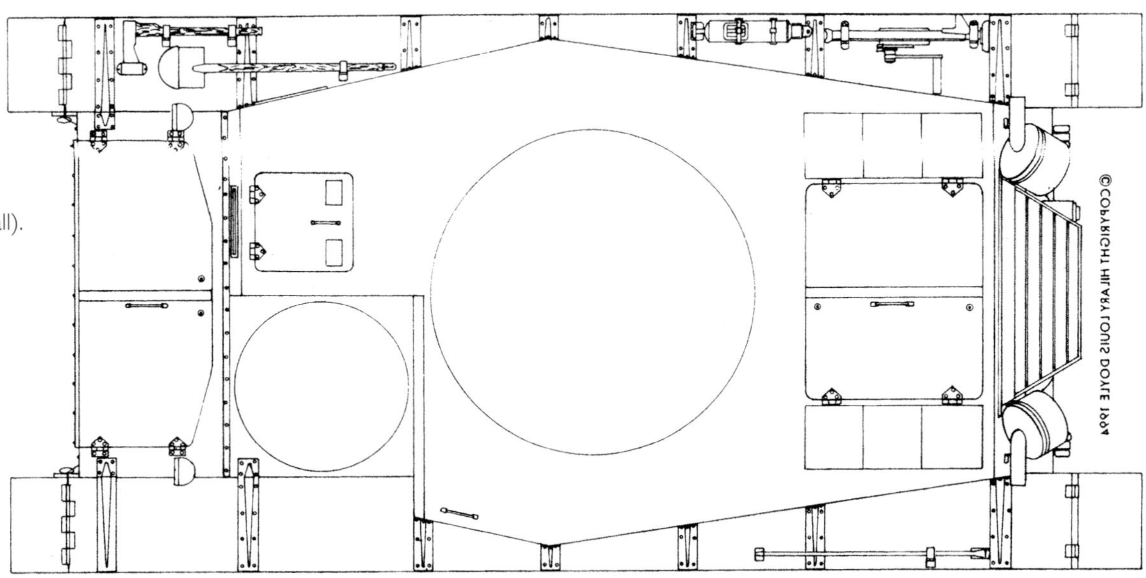

B.W. Test Chassis (Rheinmetall).

Panzerkampfwagen IV
(7.5 cm), (Vs.Kfz.622), Type A

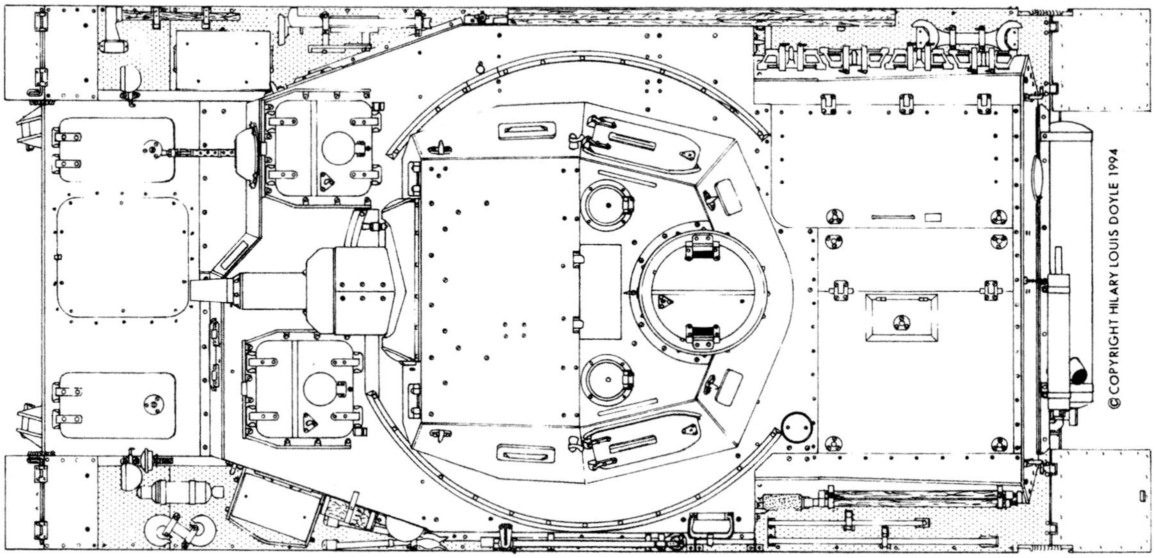

Panzerkampfwagen IV
(7.5 cm), (Vs.Kfz.622), Type B

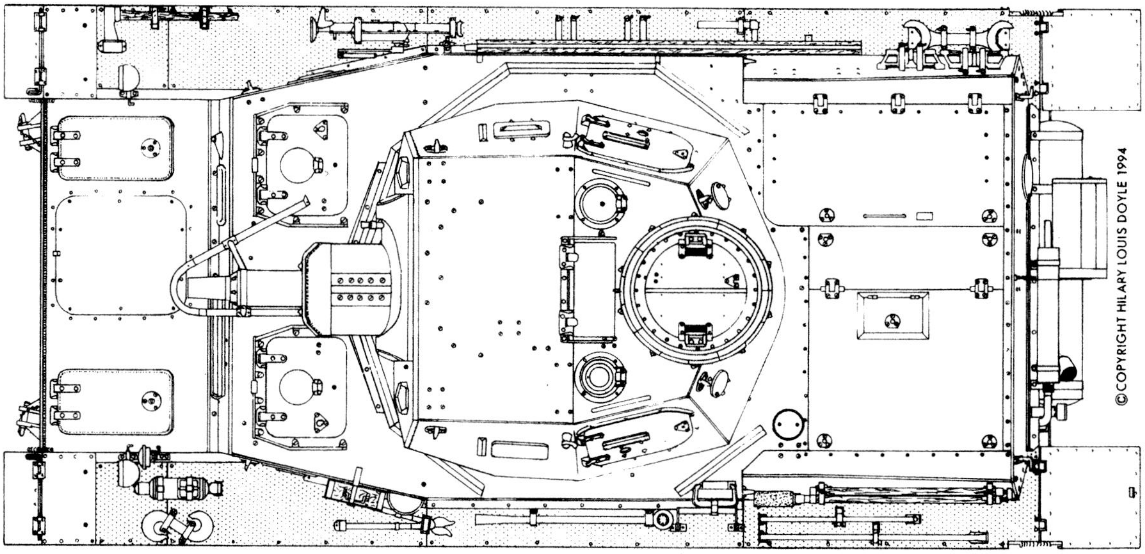

Panzerkampfwagen IV
(7.5 cm), (Vs.Kfz.622), Type C

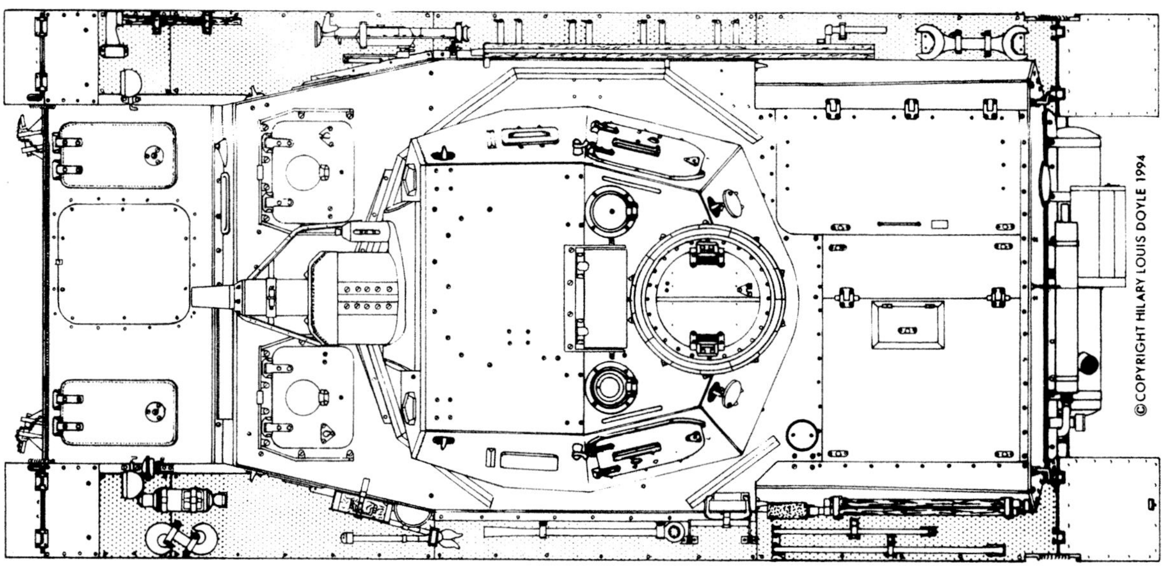

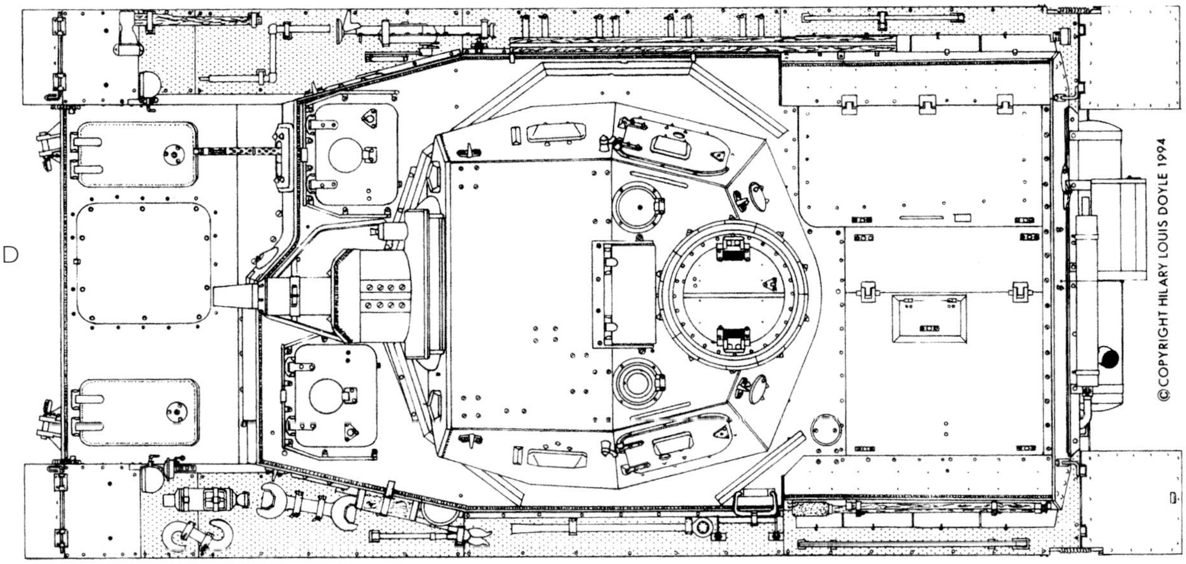

Panzerkampfwagen IV
(7.5 cm), (Sd.Kfz.161), Type D

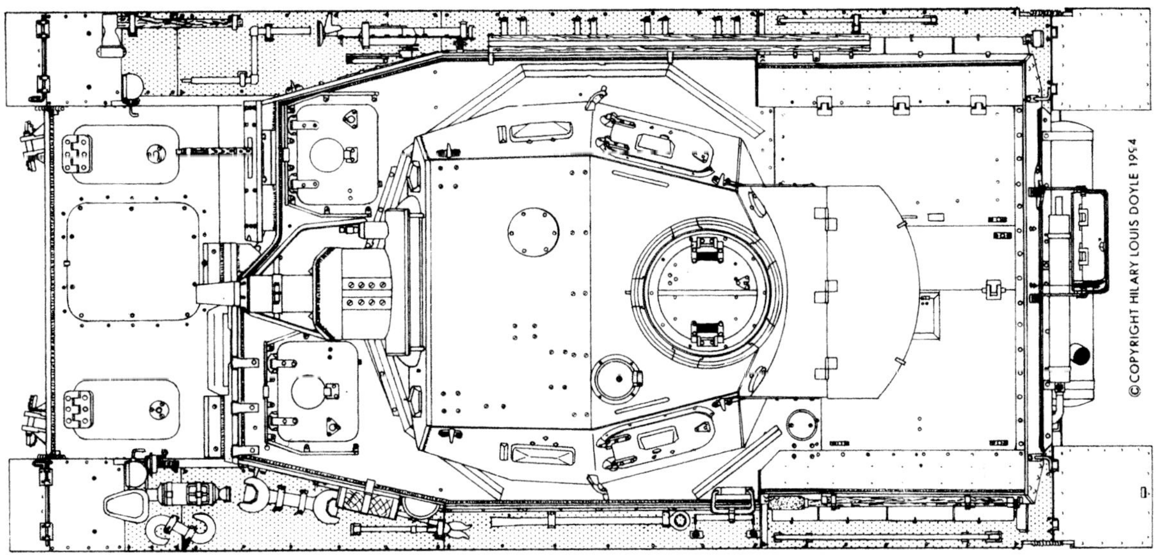

Panzerkampfwagen IV
(7.5 cm), (Sd.Kfz.161), Type E

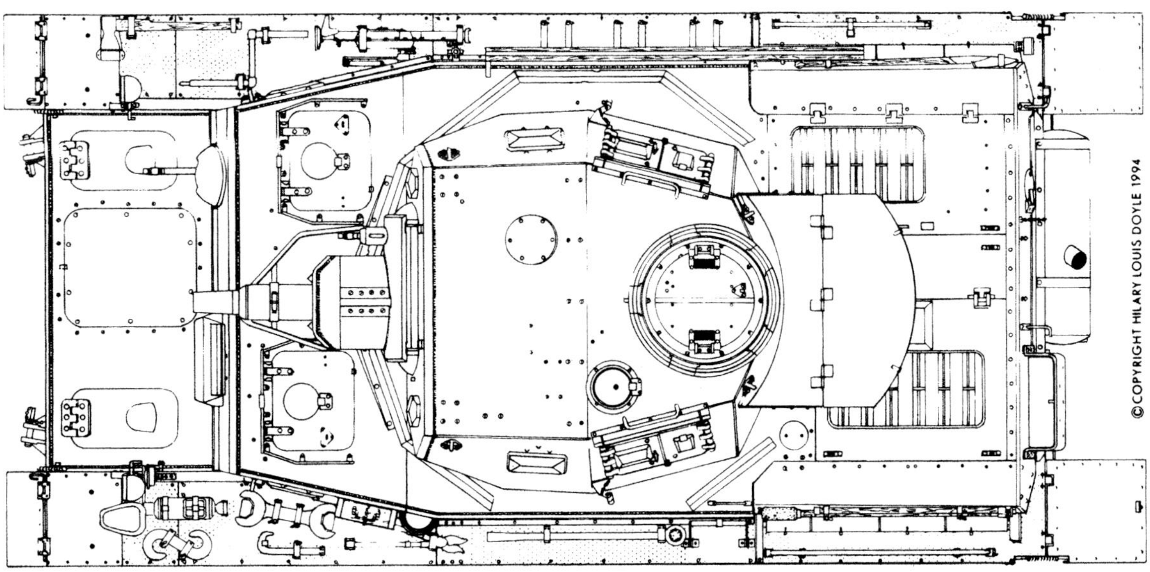

Panzerkampfwagen IV
(7.5 cm), (Sd.Kfz.161), Type F

Panzerkampfwagen IV, Type F, with Front Armor.

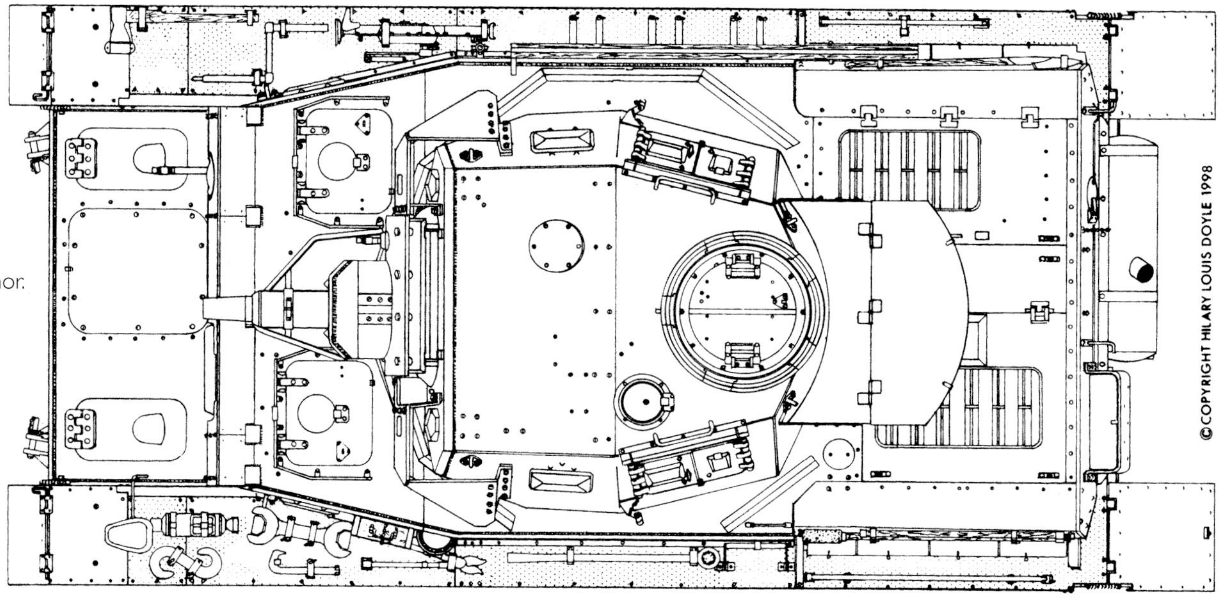

Panzerkampfwagen IV, Type G (earlier than Type F2)

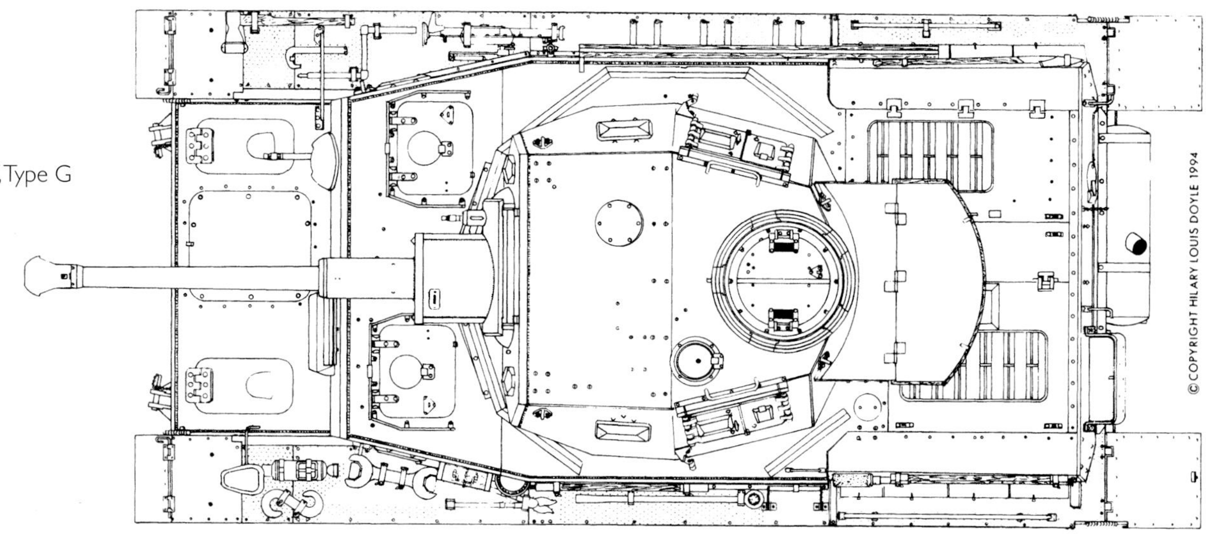

Panzerkampfwagen IV (7.5 cm), (Sd.Kfz.161), Type G, With welded additional armor, produced April 1943

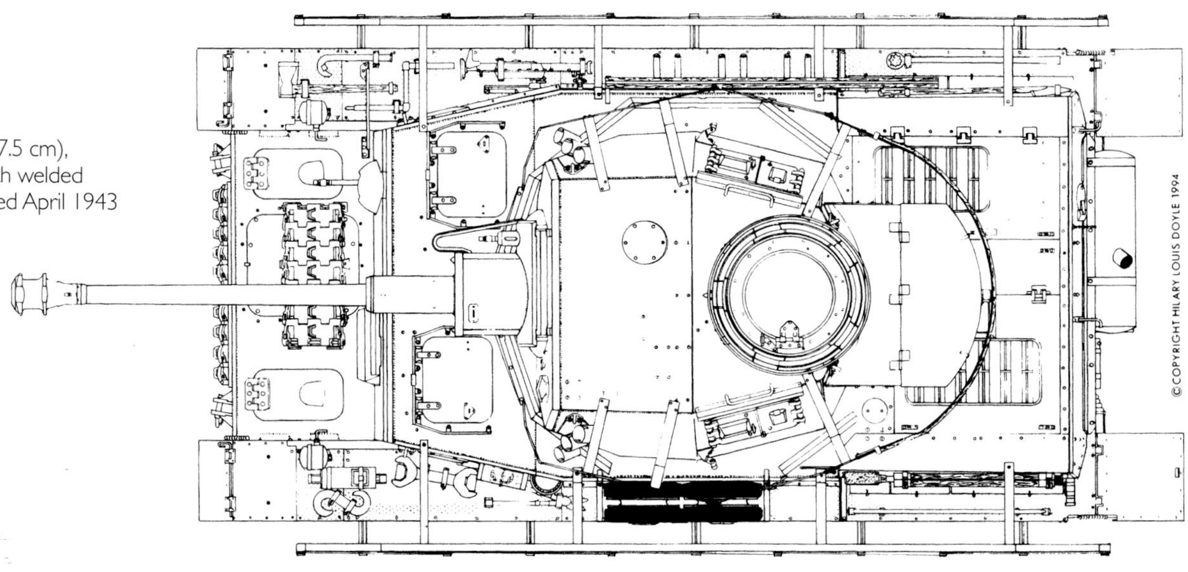

Panzerkampfwagen IV (7.5 cm)
(Sd.Kfz.161/2), Type H,
Produced as of December 1943.

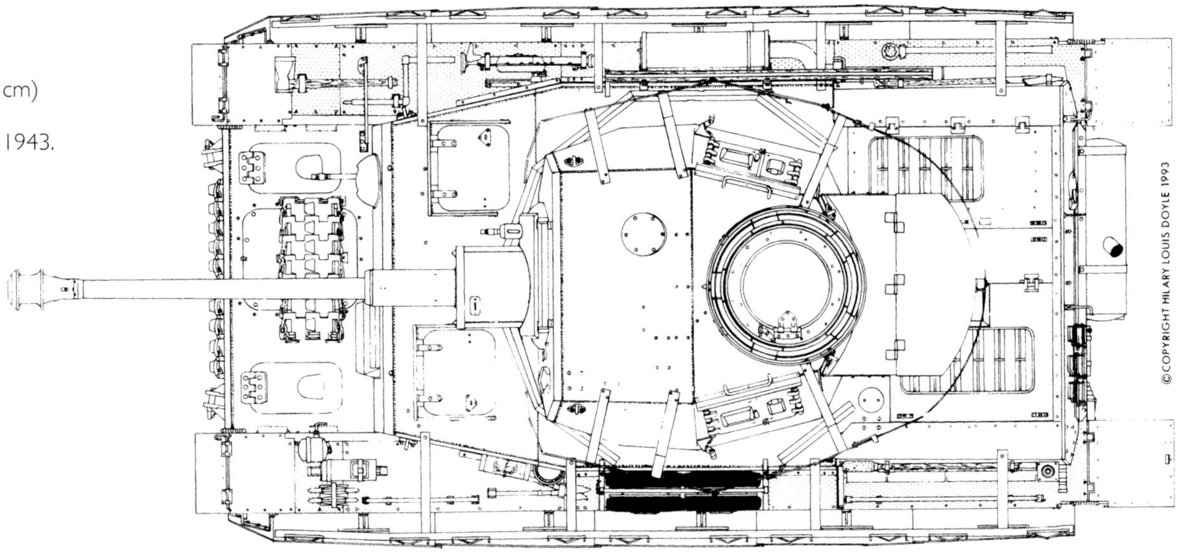

Panzerkampfwagen IV (7.5 cm)
(Sd.Kfz.161/2), Type J, As produced
at the Nibelungen Works in 1945.

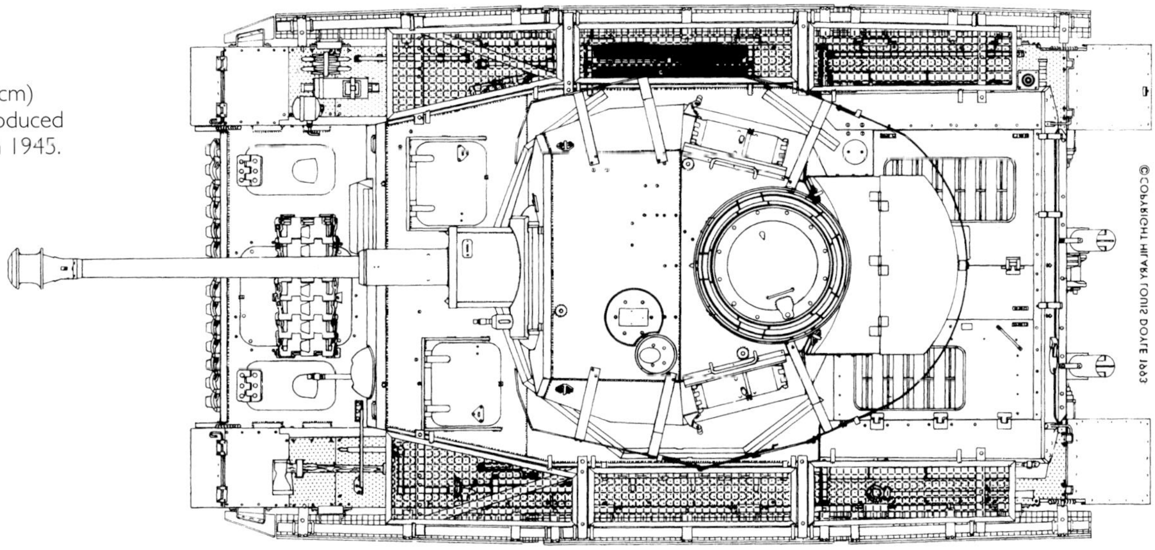

Pz.Kpfw.IV Type J, rebuilt as an armored
reconnaissance vehicle.

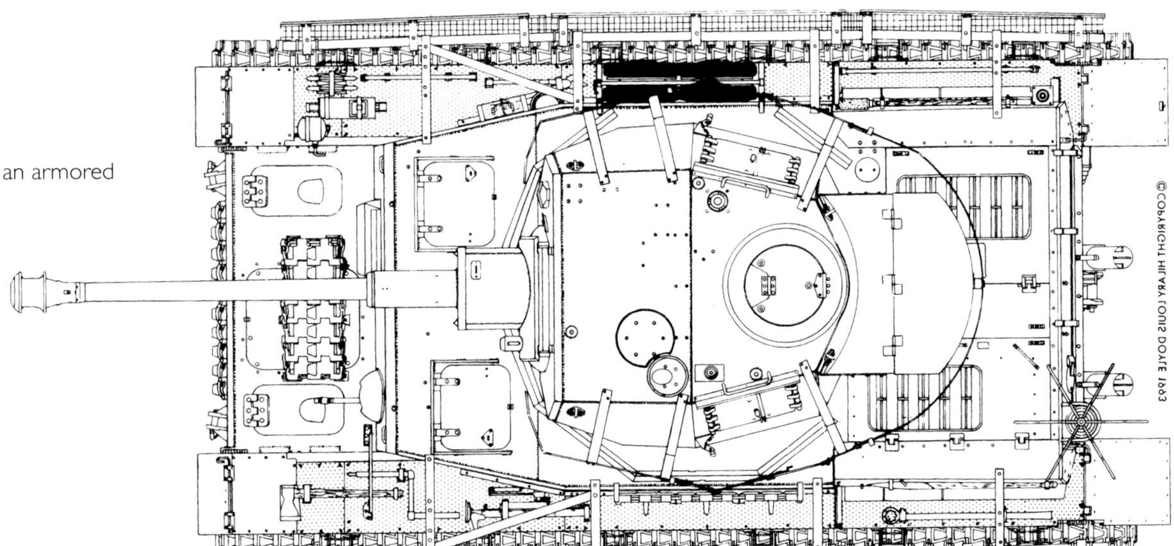

Introduction

Twenty years of intensive research, particularly by Thomas L. Jentz, have resulted in the completed history of the development of the B.W. (Escort Vehicle) series now before you. The exact chronological line of this development, plus the discovery of many hitherto unknown and development-indicative test vehicles, are the results of long and very intensive research in company records and files that have survived to this day. The following sources have proved to be of particular importance:

- Fried.Krupp A.G., files on Artillery Design (AK),
- Fried.Krupp A.G., business reports,
- Inspector-General of the Armored Troops, files,
- Service Instructions of the German Army, D653 Series,
- Collected Files of the Nürnberg Trials.

These original sources allowed an intimate insight into the fascinating developmental processes from 1925 to 1945, beginning with the creation of the large tractors (originally known as Armeewagen 20) and on to the new vehicle and the final tests in 1944 meant to house the 7.5 cm Kw.K. 42 L/70 of the Panther in the Panzerkampfwagen IV.

The way through this developmental process is indicated in the following account of the development of the B.W. series and described in detail in the individual chapters.

The key factor in the portrayal of this development, which has been overlooked to date, is the use of group numbers by the Army Weapons Office for every vehicle to be developed anew. These group numbers give the key to the exact order of development frequencies and the connections between the series vehicles and the numerous test vehicles that were suggested. The group numbers were especially important for identifying the actual processes in the development of the VK 2001 series and the Panzerkampfwagen III/IV.

Originally the group numbers were applied to the designation of all drawings of complete vehicles, components or parts and repeated in casting or engraving on the applicable parts. To rule out errors, the relevant group numbers were repeated in sets of drawings as well as in descriptions. The group number precisely identifies the vehicle for which a part was originally intended and later improved. All group numbers for armored vehicles began with the prefix 021.

The relevant group numbers that are of interest for the vehicles in this book run as follows:

Series	Designation	Group Number
B.W.I & II Kp.	Test series	021Gr8400
1./B.W.	Pz.Kpfw.IV Type A	021Gr14700
2./B.W.	Pz.Kpfw.IV Type B	021Gr3300
3./B.W.	Pz.Kpfw.IV Type C	021Gr8700
4.& 5./B.W.	Pz.Kpfw.IV Type D	021Gr8800
6./B.W.	Pz.Kpfw.IV Type E	021Gr33800
6./B.W. with box running gear		021Gr1200
VK 20.01(K)		021Gr300
VK 20.01(D)		021Gr10800
7./B.W.	Pz.Kpfw.IV Type F	021Gr100
8./B.W.	Pz.Kpfw.IV Type G	021Gr48100
9./B.W.	Pz.Kpfw.IV Type H	021Gr48400
10./B.W.	Pz.Kpfw.IV Type J	021Gr82400

The text is noteworthy, in that it does not bother with superficial designations and names. But when the basis for key decisions was to be made, important reasons had to be determined or details of original discussions turned up in the original documents, they were added to the text. These additions afforded certainty as to what in many cases was still unknown until that time. They not only clear up "why," but also questions of "when" and "how" certain decisions had to be made.

One more note as to the language in this book. One can see clearly than in the past 70 years means of expression have changed considerably. If one adds to this the peculiarities or official terminology and the language used between agencies and industries, one often feels lost in this excess of communication. The reader is asked to be understanding about this. The precise, unchanged contents of original documents was included in this book, so as not to let a distorted portrayal result. The writing style of the original documents was kept in order to portray opinions and intentions of the time.

It also proved to be necessary to portray the organization and the personalities that were responsible for development and production clearly to the reader.

They always turn up when developmental decisions are examined more closely. While many participants in important conferences certainly contributed their input to the building program, it must be determined that Senior Engineer Erich Woelfert of the Fried. Krupp AG was directly responsible for the design and development of the B.W. series.

An original document of 1943-44 indicates as follows what Woelfert's services in the development of armored vehicles for the German Army amounted to:

Development of Tanks by Chief Engineer Erich Woelfert

Mr. Senior Engineer Woelfert has been active in German tank building since 1928. At first the Large Tractor was designed and built experimentally by Mr. Woelfert. (Tested and then built in Russia as T 28).

Under his direction the La.S. (Pz.Kpfw. I Types A and B) were built. The latter was displayed to the public in 1935 as the first modern German production tank. From its testing in the Spanish Civil War, much experience was gained, and in the present war too, the La.S. has given good service. One can say that the La.S. has formed an essential basis for German tank construction, since the M.A.N., Daimler-Benz and Henschel firms were able to base their successor tanks on it.

Above all, the design of the B.W. (Pz.Kpfw.IV), which contributed essentially to the lightning war in Poland, in the West and in the Balkans, comes from Mr. Woelfert. Without the B.W. the great encircling battles would not have been possible. Above all, it is thanks to the B.W. that the German Army has penetrated to the Atlantic coast and far into the East. Finally, the B.W., still giving the best service to this day, formed the basis of the ***Hummel, Hornisse****, various Flak vehicles and other special versions.*

Besides the noted production vehicles, numerous test vehicles were built under the direction of Mr. Woelfert, including, among others, self-propelled guns that proved themselves splendidly in the East. From the test designs, the design of the 8.8 tank destroyer was taken over for series production by the MIAG.

From the above it may be concluded that Mr. Woelfert has done the basic and decisive pioneer work in the area of tank construction.

The authors owe their thanks to Messrs. Loos and Meyer of the Federal Archives/Military Archives in Freiburg for years of support in locating and studying relevant files. The same applies to Mr. Bielges, who was formerly available to us again and again at the Photo Archives in Koblenz. Our thanks also go to the staff of the Krupp Archives for providing copies of business reports that were still lacking in our materials. The support of our good friends, Colonel Aubry of the French Tank Museum in Saumur, Lieutenant Colonel Grundies of the Training Collection of the Armored Combat Troops at the Armored Troop School in Munster, Captain Kosinski of the Defense Technical Study Collection in Koblenz, and Dave Fletcher of the British Tank Museum, is recognized with thanks. Their readiness to guarantee us extensive access to tanks of the former German Wehrmacht for research and measurement contributed much to making this book so complete.

Walter J. Spielberger
Hilary L. Doyle
Thomas L. Jentz

1. LARGE TRACTOR (Gr.Tr.)

Formerly named Army Vehicle 20 (A.W.2) by Wa Prw 4

1.1 DEVELOPMENT

1.1.1. Design Requirements

The oldest documents still to be found on the development of the Large Tractors (originally known as "**Army Vehicle 20**") are found in a letter dated April 12, 1926, and sent to Dr. Ritter of the Fried. Krupp AG. It contains an addendum dated April 10, 1926, from Department 4 of the Inspection for Weapons and Equipment. This letter with addendum gives information on the first activities regarding the design and creation, and provides hitherto scarcely known details as to basic decisions.

The Jwg. Is at this time building a wooden model of an Army Vehicle 20 through which the questions of the measurements and inner arrangements shall be defined.

A wooden model of a gun shall also be installed in this model; its original version is to correspond to the enclosed requirements.

The Department requests that as to the enclosed requirements the drawings for the wooden model of the gun be determined in general terms and sent here as soon as possible.

The Department requests that it receive these drawings, through which not all the details of the gun design have to be determined, in 4 to 6 weeks. The building of the wooden model will be done by the Jwg.

The resulting costs should be forwarded in special calculations and need not to be added to the sums made available for the Design Bureau.

The design is to be handled as especially confidential within the narrow parameters of the workers.

J.W.G.4/1 **Berlin, April 10 1926**
Bd.Nr.71/26 Jwg.4/1 secret
Design Requirements for a 7.5 cm Gun in Army Vehicle 20

A. General.

1) *Army Vehicle wood model will represent a medium type of 15-ton weight. The layout surface will have a standing length of ca. 6 m and a standing width of ca. 2.40 m.*
2) *The vehicle shall, besides the M.G. armament, receive a 7.5 cm gun in a turning turret.*
3) *The turret must be kept as low as possible, its greatest height may in no case be over 1.80 m, measured from the ground surface.*

On the vehicle forms considered at this time, the turret will project ca. 60 cm over the outlines of the vehicle. The diameter of the turret is to be made as small as possible; at this time a diameter of 1.20 m is regarded as necessary. The turret shall have an upper cupola pivoting sideways to 360 degrees, with the gun barrel projecting through a port in it; the gun barrel should turn along with the cupola.

Suggestions for the form of the turret including the turning upper cupola are requested. As for the form, good mobility and sustained sealing of the gun ports and a form to be produced successfully are to be considered.

The armor thickness of the turret walls projecting from the vehicle outline 10 cm, the turret roof 6 cm.

B. Gun

1) *Barrel: 7.5 cm caliber*
   Barrel structure based on the barrel of the future mountain gun, performance: Vo = 400 m/sec, G = 6.3 kg, allowable maximum gas pressure – 3000 Atm., at + 10 degrees powder temperature, barrel length: as short as possible, but long enough so that the gas pressure does not exceed 600 Atm. Occurrence of a large fire appearance as the result of a short barrel is of no concern.
   Breech semi-automatic.
2) *Barrel recoil brake with pneumatic recuperator. : Recoil length with consideration of the available space and the condition that the shell casings are still ejected within the turret; on the other hand, barrel recoil as long as possible, to avoid excessively great shaking of the turret when firing. The components of the barrel recoil brake that extend over the armored turret must have special protection against direct hits and small splinters (5 mm armor).*
   A deflector to the left of the barrel recoil.

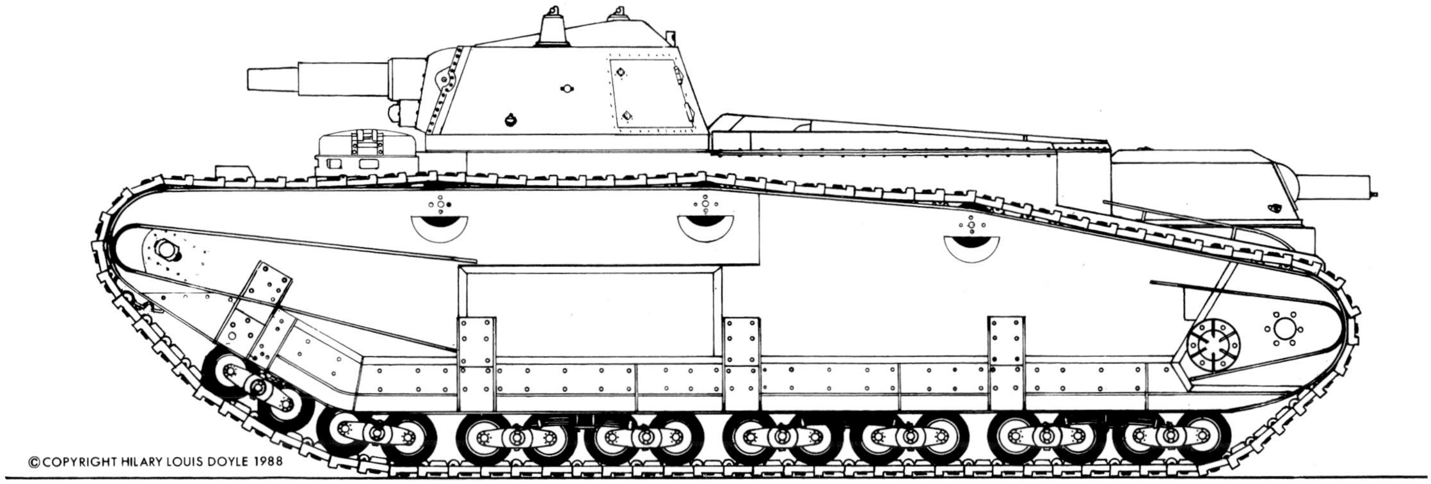

Daimler-Benz Large Tractor No. 41 and 42.

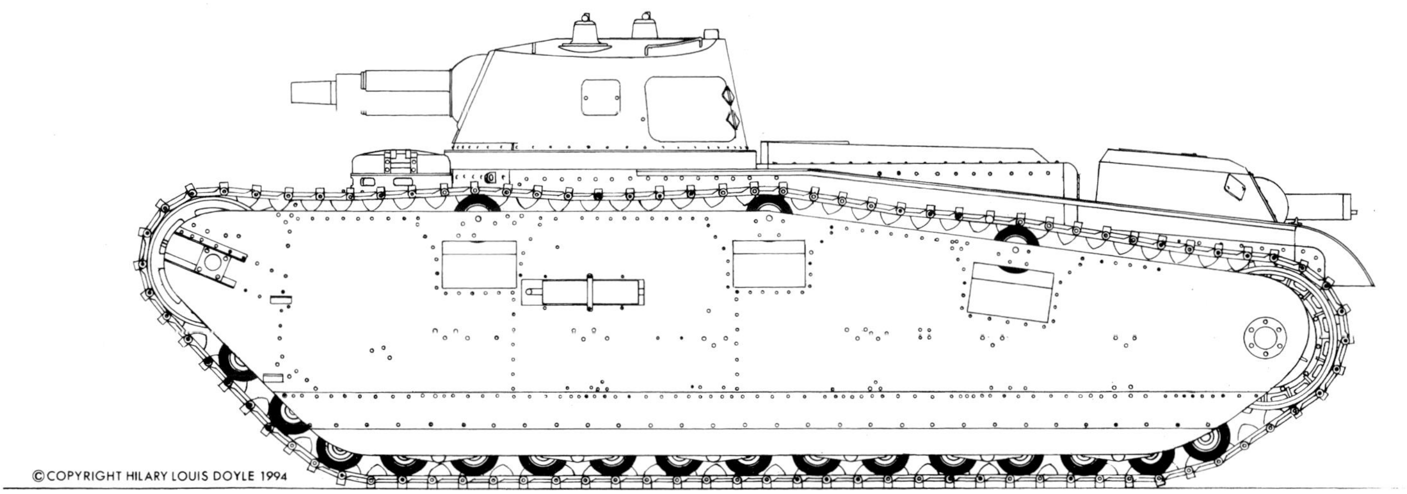

Krupp Large Tractor No. 43 and 44, original running gear.

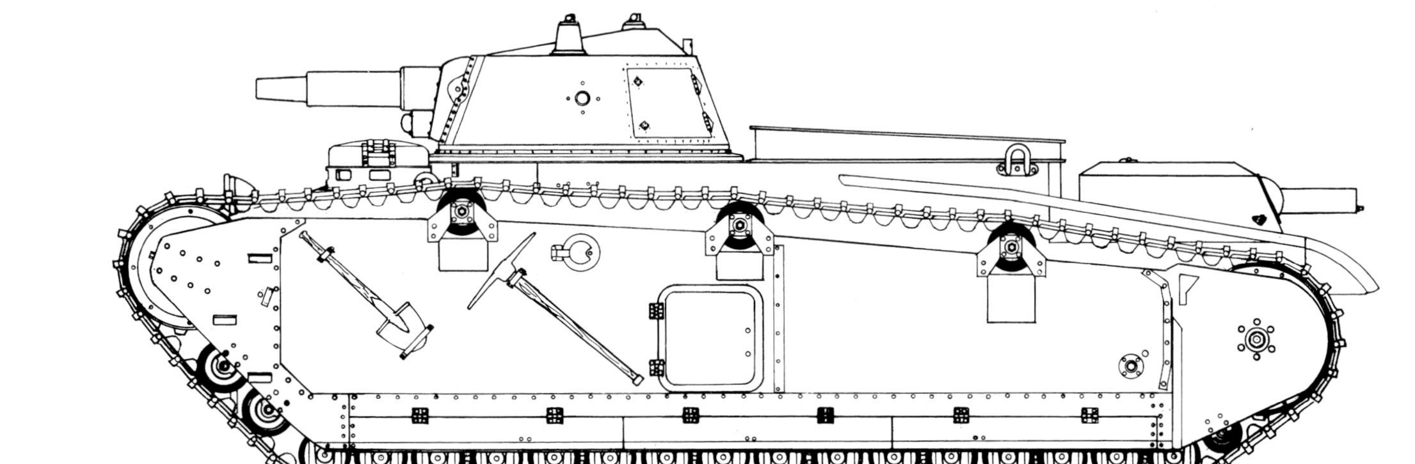

Rheinmetall Large Tractor, No. 45 and 46, original running gear.

DEVELOPMENT OF THE B.W. SERIES

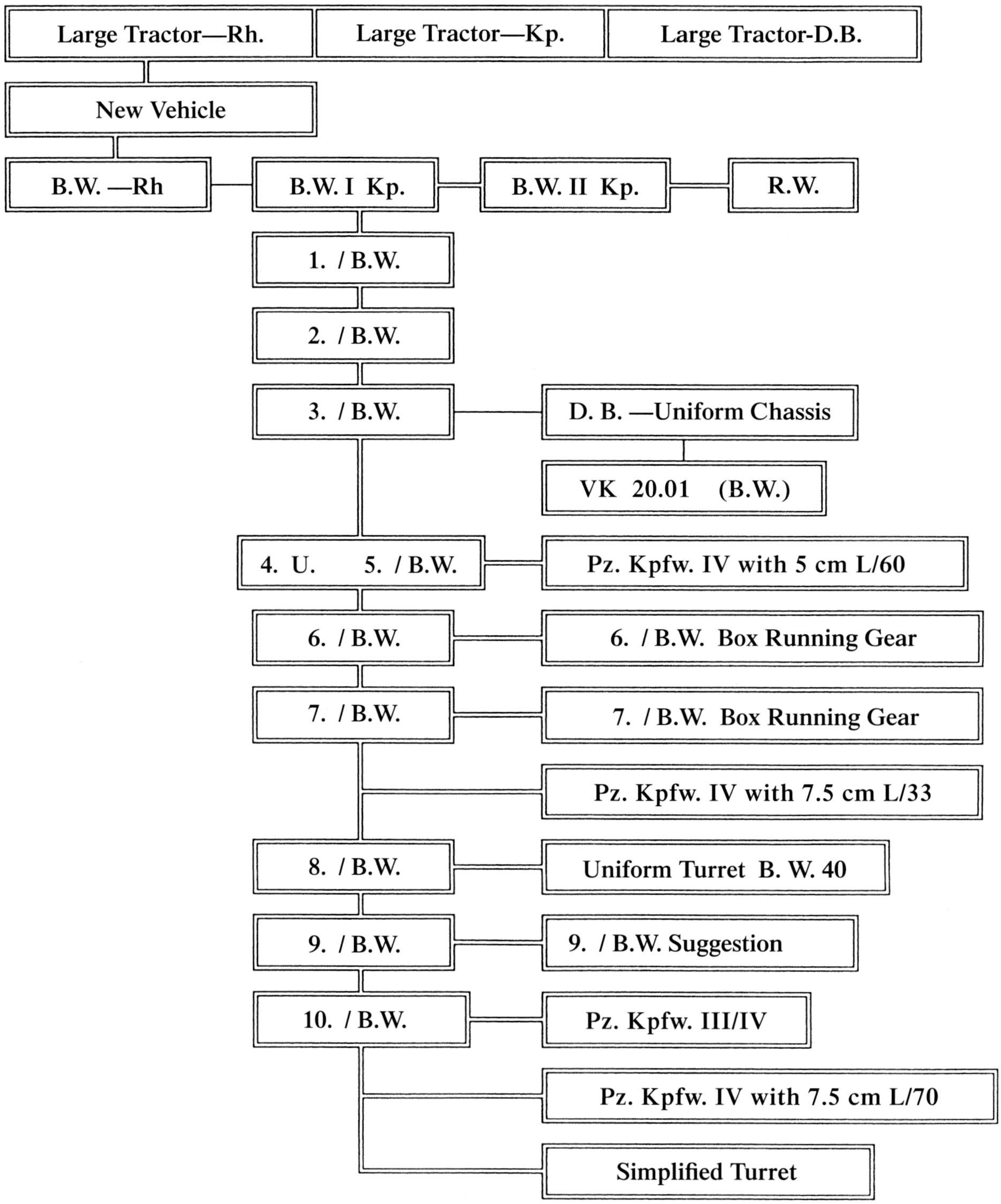

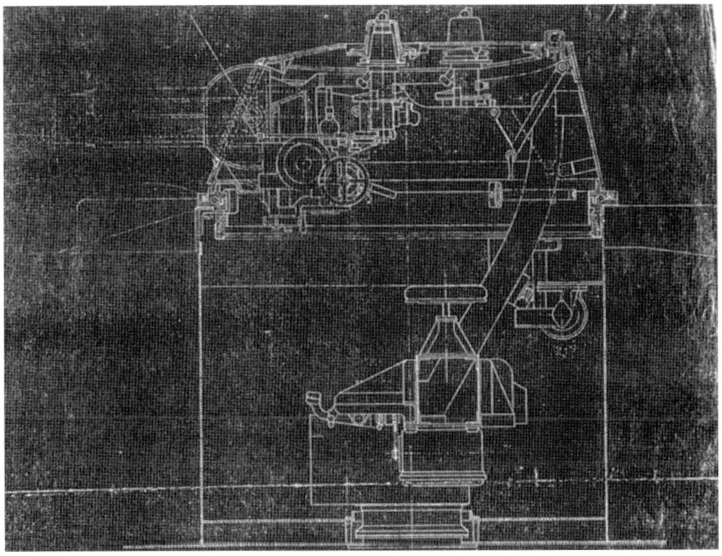

Krupp Turret, internal equipment, details.

3) *Ammunition: Shell ammunition. Types of shells are to be determined later. If need be, the separated ammunition of the future mountain gun and field guns with correspondingly smaller charges can also be used.*
4) *Rate of fire: 15 rounds per minute.*
5) *Firing height: As high as possible over the entrance step of the turret, so that the upper part of the armored cupola through which the barrel is to pass is kept as low and thus as light as possible.*
6) *Installation of a barrel-blowing apparatus: compressed air is made use of by the vehicle designer and conducted to the entry step. The gun designer arranges for further conduction of the compressed air from there to the barrel and for attachment of the switching-off apparatus. Automatic switching-off is desired. Sucked-in air is not usable.*
7) *It is desired that one or two light Dreyse machine guns of 7.9 mm caliber in fixed mounts be added to the gun barrel. These machine guns and the tank gun shall be able to be used by choice. Removal when they become unusable must be possible. If two guns can be installed, their parallel placement, coupled aiming and firing, but separate loading must be made possible.*
8) *Mounting is to be left to the gun manufacturers. They assure that the other barrel part projects into the interior of the turret as little as possible. It is to be assured that the mounting of the barrel, which must allow a higher and lower swinging within the required elevation field, must be kept as small as possible. This will require that the turning point of the barrel is located as far as possible in the turret wall.*
9) *Elevation field: for ground angles of -15 to +40 degrees, for shot-table barrel elevation from 0 to +10 degrees. Traverse field: endless, to both left and right.*
10) *Aiming machinery: For elevating and traversing machines, very easy access is required. For any machine operation, compressed air or electric power can be made available. Conduction of the compressed air or electric power to the entry step is regulated by the vehicle designer, from there on, regulation of the further conduction and type of operation by the gun designer. For the use of machine drives, hand operation is to be planned as a reserve.*
11) *Sealing of the mount: Gas-sealing of the mount is intended; the seal must be guaranteed at every elevation level of the barrel.*
12) *Visor arrangement*
    a) *For direct aiming, visibility parallel to the gun barrel with appropriately adjustable angles of elevation. Direct transmission of the vision changes with barrel movement (no indicator aiming operation). Visor to be set up optically so that the armor protection does not have a gap.*
    b) *Seeking visor over the turret.*
    c) *Elimination of angle errors for firing straight downward from a vehicle setting at an angle up to 33 degrees.*
    The visor for direct aiming must also be usable for the machine gun, if firing is to be done with the linked machine gun instead of the tank gun. No special visor is needed for the second machine gun. Parallel positioning of the second machine gun with the first one must be attained in order to be usable together against the same target.
13) *Two men (one loading and one aiming gunner) are foreseen as the crew for the gun, including the linked machine gun. Ammunition supply for the gun: 100 rounds, storage of the ammunition is determined by the vehicle designer in cooperation with Jwg.4. Machine-gun ammunition = 6000 rounds for all machine guns. The vehicle designer determines the storage.*
14) *Equalization of vibration while driving is developed by the vehicle designer. Final goal: it must thus be possible to make directed fire while in motion.*
15) *Lashing the barrel in driving position outside of battle is provided for; during battle driving without lashing must be possible.*
16) *Aiming and loading gunners' seats are to be provided. Arrangement of the seats perhaps in belts, if these appear to be purposeful.*
17) *Total weight of the gun including mounting and vehicle armor as light as possible, maximum weight 600 kilograms.*

Design conditions determined by the details and requirements for the chassis must be worked out in the same time period. Unfortunately, until now no documents could be found. But information from the original documents from the Daimler-Benz and Rheinmetall firms show that the two firms were carrying on preliminary developmental work in 1926.

In the *Developmental History of the Gr.Tr.* published by Krupp in 1935, the following chronological listing of the initial activities appears:

1926: **Carrying out Preliminary Designs**
03/01/1927: **Preliminary discussion and drawing up a contract.**
03/25/1927: **Offer for Design and Production. Wood model shown in Krawa (General Ludwig).**
03/30/1927: **Contract for design and production of two vehicles. Then design and Working out.**
04/28/1927: **Discussion of the main designs.**
Aug. 1927: **Beginning of preliminary discussions and production.**

By contract of March 26, 1927, the Daimler-Benz AG also received the green light to design and build two of these vehicles. It can be assumed that Rheinmetall received, at the same time, the contract to design and build two of these vehicles, and also for four turrets for the Rheinmetall and Daimler-Benz vehicles.

1.1.2 Krupp Design Dates

On March 19, 1927, the Krupp A. K. Department in Essen turned over the design data which it had developed to the Reich Defense Ministry, Army Weapons Department, Test Section 4 I.

7.5 cm Gun Turret A.W.20 Design Data

Barrel length	*L/20 = 1500 mm*
	Vo = 400 m/sec
	G = 6.3 kg
Turret weight with gun	*1660 kg*
Gun weight	*570 kg*
Barrel weight with breech	*230 kg (270)*
Weight of recoil mechanisms	*250 kg (285)*
Recoil distance, normal	*500 mm*
Maximum elevation	*60 º*
Lowering over center line	*12 º*
Lowering over driver's stroboscope	*12 º*
Lowering to the side	*12 º*
Lowering to the rear	*6 º*
M.G. swinging range, elevation	*+/- 5 º*
M.G. swinging range, lateral	*0 º*
Side speed	
By hand	*3 º/sec*
By electric drive	*4.8 & 12 º/sec*
Elevation speed by hand	*2 º (3.3)*
Armor thickness, sides	*18 mm (15)*
Armor thickness, top	*18 mm (8)*
Inclination of sidewalls	*82 º (78)*
Outside diameter of turning circle	
Below	*1500 mm (1550)*
Above	*1344 mm (1350)*
Height of upper armor edge	
Above ground	*2368 mm (2330)*
Above vehicle	*618 mm (587)*
Highest point of armor edge	*2463 mm (2430)*
Height of entry step above ground	*500 mm*
Firing height of gun	
Above ground	*2115 mm*
Above vehicle top	*365 mm*
Gun ammunition in turret area	*100 rounds*
MG ammunition	*3000 rounds*

The data in parentheses () are handwritten additions that define the first deviations from the original design data.

1.1.3 Final Design Conditions

On June 10, 1927, the Army Weapons Department, Test Center 4, sent a copy of the *Final Design Conditions for the Gun Turret of A.W. 20* to Krupp. Included in its full extent, this document shows in astonishing detail the progress that the armored turret development had made by 1927. Many of the basic requirements, such as exhaust of the propellant-charge gases, sufficient sight for the crew, and the prevention of entrance by poison gas were problems that still occupy the developmental engineer today.

A. General

As to making equipment uniform in view of mass production, interchange principles and easy maintenance possibility, limitation of metals to be saved, use of metric threading and consideration of the Army equipment and D.I.N. norms, those apply in the design conditions of the mountain gun.

B. Turret

1) *For general arrangement and situation of the turret, the installation sketches provided under No. 80/27 Wa.Prw.6 geh. Kdos. are decisive. [geh.Kdos = secret command matter]*
2) *Armament: 1 – 7.5 cm tank gun and 1 M.G.*
3) *The dimensions of the turret are to be kept as low as possible under the viewpoint that with a lowering of -12 degrees, firing over the front end of the tracks can be done. For firing backwards over the rear of the vehicle, a lowering possibility of 6 degrees is sufficient.*
4) *The light height: Entry step—turret roof should measure 1800 mm. The turret diameter is to be made as small as possible, maximum 1500 mm.*
5) *The turret shall have an upper cupola that traverses without end sideways, and through which the gun barrel passes in a port. The gun shall be turnable with the cupola. To achieve easy turning, the turning track is to be made as high as possible.*
6) *In shaping the turret, angled surfaces, good mobility and constant sealing of the gun port, and an armor shape suitable for manufacture, should be kept in mind.*
   The armor thickness of the walls that project over the outlines of the vehicle: 15 mm; those of the turret roof: 8 mm.
   Turret lighting is planned by the vehicle designer.
7) *In the turret walls are six glazed visors with movable shields, which allow the naked eye a quick overview of the entire environment. The shields must be made of 10 mm thick armor plate and must not interfere with the gasproof sealing in any position.*
8. *Ventilation louvers are to be built into the turret.*
9) *The outer walls of the turret are to be fitted with places to attack camouflage.*
10) *All turret openings are to have gasproof seals.*
11) *For the aiming gunner, an adjustable aiming seat, for the loading gunner a folding emergency seat and handholds are to be provided on the turret walls to hold during driving and firing.*
12) *(Only for Rh.M.) The deviant track in the turret roof provided for firing at negative heights must be adjustable by hand to the greatest height for times outside combat activity.*

C. Tank Gun and Machine Guns

1) *Barrel: 7.5 cm caliber*
Barrel structure made for the barrel of the future mountain gun; at least the loading area must be unified.
Performance: Vo = 2400 Atm. At + 10 degrees.
Powder temperature: Barrel length: ca. L/20 = 1500 mm.
Breech: Semi-automatic, meaning automatic opening after firing and automatic closing after loading are required. Replacing the gun barrel in a resting position must be possible with the weapon-master's equipment.

2) *Barrel recoil brake with pneumatic recuperator.*
Barrel recoil as long as possible considering the available space, in order to avoid too-great shaking of the turret when firing.
Parts of the recoil brake that extend out over the armored turret must have special protection against shell hits and smaller shell fragments (10 mm armor).
To the left and right of the recoil path, deflectors are to be mounted to protect the gunners, but they must not disturb the operation of the gun.

3) *Ammunition: Shell ammunition.*
The cartridge case is to fit the uniform case for the new 7.5 cm gun, according to drawing Zwg 13 E 6321.
If need be, the separated ammunition of the future mountain gun and field gun with the correspondingly smaller charges can be used. The use of this ammunition must be possible in time of need.
The ammunition supply in detail remains dependent on later decisions. The complete ammunition supply: ca. 100 rounds on hand, in the vehicle; depending on available space.

4) *Rate of fire: 15 rounds per minute.*

5) *Firing height: over the entry step of the turret as high as possible, in order to be able to keep the upper part of the armored cupola, through which the gun barrel is to pass, as low and thereby as light as possible.*

6) *The installation of an apparatus for blowing out the barrel is eliminated. The streaming on powder and poison gases into the interior of the turret is avoided by high air pressure.*

7) *A heavy machine gun is to be attached to the right side of the gun barrel. A special elevation range of the machine gun of +/- 5 degrees is planned. This range is controlled by the left gunner (aiming gunner). The machine gun is fired by the aiming gunner. The machine gun must not be in the way when the gun is loaded. Ammunition is supplied by means of the drums on board, which may be loaded only when they are to be fired. The elimination of machine-gun loading obstacles is left to the loading gunner. It must be possible to remove it if it becomes unusable or replace it by a troop machine gun.*
For the machine gun, 3000 rounds, at hand in drums as hand ammunition, are carried.

8) *The turret wall serves for mounting the tank gun and machine gun.*
It is to be kept in mind that the port for the tank gun, which must allow a lower and higher elevation within the required field of elevation range, must be kept as small as possible. This requires that the turning point of the barrel be situated as far as possible into the turret wall.

9) *Elevation field:*

Maximum elevation	*+60 °*
Lowering over the center line	*-12 °*
Lowering over driver's observation turret	*-12 °*
Lowering to the side	*-12 °*
Lowering to the back	*-6 °*
MG special swing range, elevation	*+/- 5 °*
Traversing range	*endless*
MG special swing range laterally	*0 °*

In case the commander's and driver's observation turret becomes lower than 300 mm, correspondingly greater lowering is desired.

10) *Aiming machinery*
Easy access to the elevation and traversing machinery is required. For traversing: electric drive at various speeds and hand-operation are to be provided for; reciprocal addition possibility is desirable.
Elevation: electric drive and hand-operation, the former with higher speed than hand-operation.
The apparatus for firing is to be designed so that the aiming gunner can activate the gun and the machine gun for firing during aiming and without removing his hands from the aiming machinery.

11) *Sealing the port. The port is to be provided with a gasproof seal; sealing must be guaranteed in every elevation position of the gun.*

12) *As a visual setup a periscope going through the turret roof, without a stroboscope, is to be planned. Its development is to be carried out by Wa.Prw.1 in cooperation with Wa Prw.4 and 6 and the Zeiss firm.*
   a) *The part projecting out of the turret is to be fitted with armor protection. The easily damaged prism shall be replaceable by a simple hand operation.*
   b) *The periscope's field of vision shall include ca, 58 to 60 degrees. 1.25 fold enlargement should suffice.*
   c) *The periscope must be endlessly traversable by means of a handwheel. For the attained zero setting (being parallel with the gun barrel), a clearly tangible rest is to be provided. The rest shall also be tangible when the turret is turned along with the periscope.*
   d) *According to the height, the traverse must be matched with the view so that targets within an angle of -12 to +30 degrees to the ground can be seen. A lowering to -6 degrees to the right over the turret roof suffices.*
   e) *For viewing of the setup angle, an adjustment of the hair cross up to -7 degrees is to be provided. This corresponds at Vo = 400 and G = 6.3 kg about to a shot range of 3000 meters. For 100 to 100 m touch rests are wanted, so that the aiming gunner can correct his shot variances by feel, without having to look at the range gauge.*
   f) *For firing with machine guns, a special range scale, going up to 2000 m, is to be installed.*
   g) *The principle of the panoramic telescope is to be combined with the periscope, so that out of a side position an indirect aiming can be made.*
   h) *An apparatus to rule out an unbalanced wheelbase is eliminated, but at the suitable place inside the turret a*

wheelbase level and correction table is to be attached to work out the wheelbase error.

j) The periscope as a whole must be easy to remove. It is done from the inside out. Replacing shot-out prisms must be able to be done by the crew.

k) For aerial observation, a second, exchangeable periscope with an elevation field up to 60 degrees is to be developed. Because of its greater vulnerability, though, it should be replaced by the ground-target periscope for tank battles. Ground and aerial target periscopes must have the same diameter, so that they fit into the same turret opening.

l) An emergency visor is to be developed:

1. *A horizontally positioned telescope, for terrain angles +/- 10 degrees, setup angle +2 degrees (corresponding to ca. 1000 meters of shot range). Field of vision about 40 degrees. Aperture shieldable with a plate. When not used it is to be swung aside or pushed together.*
2. *An optical visor extending sideways out of the turret under the same conditions as before. But it will not be possible to provide a shield here. The emergency visor must be able to be made ready for use with a simple hand motion. The decision of which of the two emergency visors to put into use will be made later.*

m) All visors are to be fitted with strong eye padding, and the periscope also with a forehead pad.

n) For the loading gunner there will be no optical visor; he should have the possibility of observing part of the terrain through the vision slit.

13) As a crew for the gun including the attached machine gun, two men (one aiming and one loading gunner) are planned for.

14) Used cartridge cases should be able to be removed from the turret in the shortest time. The ground clearance of the vehicle may not be influenced by the location of an automatically closing flap of the shell-ejecting apparatus. Thus, they may not reach down over the entry step.

15) The necessity of quick ammunition loading is mentioned; for it, the equipment of the ammunition vehicles can include a slide, which can be used advantageously in loading ammunition.

16) Between the commander and the aiming gunner, straightforward and clear yet simple communication must be created to select targets, even in the noise of battle, by utilizing point choices. (Handwritten note: Why not telephone?)

17) Absorbing shocks while driving is dealt with by the vehicle designer. Final goal: It should be possible to make accurate shots while driving.

18) Lashing the barrel in driving position outside of battle is to be foreseen; during battle driving without lashing must be possible.

19) Total weight of the tank gun including mounting and vehicle armor must be as light as possible. Maximum weight of the turret including the gun ca. 1650 kg.

20) The observation turrets of the commander and driver must be furnished in advance with protective panels for protection against the pressure of outflowing powder gases.

Note: For the first test turrets, production with sheet iron is to be foreseen. Here, though, the same weight as that of the later armored turret.

1.1.4 Armament Program

A discussion at the Troop Department (General Staff) on 3/14/1928 on the motor vehicle building program resulted in the following situation as to the large tractor: The delivery of the Large tractor prototypes (15 ton) should take place in the summer of 1928. Testing from 1929 to 1930 was planned. The production of the series should begin in 1931. At first the equipment for one company (17 vehicles) was foreseen; then others after the release of available materials. The price of equipping one company was set at some 2.5 million RM.

1.1.5 Other Possible Uses as Excavators

From a discussion on 4/25/1928 at the Krupp firm in Essen-Ruhr, with Gassert, Austmann, Pirner, Kniepkamp (of Wa Prw 6) and Ritter, Hagelloch, Woelfert (of Krupp):

The firm reports that the testing of the Gr.Tr. chassis as excavators had been done, that they were much too light and therefore were not considered for such uses. But in the opinion of the firm the chassis is usable for purposes of the Reich Railroad Central Office. In an orientation about the state of production work, the firm showed a finished armored body.

1.1.6 Secrecy And Camouflage as Tracked Towing Tractorsa

By Lt. Col. Gaissert, Department Head of Wa Prw 6, to the Commander of Wa Prw, September 8, 1928, subject: Gr.Tr.

The repeated discussion with several offices of the entire house and the talk by the Chief of the Army Command on 8/28/1928 have led me, as to the further development of the Gr.Tr. question, to the following considerations and attached suggestions:

The purpose of the designing and building of the Gr.Tr. was to create models from which, after thorough testing, a final model for later introduction and mass production was to arise.

In the scope of the task, which broke up into many parts, this goal can be reached only if the tests with the models now being worked on are undertaken with the closest collaboration of all involved persons in Wa Prw, plus the designers of the various firms. But in the testing now planned for outside the country (R), this will not be possible in any case, whether it be Kama, Lgrd. Or Mak. The reason for this is in the following points:

1.) Personnel of Wa Prw cannot, for political reasons, be active uninterrupted for so long or repeatedly for a shorter time in R. as necessary. Aside from that, such long absence—such as by members of Wa Prw 6—would considerably limit the advance of other important tasks (Motorizing).

2.) It seems more than questionable whether all firms—and in them again the individual persons—will be ready to take part in the testing in R. after the last events in R.

3.) It is very questionable whether the persons whose collaboration is necessary can always be together in R.

4.) The transport of the vehicles to R. may not be simple and contains many dangers of an internal and external political nature.

5.) Partial tests, changes and new designs of individual parts can be carried out only by the German production firms or design bureaus. Frequent transport and considerable loss of time will be the results.

6.) *We are revealing our newest and most secret technical problems to R. without having the guarantee that we ourselves will have the sufficient use of them.*

7.) *Since the same travel and residence costs for the personnel of the various firms will also have to be borne by the Rw.Min., testing in R. would cause very extraordinarily high costs if they are to be carried out purposefully.*

How can we now get the best results from the work to date and the large applied means amid the internal and external forced situation?

Of the many problems involved in the Gr.Tr., this one is and remains the basis and main one—the running gear.

All other tasks—gun turrets, machine-gun turrets, means of observation, F.T. system, gas sealing, air supply, armor etc.—remain valueless if the main problem—running gear—is not solved. Outside the country too, this solution has not yet been found to this day, and is placed in the foreground as the most important problem, as opposed to all others. The other remaining tasks are of course difficult, but do not form major problems in the sense that the running gear does.

This leads me to the following suggestions:

Of the three models (two examples of each), one of each will be finished at Unterlüss. None will leave the assembly halls. They should be used only to test the whole structure, as far as that is possible in the halls. After a general impression has been gained, the three finished models can be disassembled again as soon as possible. Individual pieces form no danger. The security in Unterlüss is described as good.

The assembly of the other three models will be halted at once. Everything will be disassembled down to the running gear. Of each of these vehicles, the body will be cut up lengthwise. Only the running gear will remain, open at the top and holding only the motors and transmission. The vehicle will then have the appearance of a tracked towing tractor, such as have been running for a long time at the training command of Wa Prw 6 in Kummersdorf, with somewhat larger dimensions. It will then appear as a heavy towing machine for targets.

In action, targets will actually be towed. In addition, the vehicle will form a counterpoint to a fixed tracked-vehicle test bench in that it is used as a driveable track test bench for towing tests. In the vehicle testing activities of Wa Prw 6 in Kummersdorf the vehicles are not eye-catching, while this could be the case at Unterlüss—where such activity is not constant.

All other parts will be dismantled individually at Unterlüss—partially also at Kummersdorf. The vehicles themselves will be packed up and taken to Kummersdorf. Thus, the running gear can be tested best, fastest and most cheaply. The further development of the other parts can be carried on thoroughly in individual tests. For floating tests there must be postponement at first, which can be taken in the bargain without concern in comparison to solving the total problem in view of the small-model tests in Hamburg.

I believe that in this way we will attain the best possible results that are possible at all under the present conditions. The sources of danger in internal and external political activity should not be greater than before, and rather less than with testing in R.

Since according to the latest instructions at the time the assembly in Unterlüss has been carried on as quickly as possible, I ask for the soonest possible decision and instructions to Wa Prw 6, at first to have one vehicle prepared in the suggested form, so that if possible it can be displayed to the Chief of the Army Command at Unterlüss for the purpose of further decisions at the beginning of October, if possible.

1.1.7 Final Assembly At Unterlüss

The following information from the Krupp firm, dated 1935, in the *Developmental History of the Gr. Tr.,* shows the decision to produce the "camouflaged vehicle" again in its original form and mention the time period in which the *Grosstraktoren* were finished and received by Wa Prw 6:

8/1/1928: **Beginning of final assembly in Unterluess.**
9/15/1928: **Contract for a chassis to be camouflaged as a commercial vehicle. (With removal of cover, better access in testing.)**
1/31/1929: **Contract to modify camouflaged vehicle again.**
2/8/1929: **First driving test in hall.**
6/23/1929: **Acceptance of the two vehicles.**

All six *Grosstraktoren* (two each from the Daimler-Benz, Krupp and Rheinmetall firms) were assembled under the greatest secrecy at the Rheinmetall works in Unterlüss. The acceptance by Wa Prw 4 in Unterlüss of the Gr. Tr. gun turrets by the Rheinmetall firm took place on May 8, 1929.

The report of the Artillery Design Department of the Fried. Krupp AG on the *Development of Army Guns from November 1918 to 1933* describes the development of the 7.5 cm L/20 gun for the Large Tractor by the Krupp firm:

Along with the complete "Grosstraktor" battle tank, a 7.5 cm L/20 gun with 6.65 kg shot weight and Vo = 425 m/sec performance was developed. Semi-automatic vertical breech, 2 barrel brakes in diagonal position, in the other diagonal a recuperator and a fixed machine gun in the cradle. Electromagnetic firing from handwheel of elevation machine. Safety switch for the loading gunner. According to previous studies, the design was presented in April 1927 and the gun with turret completely designed by November 1928. The test device was delivered in July 1929.

1.2 EQUIPMENT DESCRIPTION

This document describes the Large Tractors at the time of their finishing and delivery by Wa Prw 6 in June 1929.

Improvements and changes carried out after 1929 will be described later. Setups of the particular component groups were made by the individual firms, to make sure that even on the basis of only six vehicles an optimum of experience in the formation of such an aggregate could be gained. The main differences between the Daimler-Benz, Krupp and Rheinmetall Grosstraktoren will be described for all components.

The Grosstraktor consists of the chassis with main turret and a machine-gun turret behind it. In the main turret a 7.5 cm tank gun and a Dreyse machine gun are present.

The crew consists of six men with the aiming gunner and loading gunner (or commander) in the turret, the driver at left

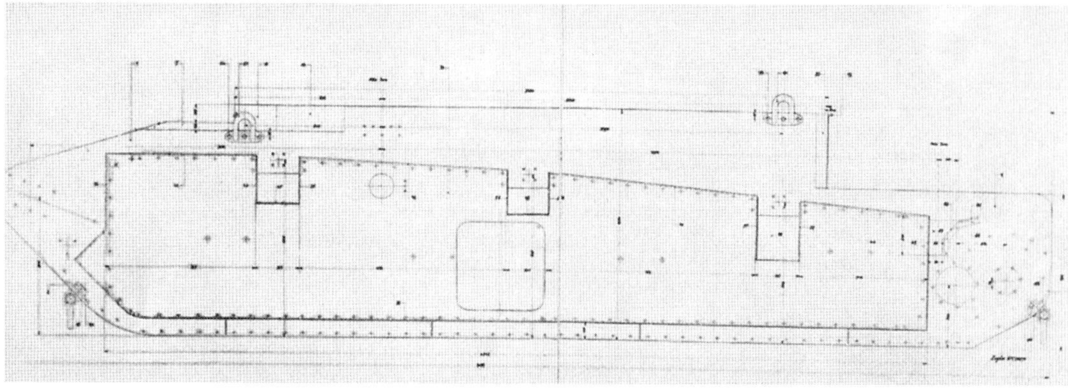

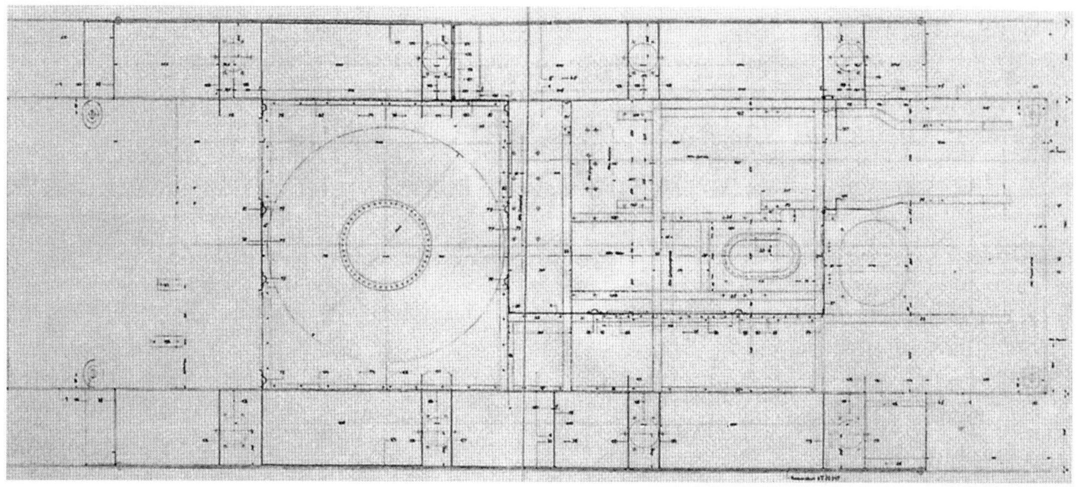

Design drawings of the hull of the Rheinmetall Large Tractor.

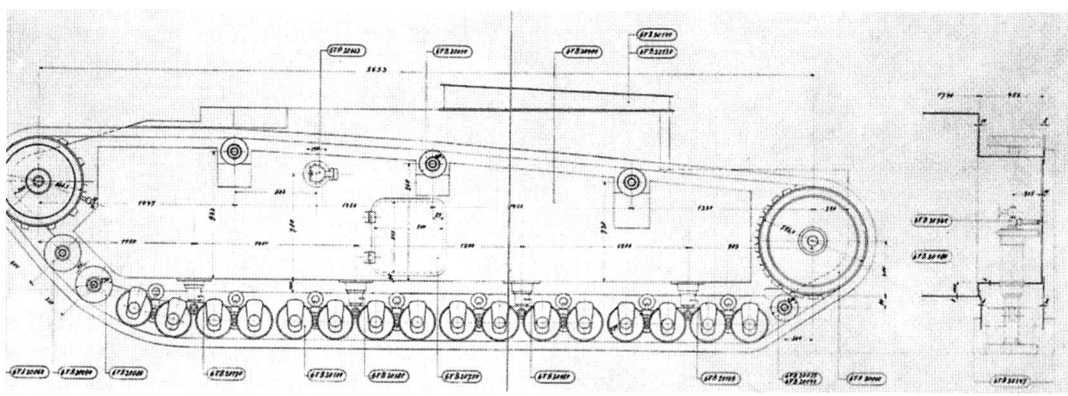

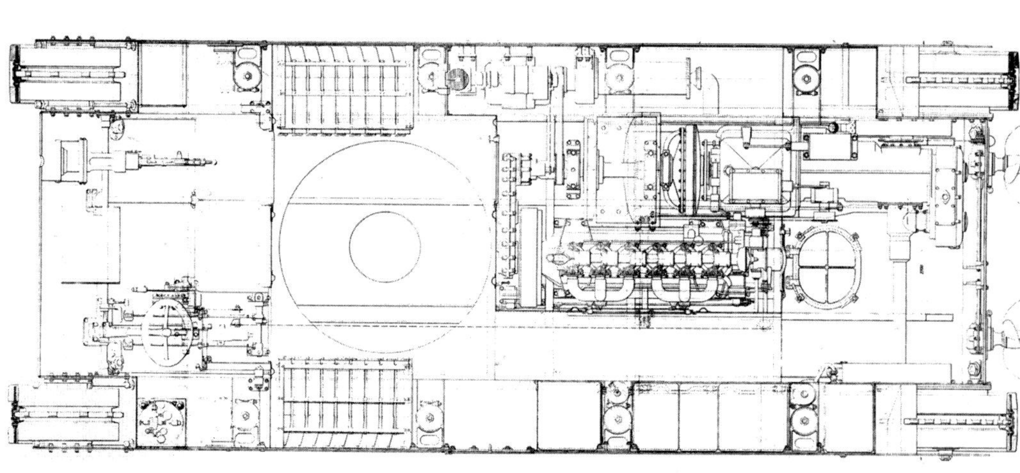

Design drawings of the chassis of the Rheinmetall Large Tractor.

The finished, assembled chassis of the Daimler-Benz Large Tractor. At the bow of the vehicle is a headlight protected from fire.

The Daimler-Benz Large Tractor seen from behind, with the installed drive wheels. On the upper rear panel is the opening for the rear machine gun turret.

The state of assembly of the Daimler-Benz Large Tractor in mid-January 1929 at Unterlüss, near Celle (Munster Camp).

Front view of the Daimler-Benz Large Tractor. The bow machine gun was installed later.

The engine-room cover with the side passage—seen from front toward rear—of the Daimler-Benz Large Tractor.

The Daimler-Benz Large Tractor at Unterlüss after the assembly was finished.

The right side of the Daimler-Benz D IV b aircraft motor (Type 182 206), weight without wheel box and adjunct motor: 604 kg.

A view of the open planetary gearbox of the Daimler-Benz Large Tractor.

The left side of the Daimler-Benz D IV b aircraft motor (Type 182 206).

The planetary gearbox of the Daimler-Benz Large Tractor in a closed condition.

front, the radioman at center front, the commander at right front in the chassis and one machine gunner in the machine-gun turret. For entry and exit there are a hatch at the right front of the body plus openings in the stroboscopes, in the gun turret, at the bottom of the rear space and in the roof of the rear machine-gun turret.

1.2.1 Chassis

The chassis consists of the following main parts: armored hull, motor with cooling system, driveshaft, main clutch, transmission, steering drive, running gear, fuel tanks and ammunition storage.

1.2.1.1 Armored Body

The armored body serves to contain the main parts, is fitted with track housings on both sides, around which the wheels revolve, and to which the drive and steering wheels and road and support wheels of which, with their suspension system, are attached.

At the front end of the armored body sit three men side by side, being the commander, radioman and driver. The commander has a machine gun in a watertight ball mantlet before him.

The gun turret is 775 mm ahead of the central bulkhead. Behind the latter is the motor and gearbox space, which unlike the crew space is enclosed airtight and under low pressure. On one rear offset of the armored body rests a turnable machine-gun turret. To the left of the motor space is a connecting passage between the fighting compartment and the rear machine-gun space, from which the motor can be serviced. The six prototypes were welded together of soft iron sheet metal. According to specifications, the armor was to be made of 6 to 13.5 mm sheets secure from steel-core bullets (S.m.K.).

1.2.1.2 Motor

Daimler-Benz installed a former Daimler DIVb (Type 182 206) aircraft motor, which at 1450 rpm gave a maximum power of 260 HP. This six-cylinder gasoline engine with upright individual cylinders had a displacement of 31.2 liters. A DKW two-cylinder two-stroke 10 HP motor was planned as starter. Krupp and Rheinmetall installed the water-cooled BMW Va gasoline aircraft

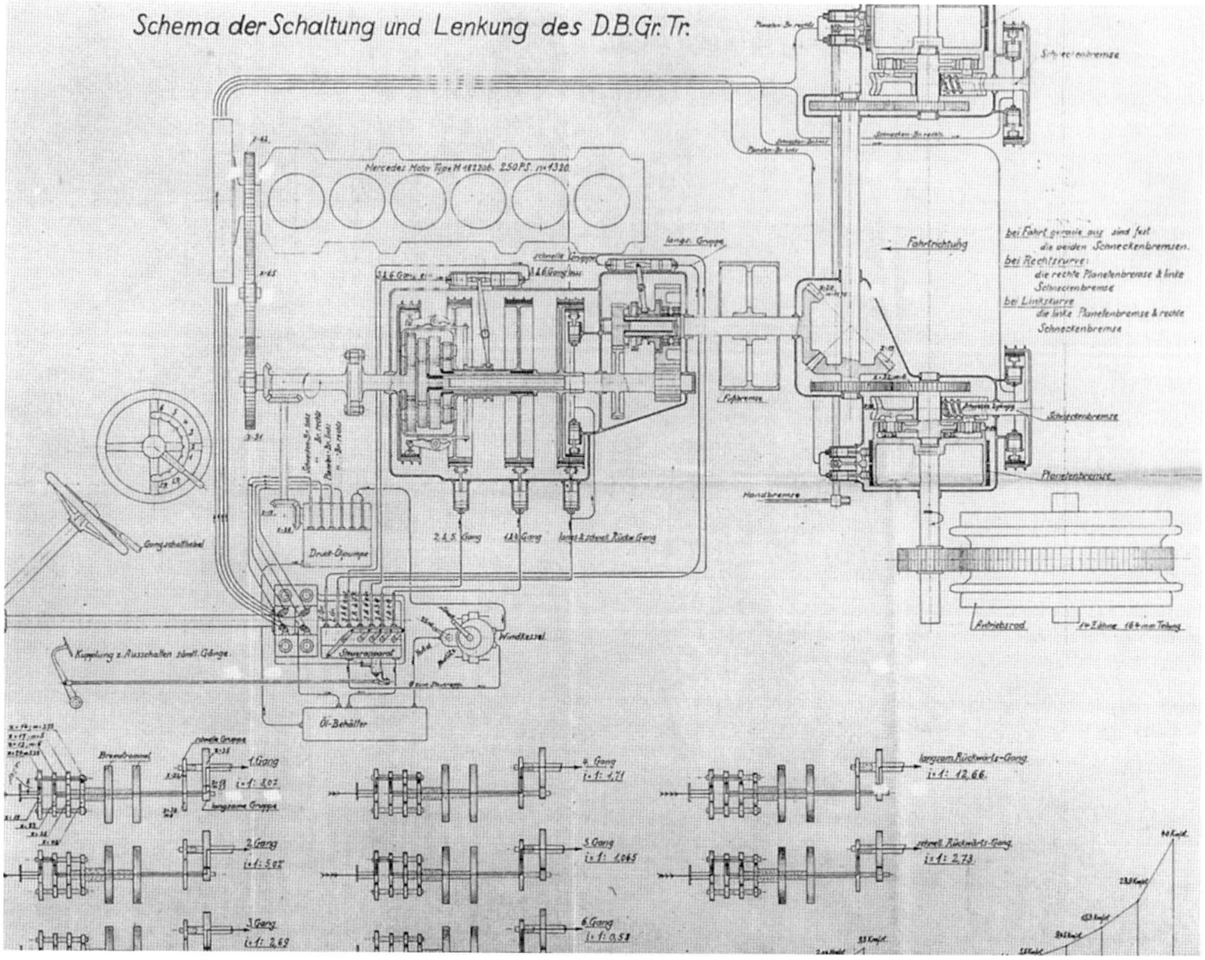

Layout of the transmission and steering of the Daimler-Benz Large Tractor.

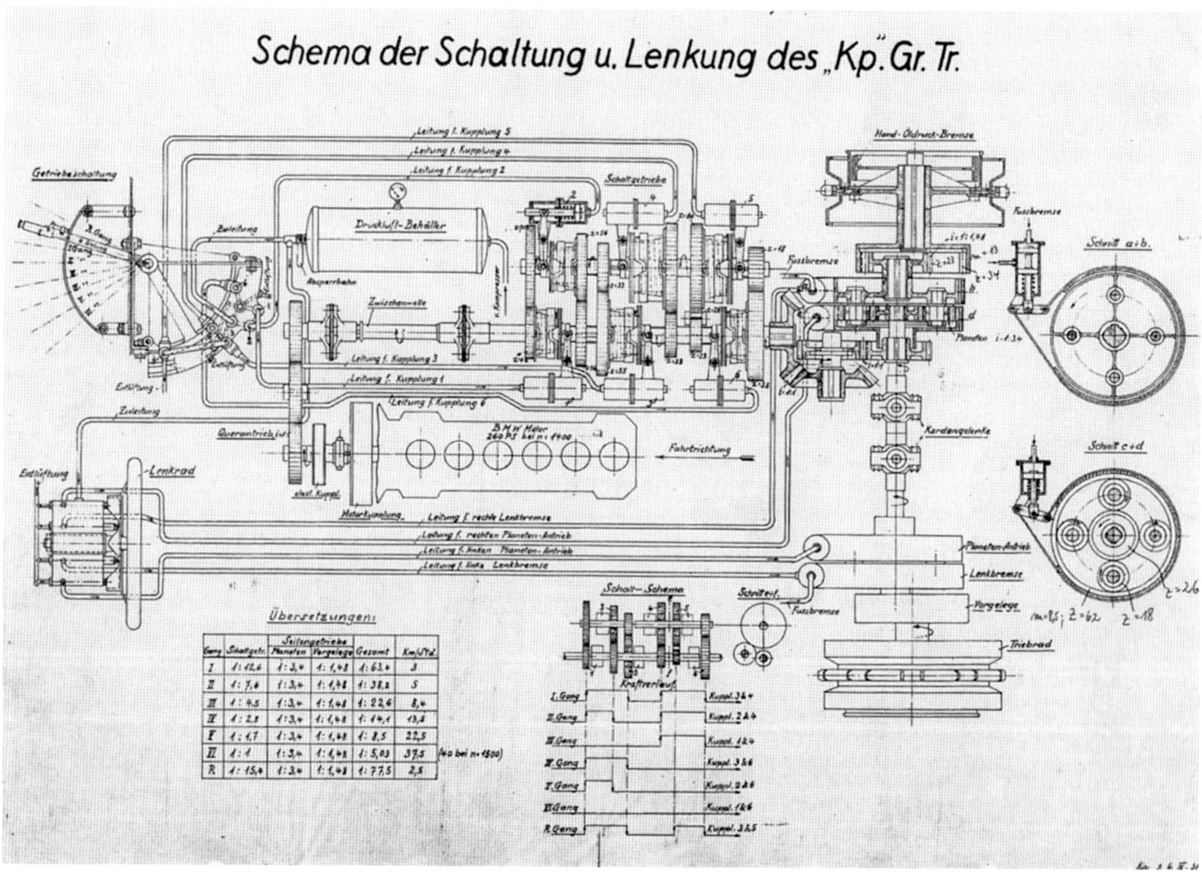

Layout of the Steering of the Krupp Large Tractor.

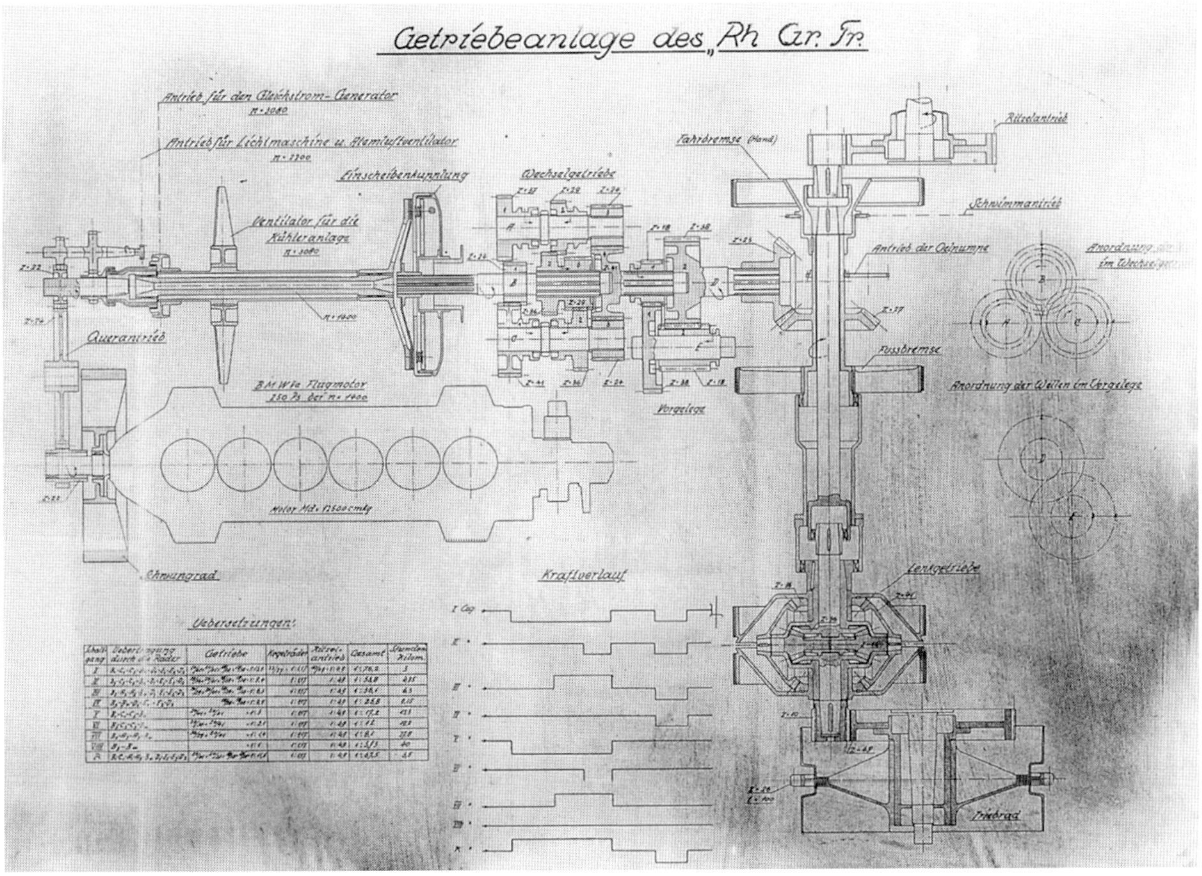

The transmission of the Rheinmetall Large Tractor.

motor with dry-sump lubrication. This was an in-line six-cylinder motor which, with a displacement of 22.9 liters, produced 250 HP at 1400 rpm.

1.2.1.3 Cooling System

Water and air cooling were separated. To remove the cooling-water warmth, a two-part radiator with a ventilator in between was used, while a special air exhaust system that drove fresh air over the motor was planned to remove the radiant heat of the motor.

1.2.1.4 Transmission

Daimler-Benz installed a planetary drive with band brakes and three-speed pre-selector (6 speeds forward, 2 speeds reverse).

Krupp used a spring clutch and group gearbox with compressed-air shifting (6 speeds forward, one reverse).

Rheinmetall had installed a group gearbox with pre-selection, electric transmission (8 speeds forward, 1 reverse).

1.2.1.5 Steering Gear

Daimler-Benz installed a planetary drive with oil-pressure brakes and clutch-steering principle.

Krupp used a planetary drive with band brakes, clutch-steering principle and compressed-air activation.

Rheinmetall had installed a differential steering with non-step steering by a P.I.V. gearbox. The P.I.V. system was a chain drive with non-step changeable drive speeds.

1.2.1.6 Running Gear

Daimler-Benz: Each side of the hull was supported by hydraulic support cylinders, each in connection with four road-wheel systems sprung on leaf springs, with four fully rubber-tired road wheels (wheel size [300] 90-203). The track drive was their own design with a rear-mounted drive wheel. The track was directed in the middle by one road whccl each, whcreby greater off road capability was to be gained. Pressure takeup by the chassis. The tracks, 380" wide, were built with individual links of pressed sheet steel.

Krupp: On each side, three upper return rollers and 16 road wheels were arranged, 12 of which were in a line four meters long. This length allowed the crossing of trenches two meters wide. Every two wheels were mounted in ball bearings on swinging arms and supported with a guide on a coil spring. This directed the power over a housing to the swinging-arm system. To tighten the tracks, the sprung leading wheels were adjustable. Since they were mounted 1.20 meters above the ground, the climbing of obstacles a meter high was possible. The rubber tracks, 380 mm wide, had inset track links and double-step guiding.

Rheinmetall: Each side of the hull was braced by two supports, each with pairs of 2 x 4 road wheels. In each group were 2 x 4 pairs of road wheels with hydraulic equalizers, four attachments for road wheels on the armored wall. The rubber tracks, 380 mm wide, had inset track links and double-step guiding.

1.2.1.7 Water Drive

For water drive, two propellers were installed. They could run forward and backward, and with a trim position of some 71 degrees, a water speed of about 5 kph was possible.

1.2.1.8 Fuel Capacity

The fuel tank for some 370 liters in the Large Tractor allowed six hours' driving at an average speed of 25 kph. The Rheinmetall

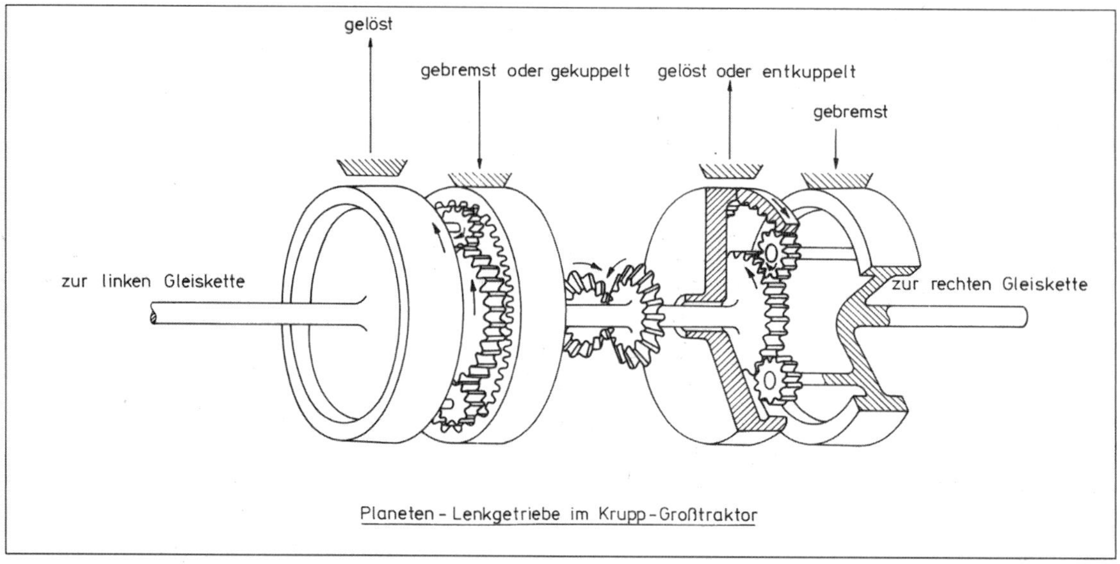

Operation of the planetary steering gear in the Krupp Large Tractor.

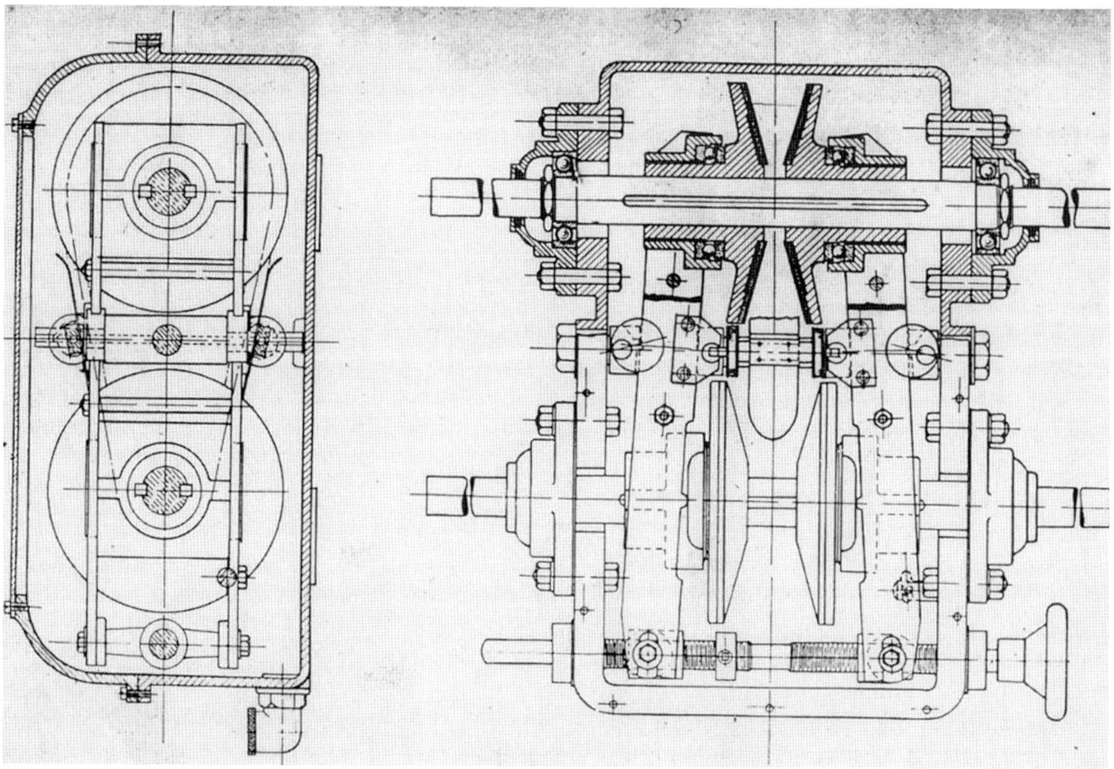

Schematic drawing of the non-step P.I.V. steering gear considered for the Rheinmetall Large Tractor.

Large Tractor had a fuel tank with 480-liter capacity and consumed 400 liters of fuel per 100 kilometers.

1.2.1.9 Ammunition Storage

The 99 to 104 rounds of 7.5 cm ammunition were kept in individual packages on the left and right sides of the fighting compartment, while 6000 rounds of machine-gun ammunition were divided between the fighting compartment and the rear machine-gun space.

1.2.1.10 Radio And Supplies

The radio set was soundproofed and suspended in front of the radioman. The foglaying apparatus was at the commander's right. The intercom, signal board, light signal, first-aid kit, fire extinguisher, tools and supplies were kept near at hand.

1.2.2 Turret

The gun turret with its ball-bearing track, on which it could be turned 360 degrees laterally was located about in the middle third of the vehicle on the reinforced top of the body. The sidewalls of the turret were inclined 78 degrees from the horizontal and were 12 mm thick; the turret roof was 8 mm thick. All the openings in the turret for the gun cradle, periscope, exit hatches and turning track were sealed gas-tight with special sealants. The weight of the turret with the gun, including the shell ejector, turning motor, turning drive, targeting devices, and machine gun added up to around 1900 kilograms.

In order to gain additional space in the lowest elevation area for the recoil of the 7.5 cm gun, the roof of the Krupp turret was raised and a recoil track was built into the flat roof of the Rheinmetall turret. The Rheinmetall turret was used on the Rheinmetall and Daimler-Benz vehicles; the Krupp turret on the Krupp tractors.

P.I.V. drive for the non-step steering of the Rheinmetall tractor.

1.2.2.1 Armament

The armament consisted of one 7.5 cm recoiling gun and one normal heavy machine gun. The Krupp gun with its vertical breech and L/20 caliber length fired shells weighing 6.65 kg, with Vo of 428 m/sec for 9200 meters. The Rheinmetall gun with a side breech fired 6.65 kg shells with Vo of 428 m/sec for 8500 meters. The barrel and the heavy machine gun parallel to it were mounted in a rolling type of cradle, the shield attachment of which was fastened to a wall of the armor cupola, so that the roller covered the opening in the cupola at all elevations. The elevation was -2 to +60 degrees, the lowering to the rear only 6 degrees. The heavy machine gun, unlike the tank gun, had a special swinging range to the height of +/- 5 degrees.

1.2.2.2 Means of Aiming and Observation

These aiming and observation means were on hand: One periscope going through the turret top and one emergency visor mounted on the sidewall for the aiming gunner, plus one other identical periscope for observation by the loading gunner or the commander.

The two periscopes were fully identical, had fixed inside and adjustable outside viewers, and could be swung independently of the gun barrel to 90 degrees high and 360 degrees laterally. The periscope heads, which projected through the turret top, could easily be replaced through small openings in the top. The outside viewer of the emergency visor in the sidewall was movable only with the gun barrel. Setting angles were set by adjusting the visor indicators. For the gun and the heavy machine gun, two different setting scales were planned.

1.2.2.3 Turning Mechanism

The elevation machinery was hand-operated; the traversing apparatus manually or electrically. The electric drive of the traversing machine had twelve speeds and was operated by two foot pedals. Differences in the traversing speeds were, if necessary, equalized by simultaneous advancing or retarding of the handwheel.

1.2.2.4 Firing The Gun

Firing the gun was done electrically by a contact on the handwheel of the elevation machine or manually on the fixed left trigger. The heavy machine gun was likewise fired by the gunner via a Bowdenex on the elevation lever of the machine gun. The shell casing automatically ejected by the breech slid into a casing catcher to the case box located on the floor of the vehicle; its two flaps were operated electrically, so that the casings could be ejected from the vehicle at once with minimal loss of air.

1.3 TESTING

On June 24, 1929, the vehicles were packed into watertight cases and sent by sea from Stettin to Leningrad and on to Kazan, Russia (capital of the Tatar Republic), where they arrived in July 1929. There they were moved from the Kazan depot to the **Kama** camp six kilometers away.

After four years of *Kama Tank Testing*, the six tractors were packed watertight at the Kazan depot and shipped to Leningrad. From there each of them was shipped in a Russian freighter to Stettin, where the last transport arrived on September 21, 1933.

The *Large Tractors* were transported to the Daimler-Benz works in Berlin-Marienfelde, where they were prepared for further use in the following year. Only the four Krupp and Rheinmetall *Large Tractors* were modified further, tested and used. The Daimler-Benz vehicles were mustered out and turned over to the headquarters of the new armored troops for use as monuments.

In November 1933, in the parameters of the Motor Vehicle Instructional Staff in Berlin-Moabit, among others, the firing training was prepared, being activated in April 1934 with the *Heavy and Light Tractors* brought back from Kama and the appropriate personnel as the *Special Motor Vehicle Training, Wustrow*. There the second half of the training course for German trainees, begun at Kama in 1933, was continued. The Wustrow peninsula in the vicinity of the seashore resort of Alt Garz, Mecklenburg, was set up in 1934 as a Flak-gun school.

The completely installed running gear of the Daimler-Benz Large Tractor (as of January 1929).

One of the road-wheel trucks of the Daimler-Benz Large Tractor running gear.

At left, an external track tightener; in the center, a drive-gear shaft; at right, an internal track tightener of the Daimler-Benz Large Tractor.

At left the leading wheel and at right the drive wheel of the Daimler-Benz Large Tractor.

The water propeller of the Daimler-Benz Large Tractor could be turned off.

Krupp Large Tractor (1932)

Layout of the drive aggregate in the Krupp Large Tractor

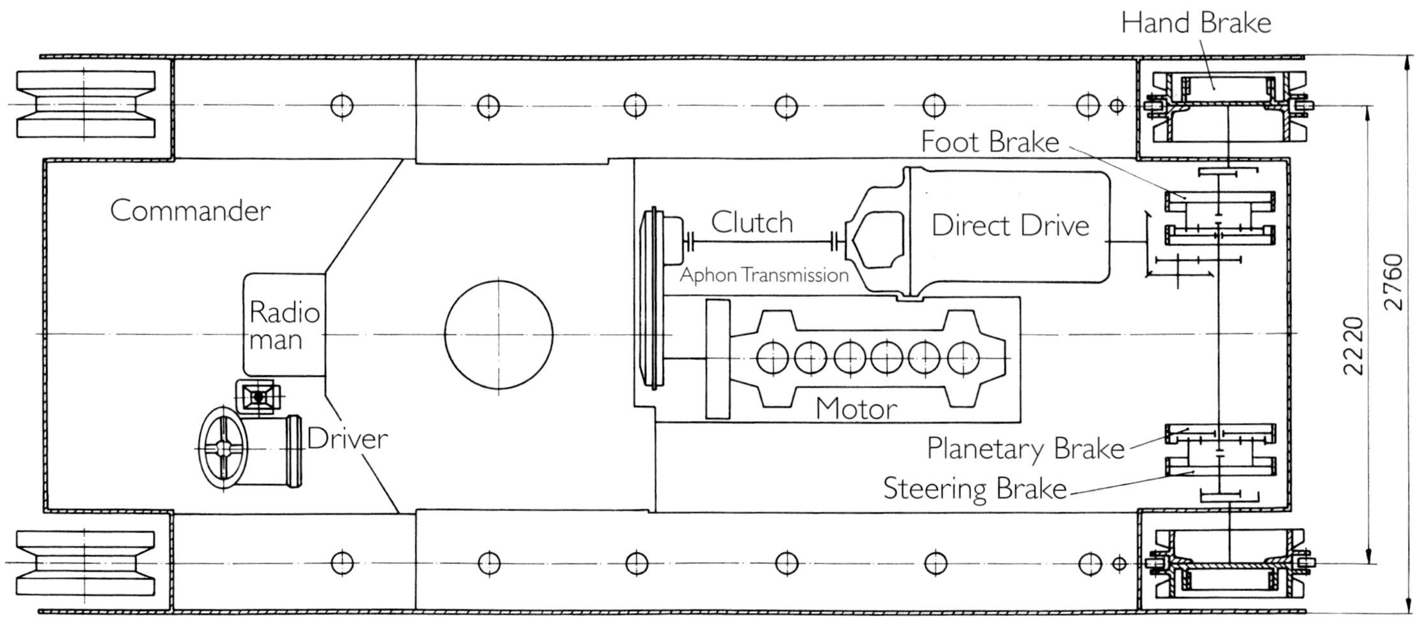

Layout of the Drive Aggregate of the Rheinmetall Large Tractor

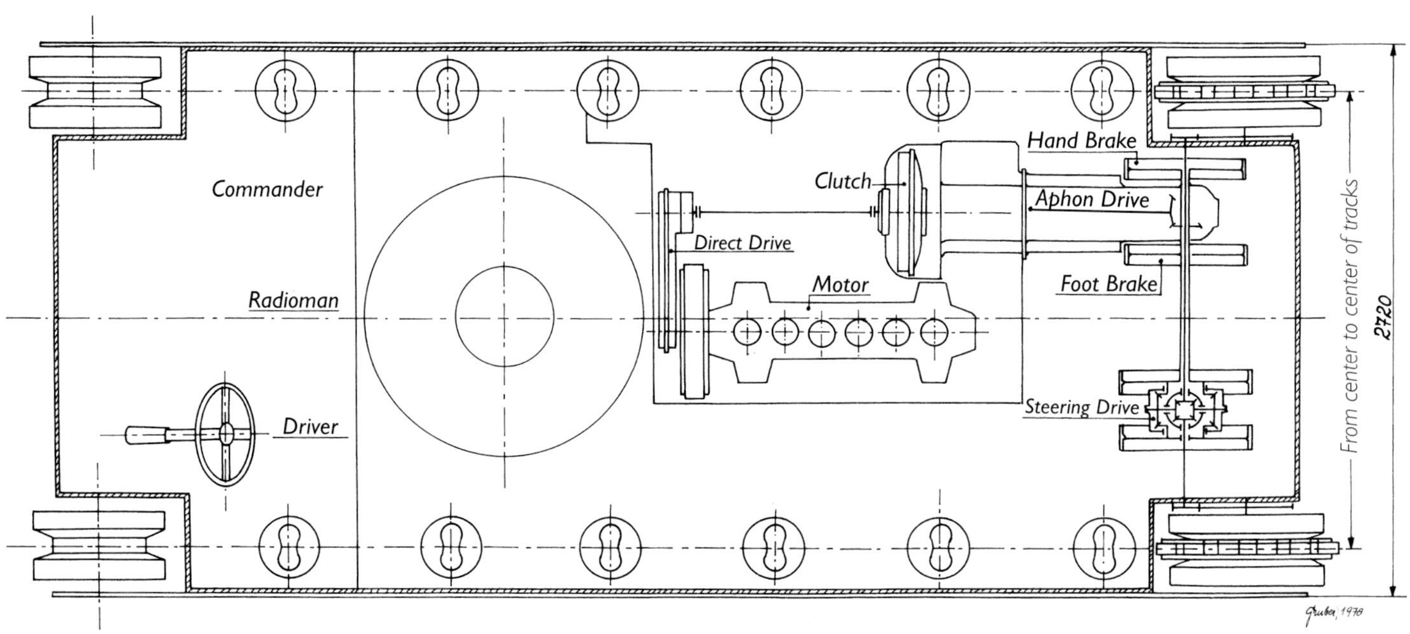

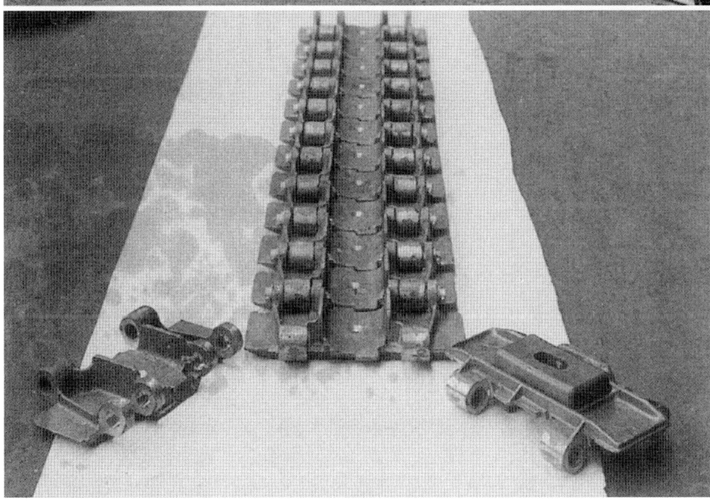

The track of the Daimler-Benz Large Tractor, Type MK 6/380/160, with rubber pads.

The firing courses were guests there. Transporting the vehicles was done under the strictest secrecy, in the Kama shipping cases, by rail to Kröpelin, and from there on Culemeyer "House-to-House" low loaders of the German Reichsbahn to Wustrow.

During the winter months of 1934-35, the *Tractors* were at the Döberitz-Elsgrund training camp to prepare for service in the large-scale armored training in the summer of 1935 at the Munster-Lager training camp. After the conclusion of this training they were moved to the newly founded Panzer Shooting School of Putlos, near Oldenburg, Holstein.

At the Putlos shooting school the *Large and Small Tractors* were used for training and comparison testing for several more years. Thus, these vehicles gave valuable service in the framework of further development of armored vehicles until they were mustered out.

1.4 IMPROVEMENTS AFTER ACCEPTANCE

1.4.1 Daimler-Benz Large Tractor, No. 41 and 42

Because of a design error in the planetary gears, which could not be corrected quickly, both Daimler-Benz vehicles had to be abandoned. The original tracks of pressed sheet steel were replaced by Mk 6/380/160 tracks with insertable rubber pads. The running performance of the two Daimler-Benz tractors in the following years was unimportant. In 1930 they were driven 25 km, and in 1931 31 km. It is incomprehensible today why Daimler-Benz no longer made any effort to correct the obvious design error. After 1931 the vehicles were no longer put to use.

1.4.2 Krupp Large Tractor, No. 43 and 44

Sept. 1929: Beginning of driving tests in Ka.
There only a few kilometers could be covered every autumn, since the other seasons were devoted to designing and building the improvements in Essen, the long transport route and installation in Ka. Difficulties with the compressed-air system delayed the tests. Road tests showed that the machinery ran quietly, the running gear was well situated and shifting was easy. Air pressure was not sufficient. The Krupp tractors showed strong signs of wear on the spring clutches of the gears. In addition, the compressed-air consumption for shifting and steering was considerably higher than expected.

1930: 1929 road tests showed improvements and strengthening of the compressor and cooling systems, as well as sound reduction.
Improvements installed.

Oct. 1930: Visit of Chief Engineer Hagelloch to Ka.

1931: 1930 road tests led to the following improvements: New tracks (Kgb 60/376/160 with roller bearings), plus drive wheels and road wheels. Change of the lever bearings of the running gear, improvement of the compressor system. Installation of hard rubber sleeves and new spring bands in the gearbox. Further road tests.

Sept. 1931: Visit of Chief Engineer Woelfert to Ka.

1932: 1931 road tests showed: Changes to the compressor system. Designs for new running gear. Temporary interruption of the road tests.

Damier-Benz Large Tractor

Oct. 1932: Maintenance and road tests.
Visit of Chief Engineer Hagelloch to Ka. The extent of road testing of Krupp Large Tractors done by 1932 added up to:

Tractor	1929	1930	1931	1932	Total
No. 43	3	--	33	154	190 km
No. 44	60	10	39	--	109 km

1933: 1932 road tests showed: Installation of auxiliary compressors in track box. Installation of new tracks and running-gear parts. Improvement of the cooling system. Installation of synchronized 5-speed Aphon SFG 280 gearbox. Movement of the clutch to in front of the gearbox.
Closing of Ka. Transport back from Ka. To Berlin.

1934: 1933 road tests showed the installation of the following parts by the Daimler-Benz firm in Berlin-Marienfelde: Limits to the stroke of the spring columns. Installation of a swing panel on the motor. Strengthening gearshift and steering brakes. Overhauling the Aphon gearbox by Z.F.
Knorr firm improved air activation of steering brakes.

8/21/1934: Transfer to Wustrow.

8/22/1934: Road tests: Good results from all problem areas in tracks and road wheels.

10/2/1934: Visit of Chief Engineers Hagelloch and Woelfert to Wustrow.

10/29/1934: Transport back to Berlin.

Nov. 1934: Suggestions for new steering gear.

1935: Overhaul by the Borsig firm. Road tests of 1934 showed the following improvements: Mounting of drive wheels on roller wheels. Installation of strengthened brake cylinders and bands, plus new road wheels.
Transfer to Putlos.

Krupp Large Tractor (as of 1933).

The Krupp Large Tractor from the rear (as of 1933).

The Krupp Large Tractor as a display model at the camp of Armored Regiment 5 in Wünsdorf near Berlin.

The Krupp Large Tractor with modified running gear (as of 1933). New running gear, larger front and rear road wheels, larger return rollers, improved ventilation over the motor.

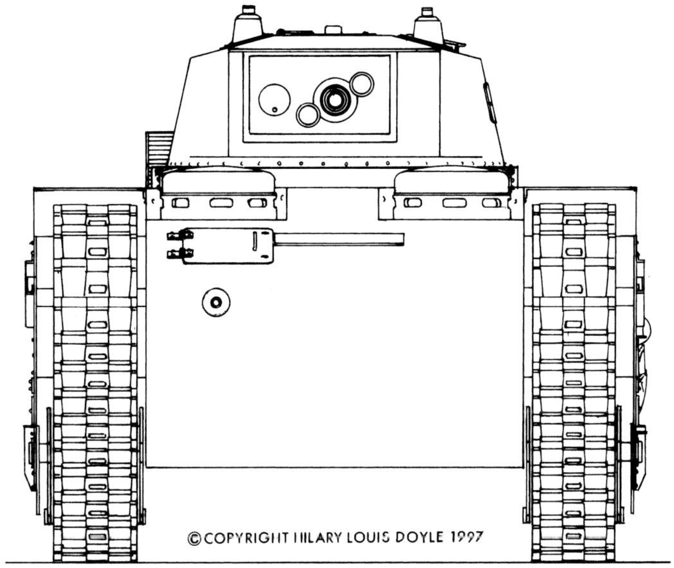

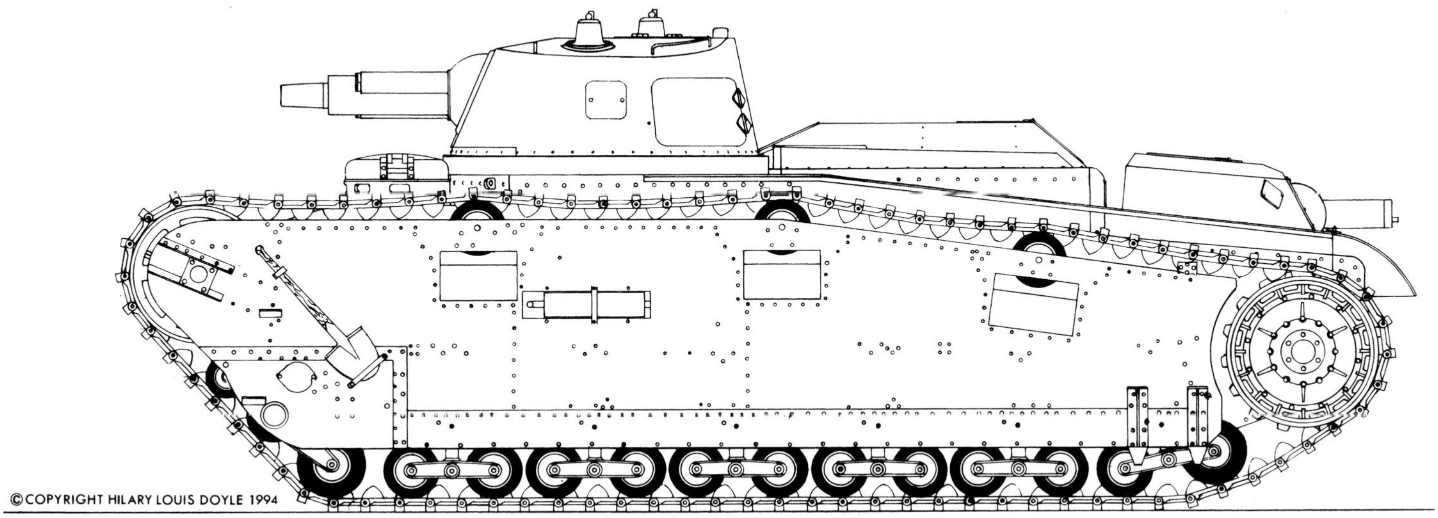

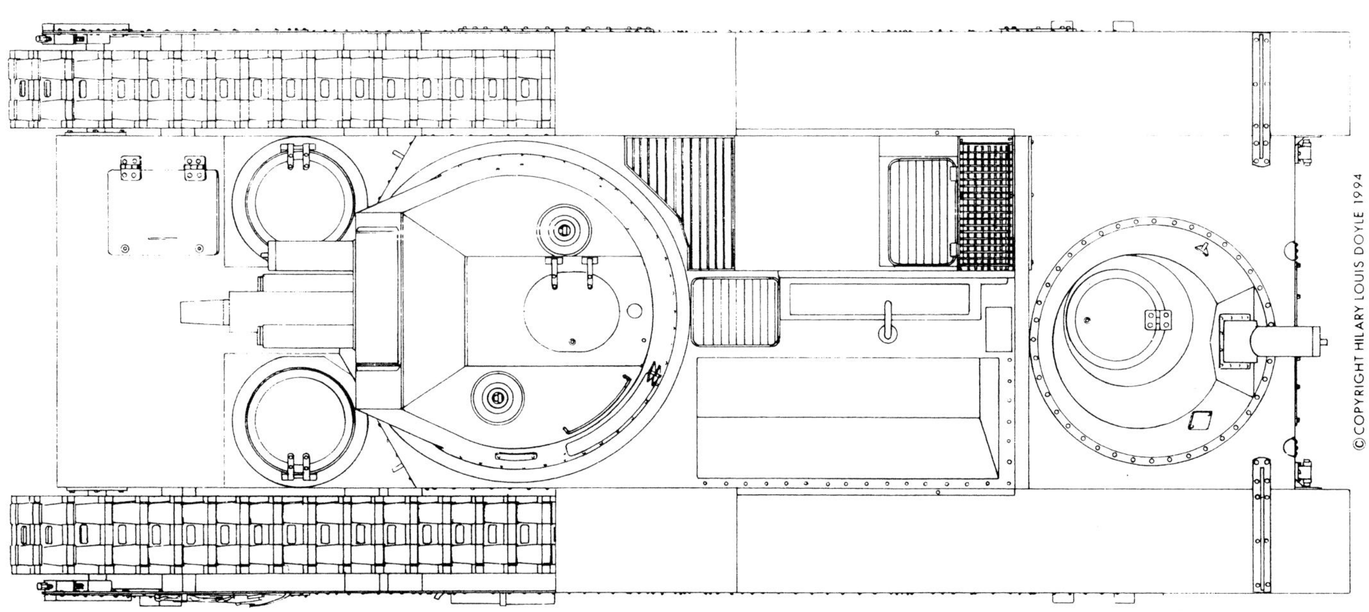

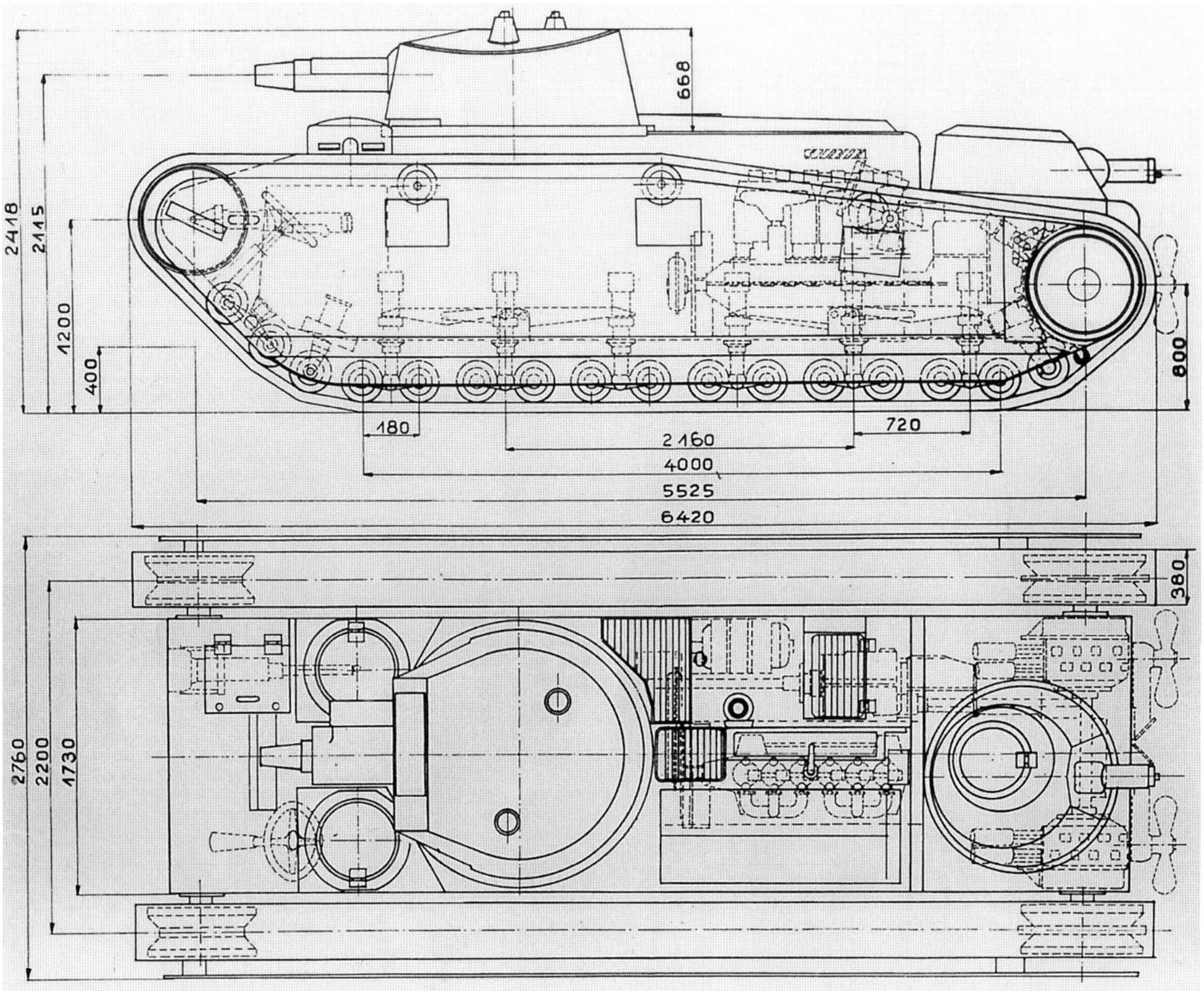

The Krupp Large Tractor, details of its internal setup.

A good view of the Rheinmetall Large Tractor with the original running gear.

Four Large Tractors (two Krupp and two Rheinmetall) and one new vehicle are seen with the newly established Medium Armored Company for the first drill of a German armored Division at Munster in August 1935.

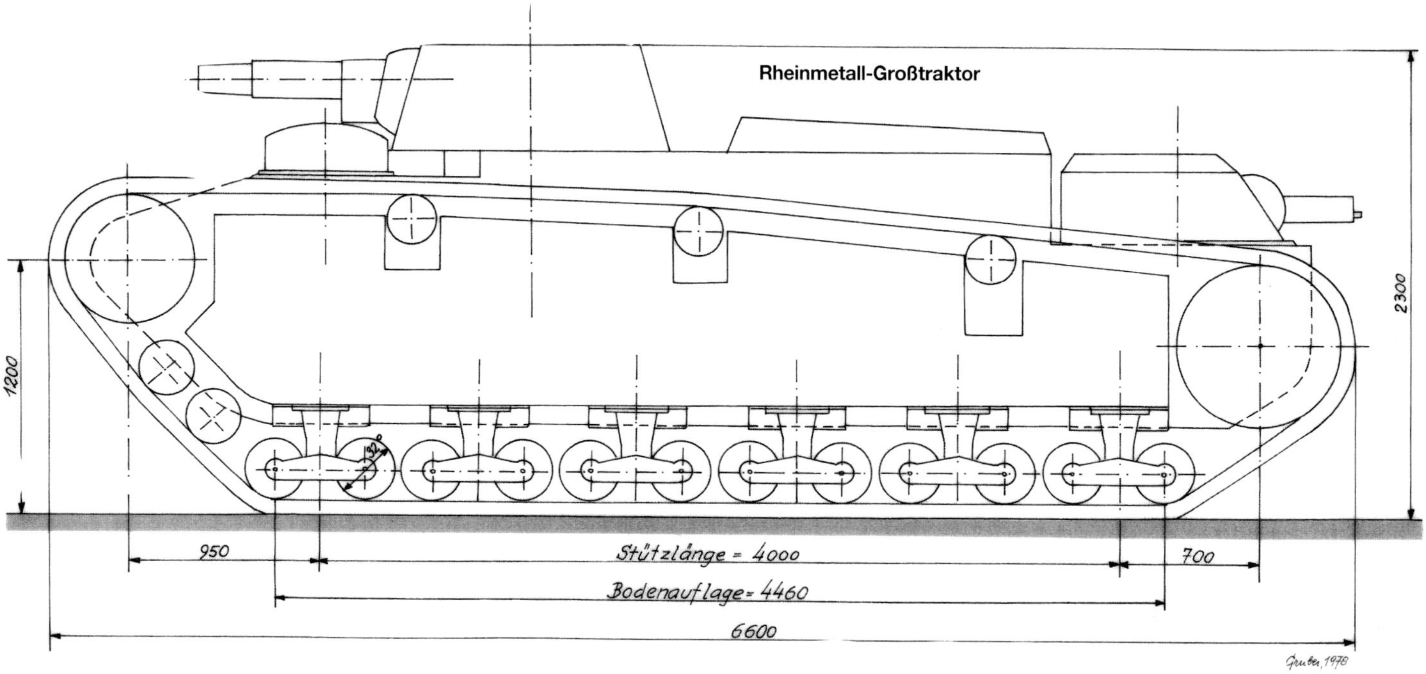

Rheionmetall Large Tractor

View of the Rheinmetall Large Tractor with the original running gear.

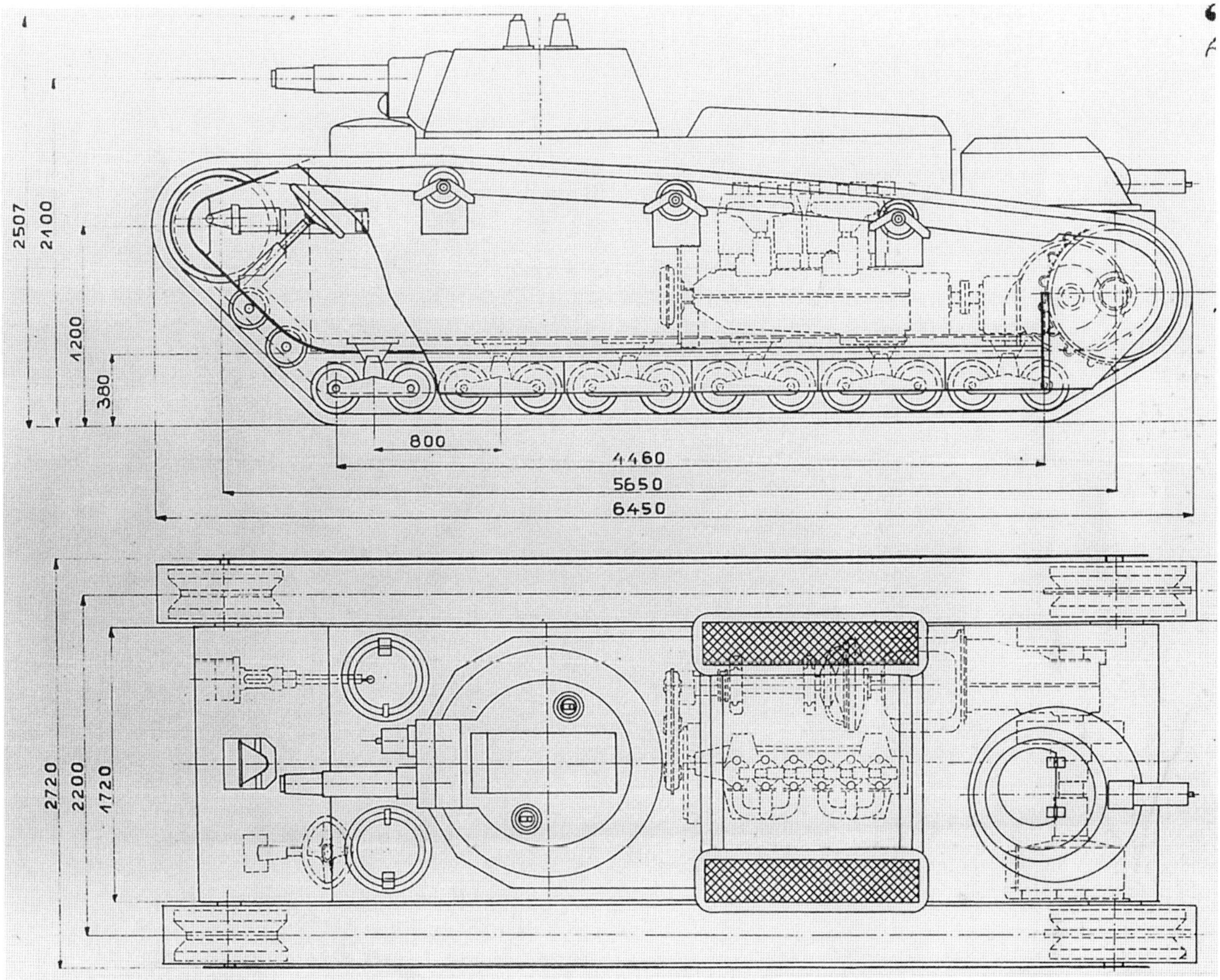

The Rheinmetall Large Tractor, details of the interior layout.

1.4.3 Rheinmetall Large Tractor

The P.I.V. drive used in the differential system of the Rheinmetall Large Tractor was unable to handle the existing demands and had to be removed. The differential steering of a simple kind remained, with all its disadvantages. But the driving possibilities were proved. The P.I.V. gearbox was supposed to bring about a smooth regulation of the differential's revolutions, thus adding a superimposed drive to the differential.

A first amphibian test of a Rheinmetall Large Tractor on October 30, 1929, on the Kaban-See, near Kasan, was a failure. While moving backward, the vehicle became unbalanced, filled with water and sank. Works Master Kerres of the Rheinmetall firm, in charge of the water drive, lost his life, although the vehicle could be salvaged after a short time. The differential steering drive installed in a Rheinmetall Large Tractor was replaced in 1930, after the P.I.V. drive had fallen out in 1929, with original Cletrac steering gear in bevel-wheel form, which proved itself. After the vehicle was brought back from Kama to Germany, the second Rheinmetall vehicle was also fitted with it in 1933-34.

The Cletrac steering gear proved itself until the vehicles were retired. As with the Krupp vehicles, synchronized ZF 5-speed Aphon SFG 280 transmissions were installed in 1931.

The rubber-belt track links used by the Rheinmetall Large Tractors proved to be unsuitable, as they inclined toward the drive wheels from steering to climbing and thus tended to derail. In 1931 they were replaced by Kgw 60/350/160 steel track links with roller bearings.

In 1930 the original running gear was already replaced by a type with 4 x 4 road wheels on each side. In 1932 this was replaced by a new type with 6 x 2 320 mm road-wheel groups. The hull was then supported with coil springs at six points.

By 1932 the Rheinmetall Large Tractors had covered the following distances:

Tractor	1929	1930	1931	1932	Total
No. 45	52	—	215	230	497 km
No. 46	81	215	181	290	767 km

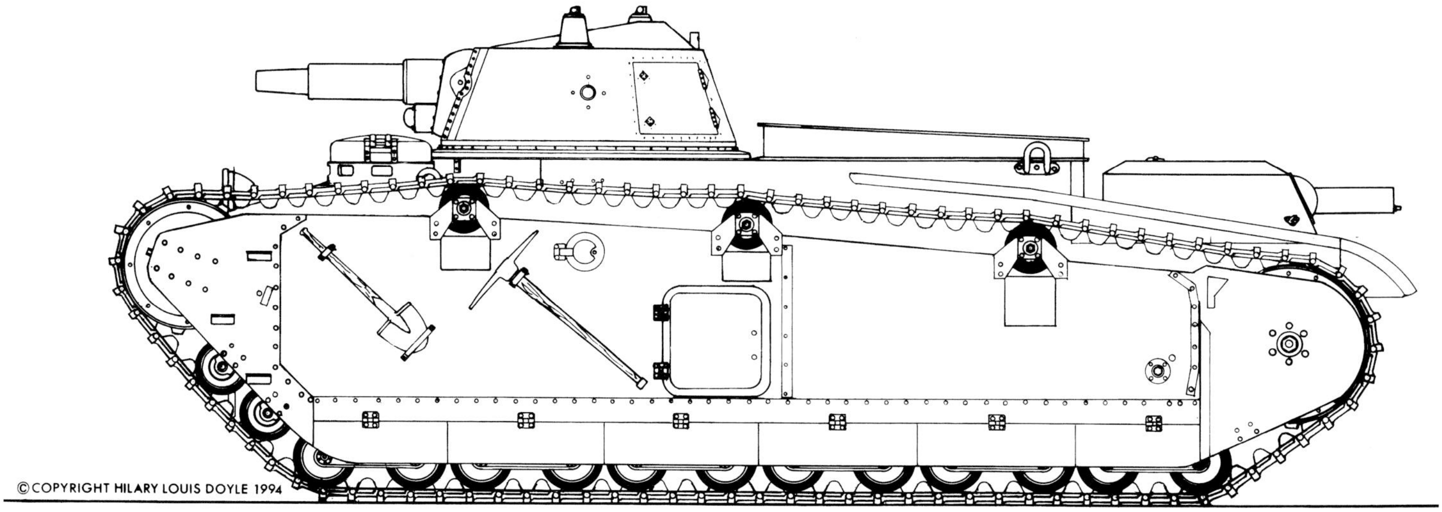

The Rheinmetall Large Tractor with changed running gear (as of 1932) had six sprung units with larger road wheels.

The Rheinmetall Large Tractor turret on a shaking stand for firing tests.

In a discussion at Kama on October 12, 1932, General Lutz reported that the Rheinmetall Large Tractor had led the way and the Rheinmetall firm would receive a contract for more of the vehicles. Lutz also wanted to order all the turrets from Rheinmetall, so as to make just one manufacturer responsible for claims.

2. NEWBUILT VEHICLE (Nb.Fz.)

Previously called by Wa Prw 6 *Medium Tractor. Gr. Tr. Copy, M.Tr.Newbuilt* and *Gr.Tr.Newbuilt.* As of 1935 as *3.7 and 7.5 cm Pz.Kpf.Wg.*

2.1 DEVELOPMENT

The first indication that a new design class for armored combat vehicles was planned by Wa Prw 6 is contained in the following excerpt from a letter from Mr. Hagelloch (Department A.K.) to Captain Steich (WaPrw6) on July 13, 1932:

Recently Mr. Blasberg (WaPrw6) spoke of a new turret with side-mounted gun and centrally mounted machine gun. I have considered the matter and also discussed it with our artillerymen. A 7.5 cm gun would be too big and heavy. If such a turret is to be designed, I would recommend using a gun that had a maximum shot weight of 3 kg. That would, as far as I can see, be a 6 cm gun. This gun would break through all tanks' armor and you would have at least twice as many shots available as with a 7.5 cm gun. Unfortunately I cannot have such a design made without having an order for it. Perhaps it is possible for you, in view of the coming medium tank, to release a sum in a department for the design of such a turret.

Further discussions about decisions for main components followed before pre-design contracts were given to the design firms by Wa Prw 6. On October 12, 1932, Dr. Stade (Wa Prw 6) decided in discussions with the Krupp firm and the Friedrichshafen Gear Factory in Essen on the M.Tr.'s main powerplant:

For the air-cooled 250 HP NAG motor supplied to the Krupp firm, cooperation between Krupp and Z.F. as to the transmission used for it. A total ratio of 1:8 to 1:9 is wanted.

In a discussion with General Lutz (In.6) in Kama, Russia, on October 12, 1932, Dr. Hagelloch reported:

The Gr.Tr. turret and gun building is still completely open. Advantages of 5 cm were made clear. General Lutz also wants to discuss this with Dr. Ritter (Krupp). For further development of the M.Tr. General Lutz wants to call in several firms if possible, for reasons of competition.

Two firms (Krupp and Rheinmetall) had taken up this competition. Unfortunately, this story is attested only partly by files from Krupp and partly from drawings and pictures from Rheinmetall.

2.1.1 Basic Viewpoints

On October 15, 1932, Wa Prw 6 continued the further development of the medium tank through the following offer to the Fried.Krupp A.G., Department A.K., in Essen:

The Krupp firm is hereby requested, according to the viewpoints in the enclosure, to enter into the design study about a new tractor—called in the future "Medium Tractor" (M.Tr.).

Basic viewpoints for the design are:

1) Upgrading the armament to 1 cannon and 4-5 machine guns.
2) Total weight not over 15 tons.
3) Armor at most secure against 2 cm weapons.
4) Creation of a vehicle of about the overall length of the existing Gr.Tr., otherwise decreasing the body, to offer no large target, and keeping armored surfaces small.

Before individual questions (extending specifically to the components) are taken up, it should be determined to what extent the established main requirements can be achieved (space and weight questions). Submission of a design offer is requested.

The following enclosures of October 12, 1932, were added to this offer:

On the basis of the previous (not yet completed) firing tests of armor plates, protection is offered against 2 cm shots.

Plate Mm	Thickness kg/mm²	Resistance meters	Distance Angle degrees
40	*130*	*100*	*75*
40	*120-140*	*500*	*90*
30	*120-130*	*100-500*	*70*
30	*120-130*	*800*	*90*
20	*120-130*	*100*	*40*
20	*120-130*	*800*	*65*
15	*180*	*800*	*90*

Neubau Fahrzeug (Newbuilt Vehicle) No. 3 to 5 with Rheinmetall chassis and Krupp turret.

Since the resistance of the plates made by Krupp is determined in a different way than by the Deutsche Edelstahlwerke (DEW), different values exist for plates of the same resistance; for example, the resistance of the 120 kg/mm² plates made by Krupp seems to correspond to that of 130 kg/mm² from DEW.

As for plates of other thicknesses and resistances, in part, no results are yet at hand.

Viewpoints for the first inquiry for the design of an M.Tr.

1. *Weight:* *Not over 15 tons.*
2. *Length:* *Apparently somewhat shorter than Gr.Tr. but longer than L.Tr.*
3. *Width:* *Presumably based on the turret diameter, not yet entering into weight questions. Otherwise favorably large width for steering Ability.*
4. *Belly height:* *40 cm*
5. *Motor:* *250 HP, air-cooled.*
6. *Speed:* *Top speed 35 kph, favorable average speed wanted.*
7. *Transmission:* *Synchronized Aphon gearbox, mechanically operated.*
8. *Steering gear:* *Clutch steering as in Gr.Tr.Kp.?*
9. *Running gear:* *Specifications not yet decided. Compare Kama report of September 1932.*
10. *Powerplant:* *2 projects*
    1) front wheel drive
    2) rear wheel drive
11. *Climbing ability:* *Will have to be subordinated to main questions.*
    To try for: Not much below the Gr.Tr.'s performance.
12. *Sealing:* *Engine room isolated from fighting compartment in terms of Fire, heat and noise. Fuel tanks separated from fighting Compartment.*
13. *Equipment:* *Turret: Cannon (7.5 or smaller caliber, 2 projects) and one Machine gun, compare Ka protocol.*

	At front and rear, two twin machine guns each (traversing Ability to 200 degrees to strive for)
14. Crew:	*Depending on the equipment: Turret: Commander, [Kan.], Machine gunner, 1 driver, 1-2 machine gunners (1 also as Radioman?), otherwise 1 radioman / 6 men (7)?*
15. Armor:	*Strive for security against 2 cm automatic, at least bow and Turret. Compare special setup*
General:	*Simplicity of arrangement, accessibility and easy removal of Individual aggregates are important.*

Beyond this, the discussions in Ka. can serve as a basis on many points.

Further, Wa Prw 6 declared to Krupp on October 25, 1933, as to M.Tr. development:

After the talk by the Department Chief, the following contract extension was telephoned to Krupp (Dipl.Ing. Woelfert) today. The space and weight developments for the M.Tr. are to be extended, so that besides the development as to 15 tons, the total weight can be increased up to 22 tons without changing the other main requirements. The higher weight could depend on, firstly, that in the limited measurements the included armament could be housed fully effectively, and secondly, that the weight for full 2-cm security would not suffice.

It could also be determined what weight would suffice for security against 3.7 cm caliber: a. of the entire vehicle, and b. mainly the turret and the bow (purely weight-based determination). (Note: According to the experience gained, plates of 120 kg/mm^2 protect against 3.7at 100 meters, 40 mm at 70 degrees, 30 mm at 35 degrees)

2.1.2 Further Decisions from the Development Period

In a discussion on February 28, 1933, between Dr. Ritter (Krupp) and Captain Streich (Wa Prw 6) about the M.Tr. it was decided:

As for the weapons, as I have now heard from Zwengauer, there is now full clarity that we are now to equip the turret with 7.5 cm Vo 450 m, and the introduced 3.7 cm with 760 m Vo. Both positions, Zwengauer's and Streich's, stressed that they would like, if we can design it so, that later the 3.7 cm barrel could be exchanged for the 5 cm barrel. A. 1 has explained, though, that the development of the 5 cm barrel will take so long that the M.Tr. cannot wait for it. While Zwengauer is completely in favor of 5 cm, Streich considers the better performance excessive. Thus we must see that the A.6's superiority can be proved in the course of time.

As for the body of the M.Tr., In 6 (Inspection of the Motorized Troops) still makes difficulties. They base them on the long delivery times that Rheinmetall requests, and keep trying to push the building of the Gr.Tr.Rhm through. Essential here is that we stick with our short delivery times and also limit them.

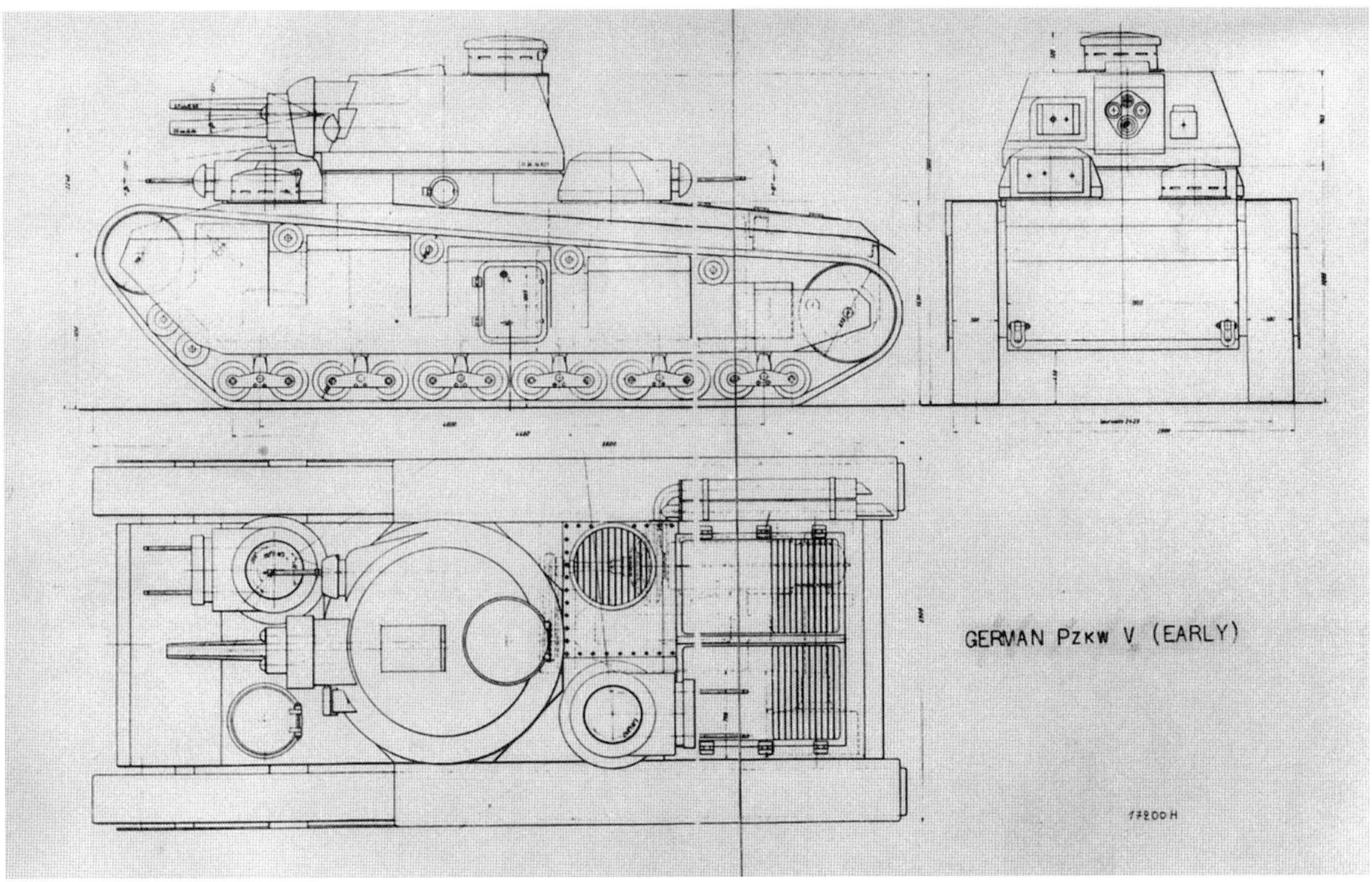

Rheionmetal design sketch of the running gear, taken over from the Large Tractor.

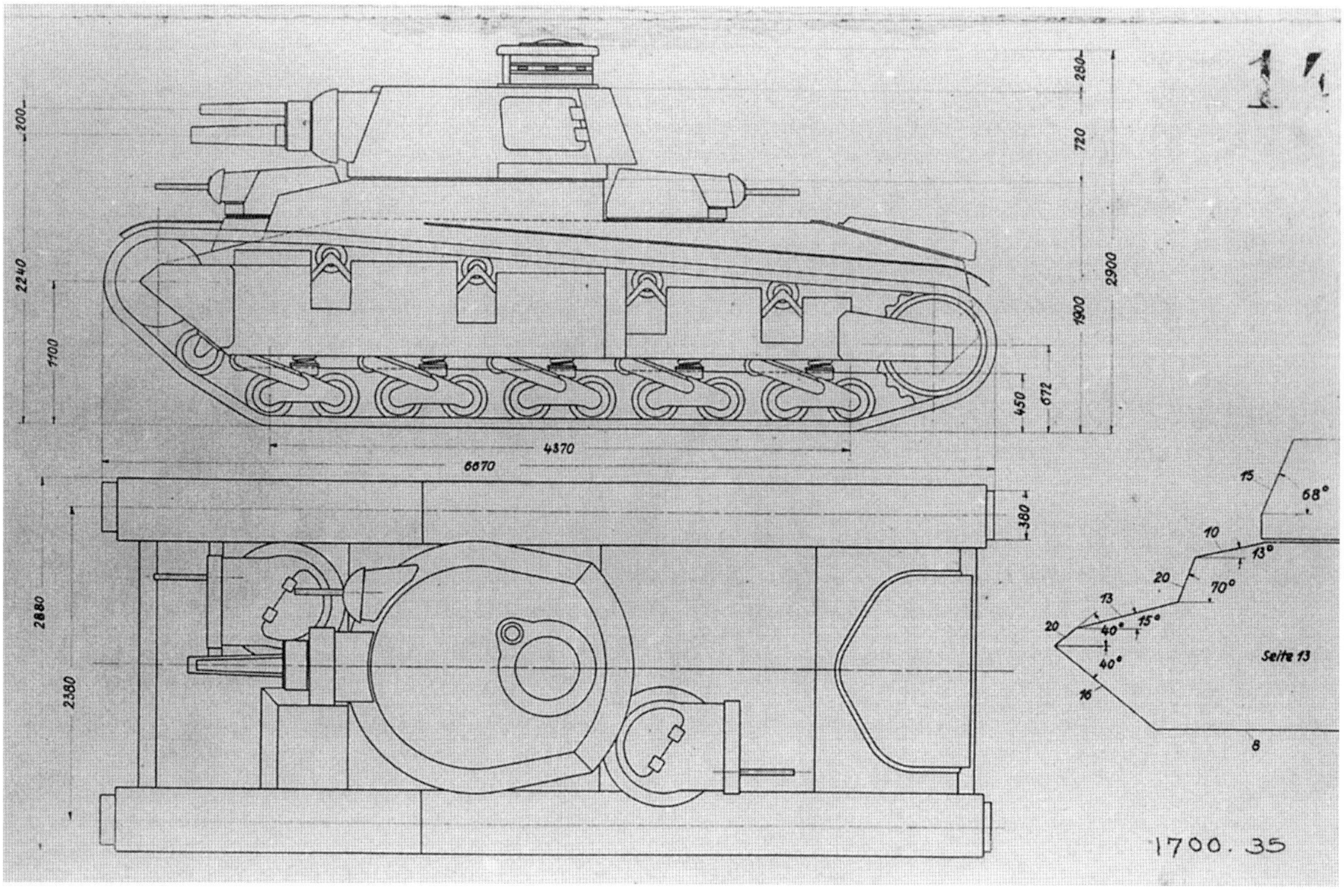

Design sketch with Rheinmetall turret and changed running gear.

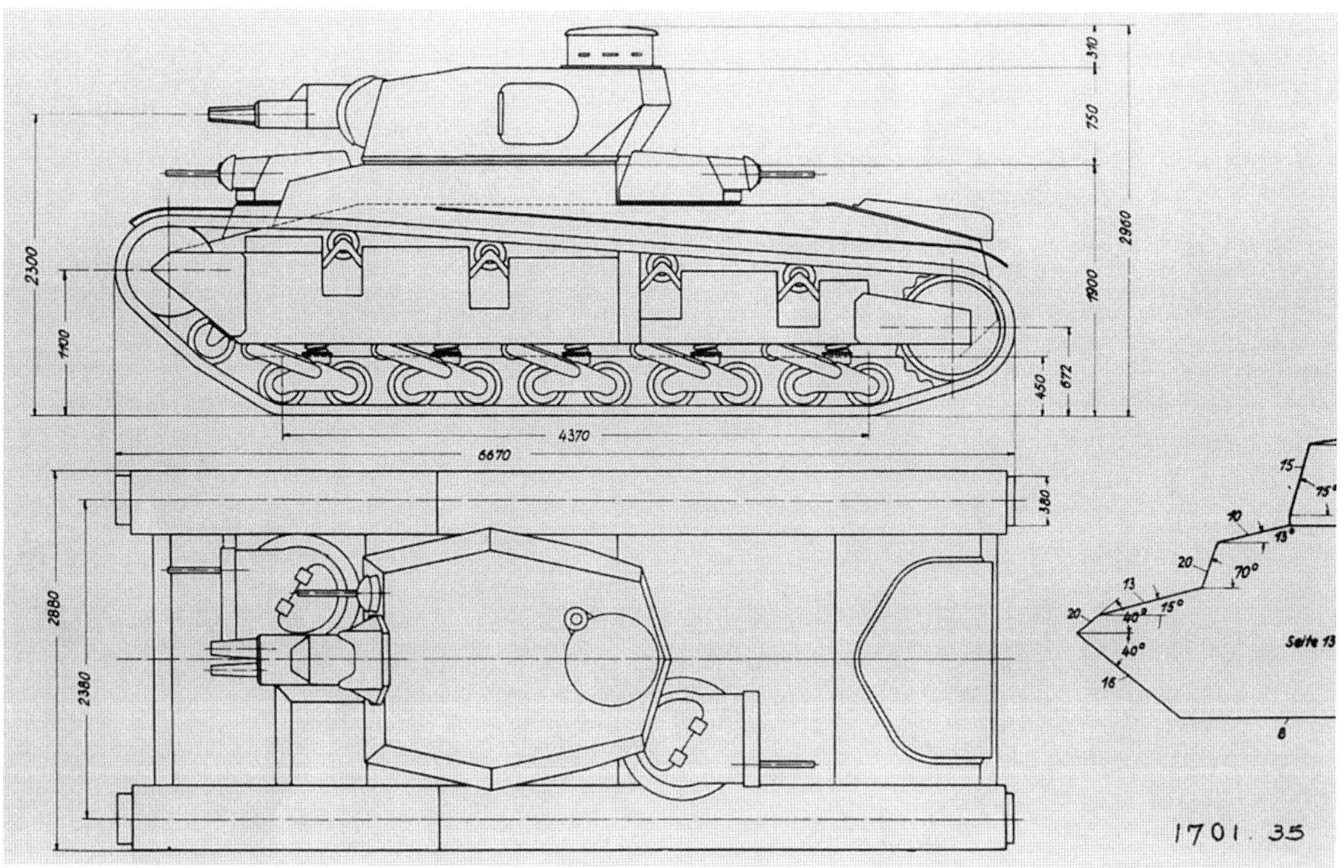

Design sketch with Krupp turret and changed running gear.

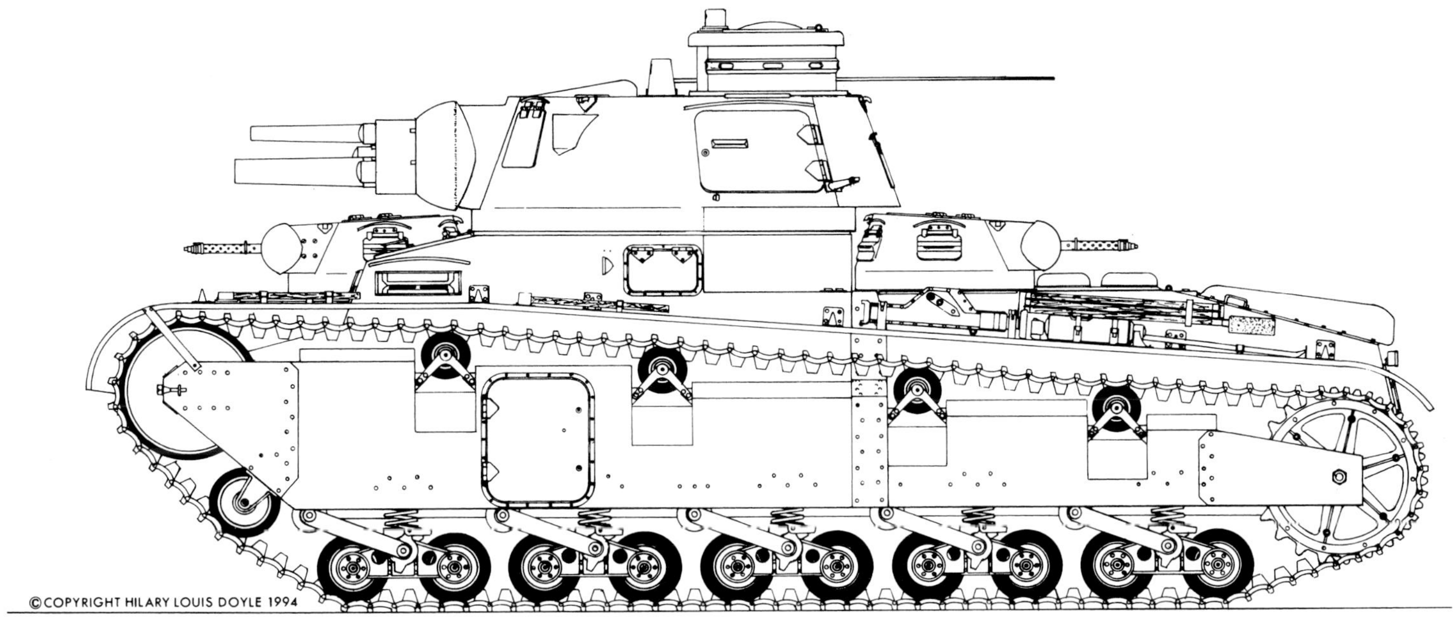

Newbuilt Vehicle No. 1 with Rheinmetall chassis and Rheinmetall turret.

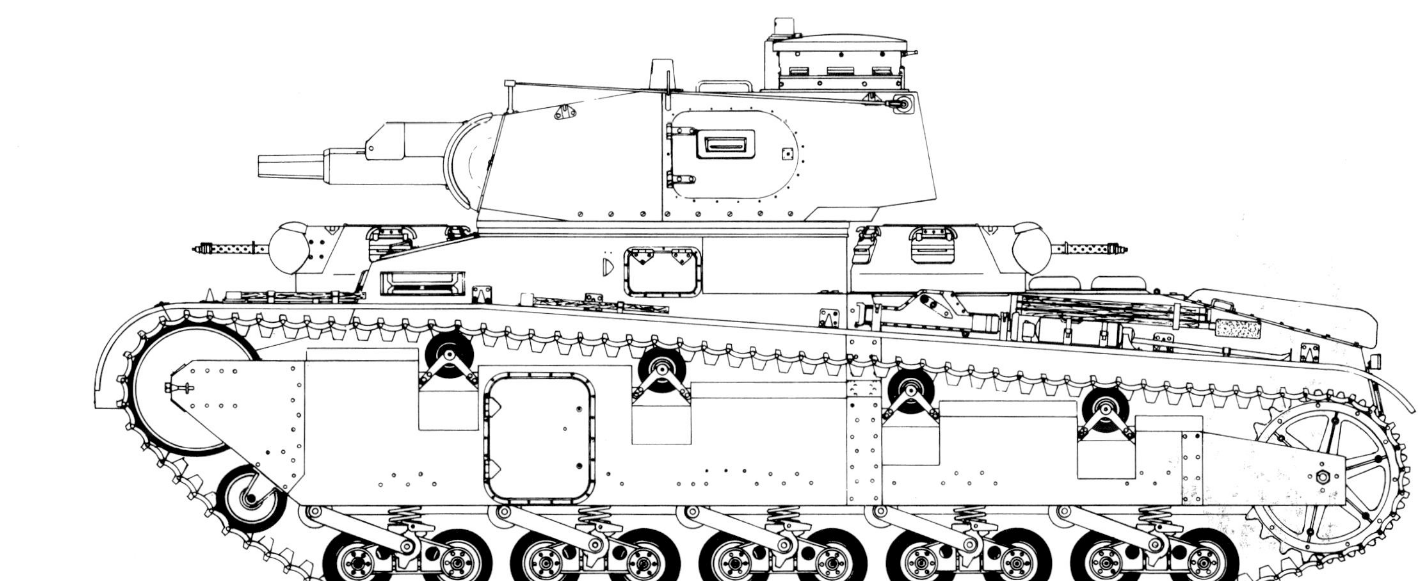

Newbuilt Vehicle No. 2 with Rheinmetall chassis and Krupp turret, modified machine-gun turret.

Streich thus recommended beginning strongly with the basic body of the M.Tr. and for the time being request information from A.6, so that we can fairly surely start work by April 1.

For this, it was necessary to speak with the Krefeld Stainless Steel Works first of all, as to how they wanted the armored body according to their concept. I have told Streich that we wanted to make it out of three pieces: front and rear ends welded, middle piece screwed on, with its plates sloping toward the strong support frame for the turret. The Stainless Steel Works must be asked above all how the front frame surfaces are to be built up, so that no blows from the back result. Streich thought one had to screw on the foremost surface, which takes up rammings, so as to be able to replace it.

As to the weight, Streich commented:

In the first calculation of the armor thicknesses, 17.5 tons of total weight, fully equipped, is to be the basis. According to his calculations, 20 mm in front, 13 mm on the sides can be attained, while the roof and floor can be kept thinner.

Thus he calculates with a 5% reserve: 500- kg—18 tons. The remaining 500 kg of the 18.5 tons should be left until last, when the total weight is to be determined finally and certainly and extras can be built into the armor thickness, especially in the middle part, thus in the sidewalls and turret armor. To be able to make the side plates thicker later, space must then be left for the running gear. He calculates that perhaps 10-15 mm would suffice. The thickenings could then be welded onto the body in one piece.

On April 20, 1933 it was further explained by Wa Prw 6:

In tractors of the weight class of the M.Tr., the crossing of streams causes problems, because the weight of the vehicle exceeds the load limit of war bridges. Sufficient ferries require numerous vehicles to transport them. Thus the idea has come up that the M.Tr., for which the wading depth of 1.20 meters is suggested, with the use of its own propulsion, be made floatable by some attached floats.

On May 16, 1933, Wa Prw 6 reported:

The research on a "Newly Built Gr.Tr." or "Newly Built M.Tr." should now be designated uniformly as "Newly Built Gr.Tr."

In a discussion with Wa Prw 6 on May 16, 1933, about the E.M.G. (*Uniform Machine Gun,* later called *M.G. 34*), in the turret of the Gr.Tr.Nb., it was decided:

Two wooden models: One model with belt loading, one with magazine loading of the E.M.G. would be sent quickly to Krupp in Essen. After the production of the first E.M.G. on June 1, 1933, a thorough examination of the weapon by Wa Prw 6 will be suggested.

From the discussions with Colonel Kuehn (Wa Prw 6) on June 30, 1933, the following was noted by Dr. Ritter (Krupp):

In the M.Tr. the body, just as in the Kl.Tr., must also be made removable, so that without the body a usable vehicle for peacetime purposes remains. Removal perhaps in three pieces: front, middle and rear.

On July 14, 1933, Mr. Griessmann (Krupp) reported:

I have thanked Colonel Kuehn (Wa Prw 6) particularly for having thought so kindly of us, although from the other side a large vehicle had been given to Rheinmetall.

It will be a very special pleasure for us to take on the large vehicle. Colonel Kuehn repeated that the decision as to the version of the large vehicle would not be made until autumn.

On October 2, 1933, Wa Prw 6 had orally requested a weight calculation for am M.Tr. in the 13-ton class from Krupp. Fried. Krupp A.G., Department A.K., replied on 7 October that the fighting weight for this version of the *M. 18,* including a five-man crew and 7.5 cm gun turret, would come to 13,000.00 kilograms.

As early as October 3, 1933, Wa Prw 6 had decided:

Neubau Fahrzeug (Newbuilt Vehicle) No. 1 with Rheinmetall chassis and Rheinmetall turret.

In the future, instead of "Gr.Tr.Nb.", the designation "Nb.Fz." (Neubau Fahrzeug) is to appear on all written communications and on all drawings.

2.1.3 Test Vehicles in Sheet Iron

Wa Prw 6 had drawn the Rheinmetall firm into the further development of the medium tank and issued a contract for the design of the entire vehicle with turret and primary weapon. The Rheinmetall turret had the 3.7 and 7.5 cm guns mounted one over the other in a port. The machine gun was mounted in a ball mantlet beside the gun port. The first design by Rheinmetall had rear drive and running gear of six pairs on roller bearings with coil springs, as taken from the Rheinmetall Large Tractor. Wa Prw 6 had declined to make a design for a Krupp chassis, and only Contract No. 66.825 of June 27, 1933, had been assigned to the Fried.Krupp A.G. of Essen for the design of the main turret structure, gun barrels and breech for the 7.5 and 3.7 cm guns of the Gr.Tr.Nb. The Krupp turret design with the 7.5 and 3.7 cm guns side by side was finished by the Department of Artillery Design in November 1933.

Contracts for production of two test chassis and one gun turret of sheet iron were assigned to Rheinmetall.

A contract to produce a main gun turret of sheet iron based on the Krupp design data of 1933, for the second test chassis built by Rheinmetall, went to Krupp in Essen. The Neubau Fahrzeug (Newbuilt Vehicle) No. 1 with Rheinmetall turret was finished in 1934, the Neubau Fahrzeug (Newbuilt Vehicle) No. 2 with the Krupp turret in 1935.

Neubau Fahrzeug (Newbuilt Vehicle) No. 2 with Rheinmetall chassis and Krupp turret.

Neubau Fahrzeug (Newbuilt Vehicle) No. 1 with Rheinmetall chassis and Rheinmetall turret.

2.1.4 Three Test Vehicles of Sheet Iron

Wa Prw 6 gave further contracts for the assembly of three chassis to Rheinmetall. On April 17, 1935, Wa Prw 6 sent an offer of a contract for a gun turret to Fried.Krupp, Department A.K., with the following explanation:

The partial completion and assembly of three Nb.Fz. gun turrets according to existing design requirements, in hard sheet metal, is to be assigned with a free hand.

A Turret consists of:

One 7.5 cm and one 3.7 cm barrel with semi-automatic breech and electric ignition.
Cradle with barrel brake and pneumatic recuperator,
Parts for the guns, brakes and recuperator, a small equipment case for each turret, for all three turrets together one more equipment case,
Turret attachments,
Electric turret-swinging machine with additional hand operation, elevating machine with hand operation,
Complete machine-gun mounts.
Bearings and drives for targeting scope, commander's optics and for the targeting rail,
Common complete targeting rail for both guns, seats for aiming gunner with 3.7 cm hand ammunition case and for commander, folding seat for loading gunner,
Lashings for elevation and traverse,
Bowdenex for machine-gun firing (foot firing), mount for entire F.T. system in turret and 2-meter rod antenna, mounts for intercom, as far as it concerns the gun turret, turret lighting, shell-case slide, turntable, shell-case ejector with foot operation and sliding-ring body, electric machine system with benzol motor including bearing mounts.
All cables and connections for turret turning, F.T. equipment, intercom system and turret lighting,
Drive for 12:00 indicator system and interior furnishing of the commander's cupola,

Newbuilt Chassis No. 2 with Rheinmetall chassis and Krupp turret.

Bolts, flaps, clasps, glass combinations and padding of all kinds for vision ports, openings, coverings,
Mounts for equipment present in the gun turret, fire-resistant paint.

Also delivered were:

All hard sheet-metal parts, worked, ready for mounting,
Complete optics for gun and commander, with parts as in first cast iron turret,
Machine guns; F.T. device and intercom,
Without factory and acceptance firing.

Newbuilt Chassis No. 1 with Rheinmetall chassis and Rheinmetall turret.

Neubau Fahrzeug (Newbuilt Vehicle) No. 3 to 5, details of the rear.

This contract for part production and mounting for three Nb.Fz. gun turrets of sheet metal with delivery date of January 1, 1936, was issued by Wa Prw 6 to Fried.Krupp A.G., Department A.K., Essen, on April 17, 1935. In the Krupp contract was specified:

If only chassis—the bodies being produced by other firms—are made, then the chassis are delivered by the vehicle firm free to the factory railroad station of the construction firm. The latter, in relations with the producing firm, is to prepare the bodies contracted for from it to the extent that it is in a position to deliver the finished vehicle 14 days after the arrival of the chassis. The chassis are test-driven by the producing firm without bodies, and if they live up to the requirements, they are to be turned over to the construction firm in orderly fashion through delivery action, in order to avoid a lack of parts and equipment at delivery to the recipient. After mounting the bodies, the Army Weapons Office reserves the right to give the complete vehicle a short test drive, which extends only to the working together of the chassis and body.

These three Neubau Fahrzeuge (Neubuilt Vehicles) No. 3, 4 and 5 with chassis by Rheinmetall and main turrets by Krupp were not finished until 1936. In a discussion between Krupp and Wa Prw 6 on February 21 and 22, 1936, it was decided:

In Nb.Fz. it would have been decided that the Rheinmetall turret was definitely cheaper, but in it the two barrels unfortunately lay one over the other. So that with the greater height, the view from the commander's cupola was nullified.

2.2 EQUIPMENT DESCRIPTION

The Newbuilt Vehicle consisted of the chassis with main turret and two machine-gun turrets. In the main turret, one 7.5 cm cannon with a Vo of 760 m/sec and one M.G. 34 were available.

The M.G. turrets were mounted at the right front and left rear, each with one M.G. The crew consisted of six men with a commander, aiming gunner and loading gunner in the turret, a driver at the left front on the chassis, and one machine gunner in each M.G. turret.

Neubau Fahrzeug (Newbuilt) Vehicle No. 1 with Rheinmetall chassis and Rheinmetall turret.

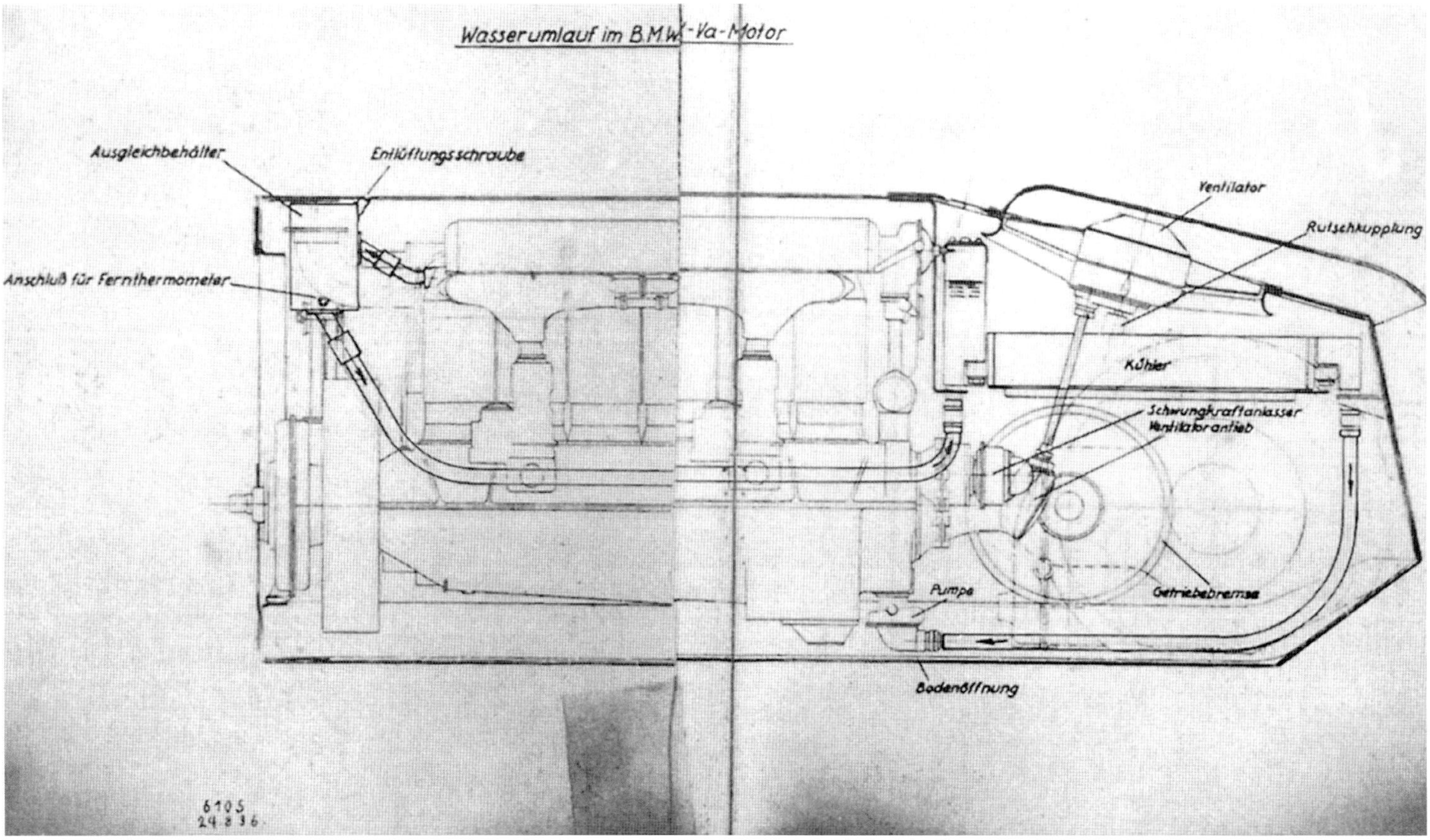

Cooling water circulation in the BMW Va motor.

2.2.1 Chassis

The chassis included the following main parts: Armored hull, motor and cooling system, lateral drive, main clutch, gearbox, steering gear, running gear, fuel tanks and ammunition storage.

2.2.1.1. Armored Hull

The armored hull was the carrier of the chassis. It consisted of armor plates welded together, with the following thicknesses and inclinations (from the horizontal).

Plate	Thickness	Inclination
Roof	10 mm	13 and 0 °
Driver's front	20 mm	70 °
Upper front wall	13 mm	15 °
Front dashboard	20 mm	40 °
Lower front wall	16 mm	40 °
Sidewall	13 mm	90 °
Back wall	13 mm	
Floor	8 mm	0 °

The front armor was supposed to be secure against 2 cm tank shells, the sides against 7.92 mm S.m.K. bullets from rifles or machine guns at ranges from 30 meters on and 90-degree angles.

In the armored hull the motors and powerplant components were mounted. The engine compartment was separated from the crew area by a bulkhead.

2.2.1.2. Motor

For the water-cooled BMW-Va gasoline aircraft engine, a dry-sump lubrication system was installed. It was an in-line six-cylinder motor, with a displacement of 22.9 liters with a governor set for 1400 RPM, giving 290 HP, and with the governor switched off it produced 320 HP at 1600 RPM. The cylinder design showed the form of a long steel box with water mantle and welded-on inlet and exhaust valve housing. The valves were dropped 60 degrees from the cylinders. There was one intake and one exhaust valve per cylinder. The crankshaft had eight white-metal bearings; the crankcase was made of aluminum. There was a carburetor for each three cylinders.

Starting the motor was done either with compressed air or by hand (starting crank). There was no generator. A.M. pumps were planned for the fuel feed, and the fuel consumption amounted to 220 g/HP/hr. The oil flow of the motor ran through the hollow crankshaft, the connecting-rod bearings had jets for the oil, which was led through the hollow rod to the piston pin. The total oil capacity was circulated 2.5 times per minute, and the oil consumption was some 3 kg/hr.

2.2.1.3 Cooling System

The cooling was done by water circulated by a centrifugal pump. There was a two-part element radiator, one element of which was used to cool the oil. The ventilator was run automatically by means of a sliding clutch. The cool air intake was in the front part of the engine compartment, the exit above the ventilator. The radiators were protected by gratings. Low air pressure in the engine compartment caused the ventilation of the fighting compartment.

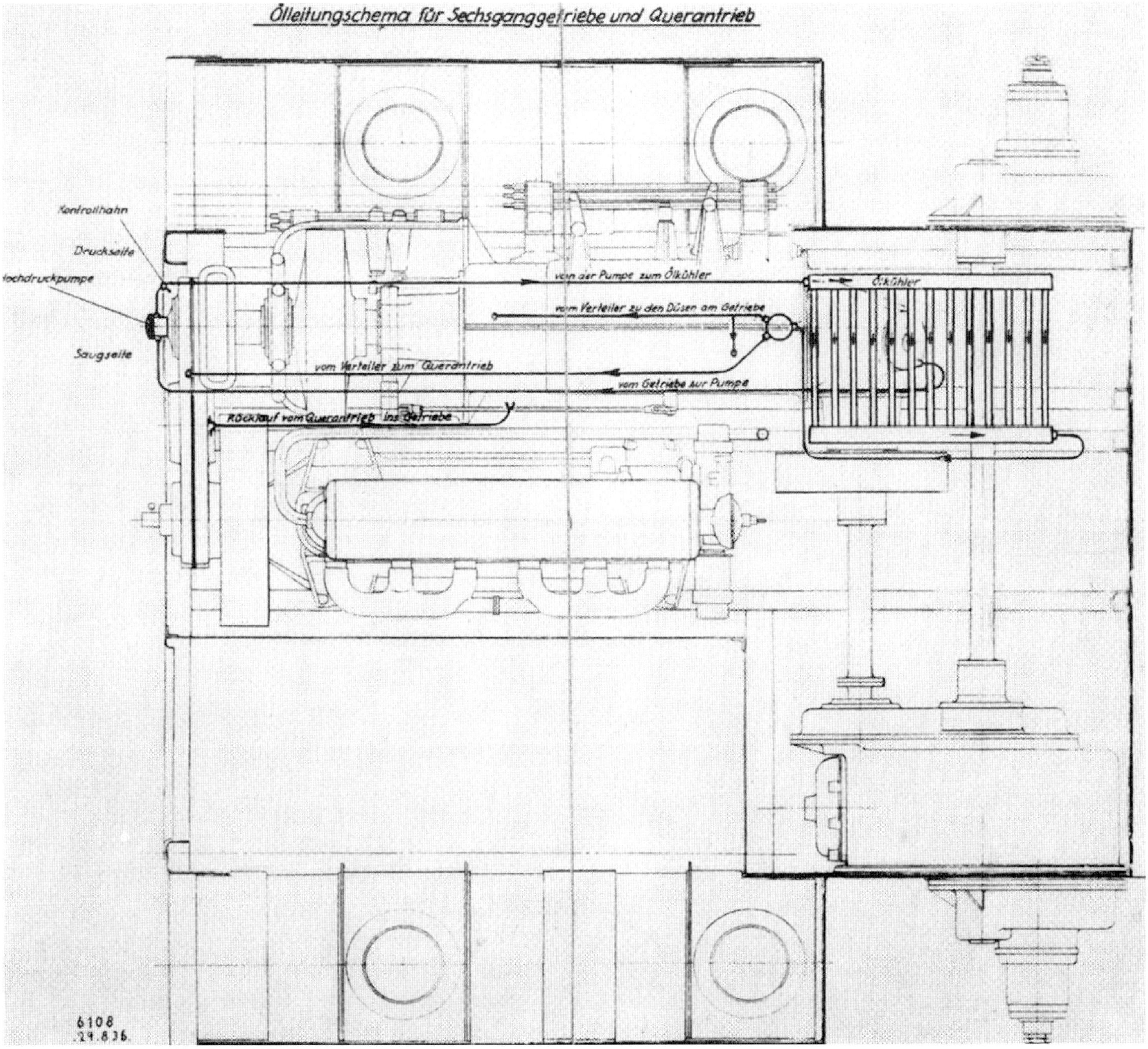

Oil circulation schematic drawing for the six-speed gearbox and driveshaft.

2.2.1.4 Power Train and Main Clutch

The gear ratio was set up so that with a cone wheel with 22 teeth at the end of the driveshaft, a speed of 30 kph at 1400 rpm was attained. From the driveshaft, the power branched out to an oil pump to lubricate the entire system. The clutch was a two-disc type laid out with ample dimensions.

2.2.1.5 Transmission

The SFG 280 gearbox made by ZF had the following ratios: 1st gear: 7.75; 2nd gear: 4.1; 3rd gear: 2.62; 4th gear: 1.75; 5th gear: 1.3; 6th gear: 1, and reverse: 7.75. Second to sixth gears were synchronized. The 22-tooth bevel wheel (v max 30 kph) could be exchanged for one with 25 teeth (v max 34 kph).

2.2.1.6 Steering Gear and Brakes

The Wilson steering gear (installed in Nb.Fz. No. 1, 3, 4 and 5) was mechanically activated. The calculation of the track pull for 15,000 kg was made with 1.2-fold security. The steering brakes also had to remain effective with light oiling. The installation of the steering gear was made on the sidewalls of the hull; they were lubricated by a pump on the driveshaft. The steering wheel was to be replaceable by a steering tiller. For testing, a Cletrac steering gear (double differential) was installed in Nb.Fz. 2. Corresponding to the ratio in the Cletrac gear (3), the outer track ran a certain amount faster, the inner track slower by the same ratio, but it did not come to a standstill. Two driving brakes, independent of each other, were used; they worked equally effectively for forward and reverse.

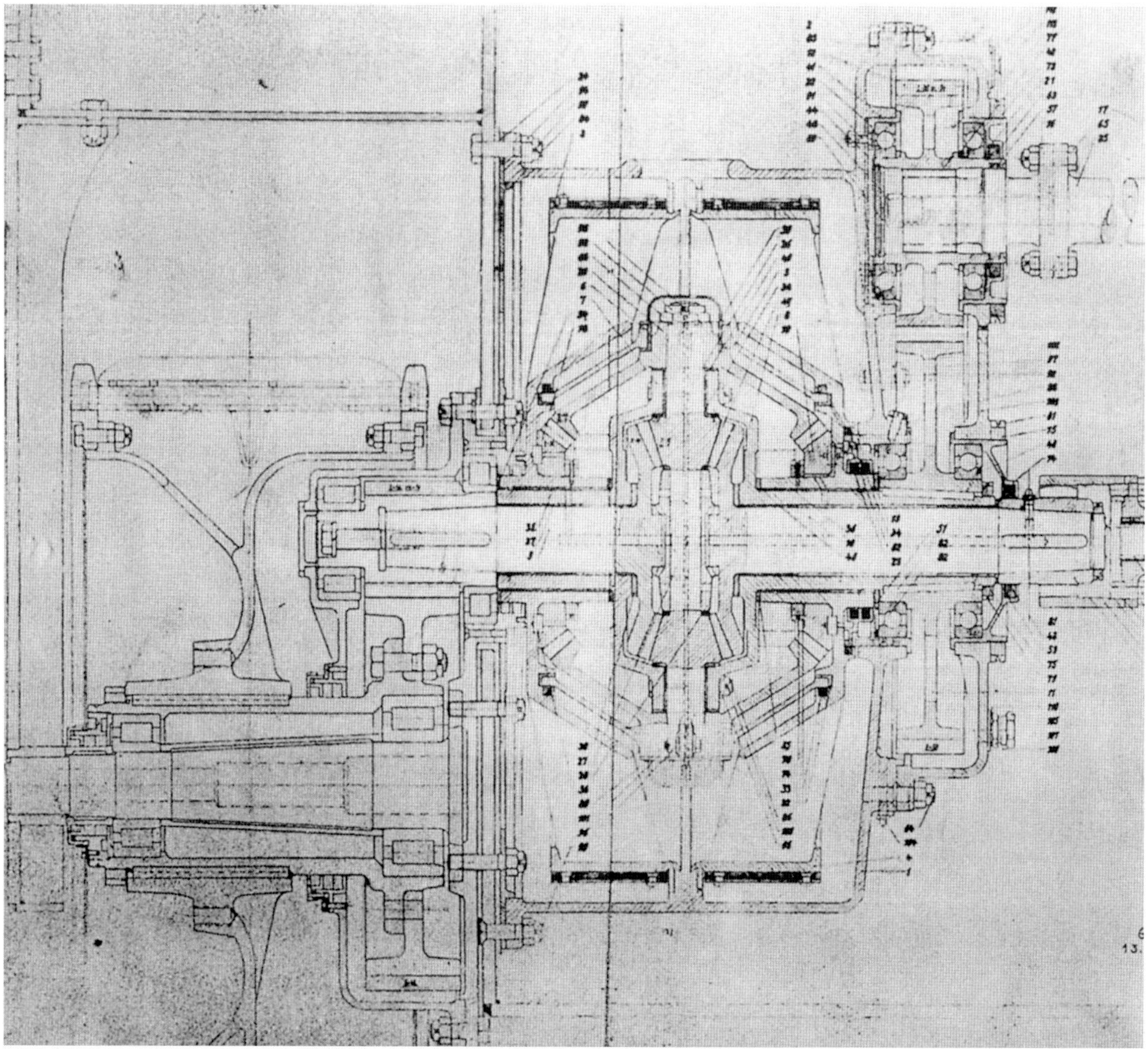

Wilson steering gear with flanged drive wheel.

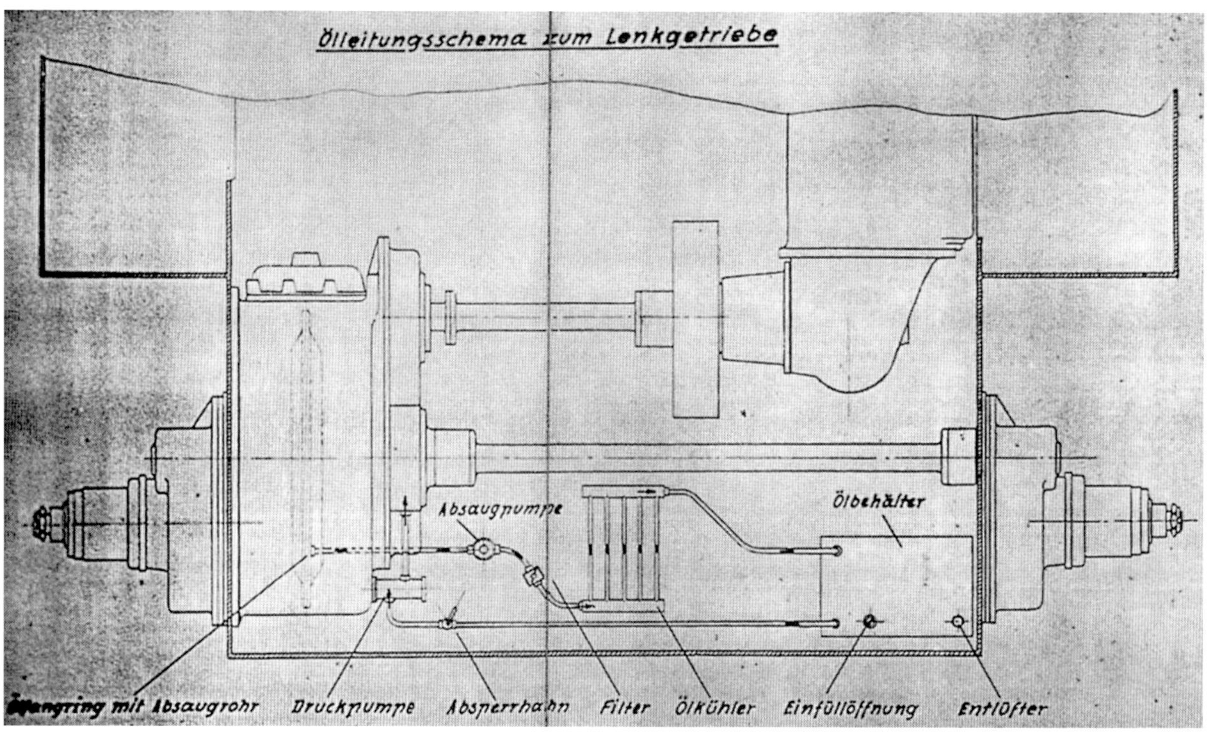

Oil circulation to the steering gear.

2.2.1.7 Runnung Gear

The rear drive was effected by a later shaft. The coil-spring running gear consisted of ten pairs of road wheels attached to the hull. Their rubber belts were made of DA Material by the Continental firm. Between the first road wheel and the steering wheel was an unsprung wheel; the drive wheel was behind it. The suspension play of the road wheels extend upward 150 mm and down 100 mm. The steering wheel height was 1100 mm.

2.2.1.8 Tracks and Track Tightener

Every track consisted of 118 Type Kgs 61/380/120 links. Being 120 mm long, they gave a track length of 14.2 meters. The track width was 2.37 meters and the ground contact was 4.4 meters. In an earlier test chassis, Kgs 65/380/100 tracks with 100 mm divisions were tested. The track width was set at 380 mm; with a fighting weight of 19.5 tons, this gave a specific ground pressure of 0.69 kg/cm2.

2.2.1.9 Fuel Supply

The fuel supply of 457 liters sufficed for 110 kilometers of driving.

2.2.1.10 Ammunition Storage

80 7.5 cm rounds, 50 3.7 cm rounds and 6000 rounds of machine-gun ammunition were carried on board.

2.2.2. Turret

The Krupp design consisted of, among other things, one 7.5 cm and one 3.7 cm barrel with semi-automatic breeches and electric firing,

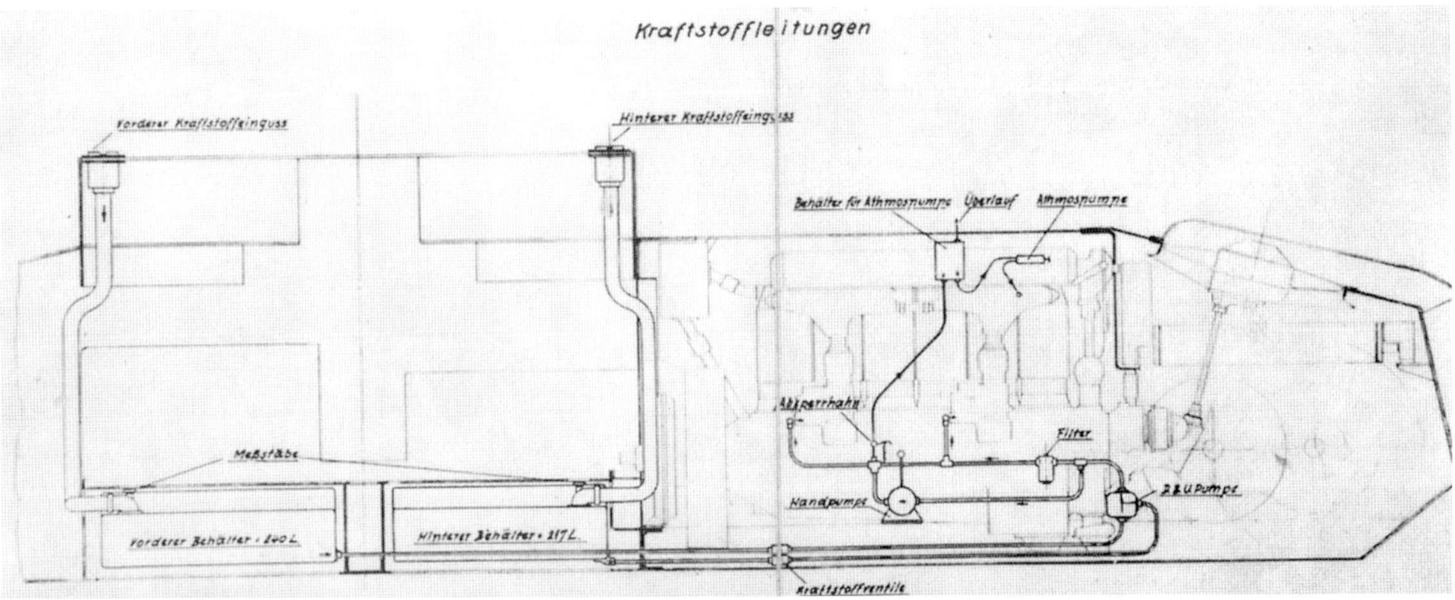

The offset fuel lines.

Details of the Newbuilt Vehicle's running gear.

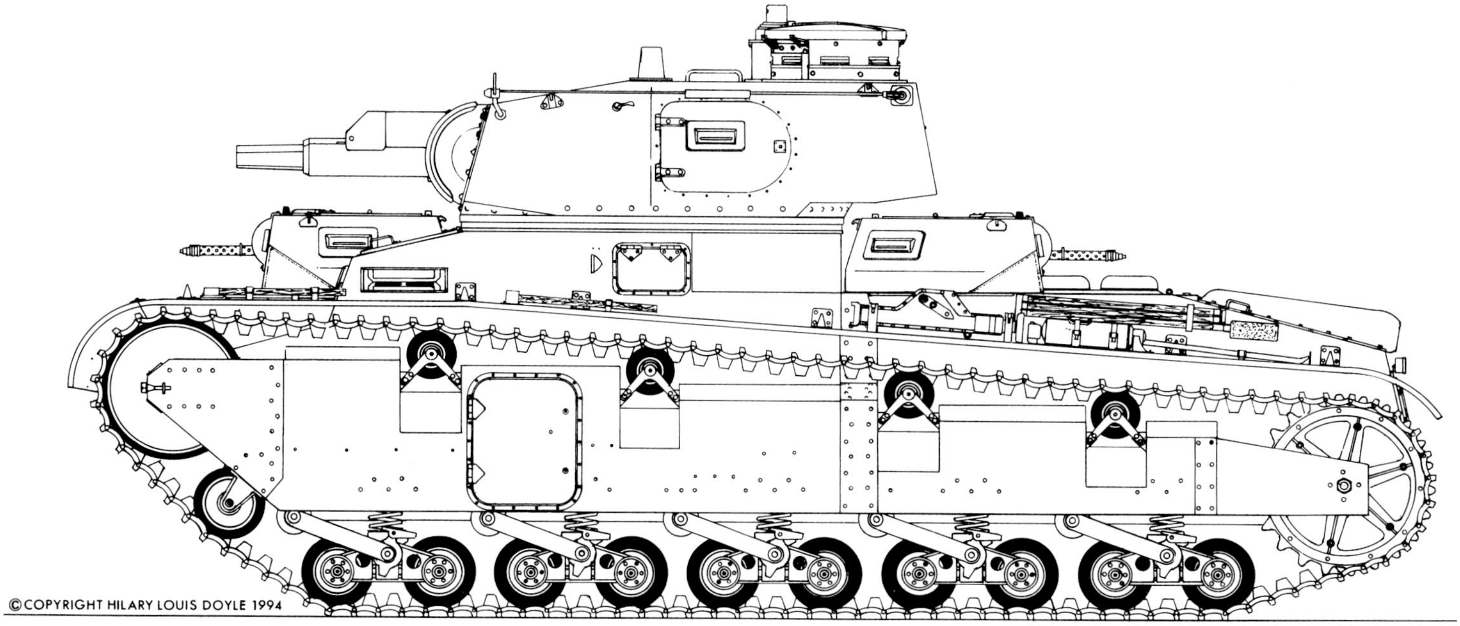

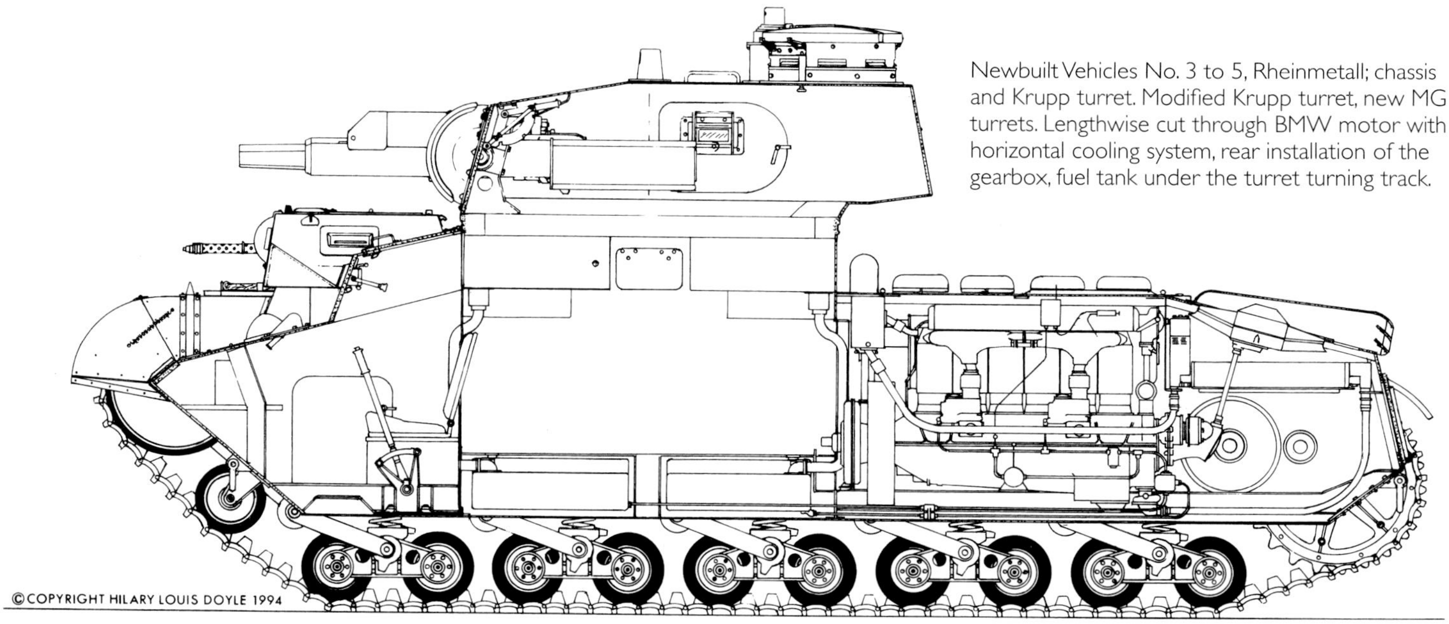

Newbuilt Vehicles No. 3 to 5, Rheinmetall; chassis and Krupp turret. Modified Krupp turret, new MG turrets. Lengthwise cut through BMW motor with horizontal cooling system, rear installation of the gearbox, fuel tank under the turret turning track.

Top view not to exact scale

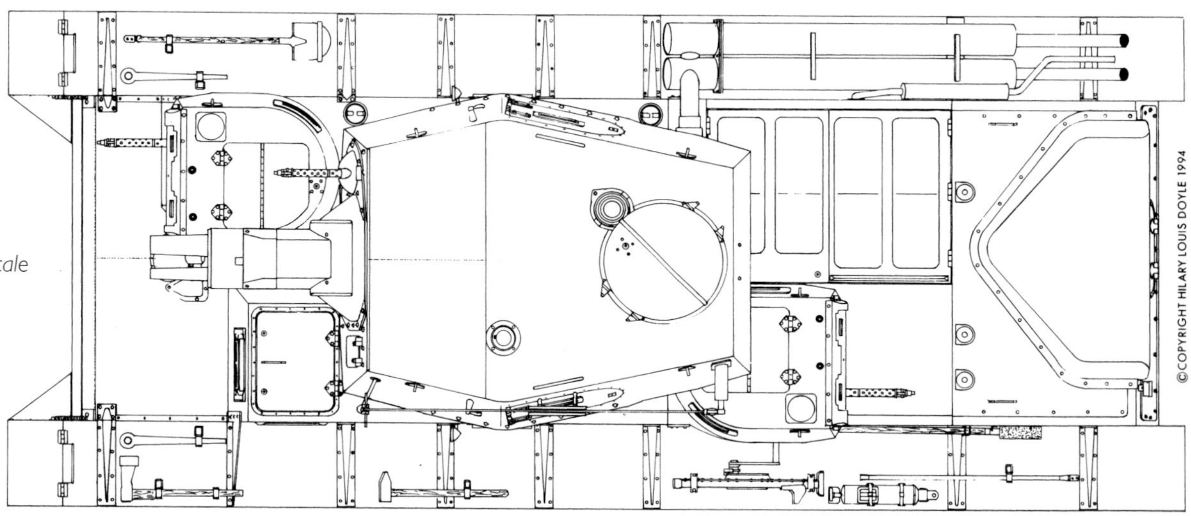

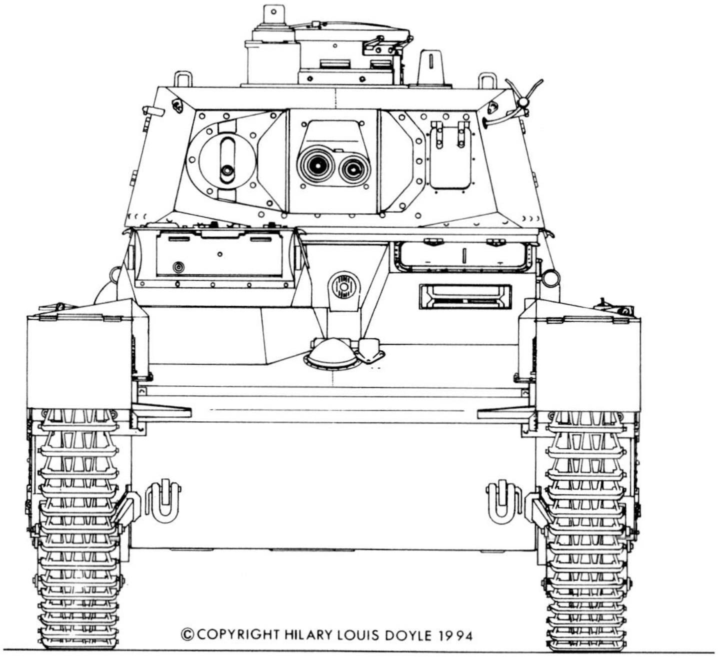
©COPYRIGHT HILARY LOUIS DOYLE 1994

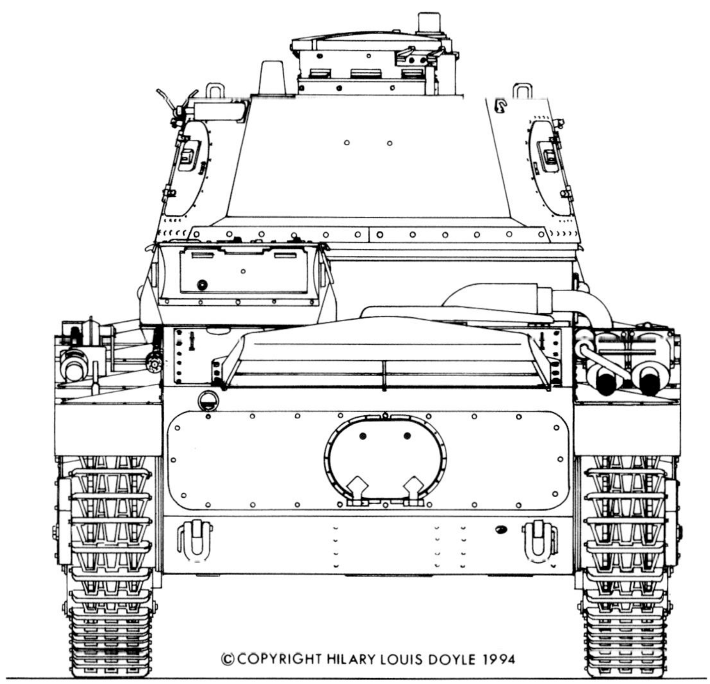
©COPYRIGHT HILARY LOUIS DOYLE 1994

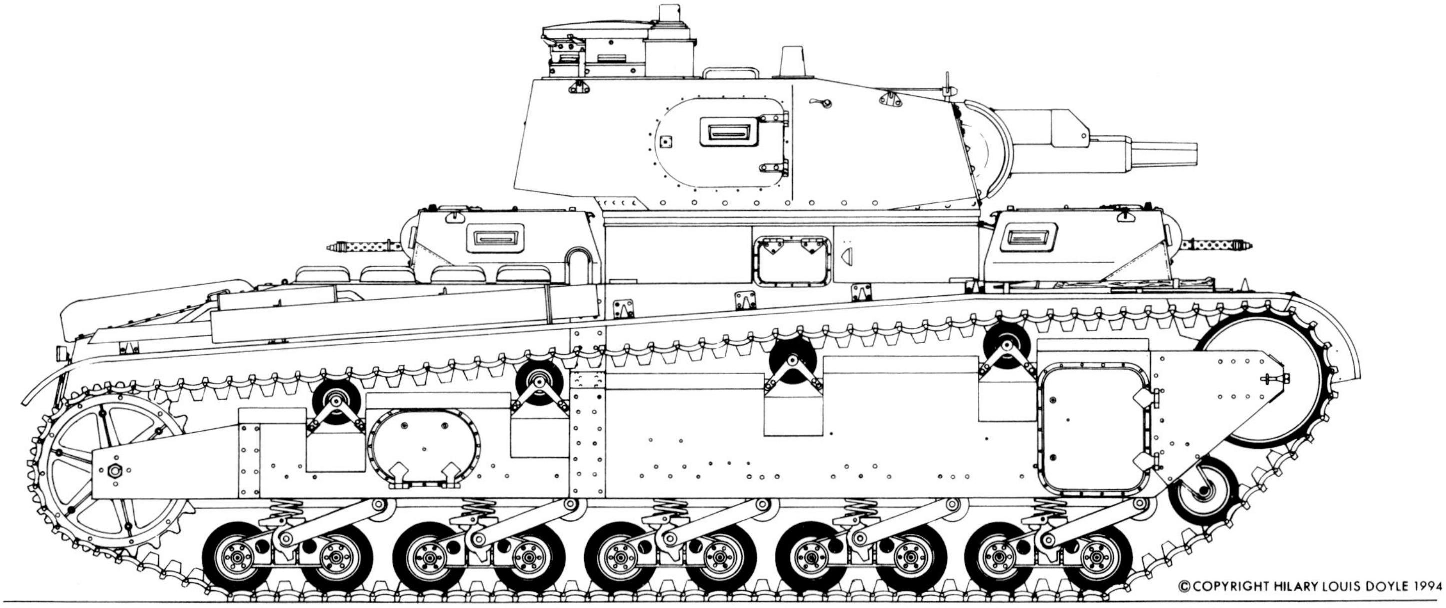
©COPYRIGHT HILARY LOUIS DOYLE 1994

Details of the driver's position, the front MG turret and the weapon arrangement in the main turret.

the cradle with barrel brake and pneumatic recuperator, an electric turret-turning machine with additional hand operation, an elevation machine with hand operation and a commander's cupola.

The turret weight of the first version was 2800 kg, that of the second 2700 kg, the light servicing circle 1660 mm. The turret armor was 13 + 1 mm all around, that of the turret roof 10 + 0.5 mm, and that of the shield 13 + 1 mm. The tilt of the turret walls was 75 degrees.

2.2.2.1 Armament

The armament was mounted in a jacket cradle in the loophole flap. The 7.5 and 3.7 cm barrels had a common base. The EMG 34 was laid in a ball loophole, coupled with the cradle or free swing range, -10 to +25 degrees in elevation, 14 degrees to the left and 20 to the right. The firing height was 2290 mm. The gun barrels made by Krupp were formed as full barrels and had self-acting breech locks. This lock required the arrangement of the 7.5 and 3.7 cm guns side by side, while the side breech built by Rheinmetall made mounting the two guns one above the other necessary.

2.2.2.2 Targeting

Two different targeting systems were planned for the Krupp turret:

1. Version: Aiming gunner: angled aiming scope and emergency targeting scope with fixed viewers, coupled with gun cradle, 12 o'clock system for sides. Commander: Panoramic scope with fixed viewer and 12 o'clock indicator inside the scope.
2. Version: Aiming gunner: Angled aiming scope with fixed viewer, coupled with gun cradle, 500 mm long aiming rail. 12 o'clock system for sides. Commander: as in type 1.

2.2.2.3 Swinging Mechanism

In the swinging mechanisms, the elevation wheel turned 1.84 degrees, the traversing wheel 1.87 degrees. The number of turns for the electric traversing machine was governed without steps. Thus, the main turret could be swung from 0.1 to 2.5 turns per minute. An electric machine with a benzol motor including bearing blocks was on hand.

2.2.2.4 Radio Equipment

The entire radio system was located in the turret with a two-meter rod antenna; the intercom was also installed in the turret. The radio equipment included a 20-watt sender and an ultra-short-wave receiver with a range of 4 to 6 km.

2.3 TESTING AND USE

The main disadvantages that appeared in testing were caused by the use of an aircraft motor. No other ready-made motor of 300 HP was available at the time. The aircraft motor with its low engine speed and high torque required a very heavy power train. The entire machinery was thus made too heavy. The running gear, which was changed several times, was revised on the basis of subsequent experience.

Newbuilt Vehicles were used at the Putlos shooting school and for training armored troops. The Newbuilt Vehicles No. 3 to 5 were shipped to Norway in April 1940 and used there.

Newbuilt Vehicle in action in Norway.

3. ESCORT VEHICLES (B.W.)

3.1 DEVELOPMENT

The first indication that Wa Prw 6 was interested in a successor to the Newbuilt Vehicle became clear in a letter of November 14, 1934, from Wa Prw 6/Vb to Director Ritter of the Fried.Krupp AG in Essen.

Situation: *Director Dr. Ritter has asked the undersigned of 6/Vb to be released from tractor-building tasks for the reason that because of the extensive tasks for Wa Prw 4w 6, the necessary number of designers would no longer be available.*

Wa Prw's position on this matter:

The B.W. test vehicle (Rheinmetall) with the original running gear and leading-wheel suspension.

In view of the absolutely necessary short development time, the cooperation of the Krupp firm cannot be done without at this time, since a large part of the important design elements after the troops' experiences in cooperation with Wa Prw 6 is anchored in the Krupp firm.

Removing the Krupp firm from tractor-building can only be considered in a broad view, but must, in the interest of national defense, be interrupted at once.

The present situation, aside from the chassis, involves the design of the gun turret of the Z.W. and improved Nb.Fz. (B.W.)

The request for release from the Z.W. design could have been discussed four months ago; you cannot be released now because of scheduling reasons.

Since the design elements for the improved Nb.Fz. gun turret (B.W.) in the Z.W. turret and the Nb.Fz. turret are already available in their farthest extent, the department must still withhold the decision in the developmental areas of the B.W. gun turret.

Should the gun turret of the B.W. be built by a different firm, then the following design drawings must be requested, in view of an absolutely necessary uniformity and with consideration for the designs made to date.

a) of the completed electric turret drive of the Nb.Fz.,
b) the commander's cupola of the Z.W.,
c) of the gun with cradle, recuperator and brake,
d) the turret shape of the Z.W.,
f) the hatch cover of the Z.W.,
g. the auxiliary machine-gun mounts of the Z.W.

The decision as to the wishes of the Krupp firm is not only dependent on the position of Wa Prw 4, but must in this special case be made by a higher authority according to reasons of national defense.

Thus, the Krupp firm is asked again for a written position statement on the situation.

In the report on the *state of the armored combat vehicle*, Wa Prw 6 stated that the official beginning of the design of the *B.W.* and the approval of the design on the wooden model on February 25, 1935, had taken place. A photo of the *B.W.* wooden model (Rheinmetall) shows the handwritten comment that the contract for its production was given to Rheinmetall on February 26, 1935. Krupp likewise produced a *B.W.* wooden model (chassis and upper parts) under contract no. 41/1952 Wa Prüf 6/Va Nr. 67025/35.

Both firms, Rheinmetall and Krupp, were required by Wa Prw 6 to submit a preliminary design for a *B.W.* with front drive, instead of the rear drive formerly used on the *Large Tractor* and *Newbuilt Vehicle*.

Instead of the formerly used aircraft motor, a new water-cooled motor for the Z.W. and *B.W.* test vehicles was requested from the Maybach firm, *Type HL 100 TR*. This motor was the first Maybach disc-shaft motor, the design of which was begun, according to the date on the drawings, in 1933-34. The advantage of the roller-bearing crankshaft in terms of the motor measurements (thus for the installation in the armored vehicles) becomes clear in comparing the HL 100 TR with the "conventional" Maybach DSO 8 Spez.:

B.W. wooden model (Rheinmetall).

Maybach motor, Type HL 100 TR.

Type		DSO 8 Spez.	HL 100 TR
Bore	mm	92	100
Stroke	mm	100	106
Cylinders	—	12	12
Performance	kW	110	220
Cyl. spring	mm	144	127
Cyl. Spring/bore	—	1.56	1.27
Stroke volume	dm^3	7.97	9.99
Building volume	dm^3	815	738
Bldg.vol/stroke vol.	—	102	74
Weight	kg	510	800
Weight/Bldg.vol.	dm^3/kg	0.6135	1.084

Two Values deserve attention: The relative cylinder spring (cylinder spring/bore) of the HL 100 TR amounts to only 81% of the DSO 8 value, and the specific weight of the motors (weight/building volume) as a measure of utilizing the volume taken up by the motor, that could be increased in the HL 100 to 176% of the DSO 8. Ten test motors of the HL 100 TR are to be made for installation in the Z.W. and *B.W.* test vehicles. The HL 100 TR produced 300 HP at 3000 RPM.

3.1.1 Design of an Escort Vehicle

Krupp reacted quickly to the request as to design requirements from Wa Prw 6 by presenting the following design of a B.W. on April 13, 1935. This document is extraordinary in the rare and precise love for detail with which the reasons for the component development and the advantages of their choice are affirmed.

Enclosure with A.K.No.51239 of April 13, 1935:

Design of a *B.W.*

I. Brief Characterization.

The W823 and W824 designs present in 1/10 scale a tracked vehicle of 17.2 to 18.5 tons according to armor. The vehicle has front drive of the tracks and rear engine. The top speed amounts to 30 kph. The main turret with 7.5 cm cannon and 1 MG is situated about in the middle of the vehicle, somewhat to the left side, to make up for the powerplant parts to the right. It is equipped with a vision cupola for the commander. At the right front, a MG turret is planned. The crew consists of 6 men, specifically:

The commander in the armored turret,

The driver at the left front,

The aiming gunner and loading gunner in the armored turret,

The machine-gunner in the right front MG cupola,

The radioman behind the driver.

For the armor, two possibilities are foreseen, specifically with a calculated total weight of

	17.2 tons	*18.5 tons*
Body		
Upper front panel	14	20
Middle front panel	20	30
Lower front panel	14	20
Sidewall	14	20
Back wall	14	15
Floor	6	8
Armored Body, upper part		
Front wall before driver	20	30
Sidewall	20	30
Back wall	14	20
Roof	10	10
Turret		
Front wall	14	20
Side and back walls	14	20
Commander's cupola	14	20
Front machine-gun turret	13	20

II. Chassis

On the basis of previous experience, the motor is moved backward and the track drive forward. We see the main advantages of this layout as:

1) *Even load distribution,*
2. *Main turret in the middle of the vehicle,*
3) *Less inclination to climb out of the tracks on turns,*
4) *Shortest way from driver to shifting and steering controls and to clutch,*
5) *Easy and readily visible installation and removal of the whole power train,*
6) *Good view of the terrain for the driver,*
7) *No discomfort for driver from motor heat and exhaust gases.*

a) Armored Body

The armored body, of resistant construction, is welded together of special steel plates of varying thicknesses, with appropriate reinforcements and lateral walls in the interior. Strong towing hooks at front and back serve for towing.

Through several hatches with closable covers, the gearbox is accessible for installation and maintenance.

The body floor is reinforced by thoroughgoing carriers. A large, well-sealed opening in the floor, under the machine-gun cupola, serves for exit if necessary and for letting messengers out.

For outer lighting the body has two lights on the track aprons, one large spotlight on the front slant and one taillight with stop light in back.

The motor and cooling system are protected by armored gratings.

b) Motor with Cooling System

The water-cooled 12-cylinder Maybach HL 100 motor, with a maximum of 300 HP, is planned as the powerplant. It is housed at the end of the vehicle with the cooling system and electric aggregate for turret turning in an area closed off from the gun space. Besides the electric starter, a hand crank serves for starting. From the exhaust manifold on the motor two curving ducts take the shortest route to the muffler, which hangs on the rear wall. Since the motor is equipped with dry-sump lubrication, it works problem-free in connection with an additional oil reservoir in larger crosswise situations.

Special value was placed on a sufficient cooling system with large openings for intake and exhaust of cooling air. The radiator is installed at the side of the motor. In order to prevent warm motor air from entering the fighting compartment, the air ducting was chosen so that low pressure exists in the motor room. The air is sucked away through a grating over the motor by two ventilators of 500 mm diameter pushed out through another grating. After loosening the belt drive, the ventilators can be folded up with the gratings, whereby the right cylinder-head cover becomes accessible. The maintenance gun space can be ventilated through a flap in the lateral wall.

There are three fuel tanks with a total capacity of 480 liters installed under the turret platform, and they are closed off by special protective panels for reasons of fire safety. The fillers are on the outside of the vehicle. The fuel suffices for a range of about 200 kilometers.

From the motor, the power goes through a swing-free shaft with two joints to the two-disc clutch. Motor and shaft lie somewhat to the side of the turret center to gain space for the electric cables to and shell ejector from the turret.

c) Gearbox and Clutch

The clutch is not right next to the gearbox, but mounted at some distance, so as not to block the servicing area for the machine-gun turret. Thus, the masses to be advanced in shifting are still comparatively meager; likewise the lever for activation is short.

In the front of the body is the SSG 75v gearbox, which was first equipped with 6 forward speeds (highest ratio 1:9) and one reverse speed, but finally with only 5 forward speeds (highest ratio 1:6). The gearing is aphonized and synchronized, and is driven by a double shaft from the clutch. The gearbox with its shafts, one above the other, is narrow and can be operated by the driver, who sits nearby, without the need for handles that stick easily.

A bevel-wheel drive with spiral gearing (ratio 1:1.2) is flanged onto the gearbox, from which the power is carried over the double driveshafts to the right and left steering gear.

d) Steering Gear and Side Shafts

The steering gear uses a Wilson planetary drive with two types of gear wheels:

1) *A toothed arc with inner and outer teeth, driven by the front cross shaft from the gear-wheel section.*
2. *A central sun wheel linked with a Brake Drum I.*
3) *Four planetary wheels linked through and armature with the gear-drive pinion.*

For driving straight ahead, the Brake Drums I and thereby the sun wheels on both sides are held immobile. The planetary drives act only as gearing and drive both tracks. To make a curve, the drive on the inside is first disconnected by releasing the Brake I and freeing the sun wheel; then the inside track is braked by pressing the Brake II. Changing the brake bands is easy to do. First there are hand levers provided for steering, which can be hand-operated without the additional aggregate as a result of the selected ratios. But a handwheel can also be installed. In both cases the transitional organs are very short thanks to the closeness of the steering gear.

The side gearing with 1:3.5 ratio is attached to the outside wall, protected by armor, and drives the drive wheel with 726 mm semicircle.

e) Driver's Space

The driver sits between the left steering gear and the gearbox, and so far forward that he can see the terrain in front of the vehicle very well. He can see over the front ends of the left and right track aprons. He observes to the front either through a glass combination after opening a simple flap, or through a curved telescope, which he pushes to the center from the left side, when the flap is closed. To the left side of the driver is a normal sight hatch fitted with a glass combination. A direction disc is also installed and easy to see. The driver has before him, in customary arrangement, the foot pedals for the clutch, gas and foot brake. The steering levers can also be used for braking and stopping the vehicle. Above the driver is a hatch for entry and exit. Before the driver, an instrument panel with the essential instruments is easy to see.

f) Running Gear with Drive and Tracks

The vehicle runs on tracks 380 mm wide; with a ratio of 1.44:1 between the ground length of 3,450 mm and the track width of 2390 mm, the vehicle is easy to steer. The drive wheel at the front has a semicircular diameter of 726 mm.

We suggest tests with two different running gear, specifically with eight-wheel and six-wheel running gear. Both versions have their advantages.

1) *8-wheel running gear as on drawings W 823 and 824.*
Four pairs of wheels of 420 mm diameter and hung on U-shaped units, which are built especially strong to take up the lateral forces when steering. The units are mounted in basins that are screwed to the sidewall and reinforced on the outside by diagonal sheet-metal walls. Each pair of wheels is supported by a coil spring on the sidewall. The wheels are sprung so that they swing as vertically to the track as possible, in order to prevent friction on the perimeter and greater heating. In this running gear, double wheels with inside guiding of the tracks are planned. The upper part of the track is carried by four return rollers of 300 mm diameter. The track is tightened by a large leading wheel of 700 mm diameter, mounted on an easily adjustable crank.

2) *6-wheel running gear as on drawings W 825.*
The pairs of wheels are so mounted on flying cranks that one wheel turns around the crank shaft, to which two leaf springs are screwed, which carry the second wheel. The upper part of the track is carried by three return rollers of 300 mm diameter. The track is tensed by a leading wheel of 600 mm diameter which, as in the other running gear, is mounted on an easily adjusted crank. Since in the tests to date with wide tracks and inside mounting of the double wheels the rubber bandages on the wheels wore out unequally after a short time, we have simply suggested wide wheels and outside driving of the tracks for this running gear. Thus, the operation also becomes easier. One test with simple rollers should be recommended.

III Body

a) Armor Box, Upper Part

The basic shape is based on the placement of the driver's seat at the left front and the machine-gun turret at the right front; the lateral extensions serve for ammunition storage. The roof of the upper part of the box, which bears the turret, is supported by strong carriers against the sidewalls and the bottom of the chassis, so as to absorb the recoil of the 7.5 cm weapon. At the left front of the roof a hatch for entry and exit of the driver and the radioman, who sits behind him, is installed. The back part of the upper part of the armor box is removable in itself, so that major repairs may be made to the motor and cooling system.

The slanted sidewalls have coverable gratings for the entry and exit of cooling air. The gratings can be opened and closed from the fighting compartment. In the roof there is a flap intended for filling cooling water and for examination of the carburetors. The rod antenna is mounted to the front on the left sidewall. It lies on the track apron when folded down and is thus well protected.

*b) **Main Turret***

The turret is formed as a welded structure. The side and back walls are inclined 65 degrees from the horizontal. The front wall is inclined under 70 degrees. There are two flaps in the front wall for orientation, and one flap in each side wall without a vision slit with external edges. The two entry hatch covers for the aiming and loading gunners have a vision slit in a firmly screwed-on flap. In the back wall of the turret, two smaller openings are cut, so that an MG 34 in a simple mount can be stuck through and fired by sighting. The commander's cupola on the rear part of the turret has eight vision slits of adjustable height plus a numeral ring for a 12-0'clock numeral system. The front vision slit in the cupola is fitted with a visor device for extra checking of the lateral aiming of the gun. For ventilation and giving orders, two signal flaps are fitted on the roof of the commander's cupola on the turret roof.

The fighting compartment is closed off at the bottom by a turntable suspended from the carrier ring of the turret housing by three pillars. The diameter of the turntable is 1400 mm, the diameter of the light operating circle in the turret is 1600 mm.

The turret turns on a rail with ball bearings; the weapons are mounted, vertically adjustable, in a hatch in the front wall of the turret. The aiming extent is from -10 to +20 degrees.

The armament consists of one 7.5 cm gun barrel, which is mounted in a jacket cradle with barrel brakes and pneumatic recuperators on both sides of the barrel, along with one MG 34 firmly attached to the cradle. For firing through the back window of the turret there is a spare MG 34 with double-drum ammunition planned.

The 7.5 cm barrel is L/21 length and produces a Vo of 440 m/sec with a 6.62 kg shot weight. The vertical breech is semi-automatic. The normal barrel recoil is 420 mm.

Provided as aiming devices are one simple periscope with adjustment by calibrated-plate movement. As an auxiliary device, there is an aiming rail 50 cm long.

The traversing machine contacts the toothed circle screwed onto the top of the armored box and can be switched by simple levers from hand operation to electric drive, so that after switching by the traversing handwheel, the electric turning device is activated.

The elevation machine is intended only for hand operation. To overcome dead-point locations, the elevation wheel can be rested in four positions on the handwheel shaft. In connection with the traversing machine, it has a numeral disc of the 12-o'clock indicator system.

Firing the gun is electric. The trigger is on the handgrip of the traversing machine. The MG 34 is fired by a foot pedal on the aiming gunner's seat via a Bowden cable.

Via a sliding-ring device in the middle of the turntable, the commander is connected to the radio and intercom system.

*c) **Front MG Turret***

This cannot be described more exactly, since it is supplied by a different firm.

*d) **Ammunition Storage***

The gun ammunition is stored so that ammunition is reachable with the turret in any position. Ten rounds of ready ammunition are kept in the turret, in the right extension ..., in the left one ... are stored. Thirty more rounds stand to the right. With the ... rounds stored elsewhere, 150 rounds in all are available. The ammunition lying horizontally is secured against falling out. Roll-up covers protect it from dust and sand.

4000 rounds of MG ammunition for the front MG turret and the turret MG are stored partly in front of the front gunner, the rest between the MG turret and main turret, reachable by both machine-gunners.

*e) **Other Equipment***

Further military equipment in the vehicle includes: gas masks, which hang near the workplaces, easy to reach; telephone, radio equipment with sender and receiver, plus rod antenna at the left side of the upper box body, light system with a special safety board on which the main light sources are secured separately, so as to be independent if damaged. Digging tools. Kept on the track aprons, plus tools and supplies for the gun, machine guns and radio sets, flags and baggage, easily reached in the interior.

B.W. General Information

Length	5520 mm
Width	2900 mm
Height (to turret level)	1635 mm
Height (+ turret, - cupola)	2240 mm
Height (with cupola)	2520 mm
Firing height	1915 mm
Track contact length	3450 mm
Track width	380 mm
Specific ground pressure	0.71 or 0.66 kg/cm2.
Track of vehicle	2390 mm
Contact length : width	1.44:1
Motor performance	300 HP
Top speed	30 kph
Ground clearance	400 mm
Maximum slope climbing	35 or 40 °
Total weight	18.5 or 17.2 tons
Armament:	
Tank gun	1 (7.5 cm)
Machine guns	2 (+ 1 spare)
Machine pistols	2
Armor	
Front	30 or 20 mm
Side	20 or 14 mm
Crew	6 men

B.W. test vehicle (Rheinmetall) with modified running gear and leading-wheel mounts.

3.1.2 Tasks for Test Vehicles

In the report on the results of the 1934-35 business year of the *War Material Department* of Krupp is stated:

Newly contracted for are two B.W. ***(Begleitwagen)*** *weighing some 18 tons, which are still being designed and built. Tank turrets have been contracted for from us; for our own turret design, a gun turret for B.W. with 7.5 cm K, L/45, herewith a design for an M.G. turret for the B.W.*

Wa Prw 6 issued the following contracts to Krupp in Essen for designing and building test vehicles:

Wa Prw 6/Va Contract No. 67096/36 of 7/13/1935 for one B.B. chassis "B.W. 1 Kp." Of cast iron (test version)

Wa Prw 6/Va Contract No. 67252 of 10.14.1935 for one B.W. chassis "B.W. II Kp." Of cast iron (test version)

Wa Prw 6/Vb Contract No. 67941/35 of 1/3/1936 for one B.W. main turret with armored box upper part and M.G. turret in cast iron (test version).

Wa Prw 6 also gave contracts to Rheinmetall for designing and building at least one, but probably two test chassis. The Rheinmetall B.W. test chassis are known from photos, but additional documentation as to their development and testing has not appeared to date.

In the *State of Tank Designing* of January 15, 1936, Wa Prw 6 reports:

The 1st test piece was to be finished in May 1936. Contracts for a 0 series were not yet issued.

In a discussion with Wa Prw 6 on February 21-22, 1936, at Krupp in Essen, it was reported:

Major Dr. Olbrich presented us with more contracts in prospect for the year 1937.

There is a large contract for Z.W. to be expected, since the intended ratio of Z.W. to B.W. is to be 3 : 1. The ordered B.W. turret must be ready on May 1, 1936, and the closing of acceptance shall be on May 2, 1936.

On April 30, 1936, Krupp reported on the Contract 41/2203 B.W.I vehicle: *The already introduced vehicle is to be finished and ready to drive today after the end of the second shift, as promised.*

On July 21, 1936, Wa Prw 6 (Va) reported to the Fried. Krupp firm, Department A.K., in the following letter, that Krupp was being contracted with as the only firm for the design and assembly of the Pz.Kpfw.IV. The Rheinmetall firm had already been singled out before mid-1936 for the design of the B.W.

The O.K.H. (Wa Prw 6) states that in the period from April to August 1937 a first test series of B.W. chassis is to be built at the Gruson works.

Even though the design of the chassis cannot yet be completed, and even during the production the aforementioned first test series of the further development must be halted, a postponement of the production date for this first series is not possible.

The production of the vehicles is thus dependent on the close cooperation between the design bureau and the production facilities of the Gruson works.

Since the vehicles should only be built by the design firm, the possibility remains to a certain degree of considering further information appearing in testing even during the testing.

3.2 EQUIPMENT DESCRIPTION

3.2.1 B.W. Test Chassis—Rheinmetall

The running gear showed eight double-mounted road wheels of smaller diameter mounted on either side, in the manner of the Neubau Fahrzeuge (Newbuilt Vehicles). The leading-wheel arrangement was changed several times. The road wheels had rubber bandages. Moreover, a lever design with built-in coil spring pressure was used. The track links were taken over from the Neubau Fahrzeug without rubber pads. The link bolts remained without lubrication. The drive wheels featured doubled drive-train crowns. The power was transmitted from the motor via a hand-shifted ZF gearbox with synchronization to the front-mounted drive wheels.

B.W. test vehicle (Rheinmetall) with covering over the planned place of the machine-gun turret.

B.W. test vehicle (Rheinmetall) with modified exhaust length and exhaust grating.

B.W. test vehicle (Rheinmetall) during tests with reinforced concrete barriers.

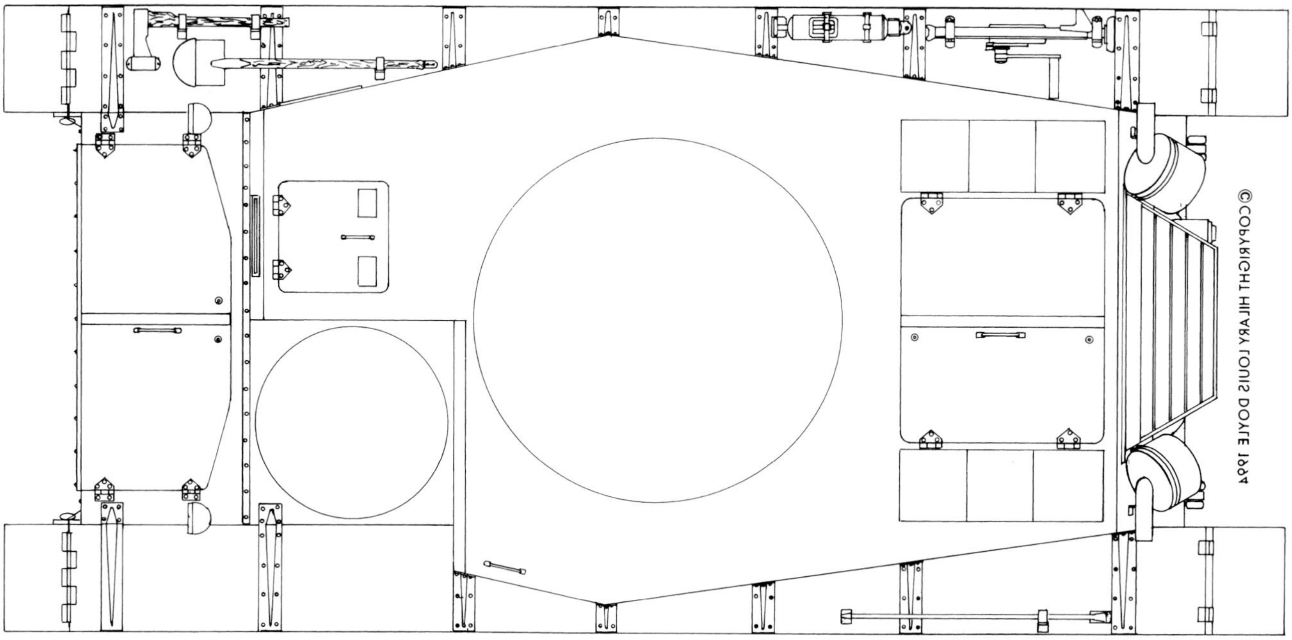

B.W. test chassis (Rheinmetall), taking over running-gear components of Newbuilt-Vehicle, top view of turret turntable for the planned main turret plus the front MG turret.

The steering gear, of the Rheinmetall type, braked the inside track from the drive. The outer track ran with the speed of straight-ahead driving. Mild curves had to be made by making the brakes rub, which led to much wear on the brake system and much heating of the gears.

On the cast-iron structure only one burdening weight was set. The place at right front on the body for the M.G. turret was closed off by a round cover. For the Rheinmetall-B.W. chassis no turrets were built.

3.2.2 B.W. Test Vehicles—Krupp

The Krupp test vehicles were completed as described in the Krupp *Design of a B.W.* of April 13, 1935, but with the following exceptions:

- *The test vehicles were welded together of 8- to 14.5 mm cast iron sheets.*
- *The M.G. turret was replaced by a machine gun located in a ball mantlet in the right upper front plate.*
- *The crew was reduced from six men to five. The radioman, who now sat at the right front, also operated the bow machine gun. Above the radioman was a hatch for entry and exit. The rod antenna was now kept on the right instead of the left sidewall of the body.*
- *The rear armor protected the engine room. On the left side of the rear armor was the air intake grille, on the right the air exit grille. The roof of the rear armor was formed without gratings for the radiator, motor and ventilator flaps.*
- *Krupp B.W.I test vehicles were built with eight-wheel running gear and B.W.II with six-wheel running gear.*

3.3 MODIFICATIONS TO ESCORT VEHICLES

After examining the first test vehicle at Krupp on June 5, 1936, reworkings and changes were requested by Wa Prw 6:

1) *Towing hook to be changed as on Z.W.-Daimler, request drawing from Daimler (maybe later).*
2) *Make rear fender end higher. Perhaps shorten it somewhat.*
3) *Adapt front fender end for folding up, fixed part should continue to be curved forward as before, so that with lower part folded up (position in combat) it offers the vehicle better dust protection (maybe later).*
4) *Driver and gunner should have handgrips on upper armor box to hold during travel.*
5) *Rubber padding for driver and gunners.*
6) *Make the viewing slit for brake adjustment lockable.*
7) *Fit flaps over the steering brakes with hinges.*
8) *Move the oil spray into the steering gear so that gear meshing is hit by the oil spray in the direction of its turning. Stronger sucking-out ducts. Maybe move pump farther forward.*
9) *Hang brake crank on lashing.*
10) *Return springs for brake bands may be changed.*
11) *Locate clutch foot pedal somewhat differently (try it out).*
12) *Install new gear circles.*
13) *Use somewhat softer springs (0.8 mm) for cap, or install the new turning-on device as in 021 D 14765 U 16. Different pinion for starting crank.*
14. *Rain shield for driver.*
15. *Attachments for leading wheel mounts, nose on toothed panel for brake lever (maybe later).*
16. *Install brake light switch (also for hand brake).*
17. *Install sliding coupling for ventilator drive, set to right turning moment. Setting-back indicator.*
18) *Remove bad road wheel and send to Conti.*
19) *Test bent running-gear lever (left front).*

Test Vehicle B.W.I (Krupp) with Test Turret V1.

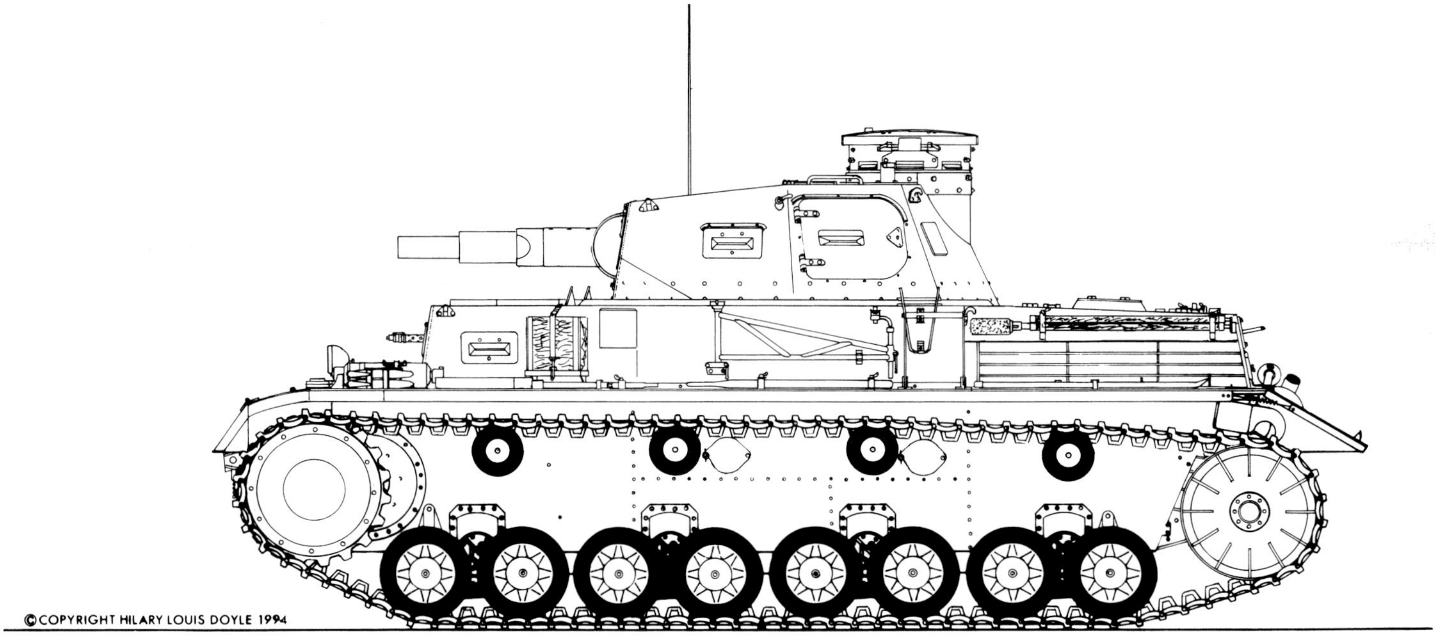

Test Vehicle B.W.I (Krupp) with test turret.

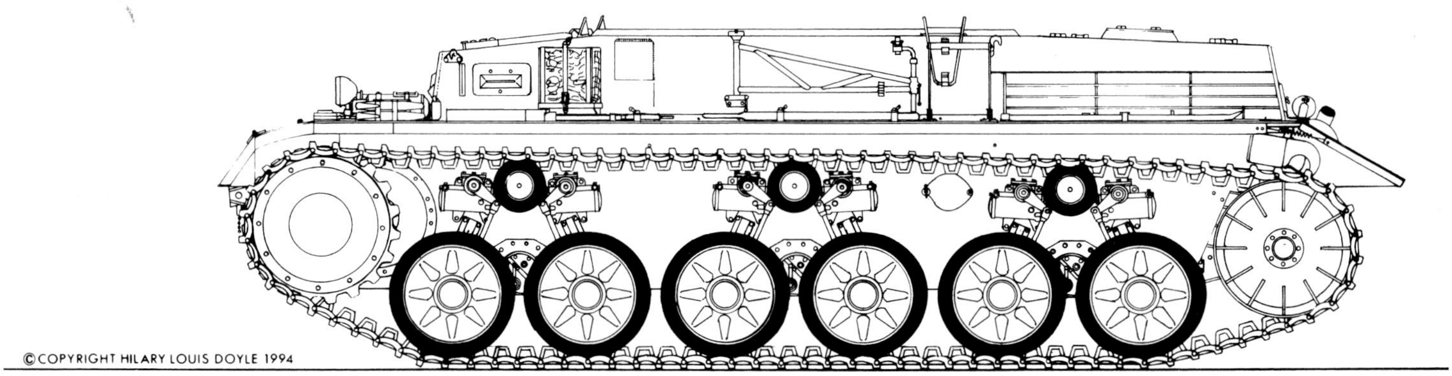

Test chassis B.W.II (Krupp) with six-wheel running gear and torsion-bar suspension. The roof armor over the driver and radioman is angled.

20) Move two pinions for ventilator cover on flap.
21) Strengthen support for flap; use stronger springs and cover hole.
22) Install new brake activation according to SK 2211.
23) Make central plate of rear armor easier to remove (hinges).
24) Most finalized driver's and gunners' flaps.
25) Divide weight of ammunition as well as possible.
26) Strengthen activation of starter flap.
27. Noise-damping exit too steeply downward.
28) Warmth protection for driver on left side gearing.
29) Remote control of hand pump and crack filter (maybe later).
30) Ventilation for carburetor (maybe later).
31) Second ladder on other side.
32) Mount spotlight in rubber.
33) Cover cable in fighting compartment.
34) Turret cover, screws should go through.
35) Install and check two batteries of 105 A-std.
36) Check air filter and move it in inside area.
37) Add S-hooks and brake rods. Towing hook on turret.

3.4 TESTING

In Krupp's technical annual report of 1935-36 it is stated:

*The design of the 18-ton **Begleitwagen,** called B.W., was finished with the choice of 6- or 8-wheel running gear (with different suspensions). After successful driving tests with the first chassis, a test series was contracted for and the design for it nearly completed. The test gun turret for B.W. with 7.5 cm K. L/24 was finished and delivered.*

The eight-wheel leaf-spring running gear was installed on the Krupp B.W.I chassis. On each side of the vehicle, eight road wheels (470 mm diameter) were united into four pairs. Each pair was attached to the hull with crank shafts. In the forward crank shaft of each pair of wheels, a leaf spring was attached. The free end of this leaf spring rolled on a wheel of the other crank shaft. When the road wheel was sprung, the effective spring length thereby became greater and the spring effect steadily lighter. Because of the self-damping of the leaf springs, no shock absorbers were installed. The effect of the leaf spring was limited by a block. For

The test chassis of B.W. II (Krupp) with six-wheel running gear and bridgelayer structure.

the front and rear wheels, a block was attached to the hull to limit the rolling. After testing five different types of running gear, this leaf-spring type was declared the best and continued to be installed on all B.W. production chassis until the war's end.

A six-wheel running gear with torsion-bar suspension was installed on the Krupp B.W.II chassis. Each road wheel was mounted on a swinging arm with a torsion-bar spring. To dampen the swinging, a knee-lever shock absorber was linked to each swinging arm. The experience with this torsion-bar suspension was so bad that the Krupp A.K. Department fought against further designs of armored vehicles with torsion-bar suspension until the war's end.

As decided by Wa Prw 6/Va on July 14, 1936, the B.W. chassis made by Krupp was tested further in Alten-Grabow.

By October 3, 1936, the B.W. No. I chassis had already covered 2043 kilometers in road tests. In road tests during the last three days, a daily achievement of 127 km at an average speed of 24-25 kph was covered.

On December 9, 1936, Krupp reported:

Re: Acceptance (taking over) of the B.W.I chassis. As is known, the B.W.I chassis, in the course of the road tests in Muensingen, was finally supposed to be taken over by the department.
Delivery receipt: Contract No. 67096/36 WaPrw6/Va
Dated 7/13/1935
Contract No. 67280/36 WaPrw6/Va
Dated 8/26/1936
Bz.No.67463/36 WaPrw6/Va
Dated 10/15/1936

We deliver to the test command of the Army Weapons Office Prw 6/Vc at this time at the Troop Training Camp in Muensingen one ***B.W.*** *chassis Krupp with HL 100 motor (300 HP), Chassis No. 1, Motor No. 14003.*

3.5 FURTHER USE

Now the Army Weapons Office had the two test chassis B.W.I and B.W.II on hand for further testing until July 1940. Then a Pz.Kpfw. IV Type D chassis, No. 80625, was delivered to Wa Prüf 6 from In 6 for testing purposes. Known tests with both test vehicles were:

- *The experimental bow armor and interior details for 2./B.W. were installed in the B.W.II test chassis between January 6 and March 30, 1938.*
- *Between April and November 1938 the B.W. turret with ammunition storage was mounted on the B.W.II test chassis.*
- *In 1939 and 1940 the B.W.II test chassis was used to test various designs of bridge laying devices.*
- *In December 1939 and January 1940 only the B.W.I test chassis from Kummersdorf was available to Wa Prüf 6 for winter testing. The B.W.I vehicle was fitted with completely new 4./B.W. running gear for these winter tests (also with ten striking blocks).*
- *On June 20, 1940, Wa Prüf 6 stated that the housing for the brake air outlet on the B.W.I test vehicle could be taken from the 1./B.W. series.*

4. Panzerkampfwagen IV (7.5 cm) (VsKfz.622) Type A, (1./B.W.)

Chassis Number Series 80101-80135

The 35 Pz.Kpfw.IV Type A were built and accepted between November 1937 and June 1938.

The Type A (Group No. 021St14700) was very similar to the B.W.I test vehicle (Group No. 021St8400) with its 8-road-wheel running gear. Because of many smaller details, almost every building group was designed anew. Only the following unchanged parts were taken over from the test vehicle:

021St8404	*Bulkhead (between motor and fighting compartments)*
021St8442	*Brakes*
021St8446	*Foot brake activation*
021St8447	*Clutch activation*
021St8450	*Lever steering*
021St8456	*Tension roll*
021St8457	*Radiator*

The following equipment description explains ongoing particulars in the design of Pz.Kpfw.IV.

4.1 EQUIPMENT DESCRIPTION

The Panzerkampfwagen IV (7.5 cm) (Vskfz.622), Type A consisted of the chassis, upper armored body and turret. The chassis and upper armored body were divided by the "separating seam."

4.1.1 Armor

The armor consisted of homogeneous nickel-free PP694 armor steel with 153 +/- 5 kg/mm= firmness. The armor was secure against 7.92 mm S.m.K. shots from rifles or machine guns at ranges of 30 meters and up and 90-degree impact.

4.1.2 Chassis

The chassis included the following main parts: armor hull, motor with cooling system and air filter, driveshafts with brake ventilation, main clutch, gearbox, bevel drive, steering gear, running gear, fuel tanks and electric equipment.

4.1.2.1 Armor Hull

The hull was the carrier of the chassis. It consisted of armor plates welded together, with the following thicknesses and inclinations from the horizontal level:

Plate	Thickness	Inclination
Upper front wall	10 mm	18 °
Forward front wall	14.5 mm	76 °
Lower front wall	14.5 mm	29 °
Side wall	14.5 mm	90 °
Upper back wall	14.5 mm	80 °
Lower back wall	10 mm	0 °

The motor and power train groups were located in the armor hull. The motor space was separated from the crew's space by a bulkhead.

The bottom of the hull was reinforced in the bow and stern by a strong transverse beam in each area, and in the central part by the boxy frame for the fuel tanks.

For servicing, removal or replacement of machine parts, the following hatches were provided:

- Unscrewable cover at the bow for installation or removal of the gearbox, for examination of the oil pump for the steering gear, the odometer and the foot brake with brake-light switch.
- Two flaps over the steering gear for the steering and support brakes were adjustable for ventilating the bow area.
- Unscrewable cover at the rear for servicing the ventilator and generator belts.
- Snap cover over the opening for the starting crank.
- Emergency exit hatch over the radioman's seat.
- Two unscrewable covers under the steering gear for checking the oil spray nozzles and oil pipes.
- Two unscrewable covers in the engine room for oil drainage and removal of the filter.

- Hand-operated bottom vents under gearbox and motor.
- Unscrewable bottom plate for removal of the fuel tanks from the bottom.
- Two unscrewable covers in the bottom plate under the fuel tanks for access to the release screws of the fuel tanks.
- Two covers, detachable via screw crossbars in the left sidewall for fuel filling.
- At the bow two towing hooks were screwed on for attaching a towline in horizontal and vertical directions or for towing with towing rods. The vehicle's winch could be attached under the towing hooks. At the rear, one trailer hitch and two towing hooks were attached.
- On both sides of the hull, rippled track aprons were attached; their front and rear ends could be folded up when driving in crater areas.

4.1.2.2 Motor

The Maybach 12-cylinder HL 108 TR carburetor motor had dry-sump lubrication. The cylinders were arranged in V form. The motor was equipped with two mechanical fuel pumps, 1 Solex hand pump, 2 Solex two-stage off-road carburetors, one electric starter and one starting crank.

The 10.8-liter motor produced 230 HP at 2600 rpm, with a torque of 63.5 m/kg.

Through the two Solex two-stage off-road carburetors it was attained that the motor received sufficient fuel on all existing slopes. Each off-road carburetor had two stages, meaning that two nozzles were activated, one after the other, depending on the burden on the motor. Each stage had two floats, which activated two vent gauges from a common shaft. For flat driving the shaft was turned by both floats; on later slopes only the lower-lying float activated the shaft and kept the right fuel level in the nozzle.

A particular starting carburetor, which was located in a housing with the main carburetor, allowed good starting for the cold motor.

For better carburetion, the suction tube of the carburetor was pre-warmed via a duct that branched off from the exhaust pipe.

Panzerkampfwagen IV, Type A, at the Krupp-Gruson Works in Magdeburg. The camouflage paint was applied at the factory.

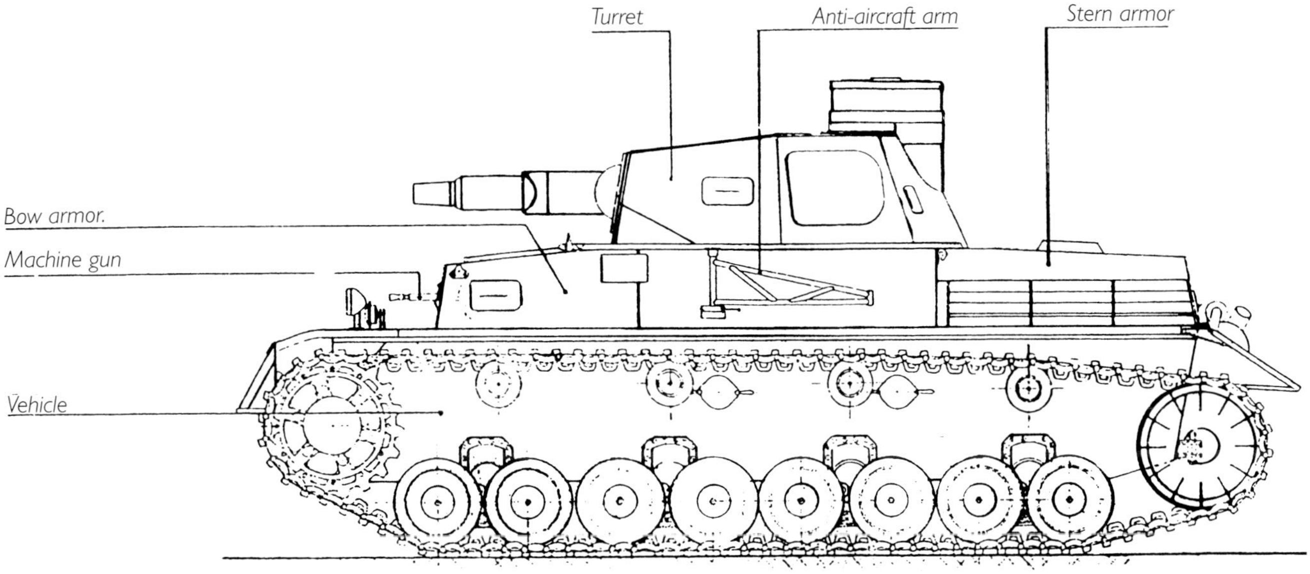

Pz.Kpfw.IV Type A (Drawing No. 021 A 14700).

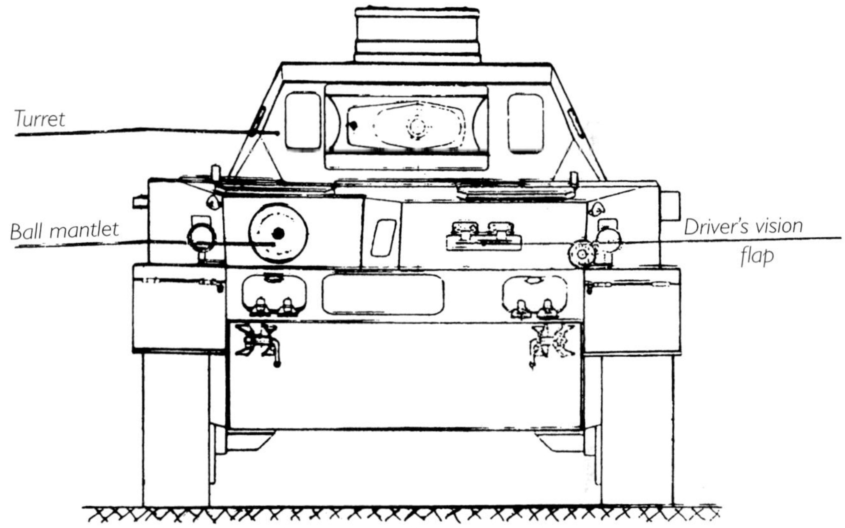

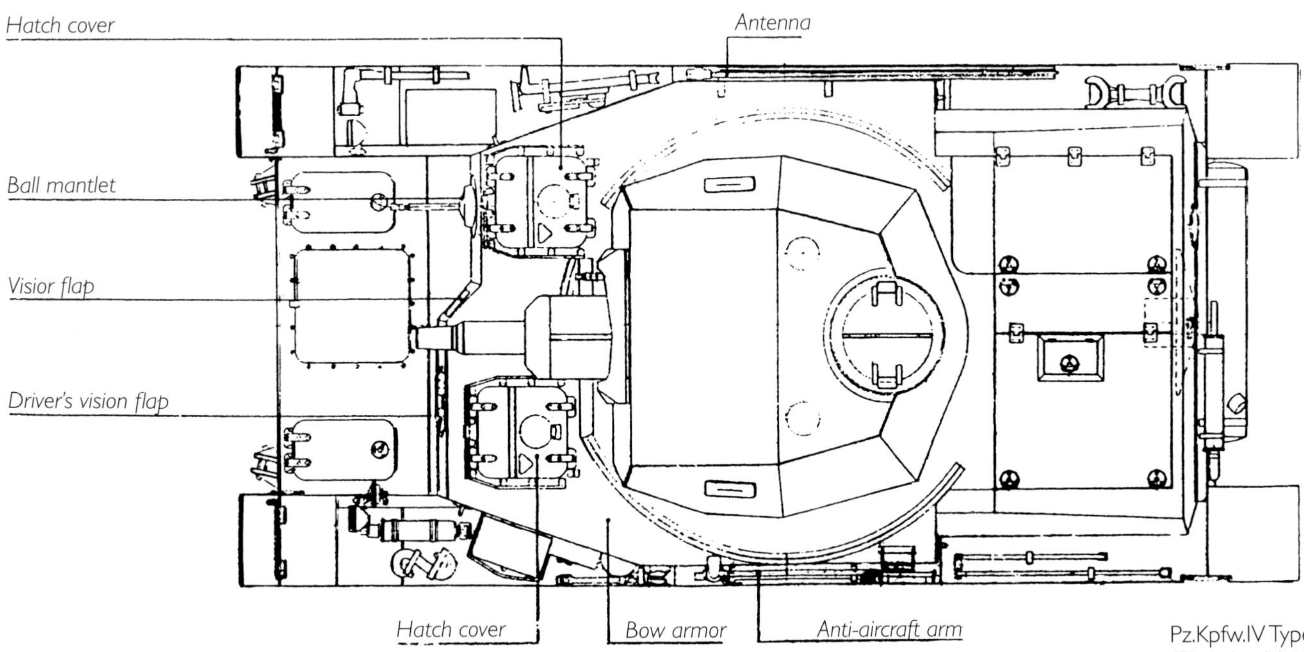

Pz.Kpfw.IV Type A
(Drawing No. 021 A 14700)

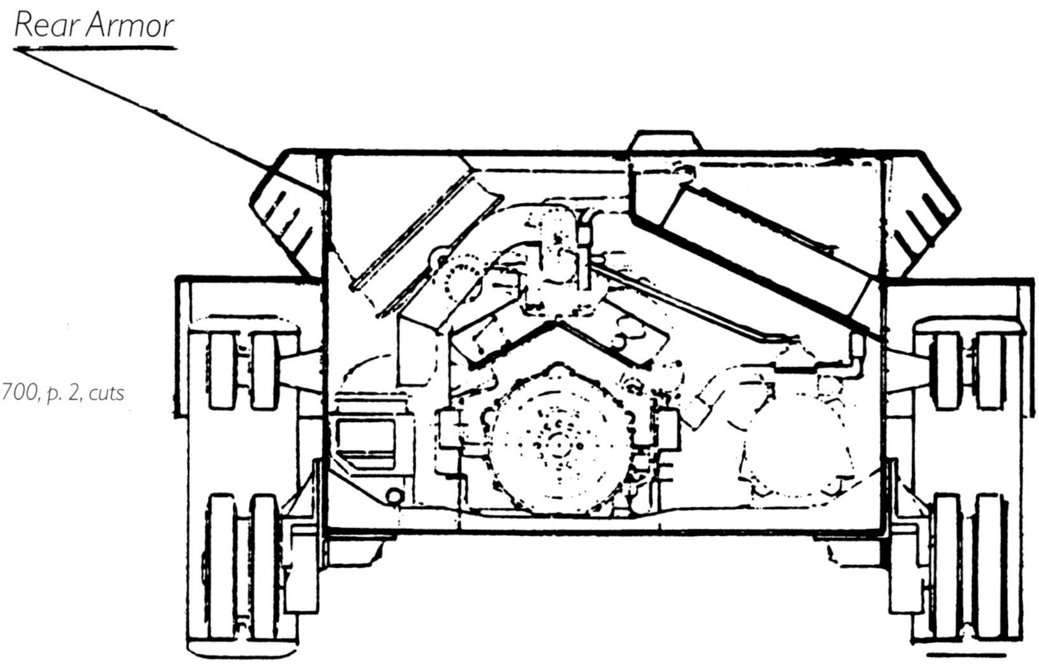

Pz.Kpfw.IV Type A (Drawing No. 021 A 14700, p. 2, cuts

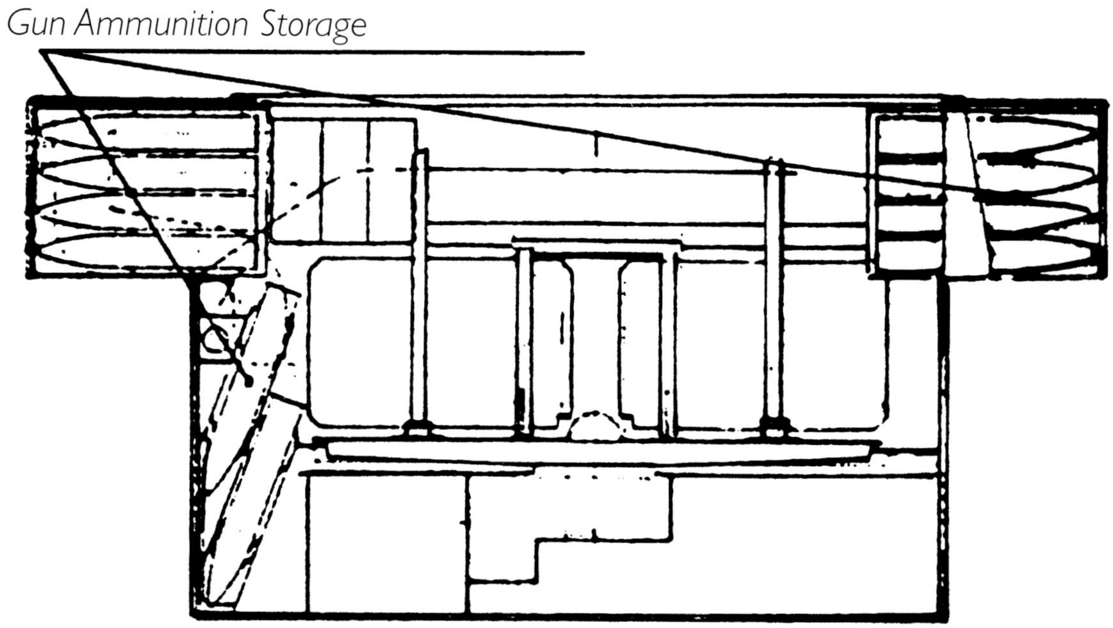

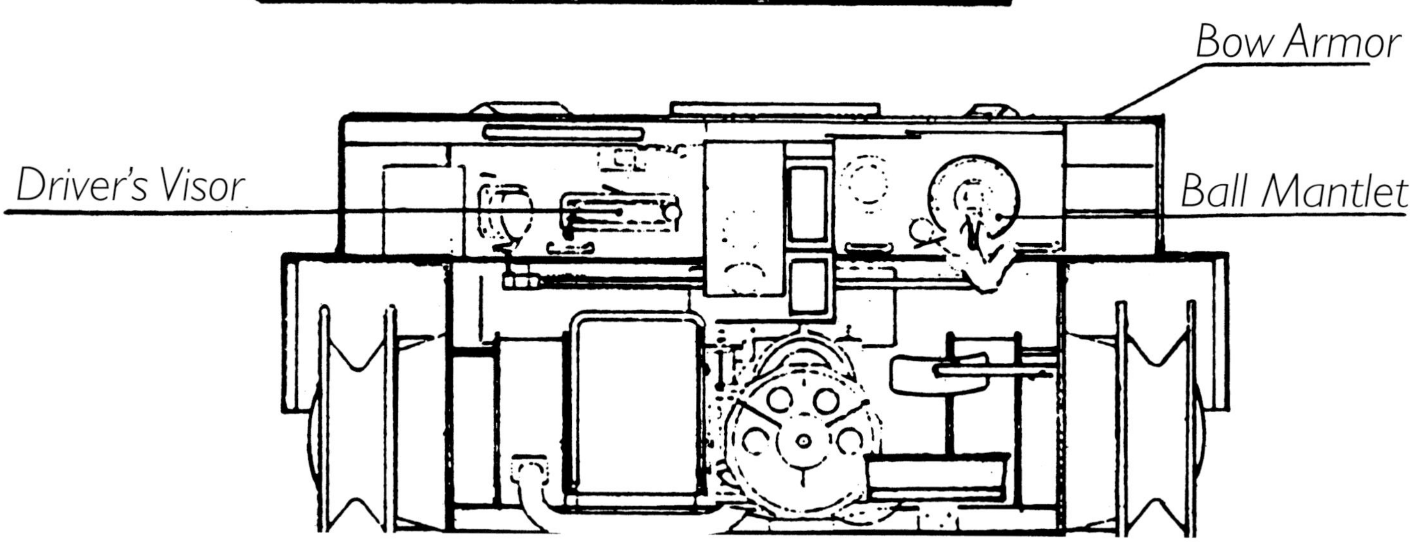

4.1.2.3 Cooling System

Two radiators were suspended in the engine compartment under the roof of the rear body. Two ventilators were suspended from the ventilator hatch in the space to the right of the motor and were driven by V-belts from the motor via a sliding coupling. The cooling air flowed through lateral openings under the roof of the rear armor on the left side, streamed through the radiators, and was driven across the motor by the ventilator fans through openings on the right side of the rear body. The low pressure in the engine compartment prevented fuel and exhaust gases from penetrating into the fighting compartment.

The ventilator flap could be opened with the suspended ventilators after shifting the sliding coupling. For the summer, the flap could be adjusted over the radiators so that the cooling air could flow directly to them.

4.1.2.4 Air Filters

Three VB3 Delbag air filters stood on the right side of the engine compartment. The air could be sucked in from outside through side openings in the body (poison gas danger), or if the dust was especially heavy, after activating a turning flap in the air intake channel, be sucked from the fighting compartment. The filters could be reached from above for cleaning after opening the ventilator flap.

4.1.2.5 Intermediate Shaft

The torque was transmitted from the motor to the main clutch through two jointed shafts, one behind the other. Between these two shafts a bearing position was attached to the lateral wall ahead. The front shaft carried the runner of a blower to cool the brakes. This sucked air from the barks housings through ducts and pushed it through a pipe and a shot-safe opening on the left side of the bow armor and out of the vehicle.

4.1.2.6 Main Clutch

The main clutch was a three-plate dry type with asbestos covering. The housing was attached to the gearbox with screws.

4.1.2.7 Gearbox

The gearbox was meant for a power of 300 HP and a torque of 75 m/kg. It was designated SFG 75; synchronized five-speed gears for 75 m/kg. The gearbox had five forward speeds and one reverse with the following ratios of the individual speeds:

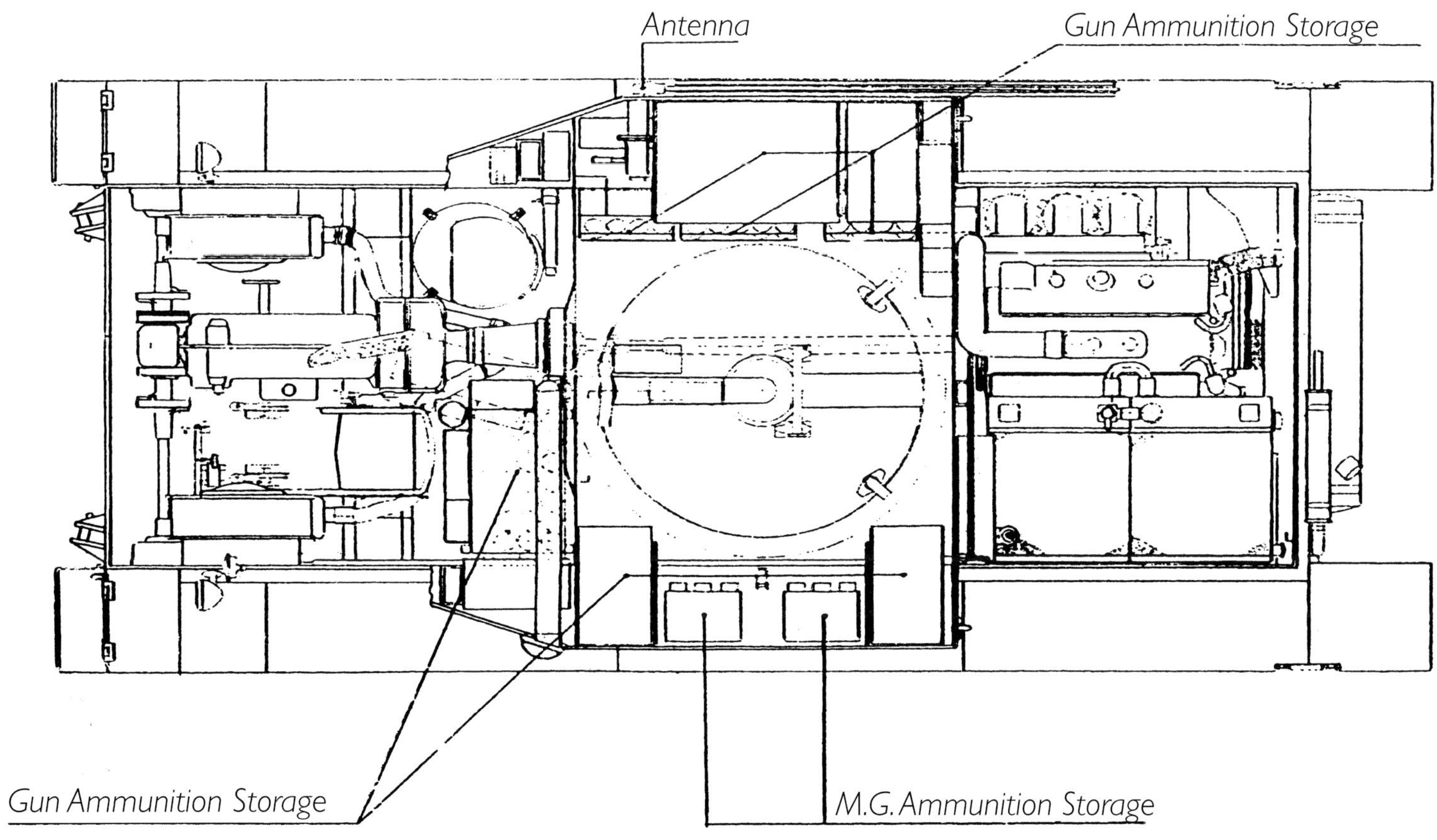

	Ratio	Speed
1st Gear	1 : 6.75	4.8 kph
2nd Gear	1 : 3.36	9.7
3rd Gear	1 : 2.12	15.3
4th Gear	1 : 1.42	22.8
5th Gear	1 : 1	32.4
Reverse	1 : 6.4	5.1

All speeds except first and reverse were synchronized. The synchronizing mechanism assured easy and noiseless shifting.

The cone drive was flanged onto the gearbox with a ratio of 1 : 1.05. It transmitted the power to the two steering drives.

4.1.2.8 Steering Drive and Brakes

The two steering drives, formed as planetary types, were enclosed in the hull and connected with the side gearing from outside. The carrier for the band brakes was screwed to the housing of the steering drives. The planetary drive served as a clutch to release the power from the braked track. The side gearing was a reduction drive and linked with a brake to brake the track when steering or stopping the vehicle. The side gearings, being outside the hull, were protected by screwed-on armor plates.

Ratios:

Gearing, wheel	1 : 2.09
Planetary drive	1 : 1.43
Side gearing	1 : 4

The steering levers right and left of the driver were connected with the support brakes and the steering brakes. By pulling on a steering lever, the support brake was opened first. The power flow to the drive wheel was thus interrupted. Continued pulling on the steering lever activated the steering brake, which slowed the track. The size of the driven curve varied according to the degree of track braking. In smaller steering movements, lightly pulling on the steering lever only released the support brake, thus the resistance slowed the track without the steering brake being applied.

4.1.2.9 Running Gear

On each side of the vehicle there were eight road wheels (470 mm diameter) mounted as four pairs. Each pair was flexibly attached to the hull wall with crank levers. To the front crank lever of each wheel pair a leaf spring was attached. The free end of this leaf spring rolled on a roller of the other crank lever. The springing of the road wheel made the effective length of the leaf spring longer and the suspension steadily lighter. Because of the self-damping of the leaf springs shock absorbers were installed. The springing of the leaf springs was limited by a striking block. For the front and rear road wheels, a block on the vehicle's hull limited the wheel's lift.

The upper body of the Pz.Kpfw.IV Type A.

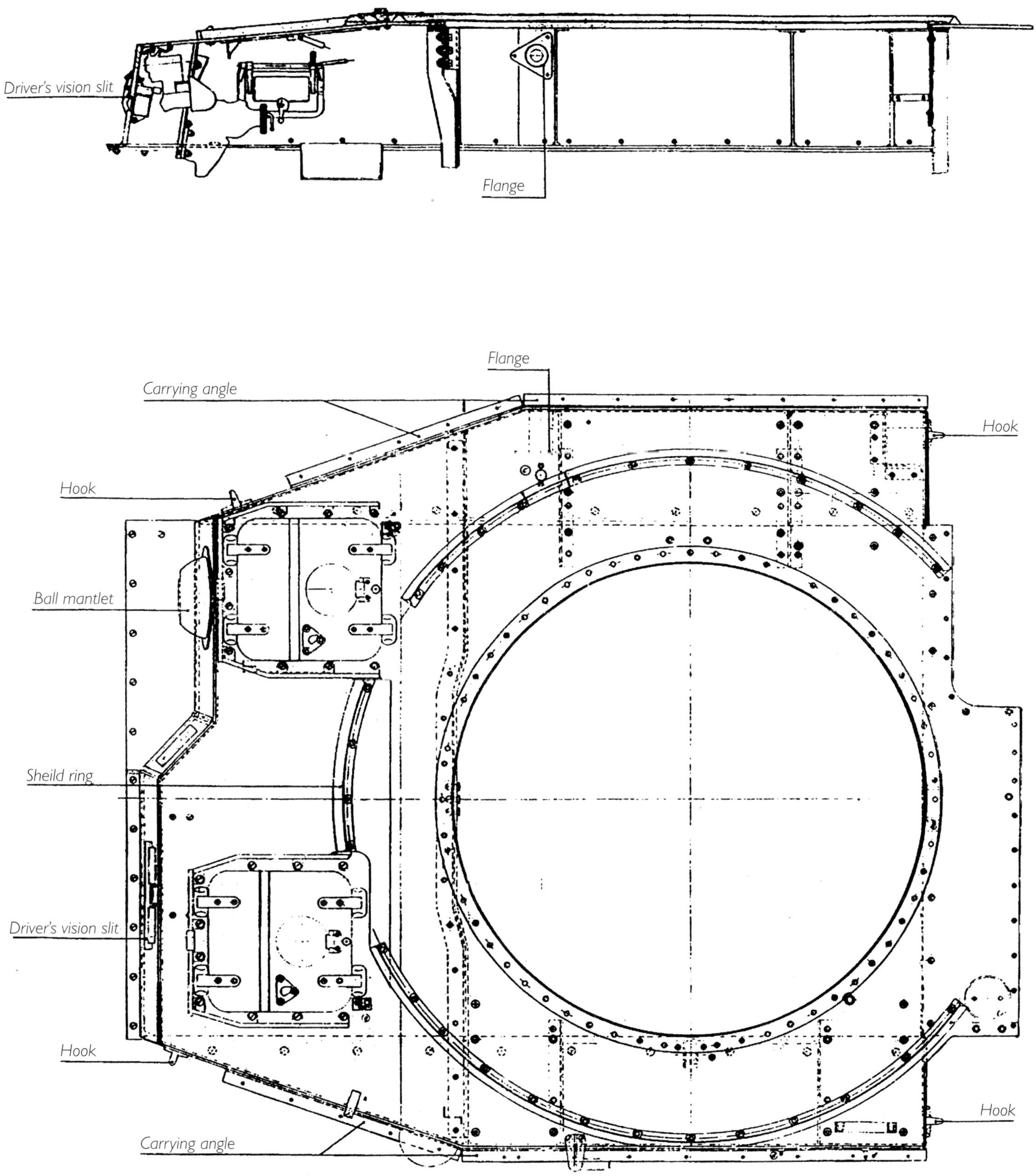

Bow armor (Drawing No. 021 A 14786 U1) for Pz.Kpfw.IV Type A.

4.1.2.10 Tracks and Track Tension

The chassis ran on two unlubricated tracks of Kgs 6110/380/120 type. Each track consisted of some 99 individual, interlocking track links held together by bolts. Their closed running surface was so great that the specific ground pressure at full vehicle weight was about 0.68 kg/cm^2.

The track was driven by a front-mounted drive wheel with a double crown of teeth. It was directed by the track crowns of the road wheels, between which the track teeth ran.

At the rear end of the vehicle the track was turned forward over the leading wheel. To prevent tracks from being thrown, the running surface of the leading wheel was made of steel. Running loose was limited by four return rollers, over which the track ran back to the drive wheel.

The leading wheel, mounted on a crank, provided track tightening. The crank's location was adjusted by two toothed plates secured by bolts and lock-nuts.

To increase their tautness, the track links had grippers. The links were made with apertures for self-cleaning.

On dry grassy ground the climbing ability was as much as 35 degrees, but it decreased on wet ground, sand or gravel. The

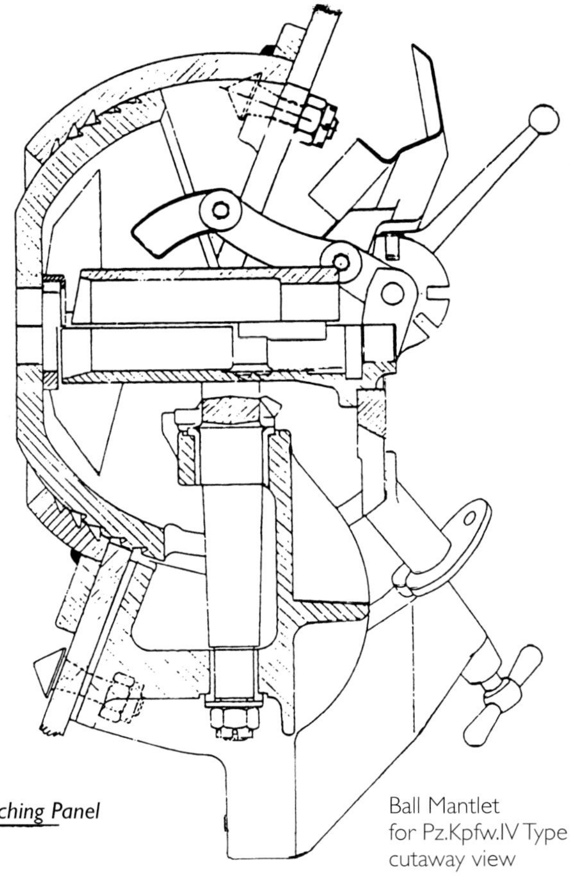

Ball Mantlet
for Pz.Kpfw.IV Type A,
cutaway view

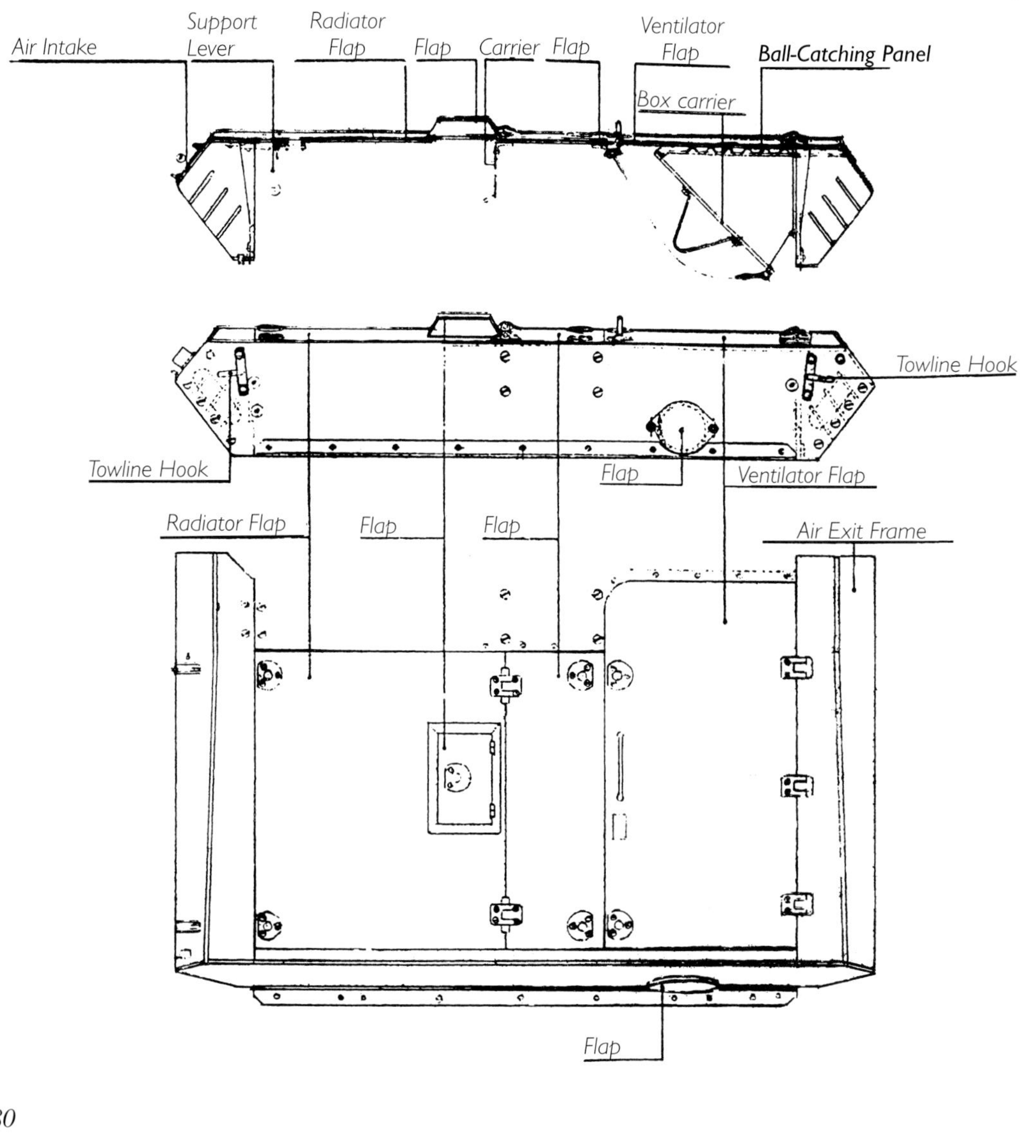

Stern Armor
(Drawing No. 021 A 14787)
for Pz.Kpfw.IV Tyoe A

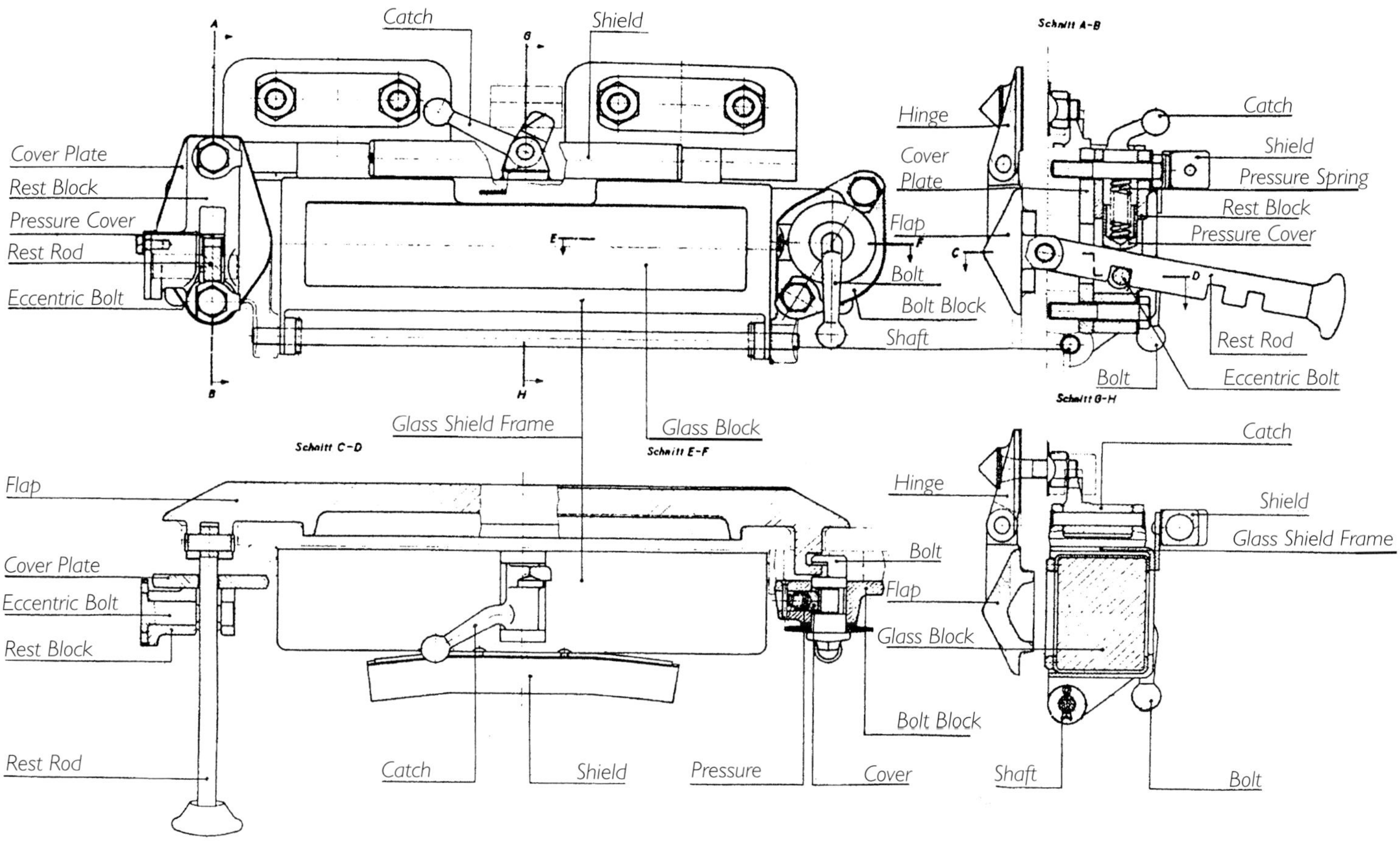

Driver's Vision Flap (Drawing No. 021 B 8960) for Pz.Kpfw.IV Type A

clambering ability (climbing over walls) depended on the firmness of the ground and the obstacle.

4.1.2.11 Fuel Tanks

The fuel was kept in three tanks, I, II and III. Tanks I and III were filled via the filler pipes. Tank II was filled from a connecting duct along with Tank III. The duct served top let air out of Tank II. The three tanks could be blocked by the vents or be connected by choice to the two fuel pumps of the motor. Capacities of the tanks: Tank I about 140 liters, Tank II about 110, Tank III about 220 liters, together 470 liters. The range (on one filling) was 210 km on a good level road or 130 km on medium-rough terrain.

4.1.2.12 Electric Equipment

To light the road, there were shieldable headlights on the track aprons. At the rear of the vehicle was a stop light. Inside the vehicle two lamps were attached to the instrument panel at right near the driver. The radioman had a light above him that could be turned off by turning it.

For a hand light with a three-meter cable there were plugs in the engine compartment and on the instrument panel.

A 12-volt, 600-watt Bosch generator loaded the batteries. The four batteries, each of 6 volts and 105 amperes, were mounted in pairs, one behind the other, to provide 12 volts of power.

The electric Bosch starter for the motor needed 24 volts of tension; this was provided by connecting the four batteries to the starter with a foot pedal.

4.1.3 Upper Armored Body

The upper body consisted of the bow and stern armor. The stern armor could be removed independently of the bow armor, the bow armor only after removing the stern armor.

The upper body was fitted with a machine gun mounted in a ball mantlet. There were special optics and three vision slits for the driver. For the radioman, who was also the machine gunner, there were an optic on the ball mantlet and a vision slit. On the right side was a folding rod antenna, on the left side an extending arm for aircraft defense. The interior space was used for storing ammunition and equipment. The main groups of the upper body armor were: bow armor, stern armor, ball mantlet, driver's vision shield and driver's optics, vision slit and portholes, antenna and radio set, extender for aircraft defense, ammunition storage and equipment storage.

4.1.3.1 Bow Armor

The bow armor protected the fighting compartment and part of the driver's and radioman's space. It consisted of welded armor plates with the following thicknesses and inclinations (from the horizontal).

Plate	Thickness	Inclination
Front wall	14.5 mm	81 °
Side wall	14.5 mm	90 °
Roof	11 mm	0 and 5 °
Rear wall	14.5 mm	90 °
Bottom	8 mm	0 °

Four sight openings were cut in around the outside of the bow armor. One was through the driver's sight hatch, and the other three were closed by flaps. The ball mantlet was mounted in the front wall. The radioman could open them and thus also use them as vision slits.

At the left front was a bulletproof opening for the brake cooling; at the right rear was one for the air coming off the motor.

On the roof of the bow armor, a pass ring was riveted on to take up the turret connection. Close to the ring, to the left rear, was the opening for the turret turning.

To cover the gap between the turret and the upper body armor, there was a three-piece shield ring. Before this ring, entry hatches for the driver and radioman, closed by two-piece hatch covers, were provided. To the left rear was an opening for fuel filling, with a cover to close it.

On the sidewalls, hooks were riveted on for the track aprons. The bow armor was reinforced inside by carriers, bridges and braces. On the sides it projected out over the width of the vehicle; the resulting space was used for ammunition and equipment storage. On the right front sidewall, a flange for the antenna mount was welded on; in back was an air channel for the motor's intake air.

On the front wall the driver's optics were attached in the left corner of the aiming arc and in front of the driver, pushable to the right side.

Between the driver and radioman under the roof there hung a frame for the sender and receiver attachments.

Four hooks were attached to the sidewalls for lifting and setting the bow armor.

4.1.3.2 Stern Armor

The stern armor protected the engine compartment. It consisted of armor plates welded together, with the following thicknesses and inclinations (to the horizontal):

Plate	Thickness	Inclination
Roof	10 mm	2 °
Sidewall (inclined)	10 mm	55 °
Sidewall	14.5 mm	90 °
Rear wall	14.5 mm	80 °

On the left side was the air intake frame, on the right was the air exit frame. The bullet-catcher plate was to catch lead fragments and small explosive pieces.

The roof of the stern armor was formed from the radiator flap, motor flap and ventilator flap. The radiator flap, if there was no danger of enemy fire, could be adjusted for better entry of fresh air. It was then held by the brace lever, which projected from the upper

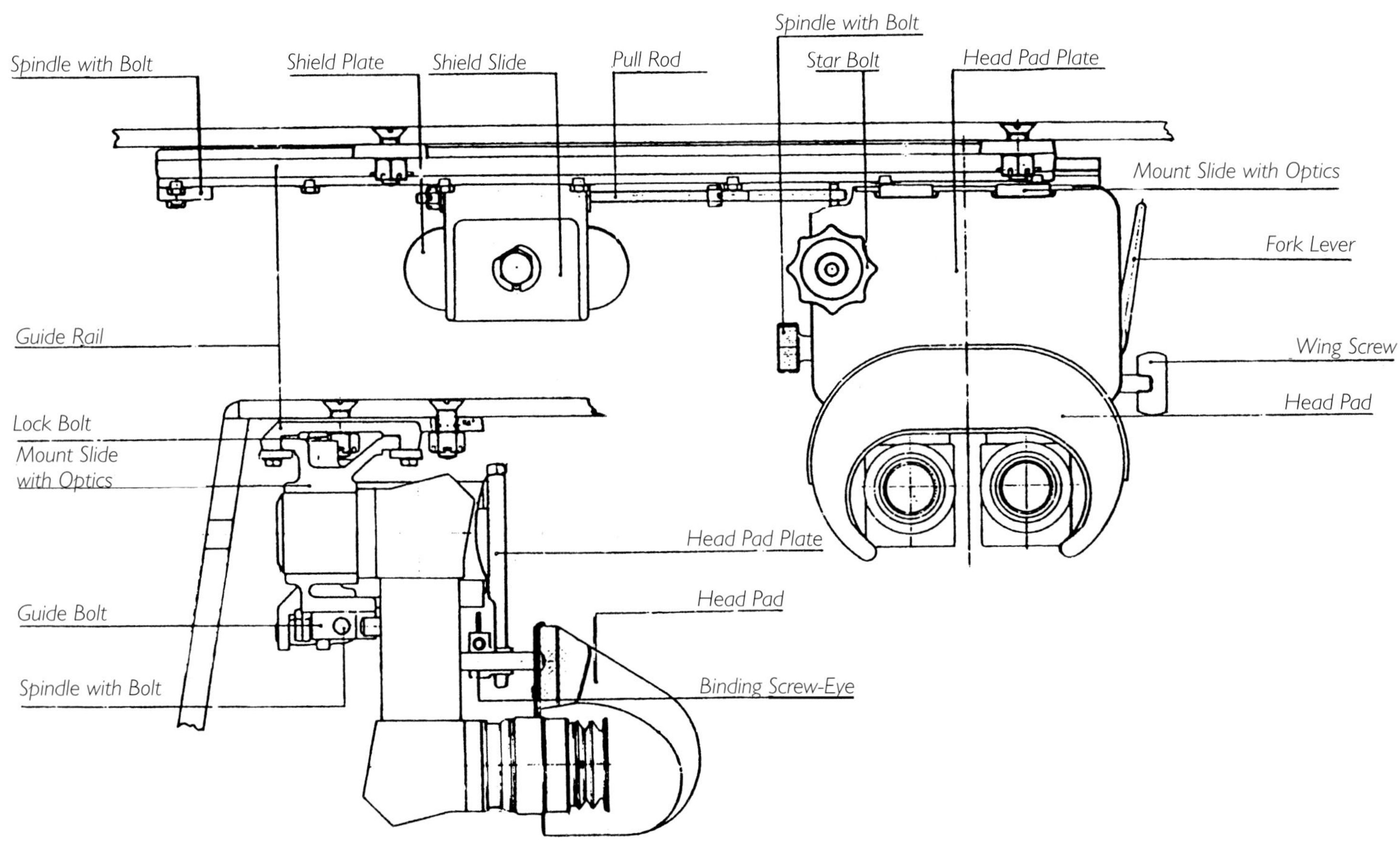

Port Cover (Drawing No. 021 A 14786 Bl.2) for Pz.Kpfw.IV Type A.

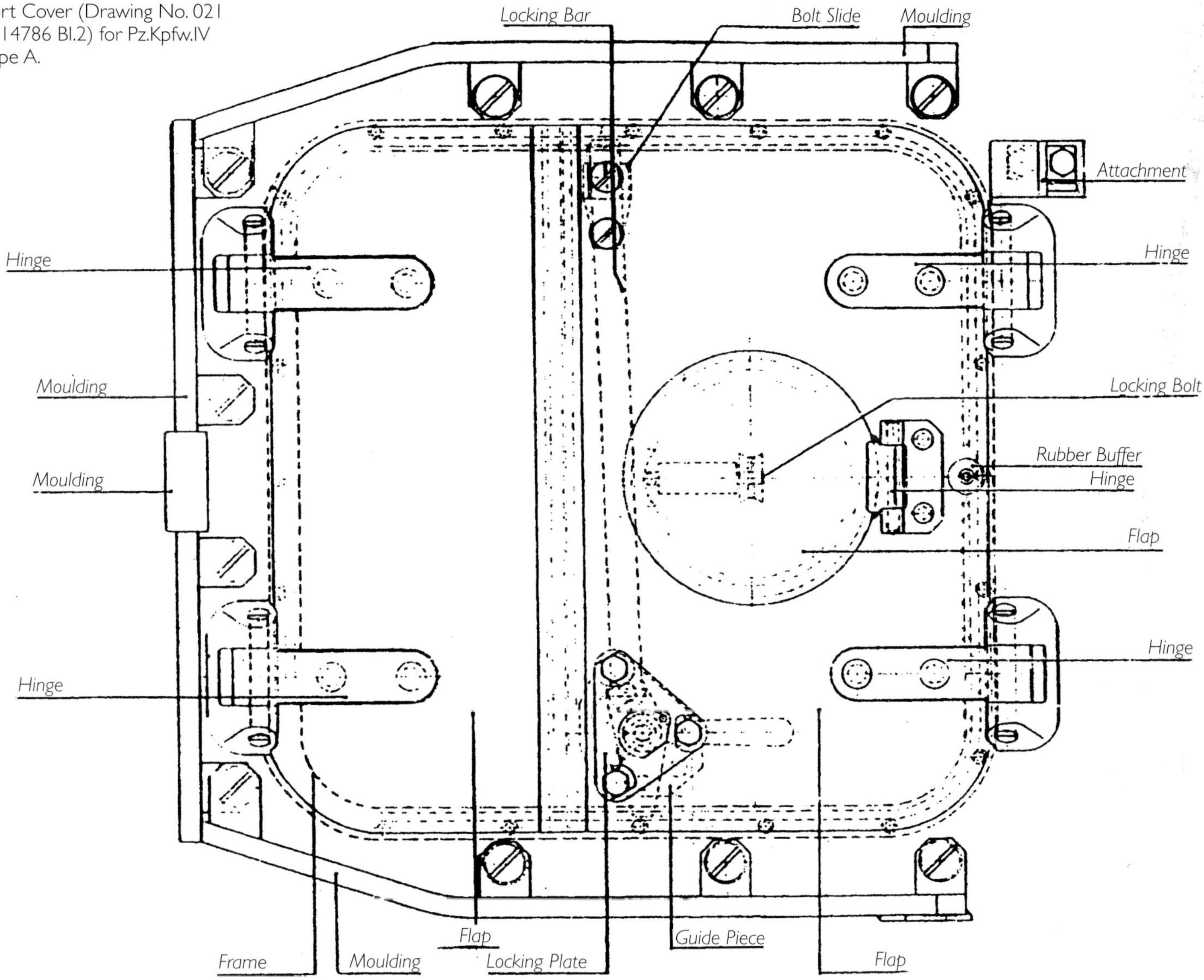

Extending Arm for MG (Anti-Aircraft)

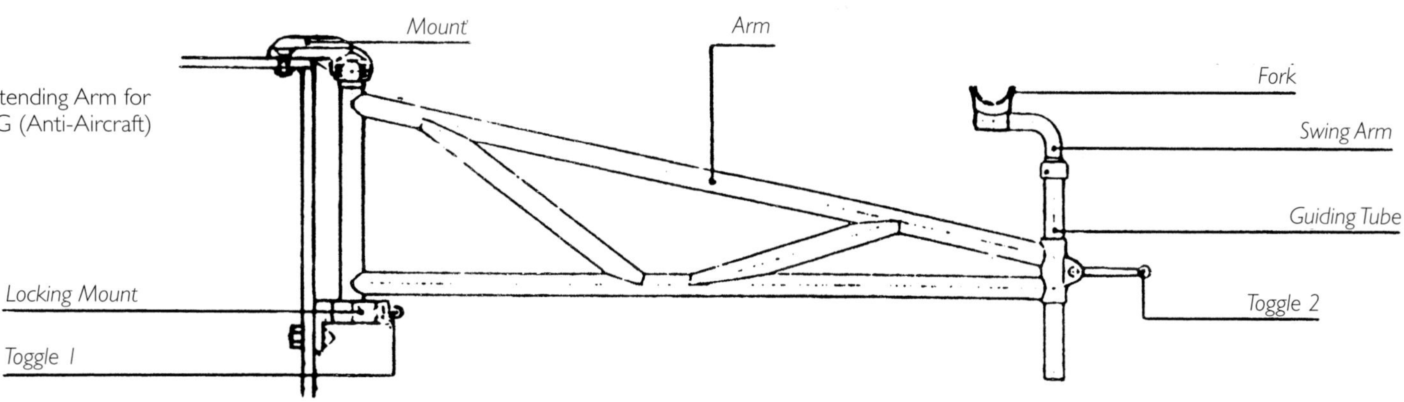

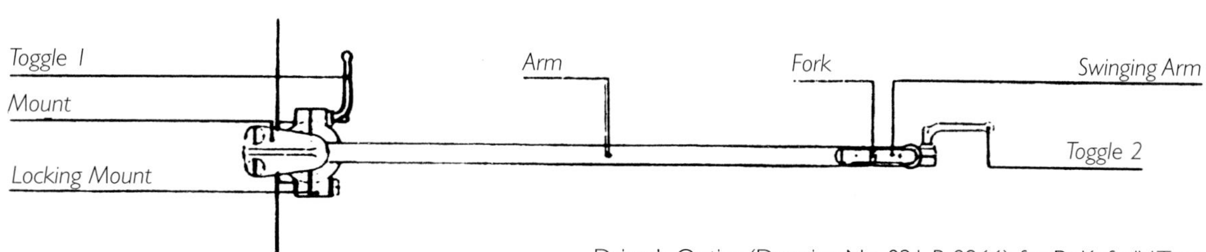

Driver's Optics (Drawing No. 021 B 8966) for Pz.Kpfw.IV Type A

edge of the stern armor and was under spring pressure. On the radiator flap a flap was fixed to allow filling water into the radiator and operating the vent for letting air out of the radiator. Before opening the ventilator flap, the ventilator drive had to be uncoupled through a round flap on the back wall of the stern armor.

The rope for towing the vehicle was carried on two hooks. Under the roof of the stern armor the carrier for attaching the radiator was located. Under the ventilator flap hung the carrier box that held the ventilator.

4.1.3.3. Ball Mantlet

The ball mantlet consisted of a round shield made of armor and movable in all directions; it was covered on the outside by a ball shield, also made of armor plate. On the inside it held mounts for one M.G. 34 and an aiming scope. The shield ball was divided horizontally in the middle into two parts; it folded up for placing the weapon and giving a free view of the terrain. The M.G. could be moved through an elevation field from -10 to +20 degrees and a traverse field of +/- 20 degrees. The ball aiming scope (K.Z.F.2) had 1.8 fold magnification and a view field of 18 degrees.

4.1.3.4 Driver's Visor and Optics

For the driver, a vision port was installed in the front wall, protected by a glass block, for direct viewing. If the port had to be closed during combat, observation was done with the driver's optics through two holes above the vision port.

The driver's sight port was a vision flap without a sight slit. The flap was fastened by two hinges outside on the front wall of the upper body armor. The flap covered the opening completely and was operated from inside via the rest rod. Behind the vision slit, the glass block of safety glass, held by the glass frame, was installed. The glass block did not protect against through shots. The shield-glass frame was attached to the front wall and moved around a shaft.

The production driver's optics (K.F.F.1, formerly K.F.F.III) made by Askania has projections for prisms on its two folding spots. The K.F.F.1 had 1.15 fold magnification and a vision field of 50 degrees. The driver's optics consisted of the guide rail, the mount slider with the optics, and the protective slider. The driver's optics were attached to the roof of the upper body armor.

4.1.3.5 Vision Flap and Ports

The sight flaps built into the upper body armor were made like those in the turret.

The ports for the entry and exit hatches were closed with hatch covers. Every hatch cover consisted of two flaps, which were attached to the bow armor with hinges. The flaps lay on frames that were riveted to the roof of the bow armor. The gaps between the hatch cover and the bow armor were filled with moldings.

With the cover closed, a fork-shaped double lever of the grip lever extended under the edge of the flap. In ventilating position, the fork of the double bolt slid into the frame. In this position the rear flap was opened some 30 mm; this position was intended for ventilation under limited shot safety.

On the rear flap a round flap was fitted with a hinge.

For longer ventilation there was a gap of 3 mm between the flap and extension. The opened flap was supported by a rubber buffer.

4.1.3.6 Antenna and Radio Set

The antenna was a rod type made for sending and receiving and mounted to be adjustable on the right side of the upper body armor. The antenna rod was a round bar that folded down 90 degrees into its resting position and lay in a wooden groove attached to the right side of the upper body armor.

Ammunition storage on the right side of Pz.Kpfw.IV, Type A, Chassis No. 80113 (finished February 1938).

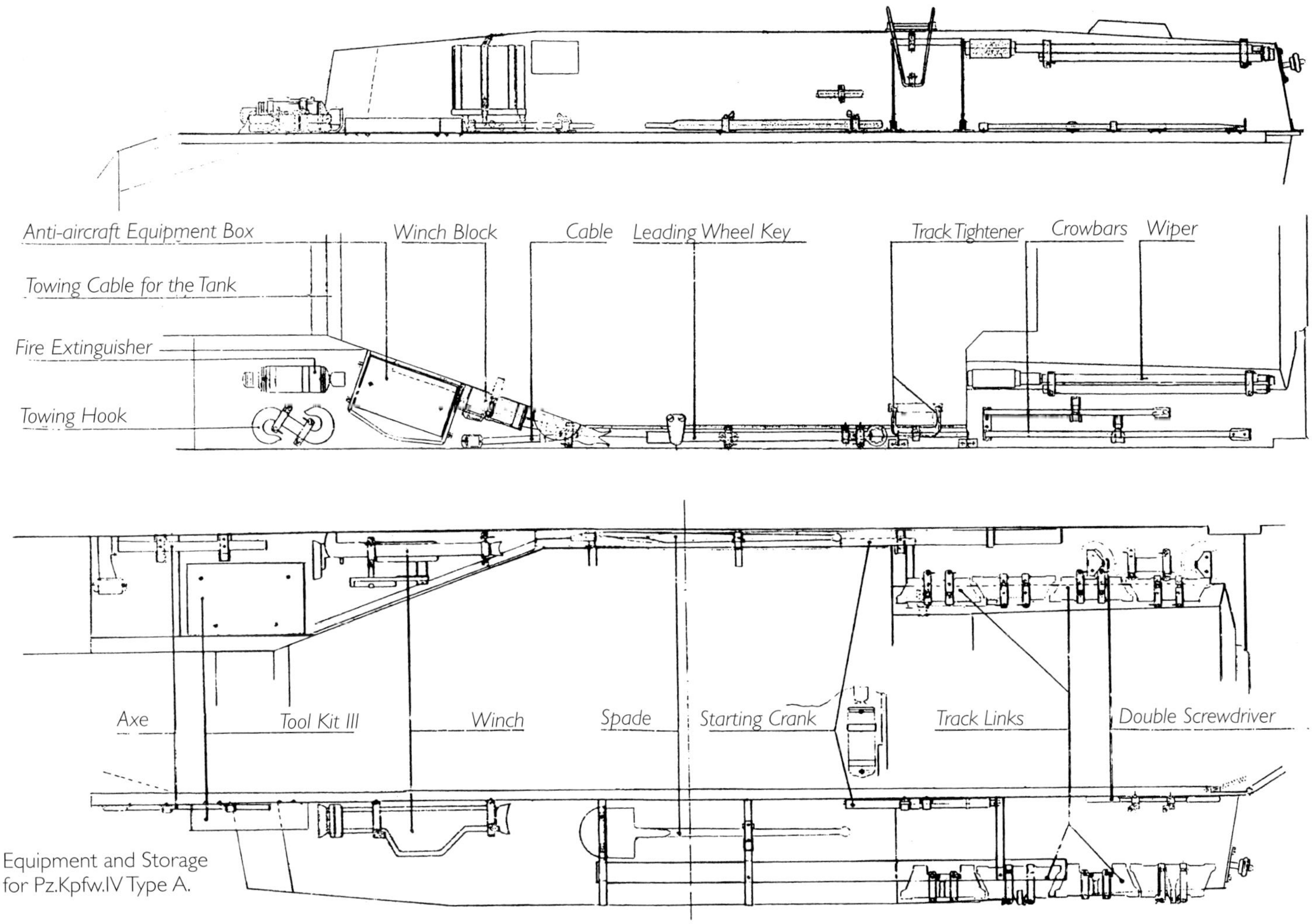

Equipment and Storage for Pz.Kpfw.IV Type A.

When touched by obstacles or by the gun barrel when turning, the erect antenna could fold to the front or rear. A turning spring righted the antenna after it was freed from the obstruction.

To the right of the machine gunner's seat the holder for the sender and receiver transformers was mounted. To the left, over the gearbox, were the frames for the sender and receiver mounts. Furnishing a company or platoon leader's tank included a radio set, an Fu 6 (SE 20 u), consisting of a 20-watt sender and an ultra-short-wave receiver. The Fu6 (SE 20 U) radio had a range of 4 to 6 kilometers (very dependent on the terrain).

The equipment for a "normal" tank had only an ultra-short-wave Fu 2 (EU) receiver installed without a sender.

4.1.3.7 Extending Arm for Anti-Aircraft Fire

The extending arm for anti-aircraft fire could be used by a halted vehicle to mount a machine gun for defense against air attacks. Onto the left side of the upper body armor a mount and a clamp were screwed, in which the arm was mounted and could turn.

4.1.3.8 Gun Ammunition Storage

In the side spaces of the bow armor there were 68 7.5 cm shells stored, in the chassis 54, making 122 shells in all. In the right side space of the bow armor there were two ammunition cases holding 32 and 12, in the left side space two ammunition cases holding 12 shells each. The shells were kept lying down in these cases.

On the right wall of the chassis there were three ammunition cases for 26 shells, and behind the driver's seat was one case holding 28 shells. The shells were kept angled or vertical in these cases, with the point downward.

All the horizontal and angled ammunition cases were covered by canvas drapes. The vertical ammunition case had a two-piece sheet-metal cover.

4.1.3.9 M.G. Ammunition Storage

To hold the drum magazines for machine-gun ammunition, eight cases were located in the upper body armor.

Of these, there were three cases on the left sidewall, one case to the left of the machine-gunner, two on the right sidewall, one under the roof behind the forward bulkhead and one to the right on the rear bulkhead, holding a total of 38 double drums. Another case with two double drums was on the turret platform.

4.1.3.10 Equipment Storage

Outside on the bow and stern armor, and on the track aprons, were the following storage places for equipment:

On the right side for: one axe, one Tool Kit III, one vehicle winch, one spade, one radio antenna (under the antenna mount), one starting crank, one double screwdriver, four track links.

On the left side for: one fire extinguisher, two towing hooks, one equipment case for anti-aircraft defense, one winch block, one wire cutter, one key for leading-wheel setting, one track tension tool, two crowbars, one wiper with handle.

On the rear wall was one cable for towing the tank, hung on two hooks.

In the chassis and upper body were the following storage places for equipment and supplies:

To the right of the machine-gunner's seat, the holder for the sender and receiver transformers, at left the holder for the M.G. barrel and one gas mask.

Behind the machine gunner's seat, one oil-can holder, at right near the seat one holder for the M.G. equipment packet, one box for glass combination supplies and vision-slit armor, one headphone box, one container for spare glasses, and below, one bottle holder.

On the back of the frame for the sender and receiver attachment were the mount for the spare glass combinations for the driver's port, the holder for the first aid kit and the scope-part kit, next to the frame one bottle holder and, behind the driver's seat, one gas mask holder.

Near the left side of the driver's seat were a plate for fastening safety kits, magnetic lamp and transformer for the aiming arc, in the upper left corner a holder for the aiming arc, and on the front wall of the M.G. ammunition case a box for headphones was attached.

The turret of Pz.Kpfw.IV Type A.

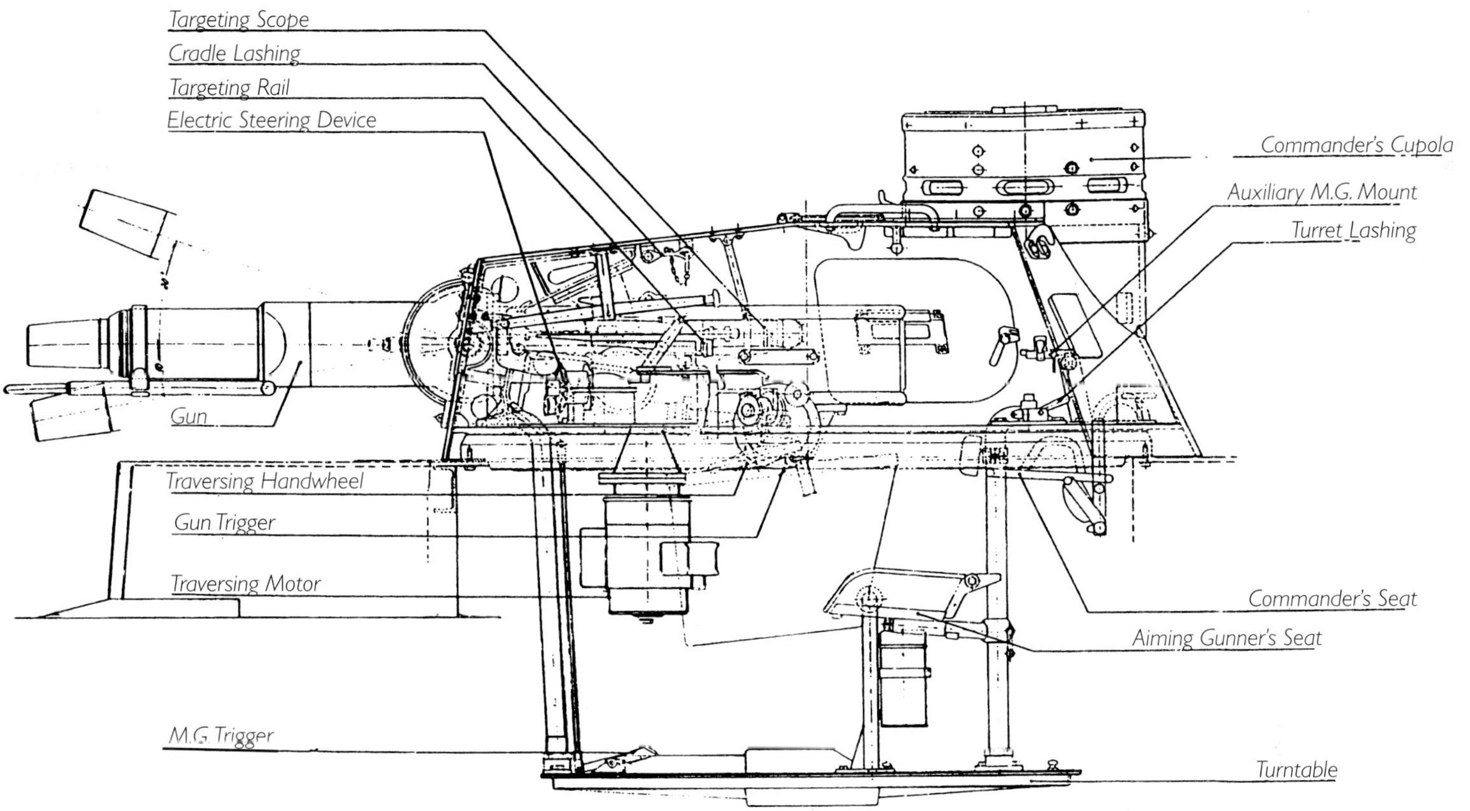

Turret (Drawing No. 021 B 14800) for Pz.Kpfw.IV Type A.

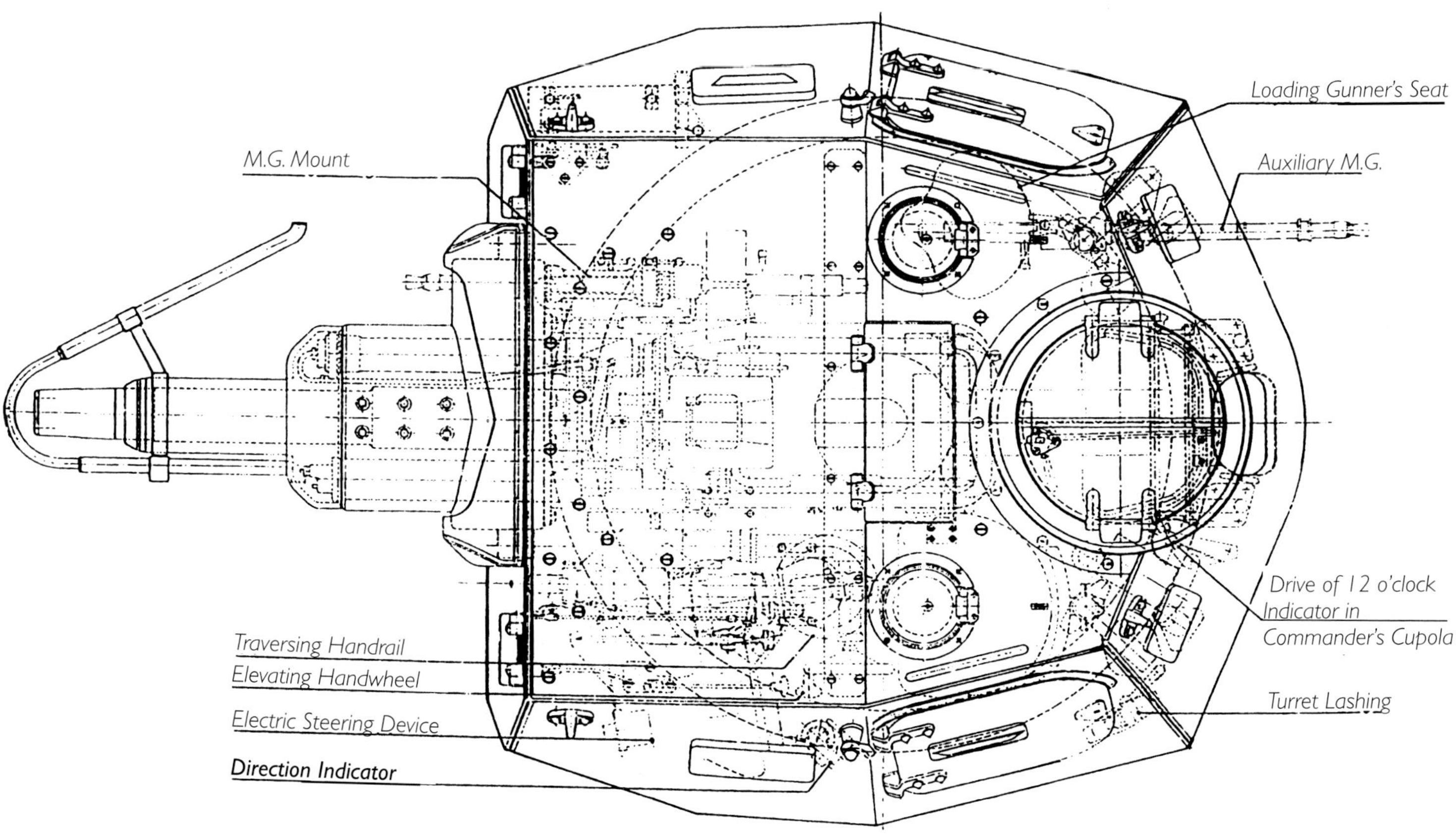

4.1.4 Turret

The turret, equipped with one 7.5 cm tank gun and one M.G. 34, was located in the center of the vehicle. The gun barrel with breechblock, barrel brake and pneumatic recuperator, was mounted in the gun cradle. The 7.5 cm tank gun and the machine gun lay side by side in a roller shield, which was installed in the front wall of the turret with lateral shield attachments. The roller shield was moved by a toothed wheel via the elevation machine by hand; the elevation field stretched from -10 to +20 degrees.

The turret could be moved around the track by the traversing machine by means of an electric motor or by hand. It was fitted with a targeting scope (T.Z.F 5 b) and a targeting rail (7.5 cm and MG 34). With the help of the 12-o'clock indicator system, traversing aiming directions could be given to the aiming gunner by the commander.

The main components of the turret were: turret and turret attachment, shield, gin, M.G. mount, targeting scope mount, targeting rail mount, turret-turning machinery (traversing machine), elevation machine, firing system, commander's cupola, vision ports and flaps in the turret, seats and equipment storage.

4.1.4.1 Turret Housing

The turret was mounted on a shoulder-high ball-bearing ring on which it could be turned all around. In the front wall of the turret was the roller-shaped shield, with a vision flap on each side. The turret armor consisted of armor plates welded together, with the following thicknesses and inclinations (from the horizontal).

Plate	Thickness	Inclination
Front wall	14.5 mm	80 °
Shield	14.5 mm	round
Sidewall	14.5 mm	65 °
Rear wall	14.5 mm	65 and 90 °
Roof	10 mm	0 and 7 °
Commander's cupola	14.5 mm	90 °

Over the roller-shaped shield a cover was screwed; it covered the gap between the shield and the front turret wall on all sides. Between this cover and the front turret wall the sealing inlay made of leather and sheet metal was clamped.

The turret was closed off above by its roof, to the back part of which the commander's cupola was screwed. Under the roof the handhold for the commander was screwed on. The back of the commander's seat was attached to the rear wall of the turret.

In the center of the roof was a ventilator flap. To the right and left of it round signal openings had been cut. The signal openings were covered by closable signal flaps.

In each side wall an entry hatch and a vision opening had been cut. The hatches were closed by the hatch covers, the vision opening in the right sidewall by the vision flap with vision slit, and the vision opening in the left wall by a vision flap without a slit. Over each entry hatch a handhold was attached on the outside. When open, the hatch covers were held by a pull bolt on the forward sidewall. Over each entry hatch a rain gutter was attached.

Two more sight openings were in the rear wall; they were covered by vision flaps without vision slits.

The turret mantle was reinforced all around on its lower rim by the carrier ring, and braced by two carriers at the front.

To place and remove the turret there were four hooks (or eyes) on the outside of the roof; they could also be used to attach means of camouflage.

4.1.4.2 Turret Attachment

The turret was linked to the upper body by what is called the turret attachment. To it there belonged:

The toothed circle, ball bearings and cover panel, the carrier ring, the turret lashing, the direction indicator (12 o'clock indicator), the extra machine-gun mount and the turntable.

The direction indicator was attached to the housing by ball bearings. The number disc, like the toothed circle in the commander's cupola, was made with divisions from 0 to 12; the indicators attached to the housing could be read.

The extra machine-gun mounts were screwed to the carrier ring behind the two rear vision flaps.

The turntable consisted of the circle made up of angle irons, covered with a plate of rippled sheet metal. The plate had a cutout in back that was closed by a flap. The turntable was hung on a brace in front and two braces on the carrier ring in back.

In the middle, the turntable was guided through the centering of the ring body and covered by a cap and a shield to protect the guide and the ring body. Behind the hood, the foot brace for the commander was screwed on.

4.1.4.3 Shield

The shield (roller shield) served to hold the gun cradle with the 7.5 cm barrel, barrel brake and pneumatic recuperator, plus the M.G. 34. It consisted of a semicircular bent armor plate that was closed on both sides by welded-on sidewalls that bore the shield attachments. Behind it stood a wall that served for attaching the cradle and the machine-gun mount. In front of the shield lay its covering of armor plate. This cover protected the gap between the shield and the front wall of the turret, and was attached to the turret wall. To increase protection against lead spray, the upper and lower shield panels were screwed to the shield, the upper cover plate to the turret roof, and the lower cover plate to the carrier ring.

4.1.4.4 Tank Gun

The 7.5 cm tank gun consisted of the following main components: the barrel with the breech, cradle with barrel brake and recuperator, electric parts, equipment and supplies.

The 7.5 cm Kw.K. had a barrel length of 1766.5 mm (L/24 caliber length). A rate of fire up to 12 rounds per minute could be attained. Optionally, the following types of ammunition could be fired.

Type	Shot Weight (kg)	Vo (m/sec)
Armor-piercing shell (K.Gr.rot Pz.)	6.8	385
Explosive shell (Gr.34)	5.74	420

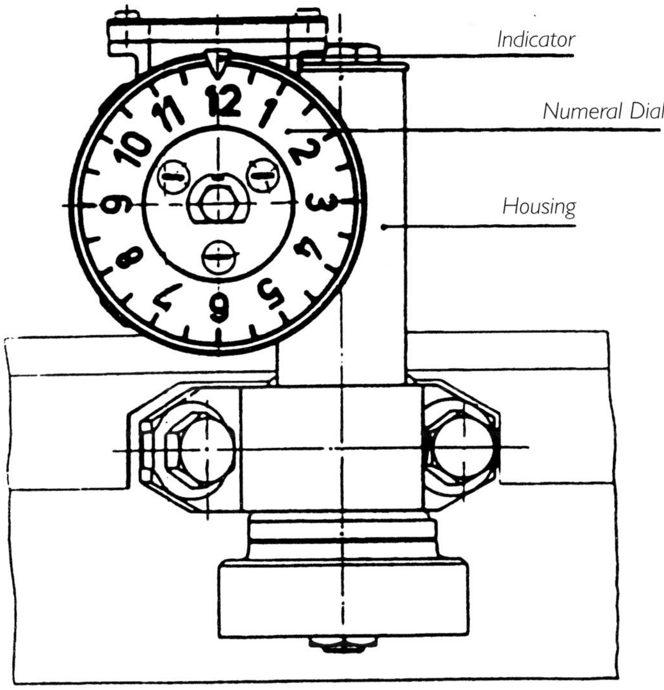

12 o'clock indicator (Drawing No. 021 C 14814) for Pz.Kpfw.IV Type A.

4.1.4.5 Machine-Gun Mount

To install the M.G. 34 there was a weapon mount, which could be aimed by an adjusting apparatus. The front part of the mount was extended toward the front shield wall, and along with the shield cover it served to protect the gun opening against shots.

The M.G.34 had a rate of fire of up to 900 rounds per minute. The ammunition load consisted of a mixture of S.m.K. and S.m.K. tracer bullets.

4.1.4.6 Targeting Telescope

The turret targeting scope *T.Z.5b* was a flexible telescope with an armor plate attached at the bending point to protect the aiming gunner. The targeting scope was used for direct aiming of the weapons in the turret of the tank.

The turret targeting scope was housed with its looking-out head in the mount; the looking-in head was hung on the carrier on the turret roof and pivoted.

The *T.Z.F.5b* had a 2.5-fold magnification and a vision field of 25 degrees (444 meters at 1000 meters).

In the head of the telescope there were two calibrated plates: one could be turned, the other's height could be adjusted. The turning plate bore small circles for shooting with explosive shells, of which the first and every other one afterward were numbered (to 2000 meters); also small circles for machine-gun fire and armor-piercing shells, of which the first and every other one afterward were numbered (to 800 meters). The highest targeted shot range for the 7.5 cm KwK was 2000 meters.

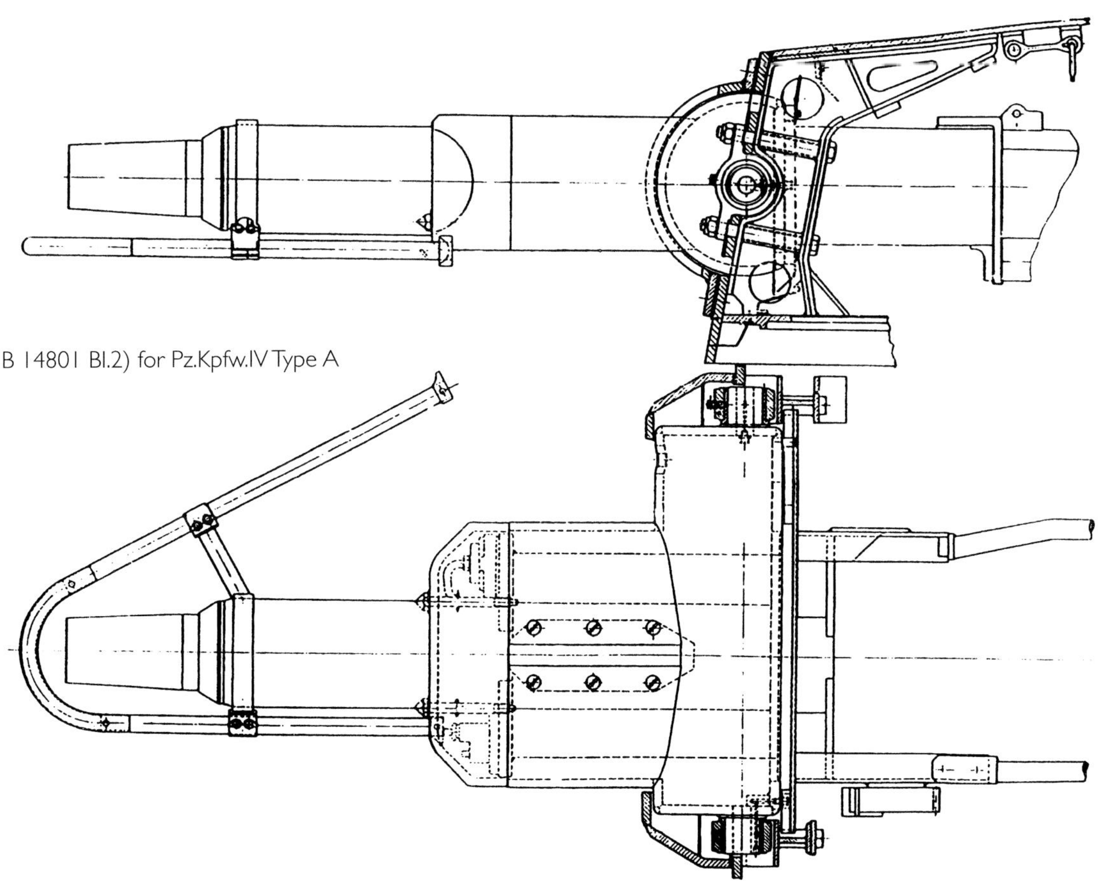

Roller shield (Drawing No. 021 B 14801 Bl.2) for Pz.Kpfw.IV Type A

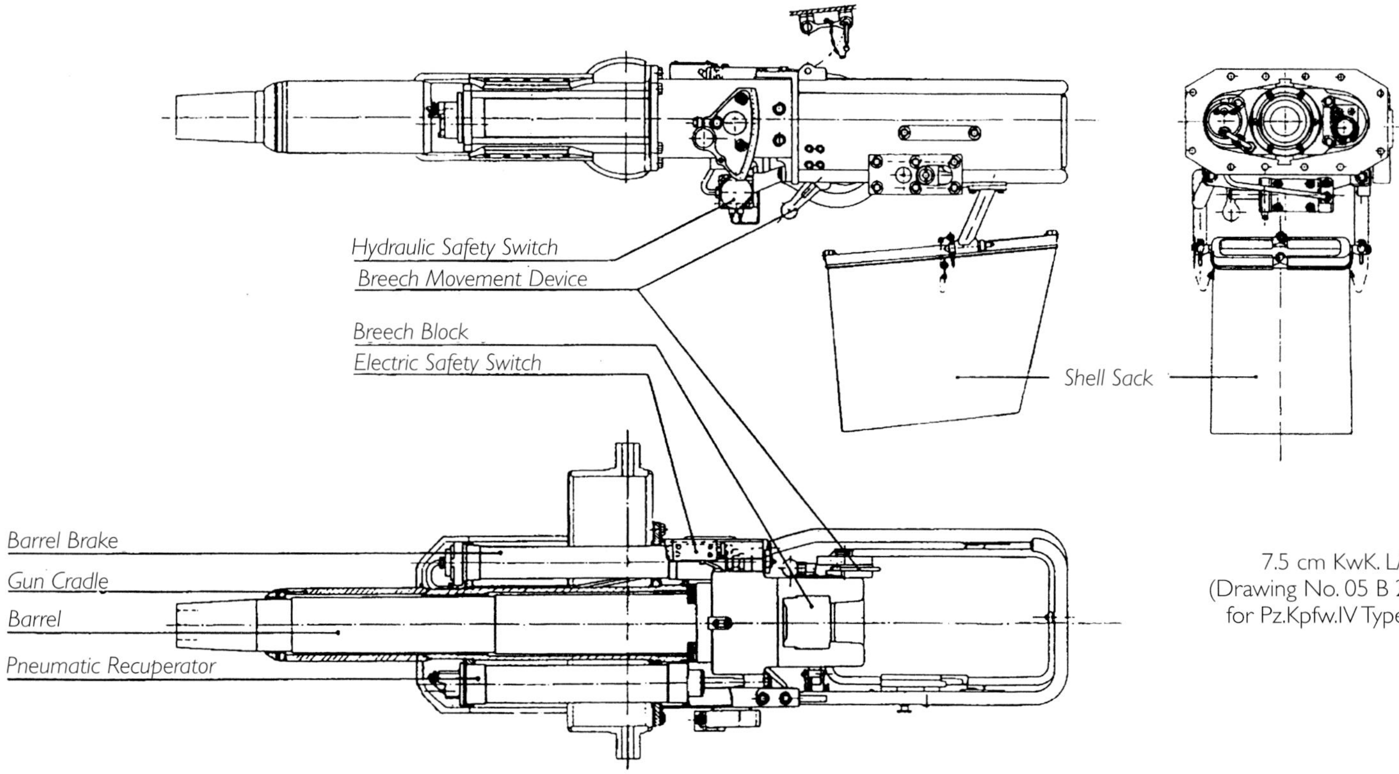

7.5 cm KwK. L/24
(Drawing No. 05 B 28)
for Pz.Kpfw.IV Type A

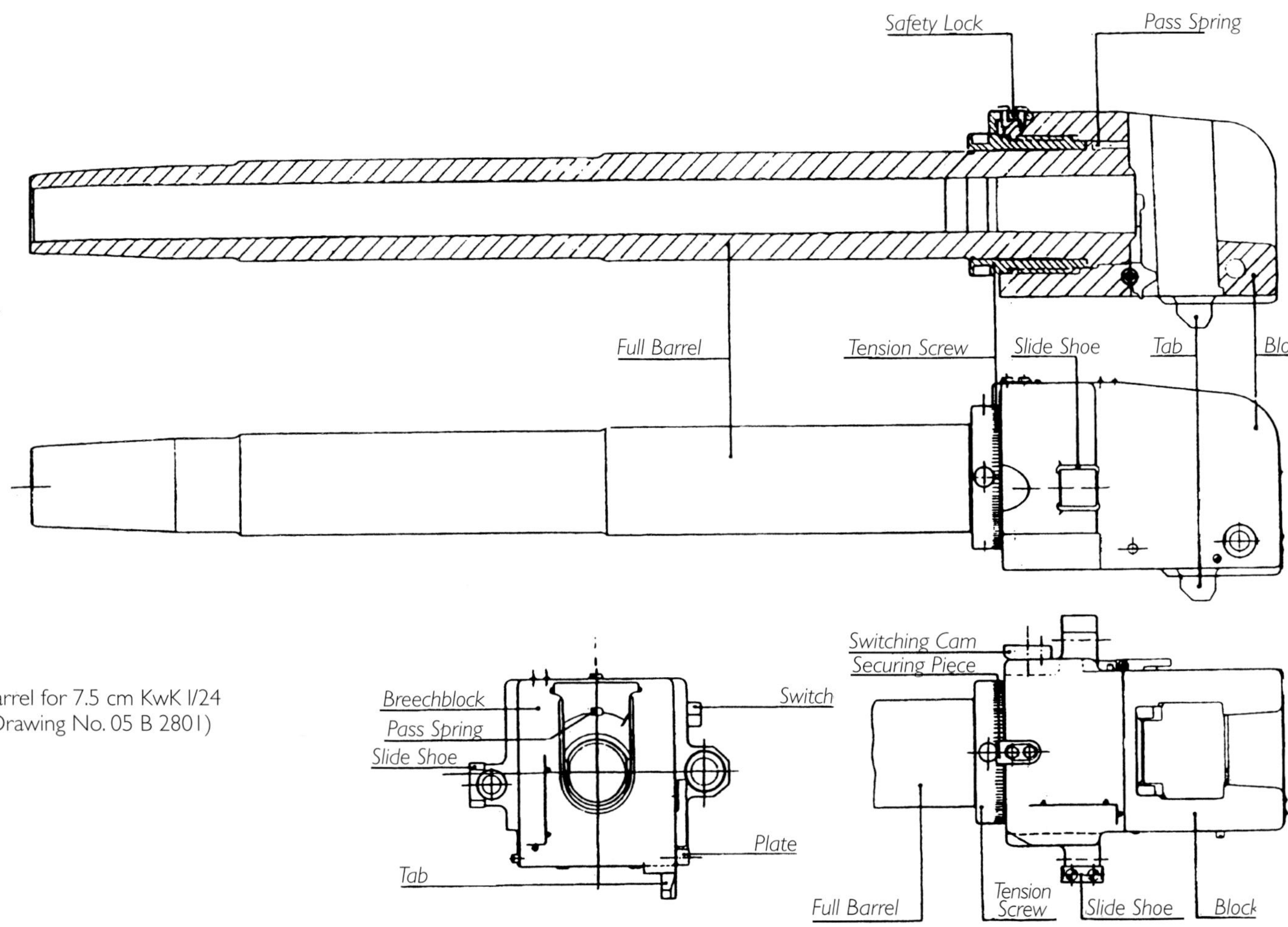

Barrel for 7.5 cm KwK l/24
(Drawing No. 05 B 2801)

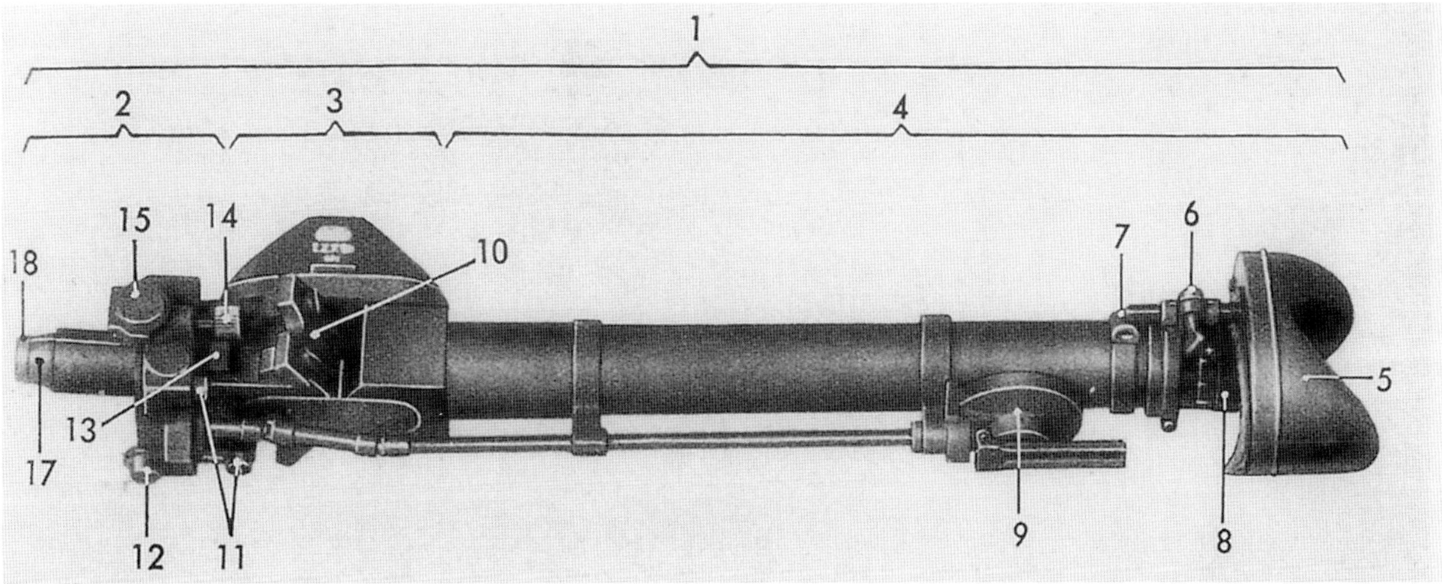

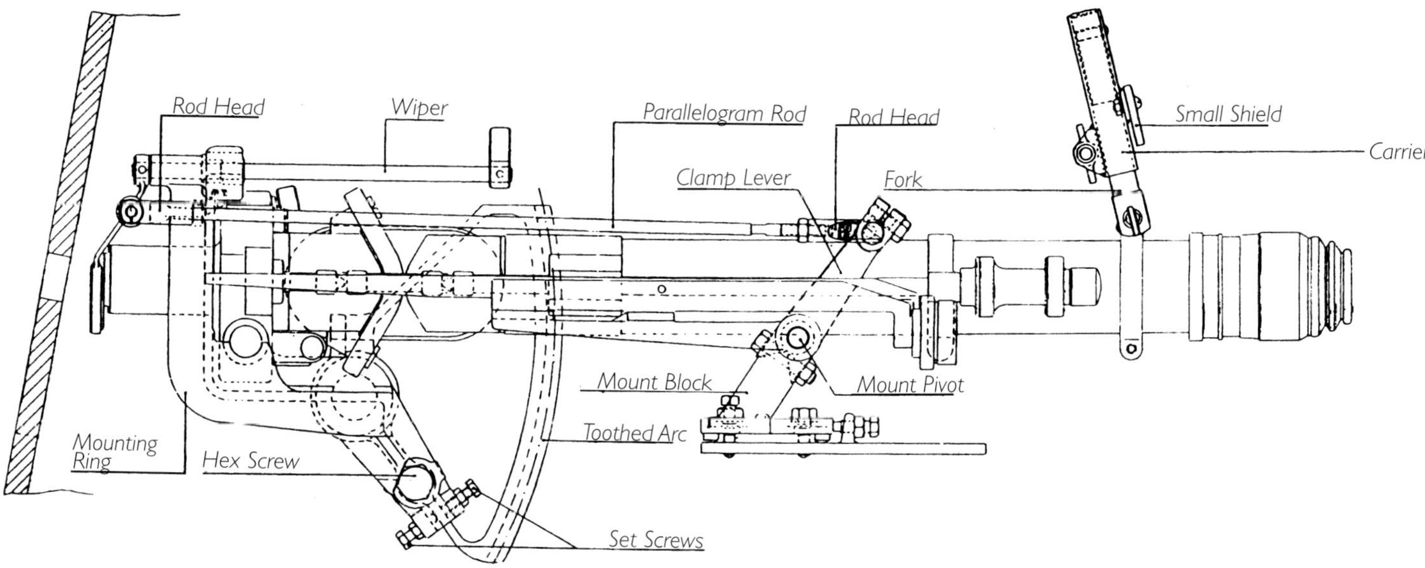

4.1.4.7 Targeting Rail

The *7.5 cm* and *M.G. 34* targeting rail could also serve as a targeting means if the targeting scope failed as a result of being hit. It allowed aiming both the tank gun and the M.G. 34 at ranges of 200, 400, 600 and 800 meters. The range was marked on the setting head.

4.1.4.8 Turret Turning Apparatus

The turret was turned by the turning apparatus, which could be operated with a turning motor or by hand. To turn the turret 360 degrees, 188 turns of the handrail of the turning apparatus were required. The turret was therefore fitted with an electric turning motor, which also allowed turning the turret if the tank was not setting level, hastened quick turning on slopes and could be used, for example, to hold moving targets steadily. By using a switch lever, turning could be done electrically or by hand.

The main components of the electric turret-swinging apparatus were the machinery, which consisted of the (Auto Union DKW two-cylinder) drive motor and the manifold generator and was located in the engine compartment, plus the turning motor by which the turret was turned electrically, and the turning apparatus by which the direction and speed were regulated.

The DKW two-cylinder motor, Type PZW 600, drove the generator. The clutch was built elastically by having a rubber disc in the middle. The carburetors were two Framo VG-V types. Since the motor was water-cooled, a Sugo pump made by Otto Hamel & Co. of Mannheim was set on the side of the generator. The pump was driven by the chain wheel, chain and other chain wheel mounted on the generator shaft. The Bosch starter, which got its electricity from the tank's batteries, started the aggregate.

The electricity was produced by a special multi-volt generator in its own housing. It consisted of a Leonard generator and constant-volt machine. The Leonard generator produced 2 kW in 20 minutes at 2800 rpm. The anchor voltage was +/- 190 volts. The generator was regulated by the control apparatus, which was attached to the traversing machine.

The turning motor was located as a flange motor under the turning machine. Its power was 1.2 kW. If a crew member came into danger of being jammed in while turning the turret, the

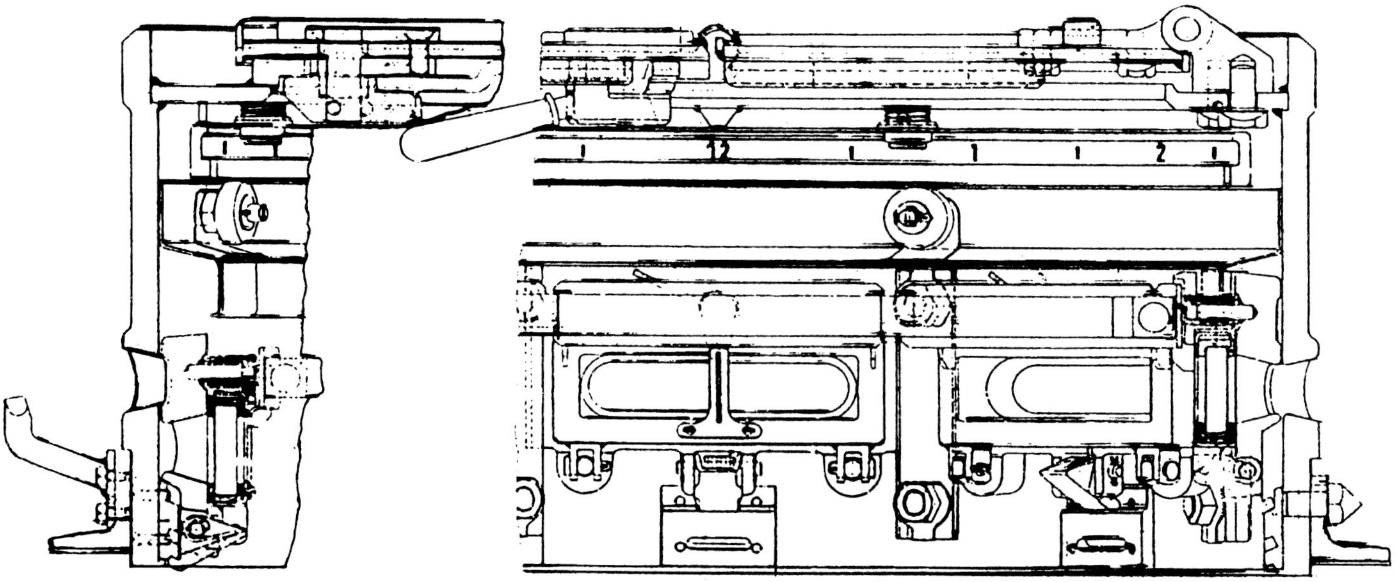

Commander's cupola (Drawing No. 021 D 15031) for Pz.Kpfw.IV Type A.

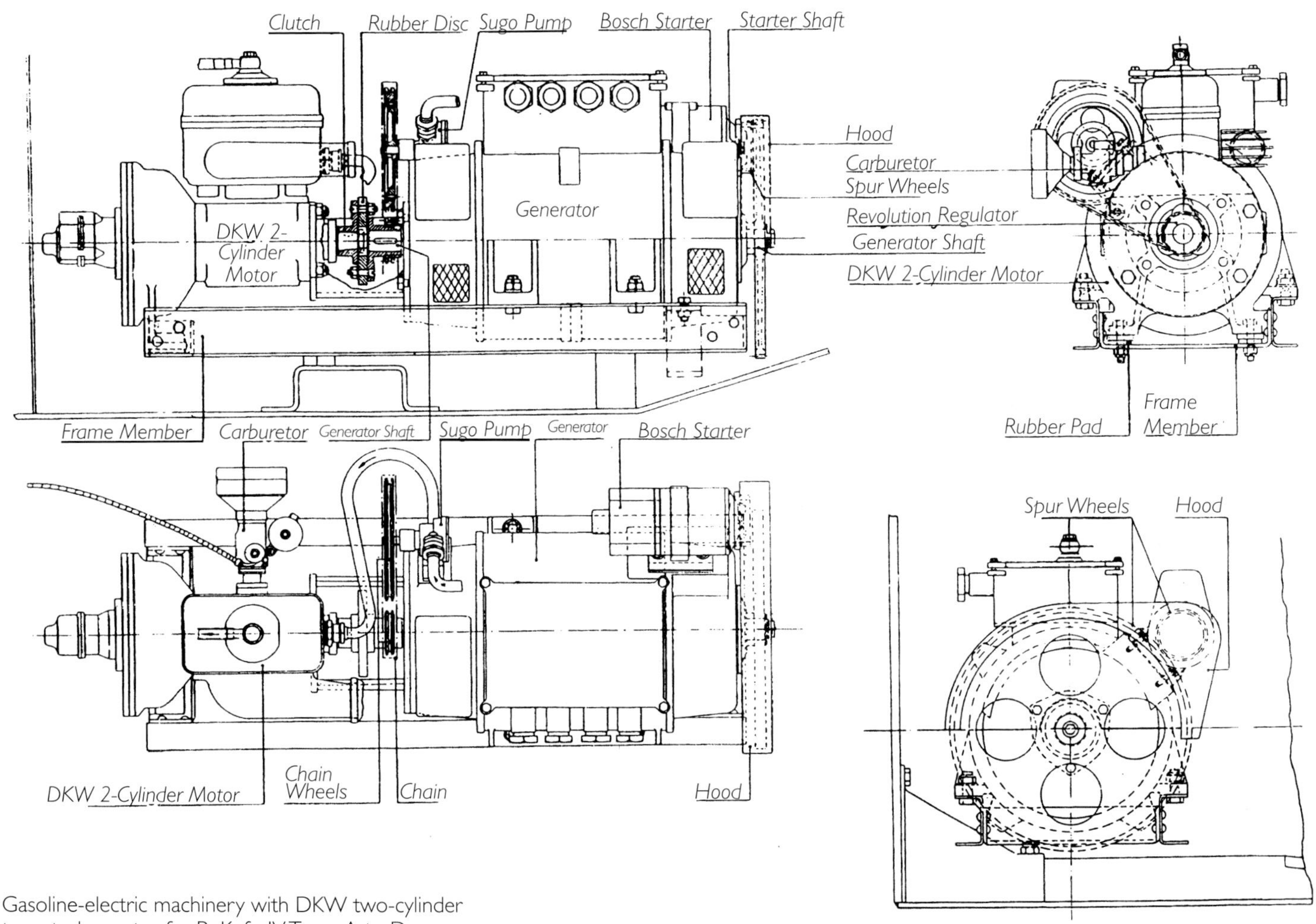

Gasoline-electric machinery with DKW two-cylinder two-stroke motor for Pz.Kpfw.IV Types A to D.

turret could be stopped immediately by activating the emergency switch.

The steering apparatus was located in a sheet-metal housing that was closed by a lid. It was attached to the horizontal flange of the turret-turning device's housing.

4.1.4.9 Elevation Machine

The shield was moved with the gun cradle by a mechanical elevation machine. One turn of the handwheel changed the elevation about two degrees.

4.1.4.10 Firing the Gun

A pistol grip for firing the gun was mounted on the handhold of the traversing machine. By pulling on the pistol grip with the index finger, the control box and, with it, the lever and bolt, were moved upward. The lever closed the contact and thus the electric circuit for firing.

The machine gun was fired by a firing lever controlled by a foot pedal.

4.1.4.11 Commander's Cupola

The commander's cupola was screwed onto the rear part of the turret. It let the tank commander get in and look out.

The cupola essentially consisted of the cylindrical mantle, the eight slides, the numeral ring and the two-piece roof. The cylindrical mantle had eight vision slits, which could be closed wholly or partly by height-adjusting levers. A lever was a rectangular plate fitted to the mantle with attachment surfaces on its sides. In the lower third there was an expandable vision slit, 138 mm wide, at the rear.

To protect the eyes against lead spray and splinters there was a protection glass, a transparent plate of safety glass, 12 mm thick, behind the vision slit. The protection glass did not protect against shots through the vision slit.

Outside on the mantle there was a direction indicator screwed on in the middle before the front vision slit. The commander could use this indicator to check the general direction of the turret.

In the upper extension of the cylindrical mantle a sheet-metal ring was welded on. On this ring the two-piece cover was mounted in two side joints. Under each half of the cover a plate of moss gum was glued as a head cushion for the commander.

The numeral disc, with four sliding rollers below and four gliding rollers above, was mounted under the welded-on sheet-metal ring. The numeral disc had a U-shaped cross-section. The lower step had 360 teeth; the middle step was calibrated in divisions with numerals 1 to 12.

4.1.4.12 Vision Hatches and Flaps in the Turret

a) Vision Hatches

In the front wall of the turret there was one vision hatch installed on each side, as also in the left and right

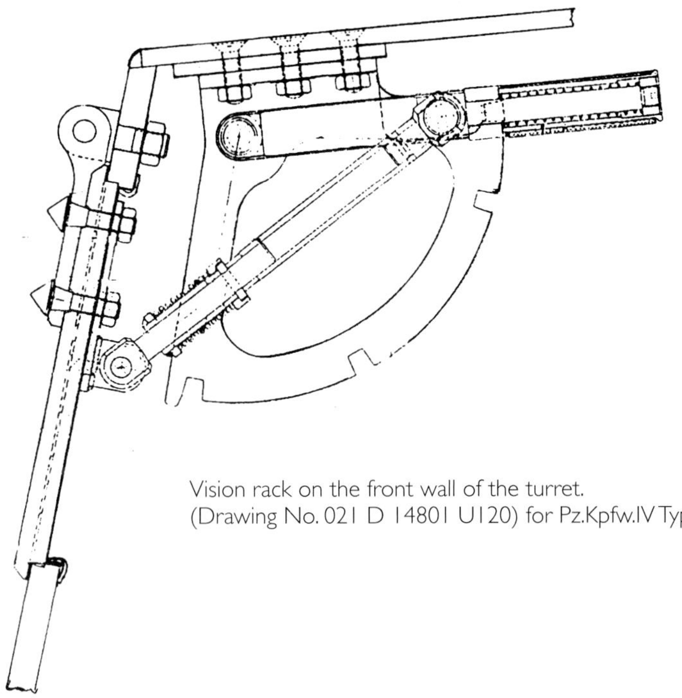

Vision rack on the front wall of the turret.
(Drawing No. 021 D 14801 U120) for Pz.Kpfw.IV Type A.

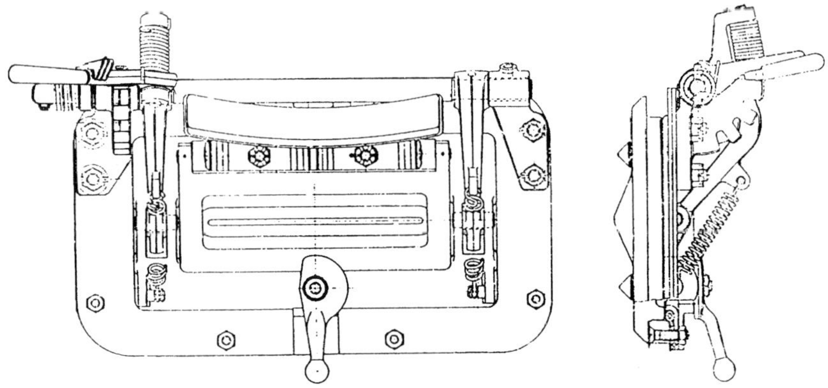

Vision rack with viewing slit in the left sidewall of the turret, for Pz.Kpfw.IV Type A.

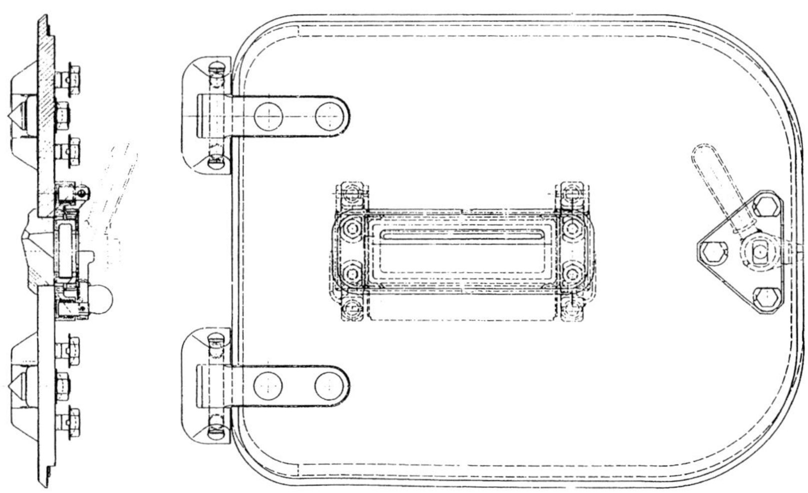

Hatch cover in the right sidewall of the turret (Drawing No. 021 D 14801 U22).

sidewalls. The hatch in the right sidewall had a vision slit, all the others did not.

The vision hatches were each screwed onto the front wall with two hinges. Inside there was a sheet-metal frame on the turret's front wall for protection against lead spray.

The vision hatch had straight surfaces, under which leather strips were glues to seal them.

The vision hatches in the left and right sidewalls were rectangular plates with straight surfaces and projecting outside frames. Every plate was cut out on the inside so that a thin sealing rim was formed around its edge. The seal lay against the rubber sealing frame during shooting.

In the right sidewall a vision slit, 4 mm wide, was cut into the flap; the slit was bordered by a rim outside to deflect lead spray and shots. To protect the eyes against lead spray and splinters, a 12 mm thick transparent plate of safety glass, called protecting glass, was mounted behind the vision slit. This glass did not protect against shots through the slit.

b) Turret Hatch Covers

The turret hatch allowed entry and exit, and was closed by two covers, which were attached to the sidewalls by two hinges. They were sealed against lead spray by sheet-metal frames attached to the rim of the hatch. The hatch cover was folded back 180 degrees when the hatch was open. In this position the cover was held in place by the turning bolt under spring pressure.

When the bolt of the turret hatch handle rested with its fork on the sidewall, the hatch cover was opened about 30 mm. This position was intended for ventilation with limited safety from enemy fire.

In the middle of the hatch cover, a fixed vision-slit plate with two mounts was screwed. The protective glass lay as on the vision hatch in a sheet-metal frame and was interchangeable with the other.

Equipment storage in Pz.Kpfw.IV, Type A, Chassis No. 80109, finished at the Krupp Works in January 1938.

A look through the left entry hatch at the turret turning machinery and turret targeting scope of Pz.Kpfw.IV, Type A.

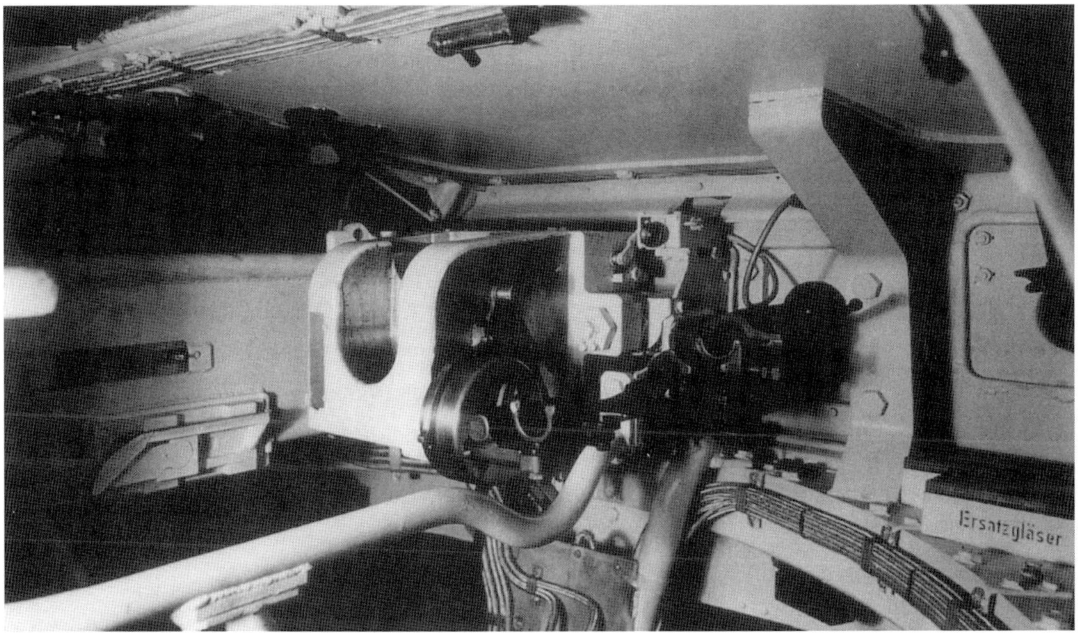

A look through the right entry hatch at the tank gun and turret machine-gun mount of Pz.Kpfw.IV, Type A.

c) Vision Hatches in the Rear Turret Wall

The extra machine guns could be mounted behind the two rear vision hatches. These hatches were rectangular plates with straight surfaces and projecting outside rims. Each plate was cut out inside so that a thin sealing rim was formed around its edge; the seal of the hatch lay against rubber sealing frames during shooting.

d) Turret Seats

At the back of the turret the commander's seat was attached to a carrier ring. The seat could be folded and fastened in this position. The seat consisted of a steel pipe frame with a welded-on seat panel. A seat cushion was attached to this seat.

The left rear brace of the turntable held the back of the aiming gunner's seat. The gunner's seat, a box type, could be folded to the side and stored on the seat mount. A seat cushion was attached to the seat.

The right rear brace of the turntable held the back of the loading gunner's seat. This seat was saddle-shaped and was welded to a shell that could be folded 90 degrees to the side on the seat frame. When not in use, the seat could be tipped to the side and swung under the commander's seat.

4.1.4.14 Equipment Storage

In the turret there were the following storage places for equipment:

- On the right side was the rack for the unloader,
- Under it on the carrier ring was the rack for spare machine-gun barrels and containers for spare glasses,
- At the rear of the right side of the turntable, above the loader's seat, was a rack for a gas mask,
- Up on the rear wall was a rack for two field flasks,
- Between the racks on the carrier ring was the belt for the extra machine-gun mount,
- Behind the commander's seat on the carrier ring was the headphone holder,
- On the commander's foot-rest was a holder for the machine-gun parts packet, and behind it the rack for two gas masks,
- Between the commander's and aiming gunner's seats, on the carrier ring, was the box for tracer bullets, and over it on the back wall the holder for the flare pistol,
- On the left rear brace of the turntable behind the aiming gunner's seat was the holder for one field flask and the flag holder,
- At the left front under the roof was a rack for the targeting rail,
- At the right front on the turntable were two boxes for machine-gun drum magazines.

4.2 FORM CHANGES DURING PRODUCTION

4.2.1 Improvements Required After Inspecting The First Vehicles

As reported in the following notes of Mr. Woelfert about the acceptance and inspection of two Begleitwagen at the Gruson Works in Magdeburg on November 30, 1937, shape changes were already requested then by Wa Prw 6.

After the two vehicles were received on November 29, 1937, by the acceptance officials in the presence of Saloman and Giesecke of Prw.6, the vehicles were thoroughly inspected by Fichtner, Olbrich and the other gentlemen.

The careful preparation by the Gruson Works was particularly noted; the easy access to the motor in the rear space after opening the upper covers, and the electric system from the bulkhead out, were noted. The following wishes for changes were expressed by Olbrich as to individual points formerly agreed on by him and his fellow workers:

1) *Attach an additional lashing on the ball mantlet of the bow M.G. (by Daimler-Benz) to ease entrance.*
2) *Strengthen the rotary attachment and push it farther into the corner,*
3) *Mount the boxes for the driver's headphones higher to protect them against any penetrating water,*
4) *Dust protection for the 7.5 cm ammunition is no longer to be attached in two pieces or with frames, but in one piece and push-buttons from the lower or spring pulls or push-buttons on the upper dust protector.*

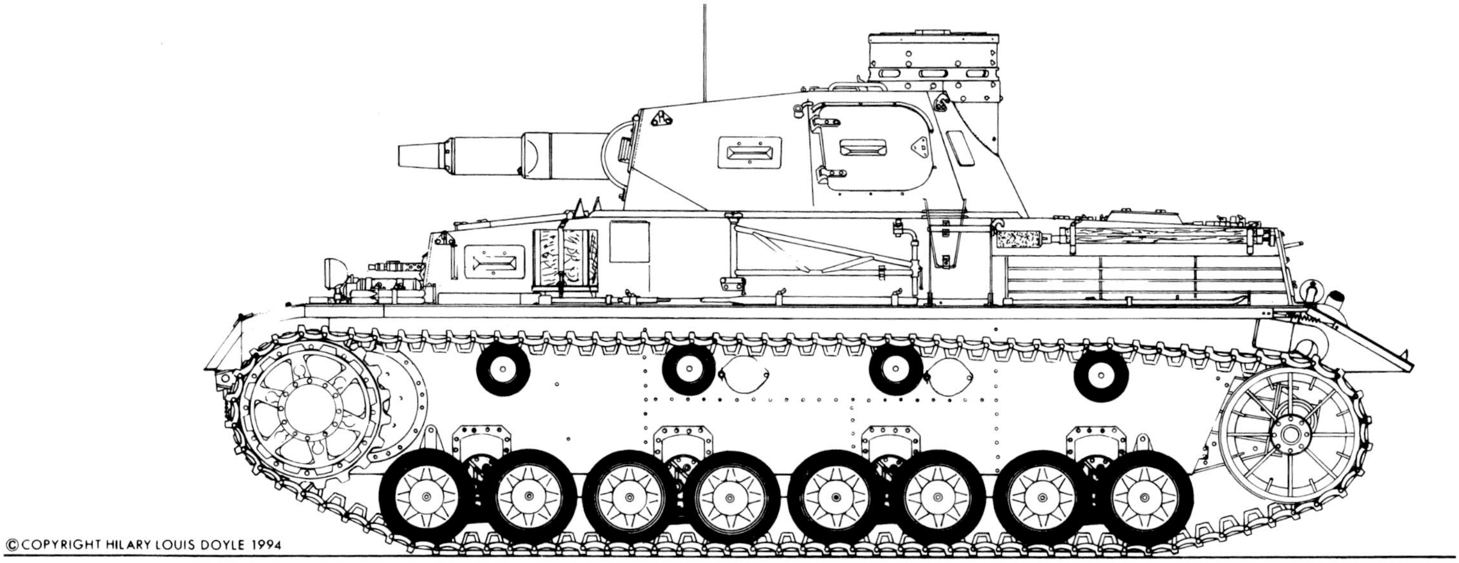
©COPYRIGHT HILARY LOUIS DOYLE 1994

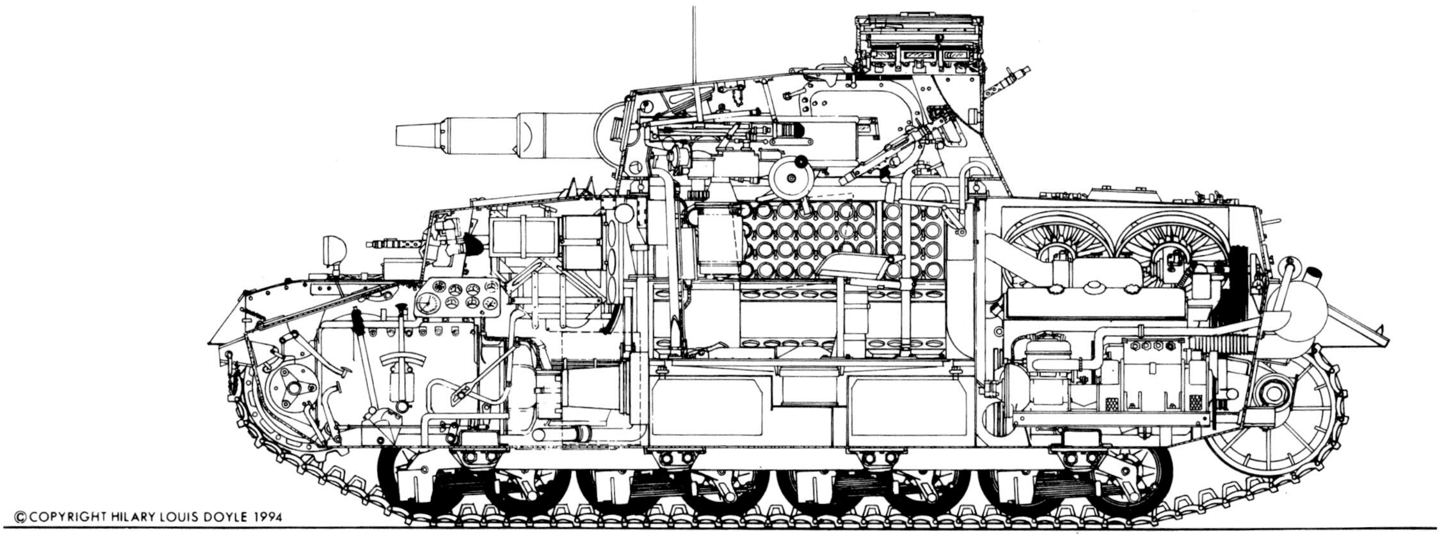
©COPYRIGHT HILARY LOUIS DOYLE 1994

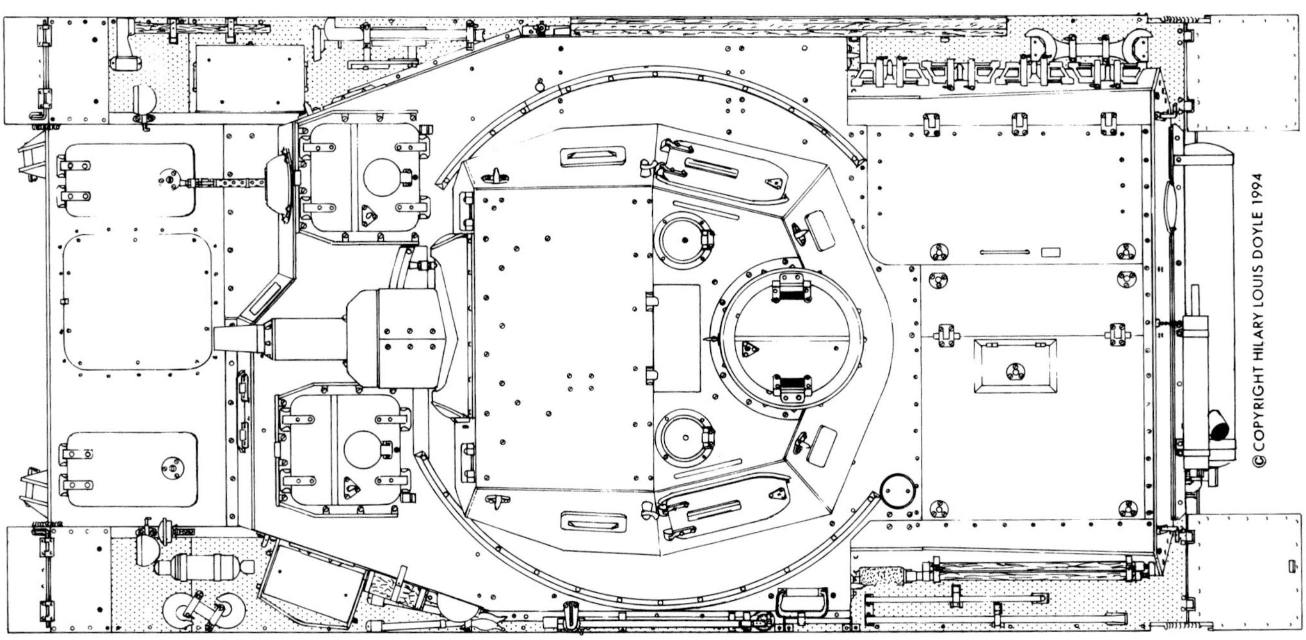
© COPYRIGHT HILARY LOUIS DOYLE 1994

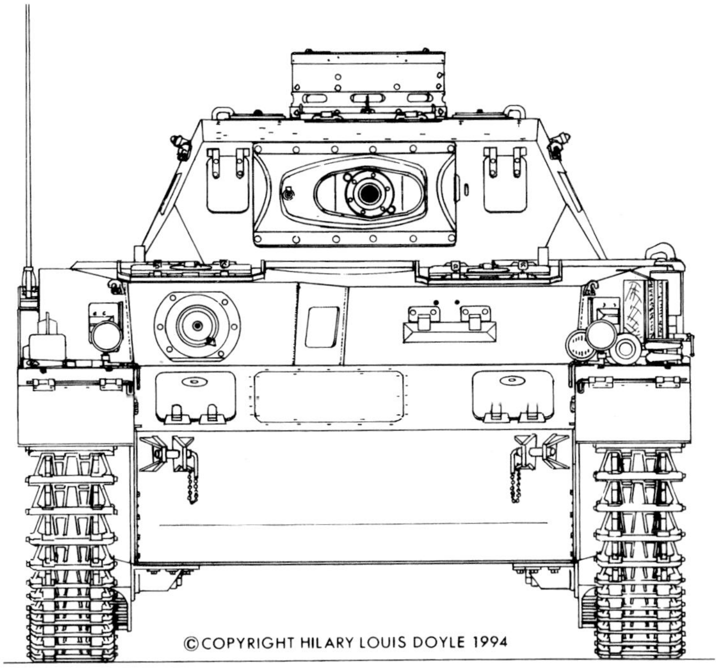

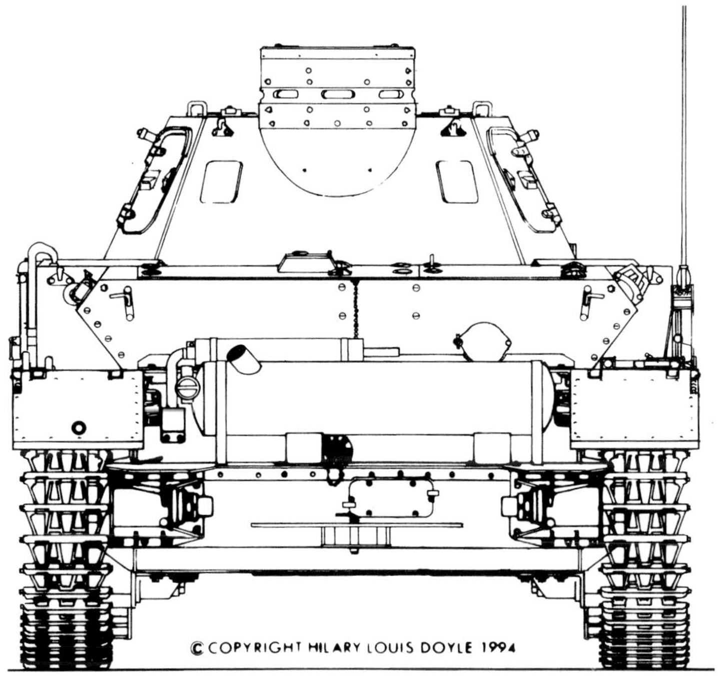

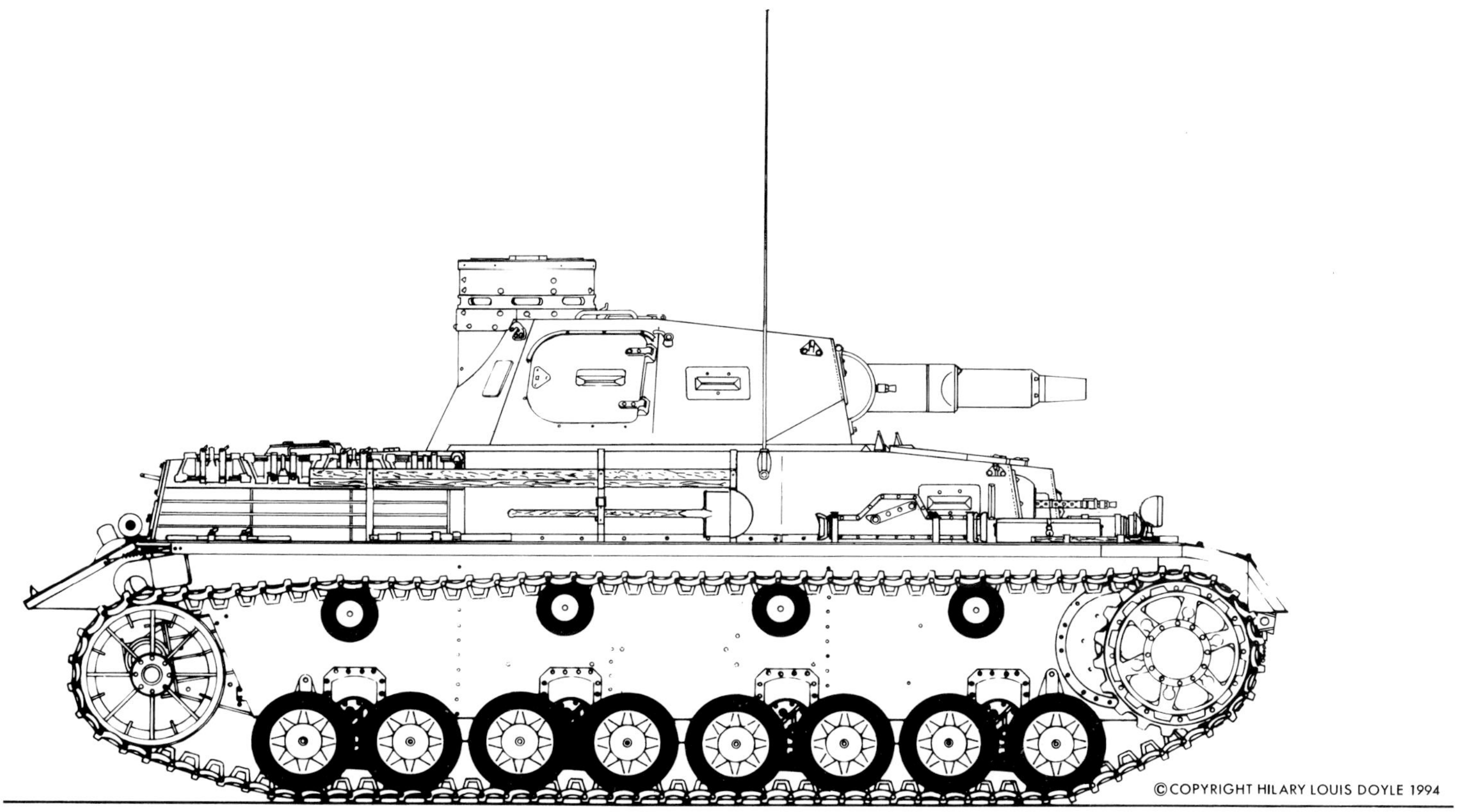

Panzerkampfwagen IV (7.5 cm), (Vs.Kfz. 622), Type A, developed from Escort Vehicle I with Escort Vehicle II body. New turret and commander's cupola. The longitudinal cut shows the SFG-75 gearbox. Storage for 122 7.5 cm caliber shells.

These wishes were discussed more precisely with Biersack after Olbrich's departure. They will be taken care of before the delivery of the first vehicles.

Olbrich indicated that the handholds for the gunners and driver were not to be welded onto the front wall, but screwed on, and that to attach the lifting eyes of the bow armor round-headed screws were to be used instead of rivets. The latter were still attached.

In the future the moldings later welded onto the bow armor at Olbrich's wish be welded on ahead of the side vision hatches before approval.

Olbrich also suggested that part of the M.G. ammunition be moved from the left to the right side of the bow armor so as to be easier for the loading gunner to reach. This must still be approved by Prw.6, since it is not possible without the elimination of numerous rounds of 7.5 cm or M.G. ammunition because of the lesser depth of the left bow armor.

WaPrw 6 (IId), Olbrich, was not as enthusiastic as Mr. Woelfert in his inspection of the first two *1./B.W.* vehicles and requested the following 30 improvements from Krupp in a letter of December 1, 1937:

The first two 1./B.W. vehicles were examined. The following shortcomings were identified:

1) The rack for the course rotor was made only sketchily. A rack must be attached that serves to hold the elastic intermediate ring for the course rotor; in addition, the rotor is to be moved some 30 mm farther forward. (As of 1./B.W.)

2) Over the switch box of the intercom system (to the right of the radioman) a step must be attached. (As of 1./B.W.)

The driver's seat layout of Pz.Kpfw.IV, Type A.

Ammunition storage on the left side of the fighting compartment of Pz.Kpfw.IV, Type A.

The radioman's seat layout in Pz.Kpfw.IV, Type A.

3) *The protective covers for the ammunition storage must be attached so that they do not lie on the floor when opened or hinder the removal of the ammunition. Instead of belt buckles, push-buttons or spring snaps are to be attached. (As of 1./B.W.)*
4) *The firing cable between the safety switch on the gun and the electric plug must be fitted with a metal duct.*
5) *The side switch does not fulfill the established requirements. The switch made by Daimler-Benz is to be installed (as of the 8th 1./B.W. vehicle. For the first seven vehicles, the switching mechanisms made by Krupp and now present may be used with the installation of two switches per vehicle.)*
6) *The cover ring of the commander's cupola must be given a higher rain barrier and be turned out more sharply at its transition. The overhanging rim on the cover of the commander's cupola may not be welded on inside. (As of 8th 1./B.W. vehicle)*
   The antenna-folding loop is to be made like the model shown at Magdeburg. (As of 1./B.W.)
8) *The leaf springs on the turret cover for fastening the signal flaps can be omitted. (As of 1./B.W.)*
9) *The M.G. ammunition box on the turntable is to be changed to hold only two drum magazines and be stored under the gun shell-casing sack. The gun equipment case formerly stored there is to be moved left to the sidewall. Two drum magazines are to be stored at the right front in the turret. (As of 3rd Vehicle 1./B.W.)*
10) *The spare barrel in the turret is to be moved farther forward. (As of 3rd Vehicle 1./B.W.)*
11) *The ball mantlet will receive a second lashing, so as to make entry through the radioman's hatch easier. (As of 1./B.W.)*
12) *The signal-lamp case for the radioman must be moved some 10 mm to the right. (As of 1./B.W.)*
13. *The case for the driver's headphones is to be placed higher. (As of 1./B.W.)*
14. *The flange of the ball-mantlet cover is to be made into a full circle (Departing from the D.B. drawing). (As of 1./B.W.)*
15) *The mounting eyes on the upper body armor to be attached with round-headed screws; riveting is not allowed. (As of 2./B.W.)*
16) *The side barrel braces at the commander's seat must be bent farther out from each other. (As of 2./B.W.)*
17) *Handholds may not be welded in place inside the armor. (As of 2./B.W.)*
18. *The armor-plate deflector before the vision hatches are to be welded on before the approval of the material. (As of 2./B.W.)*
19) *The air-escape flap in the turret roof is to be unarmored at the front and sides. (As of 2./B.W.)*
20) *The covers of gun ammunition cases must be installed in a dismantled state, so that they cannot close, or be fastened to the ammunition cases with hinges, similarly to Pz.Kpfw.III. (As of 1./B.W.)*
21) *On the head of the indicator rail the numeral field for MG34 setting is to be modified permanently with luminous paint. The "MG34" designation is to be painted on with luminous paint. (As of 5th Vehicle 1./B.W.)*
22) *To lubricate the turret ball bearings, 3 Meyrel lubricating nipples are to be divided and attached to the turret ring. (As of 3rd Vehicle 1./B.W.)*
23) *The cable drum on the front wall of the vehicle's upper armored body is to be made narrower according to specifications from Prw.7. (As of 1,.B.W.)*
24) *The cable set attached between the sender and receiver is to be removed. (As of 1./B.W.)*
25) *The insertion device for the radio button must be moved as far downward that the button can be pushed in. (As of 1./B,W.)*
26) *The moving of the cable right of the insertion device for the button is to be made according to specifications from Prw.7. (As of 1./B.W.)*
27) *The following parts were missing from the vehicle:*
    a) Container for targeting rail,
    b) Grip handle for the commander.
    The parts are to be sent by return mail.
28) *For every vehicle, a reserve rod antenna is to be delivered by the assembly firm. (As of 1./B.W.)*
29) *The rests for the vision hatches in the front wall of the turret are not right.*
30) *The constructive assembly of the electric aggregate for the turret drive is insufficient; as for operating technology, the aggregate corresponds to the established requirements.*

The Gruson Works replied on 20 December that they were not in a position to make changes to the B.W. vehicles No. 80103 (delivery planned for 12/28/37) and 80104 (delivery planned for 1/10/38), and would undertake the delivery of these vehicles without considering points 5, 7, 8, 9, 29 and 30.

4.2.2. Further Improvements

The Gruson Works reported on 8 December:

The delivery of the first two vehicles has been delayed, because according to the wishes of the troops, the oil drain of the motor had to be made easier of access afterward.

On January 11, 1938, the Gruson Works reported that the *"first two vehicles, 80101 (Motor Vehicle Combat Troop School in Wulsdorf) and 80102 (Armored Troop School, Firing Instruction, Oldendorf/Holstein) were delivered when we determined that a connecting opening exists between the vehicles' interior and the space for the fuel tank.*

Because of a change in storing part of the M.G. ammunition from the left to the right side of the bow armor, so as to be easier for the loading gunner to reach, it was decided on 10 December that the *Ammunition storage of Pz.Kpfw.IV Type A should be lowered from 140 to 122.* By 23 December the Gruson Works had built the ammunition cases on two vehicles according to A.K. suggestion, and also fitted them with push buttons according to Gruson Works succession. On January 11, 1938, these two vehicles should be finished and inspected by Major Olbrich.

On February 12, 1938, Wa Prw 6 had ordered: *The anti-aircraft extending arm is omitted for all PzKpfw.II, Pz.Kpfw.III, and Pz.Kpfw. ypes. Ordered material is to be annulled at once. Already finished extending arms are still to be installed for the time being.*

Pz.Kpfw.IV Type A with tactical number 101.

4.2.3 Five Type A Built with Type B Hulls

On January 31, 1938, the Gruson Works reported to Department A.K. in Essen:

For the last five vehicles of the 1./B.W. Series, we lack the armor material, which nevertheless has been delivered already for the 2./B.W. Series. In order to bring the 1st series to delivery, we would like to use the 2 Series armor parts. Since the structures of the 2. Series have been changed from those of the 1. Series, we ask whether the guns of the two series have the same structure, so that the hulls of the 2. Series can be used for the 1 Series without concern. This problem resulted from the lack of five armored hulls from other firms. The Harkort firm of Essen was to deliver three hulls by the end of February and two hulls by mid-March, and in fact for the completion dates for three complete vehicles in May and two in June by the Gruson Works. Since the delivery date of the hulls was not met, the Gruson works used five 2./B.W. hulls (finished by Fried.Krupp of Essen) for the last fuive *Pz.Kpfw.IV Type A* (including hulls No. 80206 for *Pz.Kpfw.IV Type A*, Chassis No. 80134).

A further question from the Gruson Works to Krupp of Essen on 4 February asked: *Are 1./B.W. guns usable for 2./B.W.?* Department A.K. in Essen answered on 5 February that *Guns for the 2./B.W. are the same as 1./B.W.. The armor covering of the gun cradle and of the rest of the turret, though, is changed from 1./B.W., as is the shape of the upper armored body.*

This question arose from the late delivery of the 7.5 cm tank guns. The last 7.5 cm Kw.K. for the 1. Series had finished its test firing in Meppen on May 11, 1938, and was installed in a Pz.Kpfw. IV Type A in July 1938.

4.3 IMPROVEMENTS AFTER DELIVERY

A series of improvements was ordered after the delivery of the Pz.Kpfw.IV Type A to the troop units. Among others:

1) As of August 1938, installing a foglaying launcher on the exhaust gas muffler at the rear of the vehicle.
2) By September 1938 the course discs for the 1. and 2./B.W. series had not been delivered at the right time for the delivery of the vehicles; thus, the course-circle manufacturer had to send these discs, on order from the department, to the troops.
3) As of spring 1940 attachment of the "Notek" night marching device on the left track apron.
4) As of July 1940, revision of the drum on the feed for machine guns and installation of cartridge-sack holders for the feed.
5) As of February 1941, reinforcing the front wall of the hull by welding on a 30 mm thick additional plate before the original 14.5 mm bow armor plate.
6) As of February 1941, installation of changes to the ventilator system and ventilator drive.
7. As of February 1941, mounting an antenna deflector at the front of the 7.5 cm Kw.K. barrel.
8) As of March 1941, mounting of a baggage rack on the back of the turret.
9) As of June 1941, installation of a coupling at the rear for a fuel trailer.

5. Panzerkampfwagen IV (7.5 cm) (Vskfz.622) Type B (2./B.W.)

Chassis Number Series 80201—80242

*The 42 **Pz.Kpfw.IV Type B** were built and accepted between May and October 1938.*

5.1 IMPROVEMENTS FROM TYPE A

Changes were requested by Wa Prw 6 in each of the main components of the second series of Panzerkampfwagen IV. The most important improvements, with descriptions, are noted in this chapter.

5.1.1 Armor

For increased safety against 2 cm antitank shells from close ranges, the front plate was strengthened. The armor plates consisted of homogeneous nickel-free armor steel with 153 +/-5 kg/mm^2 firmness for sheets up to 14.5 mm thick. For sheets over 14.5 mm thick, homogeneous nickel-free PP694 armor steel with 125 +/-5 kg/mm^2 firmness was used.

5.1.2 Chassis

The original form of Type A was kept, with the same layout of the main components. Only minor changes, if any, were made to the running gear, but other main components, such as the armor hull, motor, gearbox and side gearings were improved.

5.1.2.1 Armor Hull

The hull consisted of armor plates with the following thicknesses and inclinations from the horizontal:

Plate	Thickness	Inclination
Upper front wall	20 mm	18 °
Front wall	30 mm	76 °
Lower front wall	20 mm	29 °
Sidewall	14.5 mm	90 °
Upper back wall	14.5 mm	80 °
Lower back wall	10 mm	16 °
Bottom	8 mm	0 °

5.1.2.2 Motor

For the increased weight of the vehicle, Type B had a stronger motor, the HL 120 TR, installed. As a further development of the HL 108 TR for greater power production, the cylinders were bored out from 100 to 105 mm. Without increasing the outer dimensions or the weight of the motor, the new total displacement of 11,900 cc gave a power of 285 HP at 2800 rpm, up from the 250 HP at 2800 rpm of the HL 108 TR. Maybach Motorenbau GmbH delivered 22 HL 120 TR motors, and the Nordbau firm (Norddeutsche Motorenbau GmbH) in Berlin-Niederschöneweide built the last twenty HL 120 TR, numbered 54001 to 54020.

5.1.2.3 Gearbox and Side Gearings

From Type B on, a new gearbox designated SSG 76 (synchronized six-speed gearbox, with a torque of 76 m/kg) was installed. Compared to the gearbox of Type A, this type had six speeds and shifting rails in the housing. It had six forward speeds and one reverse. The ratios of the individual gears were:

	Ratios	Speed:
1st Gear	1 : 9.02	4.7 kph
2nd Gear	1 : 4.68	9.0 kph
3rd Gear	1 : 2.76	15.2 kph
4th Gear	1 : 1.82	23.0 kph
5th Gear	1 : 1.129	32.6 kph
6th Gear	1 : 1	42.0 kph
Reverse	1 : 7.3	5.8 kph

The 2./B.W. was harder to steer than the 1./B.W., since the ratios of the side gearings was 1 : 3.23 instead of 1.4 (because of the greater whole step of the six-speed transmission).

5.1.3 Armored Upper Body

On November 20, 1937, at the acceptance of the first two B.W. at the Gruson Works, Wa.Prw.6 requested:

Panzerkampfwagen IV, Type B.

Fichtner indicates that in individual vehicles of the second series 10-watt instead of 20-watt devices be installed, with a second receiver and transformer, in order to be able to receive different waves simultaneously, thereby making greater changes for the second series. Major Kutzki and Dr. Bruun, Wa.Prw.7, met in Essen on December 6, 1937, in order to discuss the installation of devices in the model of the radioman's space of the 2./B.W. on the next day. Models of the devices are to be prepared.

A contract was issued by Wa Prw 6 to Department AK in Essen on January 6, 1938, for "reworking the cast iron test model of the armored upper box of 1./B.W. to the smaller 2./B.W. form."

Further decisions resulted from a discussion with Wa Prw 6 in Berlin on 29 January, as follows:

AK Contract No. 86 496/L of 1/6/38 for modifying a test bow armor of a B.W.II test chassis in the state of the 2nd series will be discussed. To a return question Prw 6 stated that the unstrengthened turret carrier ring remains in existence and no protective piece for the turret gap will be added. A horizontal 7.5 cm ammunition storage, though, must be built in particular.

On 30 March Department A.L. reported:

The B.W.II with 2./B.W. interior arrangement is finished. It was inspected in Essen on 12 April by Major Dr. Olbrich and Dipl.Ing. Biersack, with the following decisions: *The ammunition storage for 2./B.W. will be examined in the B.W.II test chassis. The rack for two rounds of 7.5 cm on the back bulkhead is to be covered with sheet aluminum. Before the shell storage, curtains like those of the test version are to be added, but not before the M.G. ammunition storage. Prw 6 agrees as to the number of rounds (3000 M.G. rounds and 80 7.5 cm rounds).* These and other discussions were decisive for the following main changes to the armored upper body of Type B:

5.1.3.1 Bow Armor

The bow armor consists of armor plates with the following thicknesses and inclinations (to the horizontal).

Plate	Thickness	Inclination
Front wall	30 mm	81 °
Sidewall	14.5 mm	90 °
Roof	11 mm	0 and 5 °
Back wall	14.5 mm	90 °
Bottom plate	8 mm	0 °

The front wall of the Type B (and Type C) bow armor was made straight, with the driver's and radioman's vision hatches and the machine-gun hatch all on the same plane.

In Type B (to Type D) an intake opening for the motor-cooling air in the right sidewall in back was planned; it was protected against fire by vertical and horizontal steps.

From Type B on, the gap between the turret and the upper body armor was covered by prismatic protective plates, which protected against shots into the gap between the lower edge of the turret and the roof of the bow armor.

Under the driver's vision hatch, a lamp panel was screwed on to hold the blue warning lamps of the cross switch. The switch turned the warning light to the left or right of the driver when the 7.5 cm KwK projected to the left or right of the tracks.

5.1.3.2 Driver's Vision Port 30

The driver's vision port for Type B (to Type D) consisted of two overlapping slides. The lower slide's front edge, in a closed position, extended over the lower edge of the upper slide. For eye protection against lead spray, splinters and rifle fire that could penetrate the vision opening with the driver's port open, a glass shield 90 mm thick was mounted behind the driver's vision port and was easily replaced. To catch lead spray, a frame was installed in the space between the glass shield and the bow armor wall.

5.1.3.3. Machine-Pistol Port

In the front wall of Pz.Kpfw.IV Type B (and Type C) a machine pistol port was installed at the right next to the "vision flap before the tank radioman." The M.P. flap closed the loophole used for firing the machine pistol, which also served as a support for the machine pistol when firing. The M.P. flap was shaped like a shortened tilted cone.

5.1.3.4 Vision Ports

The vision port with vision slit was used only in Types B and C in place of the ball mantlet in the front wall. It afforded the tank radioman a view to the front. It was a rectangular plate with a stepped rim, on which the flap lay on the armor wall. To protect against lead spray, the flap was set a few millimeters into the armor wall. In the vision port a vision slit 8 mm wide was cut. Outside it was formed like a roof, so that lead spray and striking bullets were largely deflected from entering the vision port. To protect the eyes against lead spray, splinters and rifle bullets that could penetrate the vision port, a glass shield, easily replaceable, was located. The sealing frame welded onto the front armor wall from inside stopped lead spray that entered through the gap between the vision port and armor wall.

The vision port in the sidewall of the bow armor walls for Types B and C was fitted with a vision slit 4 mm wide. The vision slit was edged by ridges outside to deflect lead spray and shots. The vision port was a rectangular plate 90 mm thick with even surfaces and a projecting outer rim. Every plate was cut out on the inside so that a thin sealing surface was formed around the plate; when closed, it lay against a rubber sealing frame. To protect the eyes against lead spray and splinters, a glass shield 50 mm thick was placed behind the vision opening.

5.1.3.5 Hatch Covers

For the tank driver and radioman, there were hatches in the top of the bow armor for entrance and exit; they were closed by one-piece hatch covers. The hatch covers were mounted on two hinges to allow turning and lay on the frames set into the bow armor. The gaps between the hatch covers and the bow armor were covered by edges. On each hatch cover was a round flap for giving flag signals or other signs and for ventilation.

The SSG 76 gearbox, made by the Zahnradfabrik Friedrichshafen.

Wooden dummy (or mock-up) for 2 B.W. bodies on B.W. II test chassis.

5.1.3.6 Radio System

To the left of the radioman were the hangers for the sending and receiving sets, one over the other.

The equipment of a company or platoon leader's tank had two radios: an Fu 5 (SE 10 U), consisting of a ten-watt sender and an ultra-short-wave receiver, plus an Fu 2 (EU) ultra-short-wave receiver. The Fu 5 (SE 10 U) had a range of 4 to 6 km (strongly dependent on the terrain). The equipment for a "normal" tank had only an ultra-short-wave Fu 2 (EU) receiver, without a sender.

5.1.3.7 Tank Radioman's Seat

The seat for the radioman was on consoles attached to cross members of the hull floor. The back, formed of tubing and fitted with a pad, was attached to the seat and could be pulled out upward.

5.1.3.8 Gun Ammunition Storage

In the side spaces and on the back wall of the bow armor 26 shells were stored, in the chassis 54, making a total of 80 rounds of 7.5 cm cartridges. They were divided as follows:

In the left side space of the bow armor were two ammunition cases for nine shells each, in the right side space one case for 6 shells, and to the right on the rear wall of the bow armor one case for two shells. The shells were kept lying down in these cases.

In the chassis there were four ammunition cases on the right wall, three for eight shells each and one for two, as well as one case for 28 shells behind the driver's seat. The shells were kept standing up in these cases.

5.1.3.9 MG Ammunition Storage

In all, the Pz.Kpfw.IV Type B (to Type D) carried 32 bullet drums, each of 75 bullets, thus a total of 2400 rounds of machine-gun ammunition stored in cases.

5.1.3.10 Rear Armor

The rear armor of Type B retained the same shapes and thicknesses of Type A.

5.1.4 Turret

During the planning for the second series, it was reported as to the 2.B.W. turret on January 28, 1938:

On orders from Wa Prw 6, the ventilation flap is to be surrounded by armor protection. In the process, there are to be several holes in the turret roof. On the left signal flap a hood is to be set. The antenna deflector will be taken on anew. The left vision port will be made without a vision slit. For this reason, four holes on the left sidewall will be eliminated.

On 12 February it was explained that the 2./B.W. turret was to have two M.G. ports (back wall) and two pistol ports (hatch covers). The main size and shape of the earlier turret remained, but several other changes were made, as follows:

5.1.4.1 Turret Housing

The turret housing consisted of armor plates of the following thicknesses and inclinations (from the horizontal):

Plate	Thickness	Inclination
Front wall	30 mm	80 °
Gun port cover	30 mm	round
Sidewall	14.5 mm	65 °
Back wall	14.5 mm	65 ° & 90 °
Roof	10 mm	0 & 7 °
Commander's Cupola	30 mm + cast slide	

In the middle of the roof, the ventilation flap with its armor shield was attached. To the right and left of it, one round signal opening each had been cut. The signal openings were covered by closable signal flaps. From Type B (to Type D) a hood was set on the left flap.

The radio system, ten-watt devices, with the intercom.

The radioman's seat with the transformer for the radio system behind it.

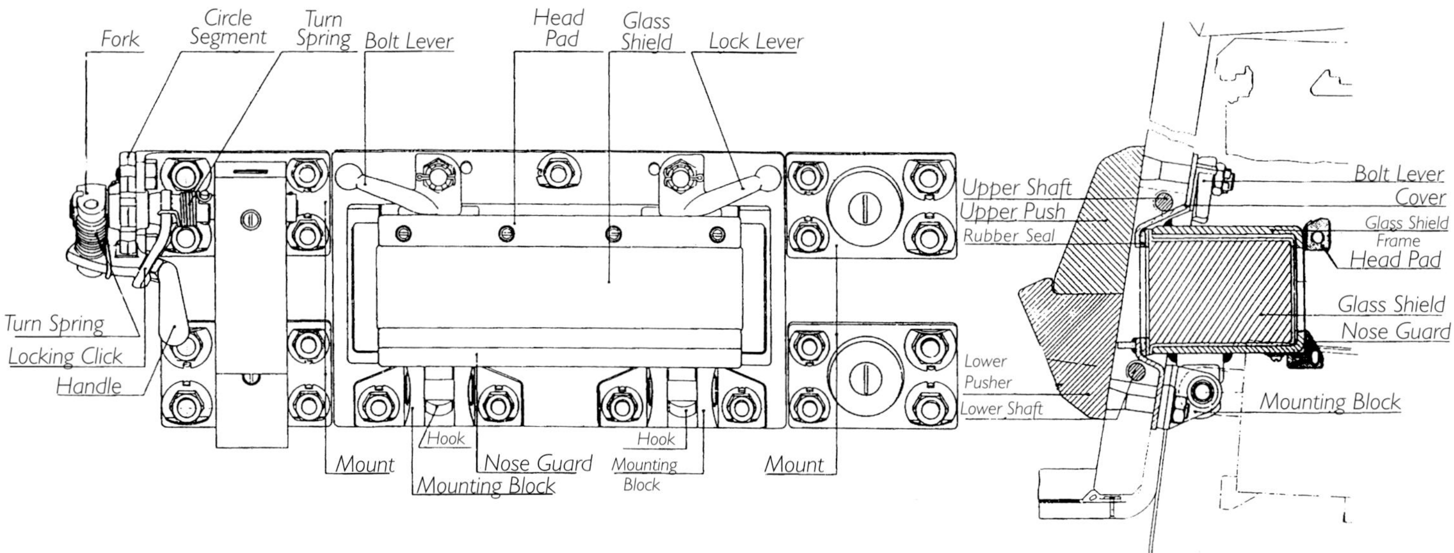

Driver's Vision Port 30 (push flap), (Drawing No. 021 C 9243) for Pz.Kpfw.IV Types B to D.

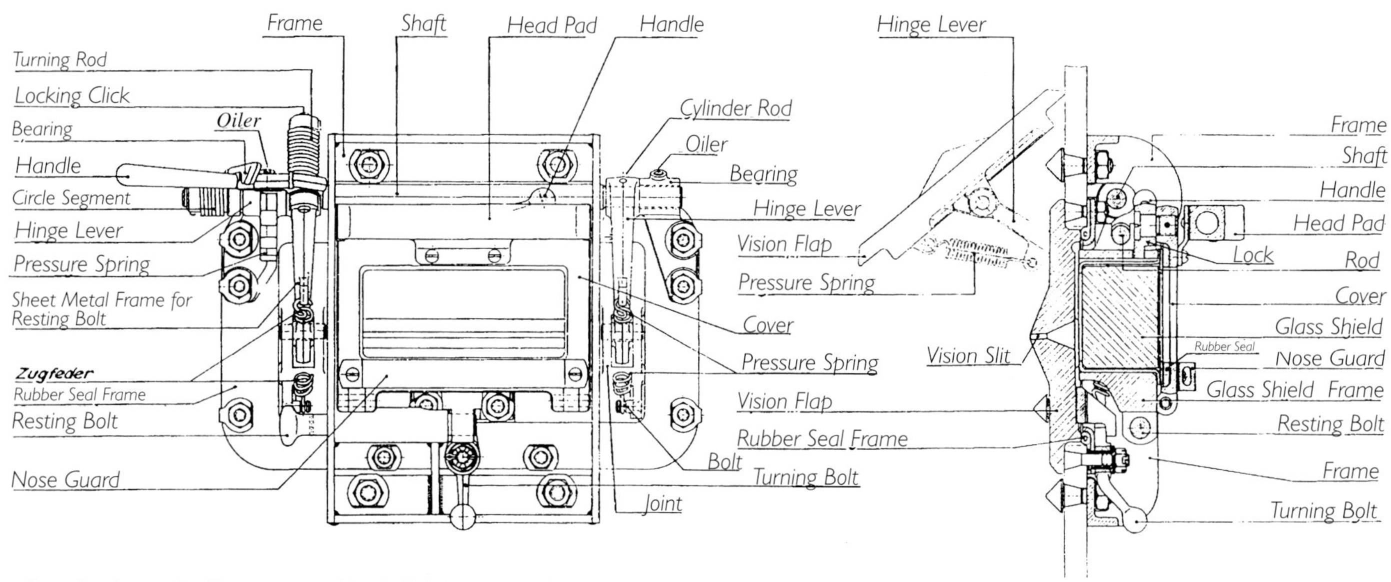

Vision Port for Sidewalls (Drawing No. 021 B 8987) for Pz.Kpfw.IV Types B and C.

5.1.4.2 Roller Mount

The roller mount of Type B (and Type C) consisted of a semi-cylindrical curved armor shield, which was closed on both sides by sidewalls which bore the shield tabs. A wall was inset behind it, and the gun cradle and M.G. mount were screwed to it. Before the mount lay the mount cover made of armor steel. The mount cover covered the gap between the mount and the front wall of the turret; it was screwed fast to the front wall. To increase protection against lead spray above and below the mount, catch plates were screwed onto the turret roof and the carrier ring.

5.1.4.3 Commander's Cupola (Strengthened to 30 mm)

As of Type B (to Type D), the commander's cupola consisted of the cylindrical mantle, five upper and five lower slides, the numeral disc and the two-piece hatch cover. The cylindrical mantle had five vision openings, which could be partly or wholly closed by the paired effect of the upper and lower slides.

For eye protection against lead spray and rifle shots, a glass shield, easily replaceable, was mounted behind each opening.

5.1.4.4 Visipn Flaps, Hatch Covers and Flaps in the Turret

a) Vision Flaps

The vision flaps in the front wall of the turret were formed so that striking shots and lead spray were deflected. To protect against lead spray and splinters, a glass shield was placed behind the vision opening and held in a glass-shield frame.

The vision flaps for Type B (and Type C) in the left and right sidewalls were rectangular plates with straight surfaces and projecting outside rims. Every plate was cut out inside so

Assembly Drawing of Turret
(Drawing No. 021 B 14851) for Pz.Kpfw.IV Type B

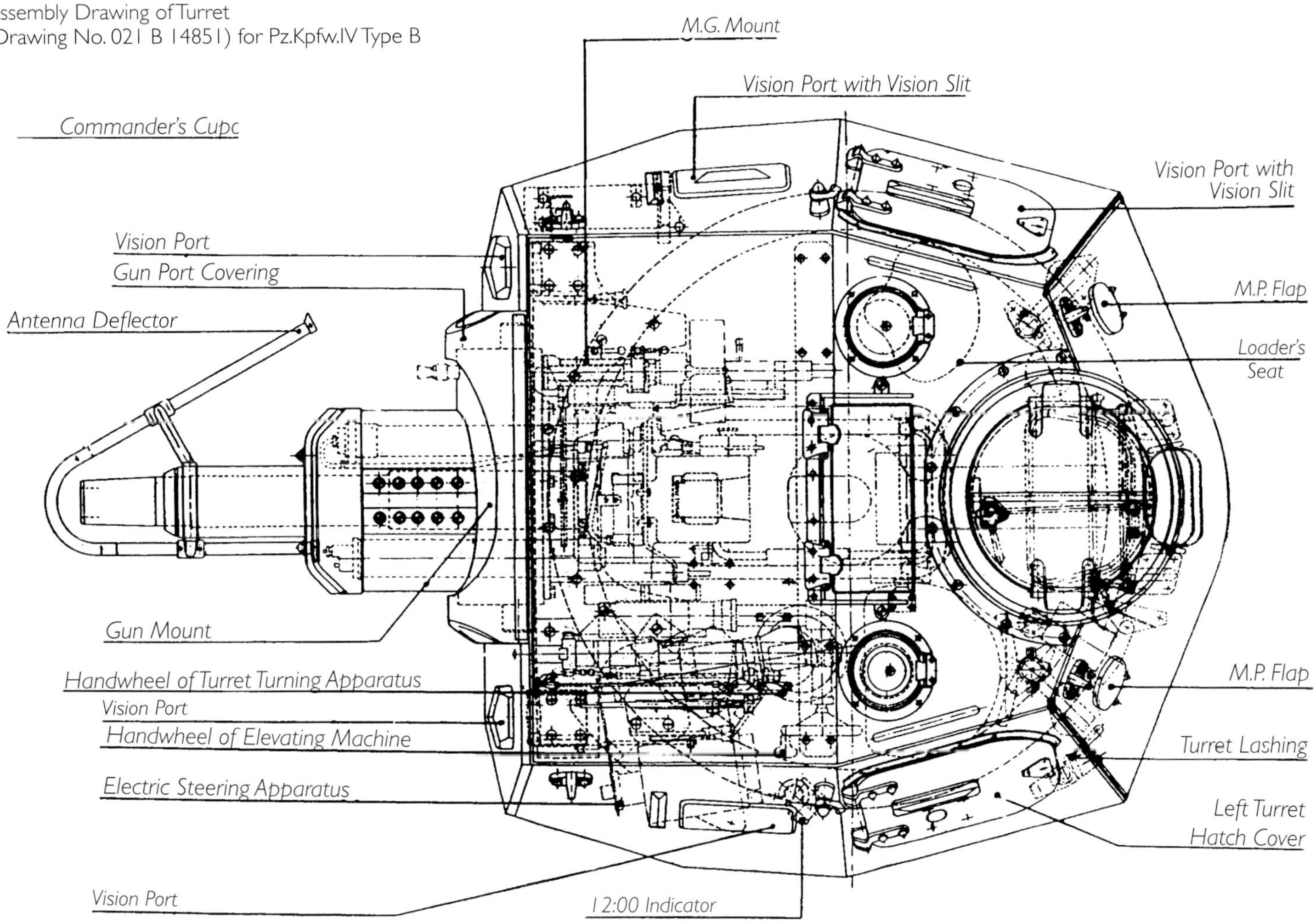

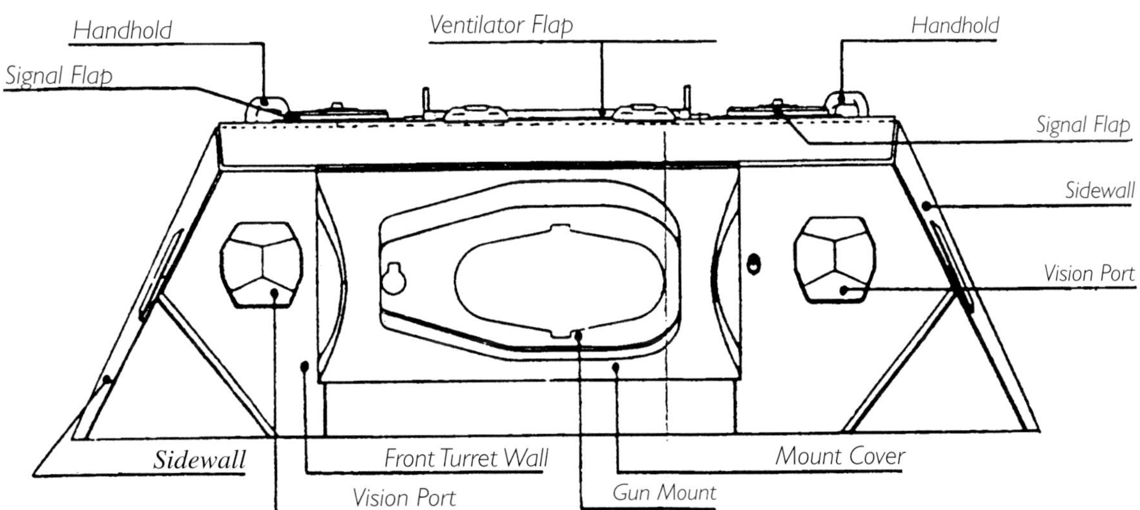

Turret (Drawing No. 021 B 14851) for Pz.Kpkw.IV Type B

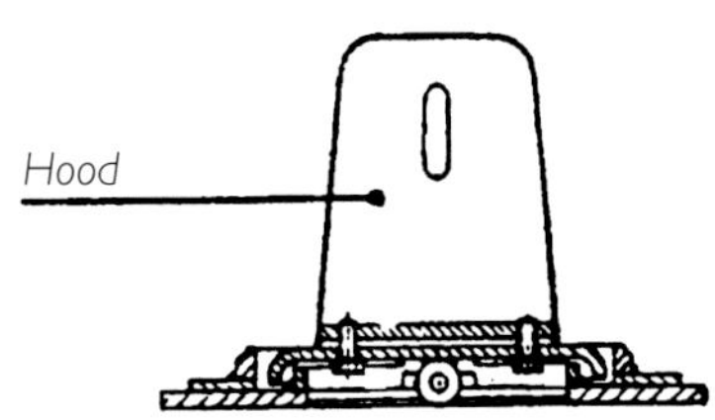

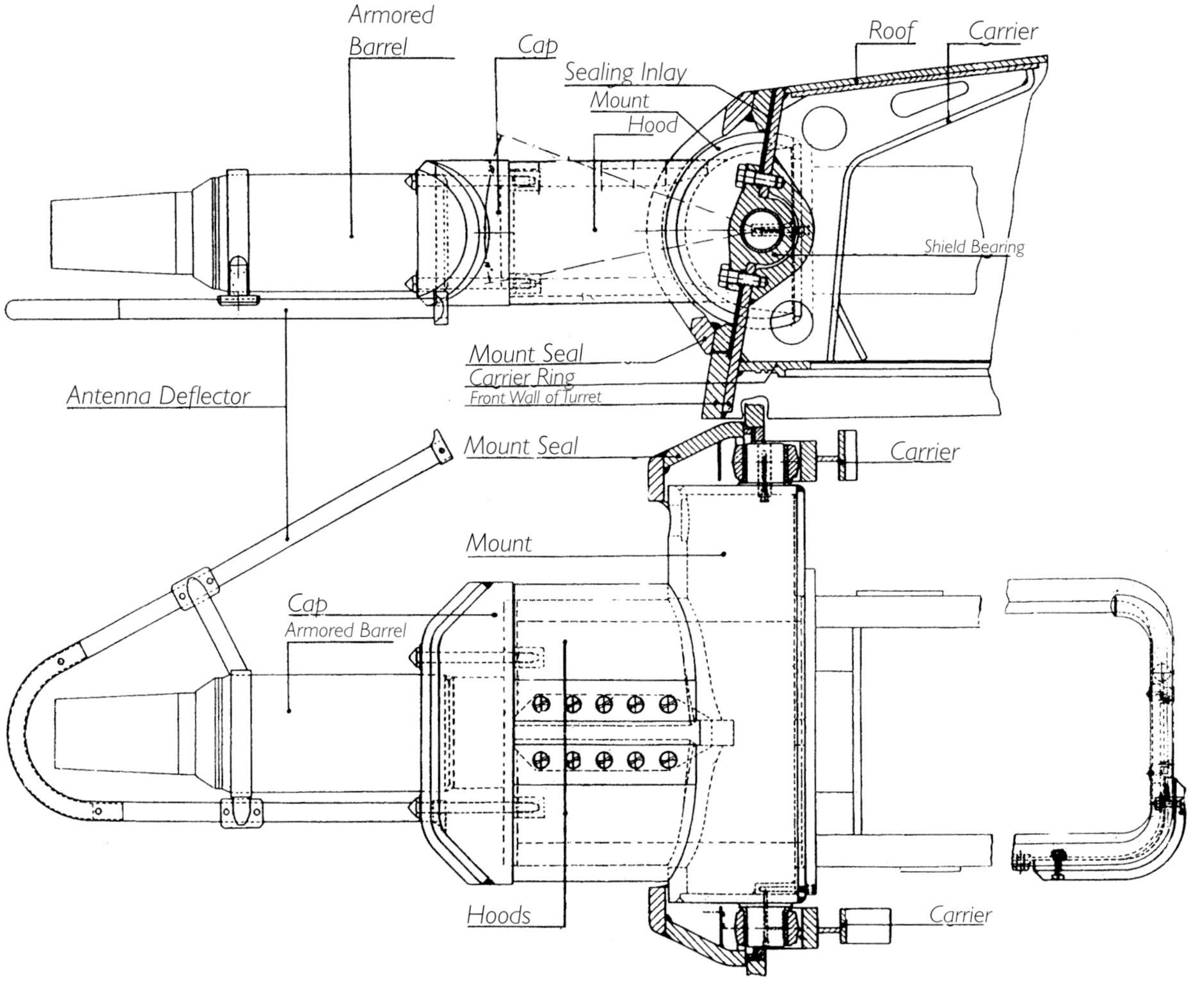

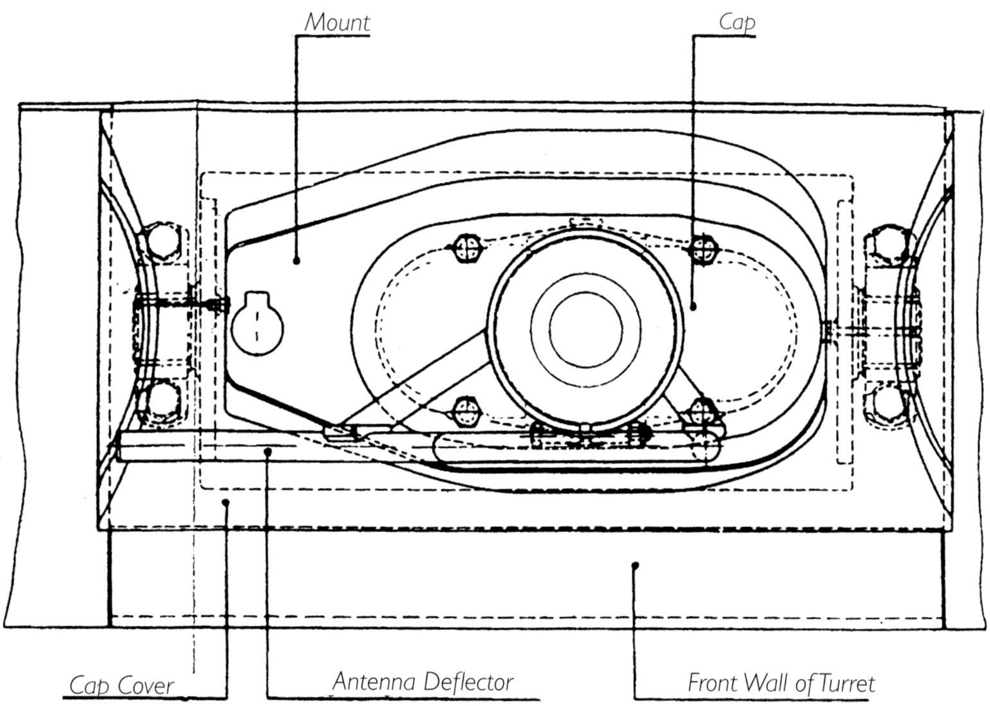

Roller Mount
(Drawing No. 021 B 14851 U18)
for Pk.Kpfw.IV Type B.

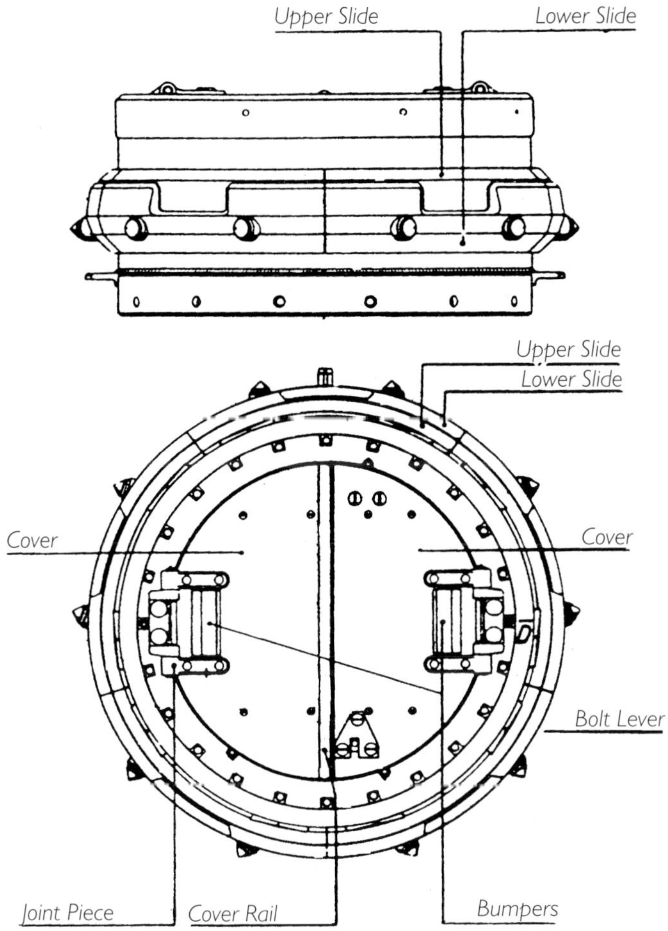

Commander's cupola (Drawing No. 021 B 9261) for Pz.Kpfw.IV Types B to D.

that a small thickening surface was formed around its rim. The vision flap in the right sidewall was fitted with a vision slit. The slit was bordered by a ridge outside, which deflected lead spray and bullets. For protection against lead spray and splinters, a glass shield was placed behind the sight opening. The glass shield, 50 mm thick, was mounted in a frame and held firmly by the cover.

b) Turret Hatch Cover

As of Type B (to Type C) a vision slit plate was screwed onto the center of the turret-hatch cover. The vision-slit plate had an 8 mm wide vision slit and was shaped outside so that lead spray and bullets were largely deflected from entering the vision slit. To protect the eyes against lead spray and splinters, a 90 mm thick glass shield was set up behind the vision-slit plate.

As of Type B (to Type E) a round pistol opening was cut into each hatch under the vision-slit plate. On the inside of the turret hatch cover a closing slide was placed before the opening.

c) M.P. Ports

From Type B on, one M.P. port each was built in to the left and right sides of the back wall of the turret. The M.P. port lay with its cylindrical part in a drilled-out spot on the back wall and covered the M.P. opening.

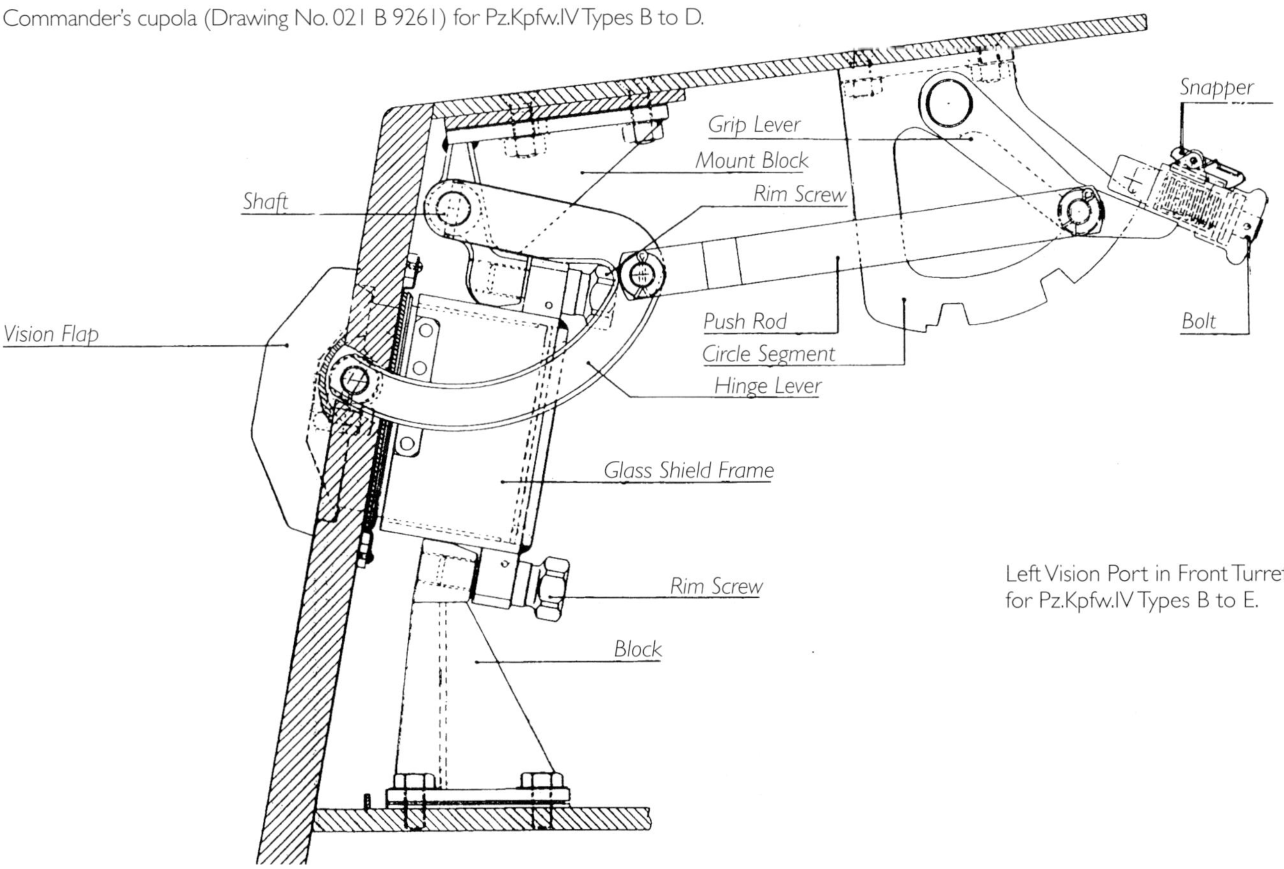

Left Vision Port in Front Turret Wall for Pz.Kpfw.IV Types B to E.

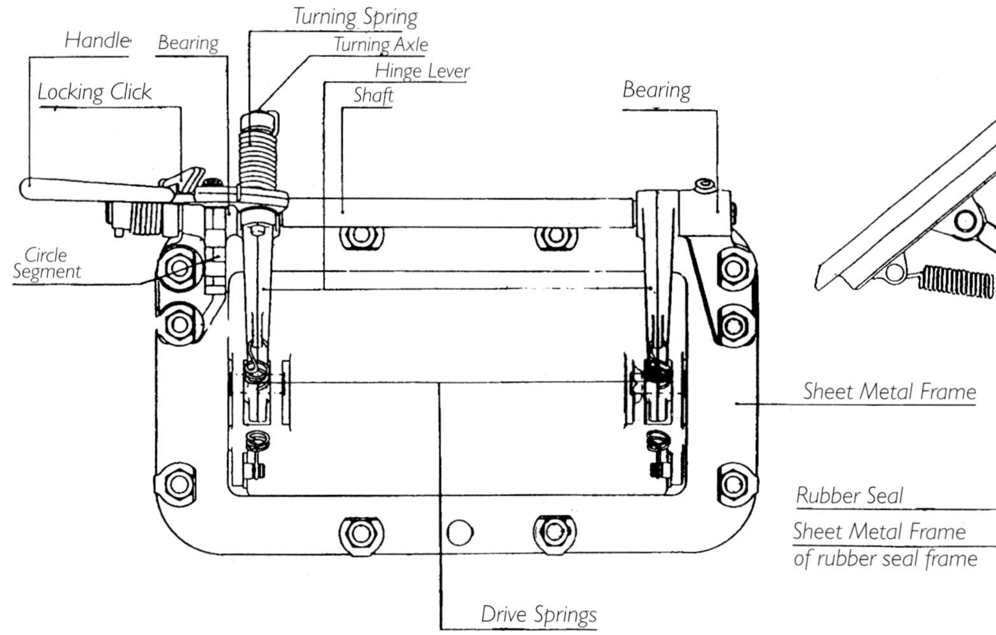

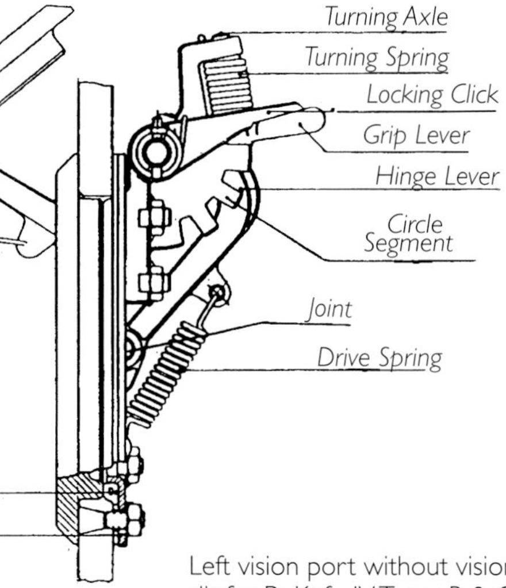

Left vision port without vision slit for Pz.Kpfw.IV Types B & C.

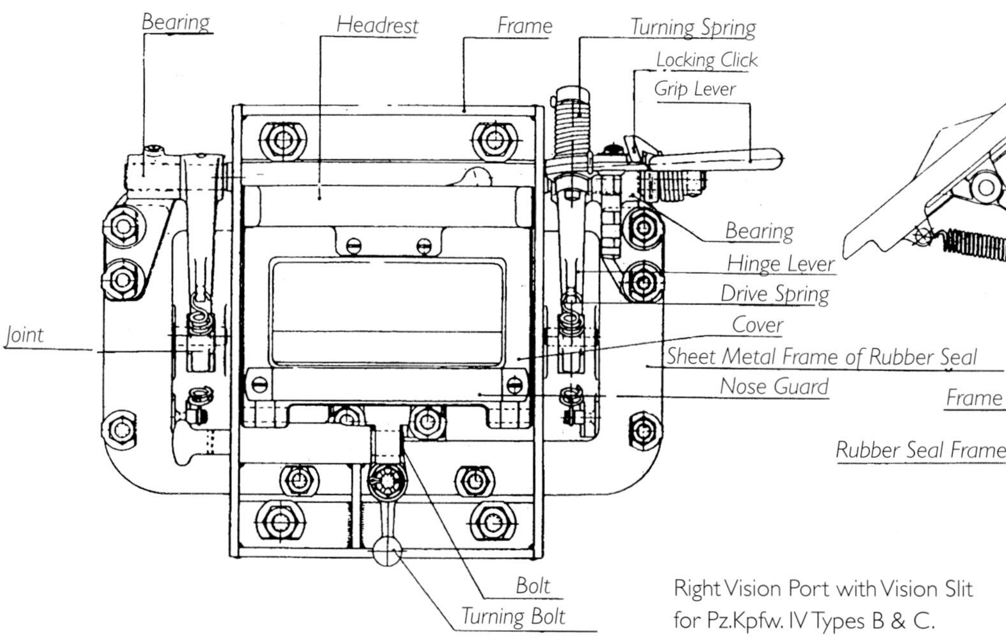

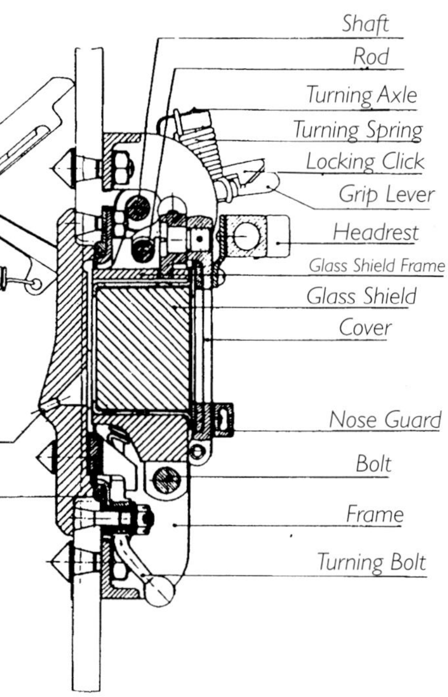

Right Vision Port with Vision Slit for Pz.Kpfw. IV Types B & C.

5.1.4.5 Equipment Storage

In Type B (and Type C) the following storage places for equipment were attached:

- In the right sidewall, the rack for the unloader,
- Under it, on the carrier ring, the holders for extra M.G. barrels and spare glass shields,
- On the right rear brace of the turntable, above the loader's seat, a gas mask holder and
- Over it, on the back wall, two holders for field flasks,
- Behind the commander's seat on the carrier ring, a holder for double headphones,
- By the commander's footrest in front, the holder for the MG tools and supply pouch, and in back, two gas masks,
- Between the commander's and aiming gunner's seats on the carrier ring, the box for flare bullets, and over it, on the back wall, the holder for the flare pistol,
- On the left rear brace of the turntable behind the aiming gunner's seat, the holder for a field flask and the signal flag holder,
- At left front under the roof, the holder for the targeting rail,
- At right front on the turntable, two cases for the bullet drums of the M.G.

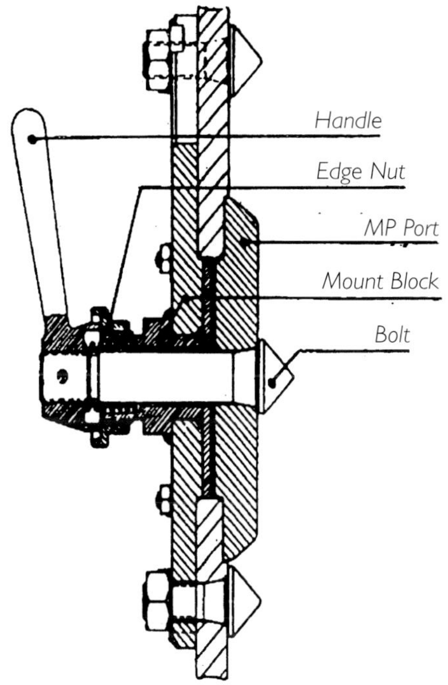

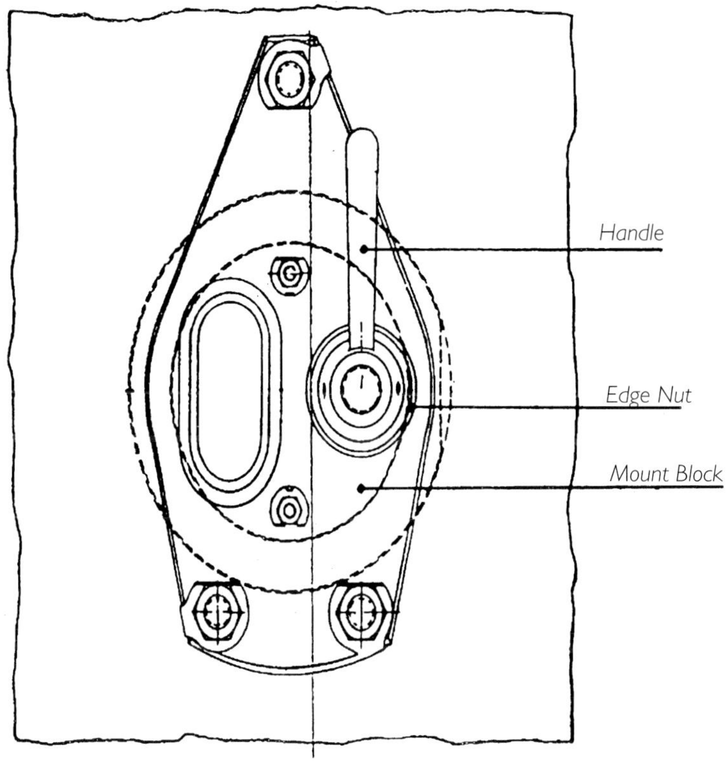

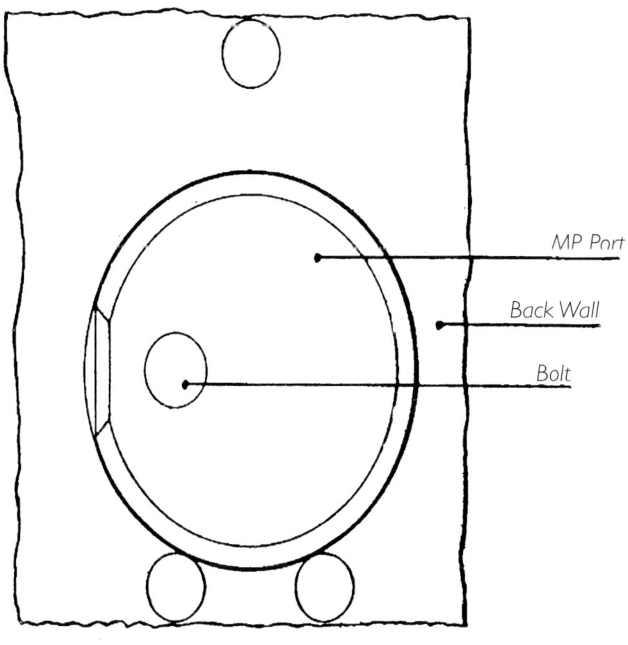

Turret M.P. Port
(Drawing No. 021 B 14851 U24)
for Pz.Kpfw.IV Types B & C.

5.2 SHAPE CHANGES DURING PRODUCTION

5.2.1 Minor Details Make Vehicles Ready For The Troops

The notes of Mr. Heerlein on discussions at the Gruson Works on May 9 and 11, 1938, explained that in the first finished main components small changes were necessary before the second series was ready for troop use:

In examining the first finished 2./B.W. turrets, the following changes were seen to be necessary:

1) *The lower block of the right front vision port receives 3 instead of 4 attachment screws, larger passage holes.*
2) *The upper block of the front vision ports receives note: "Holes bored during assembly." Turning the mount hole in back eliminated. Hole receives lubricating nipple.*
3) *Resting arch of the left front vision port to be moved 20 mm farther to the left.*
4) *Resting bolt on the front vision port to be made 10 mm longer.*
5) *Armored rim of ventilation flap in turret roof to be fitted or measurement changed.*
6) *M.P. port in left part of rear turret wall to be made in mirror image of that in the right part.*

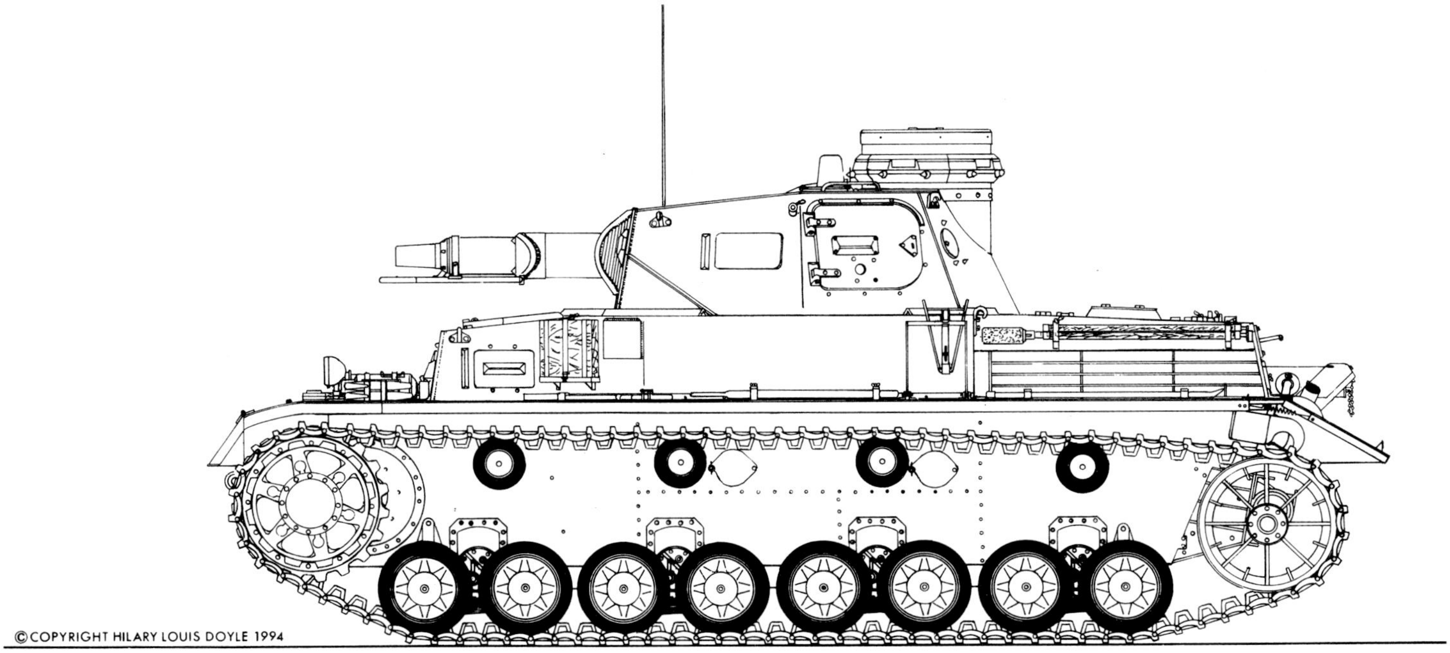

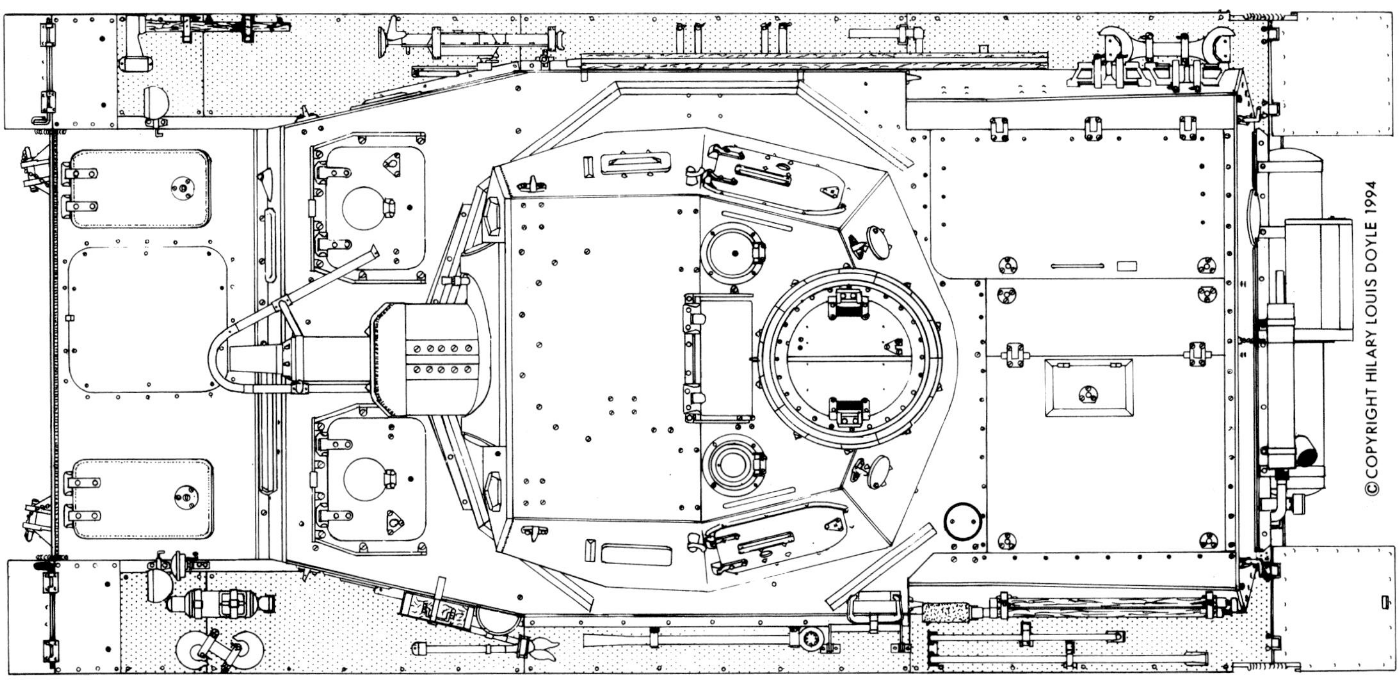

Panzerkampfwagen IV (7.5 cm), (Vs.Kfz. 622) Type B:
The driver's vision plate, stepped in Type A, was replaced by a straight plate. The structure was narrowed. Pistol ports in place of the M.G. ball mantlet, air exit at right rear and moved radio antenna. New turret with improved vision conditions, pistol opening in the back of the turret, added pistol openings in the turret hatch covers, new commander's cupola, and new inside roller mount on barrel of tank gun. Foglaying device mounted on the exhaust muffler, two replacement track links stored on the track aprons.

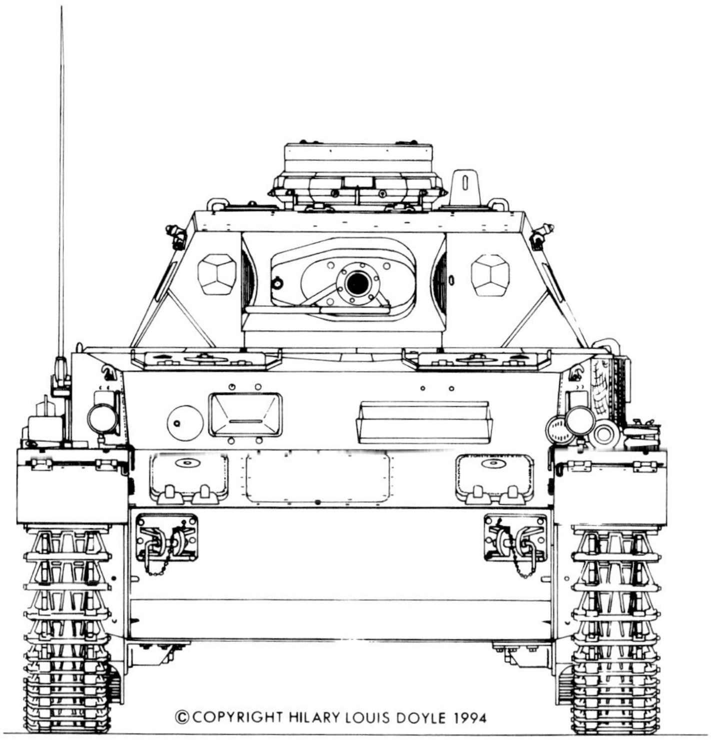
©COPYRIGHT HILARY LOUIS DOYLE 1994

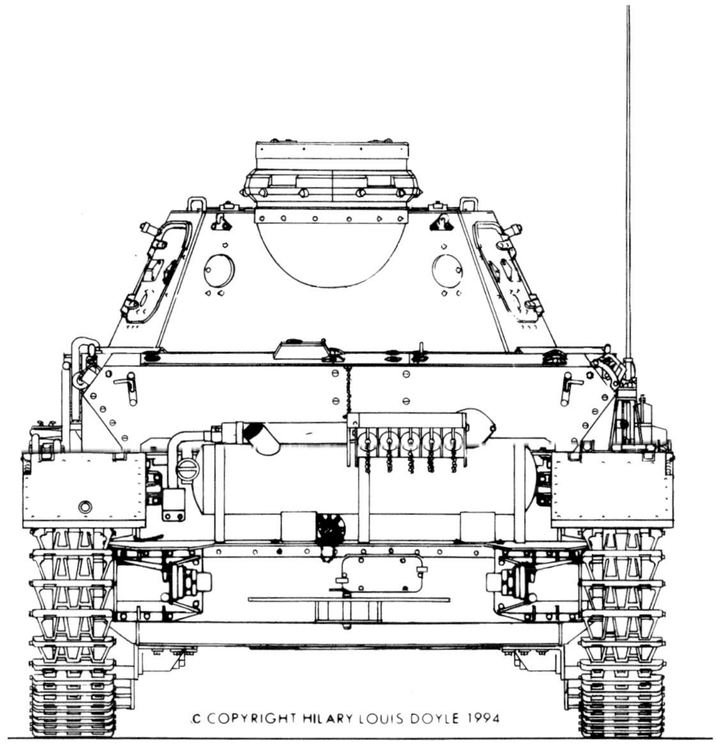
©COPYRIGHT HILARY LOUIS DOYLE 1994

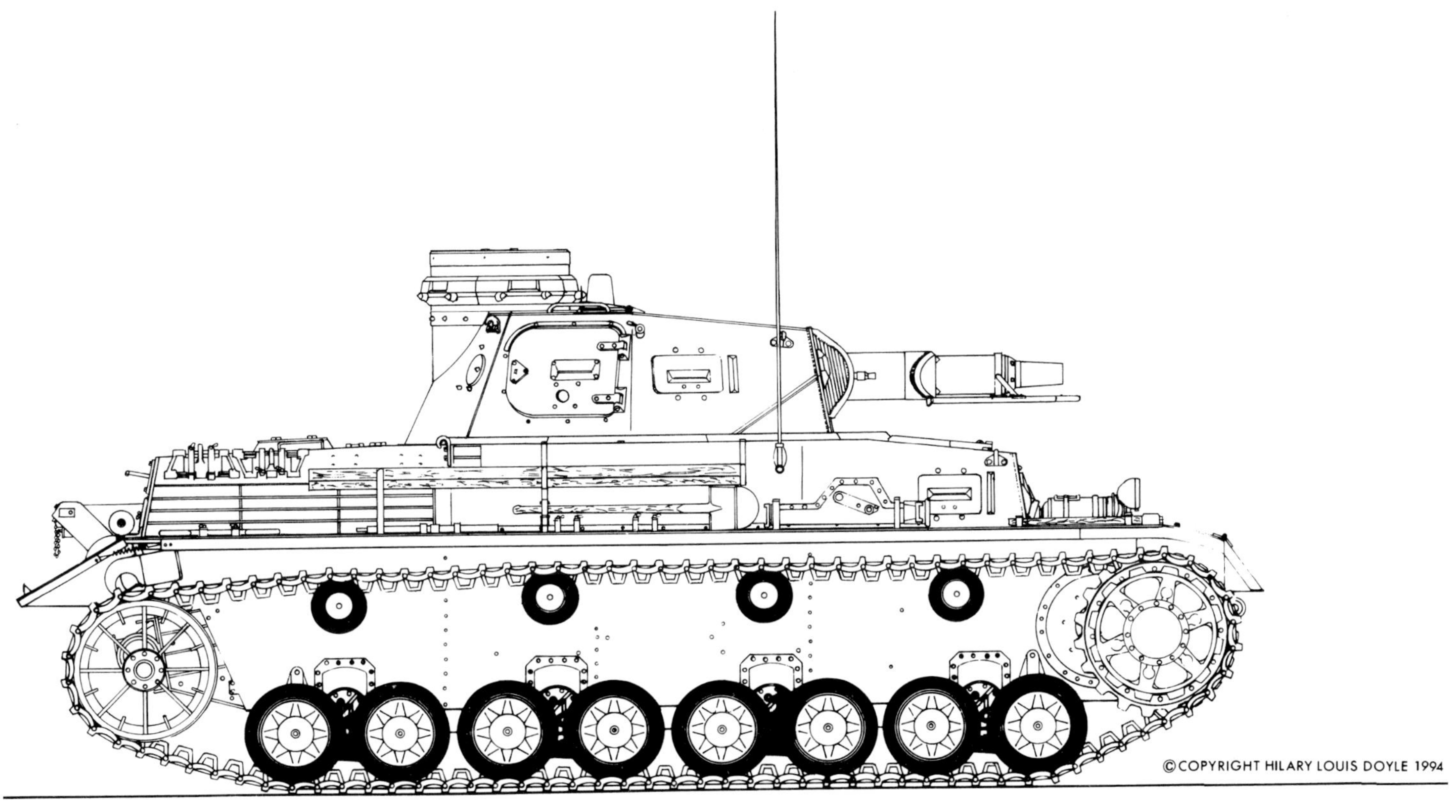
©COPYRIGHT HILARY LOUIS DOYLE 1994

7) *Commander's seat back to be moved forward some 25 mm.*
8) *Glass block frames of front vision ports to have more play in the holes for the holding screws.*
9) *Eliminate securing of holding screws.*
10) *The four boreholes closed by welded screws in each of the two turrets delivered from APPA III can be left, after a further talk with Dipl.Ing. Biersack.*
11) *Headphone plug boxes in the back wall of the turret must be moved to the sides and set higher because of the M.P. port. Cut off the previous attachment hole and attach a plate on the back wall of the turret. As of 3./B.W. use two hinged edges instead of the plate.*

5.2.2 Type B, Built on Type A and C Hulls

As already noted in Chapter 4, because of delays in delivering armored hulls by other firms, five Type A tanks were built with Type B hulls. Thus, five Type B tanks were built with Type A hulls.

The same thing happened again, as reported on September 14, 1938:

Since the firms of Harkort Eicken and Eisen- und Hüttenwerke could not deliver the hulls intended for the 2./B.W. on time, it was urged there that instead of these vehicles, those from Appa III here of the 3./B.W. production be used for 2./B.W. As a result, the hulls to be produced by the steel firms noted above had to be modified according to 3./B.W. production.

The Eisen- und Hüttenwerke had a contract for 12 armored hulls, No. 80213 to 80224, and the Harkort Eicken firm had a contract for 18 armored hulls, No. 80225 to 80242. Instead of these hulls, Type C armored hulls No. 80301 to 80330 had to be used for Pz.Kpfw.IV Type B chassis No. 80213 to 80242.

5.2.3 Improvements Resulting from Troop Experience

On September 23, 1938, Wa Prüf 6 (IId) had expressed further wishes to Department AK, as follows:

On the basis of troop experiences, the following modifications to the B.W. have proved to be necessary:

1) *For the loading gunner a back must be attached to the seat brace.*
2) *The signal-lamp box is hard for the commander to see and must be moved farther to the middle of the commander's cupola.*
3) *On the mounting block for the viewer supports of the targeting scope (on the turret roof), the splint bolt for the rack of the targeting scope must be replaced as soon as possible by a projection with a securing flap and chain, as was already suggested for 4./Z.W. (Pz.Kpfw.III Type E). It is desirable that in the B.W. and 4./Z.W. the same device is used whenever possible. The projection is also to be prepared for the already delivered vehicles.*
4) *The commander's cupola must be fitted with a visor setup on the front vision opening. The change will be made from here. The drawings will be sent by the Daimler-Benz firm.*

It is requested that the aforementioned modifications be made as soon as possible to the next vehicles to be delivered.

Side view of Pz.Kpfw.IV Type B.

Pz.Kpfw.IV Type B with crew, Chassis No. 80209, finished in July 1938.

The visor arrangement was described as follows: On the attachment ring of the mantle, the target needle was screwed on in the middle before the front vision slit. The sight block was attached to the cover of the front glass shield. With this arrangement the tank commander could recognize the targeting direction of the turret weapons and advise the aiming gunner about the targeting.

As of July 1938 a foglaying launcher was mounted on the exhaust muffler at the rear of the vehicle.

5.3 IMPROVEMENTS AFTER DELIVERY

A list of improvements was compiled after the delivery of the Pz.Kpfw.IV Type B to the troop units. Among others:

1) As of August 1938, mount a foglaying launcher on the exhaust muffler at the back of the vehicle.
2) As of spring 1939, attach a rain gutter over the driver's vision port.
3) As of spring 1940, attach the "Notek" night marching device on the left track apron.
4) As of June 1940, switch from drum to belt feed for M.G. and install bullet-bag holders for belt feed.
5) As of February 1941, reinforce the front wall of the hull by welding on an added 30 mm plate before the original 30 mm bow armor plate.
6) As of February 1941, install changes to the ventilation system and ventilator drive.
7) As of March 1941, mount a baggage rack on the back wall of the turret.
8) As of June 1941, install a coupling at the rear for a fuel trailer.

6. PANZERKAMPFWAGEN IV (7.5 CM) (vSkFZ.622) Type C (3./B.W.)

Chassis Number Series 80301—80440

Of the 140 ordered Pz.Kpfw.IV Type C. 134 were finished and accepted between October 1930 and August 1939. According to Instruction AHA In 6 (IVd) No. 156/39, secret command material of February 22, 1939, the last six Type C chassis (Chassis No. 80435 to 80440) were prepared for a special bridgelaying test series at Wa Prüf 5 and delivered in June.

6.1 ORIGINAL PLANNING

On June 1, 1937, Ober.Ing. Kniepkamp (Wa Prw 6 (Gr 3) had concluded the further development of the Pz.Kpf.Wg. chassis (B.W.) by Department A.K. as follows: *The necessity of also making the chassis of the tanks uniform compels us to give up the design of the B.W. chassis developed by you with the completion of the 2./B.W. series. For 3./B.W. the chassis of the Z. W. Model 4 (Panzerkampfwagen III Type E) being developed with an appropriate armored hull is foreseen. As a result the continuation of a special development contract for B.W. chassis is unnecessary.*

On the basis of problems with the punctual production of test vehicles, unacceptable delays in their production would have resulted if the Daimler-Benz uniform chassis had been put into production.

In a discussion with the Weapons Department on 21 June it was already mentioned that after the delivery of Series 2./B.W., further B.W. in as great numbers as possible should be produced.

6.2 IMPROVEMENTS AFTER TYPE B

Krupp had gained permission from Wa Prw 6 to carry out important design changes for the B.W. chassis. Krupp received only contracts for production of further Pz.Kpfw.IV on B.W. chassis. In the final form, there was created the Pz.Kpfw.IV Type C from a contract for the building of several basically unimproved Type B vehicles. Wa Prw 6 allowed only the following simple modifications:

6.2.1 Motor

On December 9, 1937, the Maybach firm reported:

In the type designation HL 120 TRM, the letter "M" means that the motors are equipped with snap magnets. This type first appears in the 100 motors to be delivered by us for Series 3./B.W.

The first 40 Type C (Chassis No. 80301 to 80340) should be built with the HL 120 TR motor (number series 540021 to 540060 by Nordbau). The rest of the Type C (from Chassis No. 80341 on) should be built with the improved HL 120 TRM motor (number series from 33501). This was carried out basically as planned. But two Pz.Kpfw.IV Type C, Chassis No. 80338 (with No. 33533 motor) and Chassis No. 80340 (with HL 120 TRM instead of HL 120 TR) were built.

6.2.2. Chassis

Other modifications to the chassis consisted of these improvements:

1) Exit hatches in the chassis under the radioman
2) Complete motor carriers
3) Duct connections with the air filter for HL 120 TRM motors.
4) Ventilation system.
5) Locks for radiator flap and ventilator flap on the roof of the rear end.

Thirty Type C chassis were built with 2./B.W. hulls (No. 90213 to 80242). These were to be rebuilt by the supplying firms according to 3./B.W. designs.

6.2.3 Turret

On February 4, 1938, the Gruson Works had inquired of Department A.K.: *What parts of the turret are changed for 3./BN.W.?* A.K. answered: *From 2./B.W. the M.G. mount and new turret lashing change.* For Type C the attachments for the machine gun were attached to the mount.

A further change to the turret machine gun took place, as reported during a discussion at Prüfwesen 6 in Berlin on April 22, 1938: *As opposed to our suggestion made in the previous discussion on March 24, 1938, to armor the part of the M.G. projecting from the roller mount with a shell, we have decided to protect this part completely with an armored barrel. The armored barrel is sufficiently well measured and will receive in its forward section a borehole above and below to draw off the muzzle flash.*

To better protect the gap between the roller mount and the mount covering, the front edge was made V-shaped. The permission for this had been given as follows in a discussion at Wa Prw 6 in Berlin on June 3, 1938:

The profile of the mount cutout for the 3./B.W. mount covering can be made according to the enclosed drawing 021 B-14861-5(a).

6.3 PZ.KPFW.IV TYPE C PRODUCTION WITH BRIDGELAYER CHASSIS

After preliminary test problems at Wa Prüf 6, the six bridgelayer chassis returned to the Gruson Works. Two of them were, as ordered, examined by Army Acceptance Center II (K) and released to the Gruson Workks for installation of the upper bodies, as reported below:

On March 7, 1940, the release of two Pz.Kpfw.IV, Chassis No. 80435 and 80436, from the Gruson Works took place because of the test run on March 7, 1940, at the Altengrabow troop training camp (151km by factory personnel, 20-21 km by inspector).

The one chassis (No. 80435), fitted with an infantry assault body for testing purposes, was used further. The second chassis (No. 80436) had the No.209 gun mounted on a Type C turret and was finished as a complete Pz.Kpfw.IV Type C and sent to the Army Equipment Department in Magdeburg-Buckau on April 1, 1940.

As reported in the acceptance report, the release of four Pz.Kpfw.IV, Chassis no. 80437, 80438, 80439 and 80440, from the Gruson Works took place on February 23, February 29, and March 7, 1940. The temporary receiver was the Gruson Works for the mounting of the ladder structures for Bridgelayer IV, according to OKH Wa Prüf 6 IIf Order 76g 1015 Bd.Nr.5501/139 of June 27, 1939.

The next receiver of the finished Bridgelayer IVb was the Engineer Office of Wa Prüf 5 in Sperenberg (County of Teltow) for testing and further use with an armored engineer battalion.

After combat testing of the Bridgelayer IVb in the western campaign, the Weapons Office in June 1940 had decided to rebuild several bridgelayers into normal Pz.Kpfw.IV tanks. Three Pz.Kpfw. IV Type C were finished and delivered to the Army Equipment Department in Magdeburg-Buckau as follows:

- Gun No. 205 on Chassis No. 80439 on July 29, 1940
- Gun No. 207 on Chassis No. 84037 on August 1, 1940
- Gun No. 208 on Chassis No. 84036 on August 30, 1940.

The fourth bridgelayer, with Chassis No. 80438, was used further for testing and combat experience with an infantry assault body.

6.4 FORM CHANGES DURING PRODUCTION

6.4.1 Commander's Cupola

Testing 6/IId reported to Krupp on April 5, 1938:

As ordered, we notify you that the Wegmann commander's cupola for Series 3./B.W. is being delivered after the goal was set like 2./B.W. and 3./Z.W. Commander's cupola 021St9261.

Panzerkampfwagen IV, Type C.

Maybach HL 120TRM motor, generator side.

Maybach HL 120TRM motor, oil cooler side.

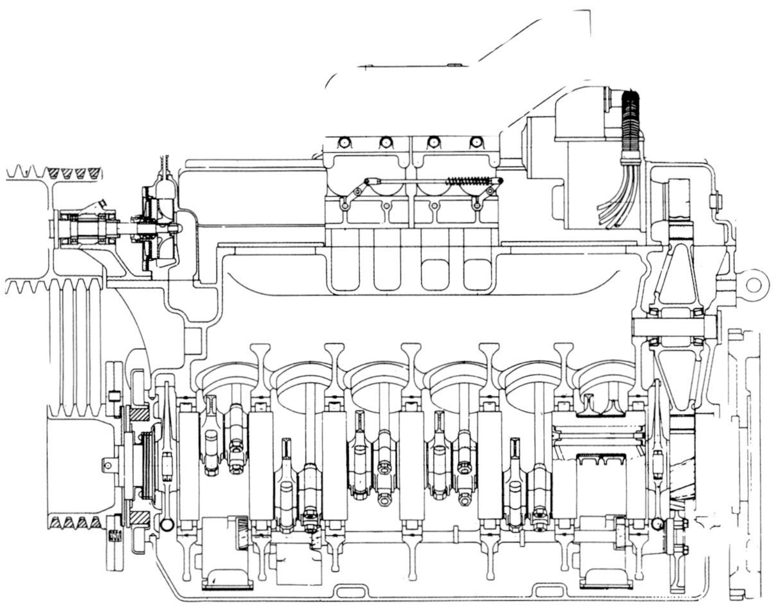

Longitudinal cut of Mayback HL 120TRM motor.

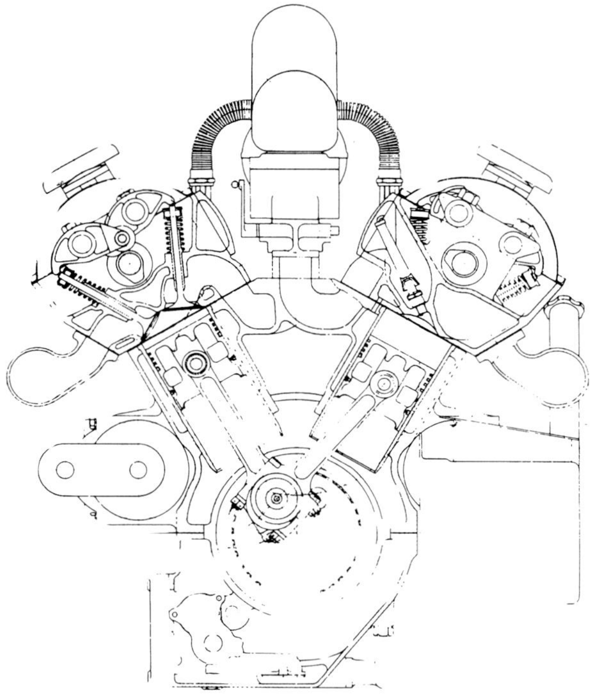

Transverse cut of Mayback HL 120TRM motor.

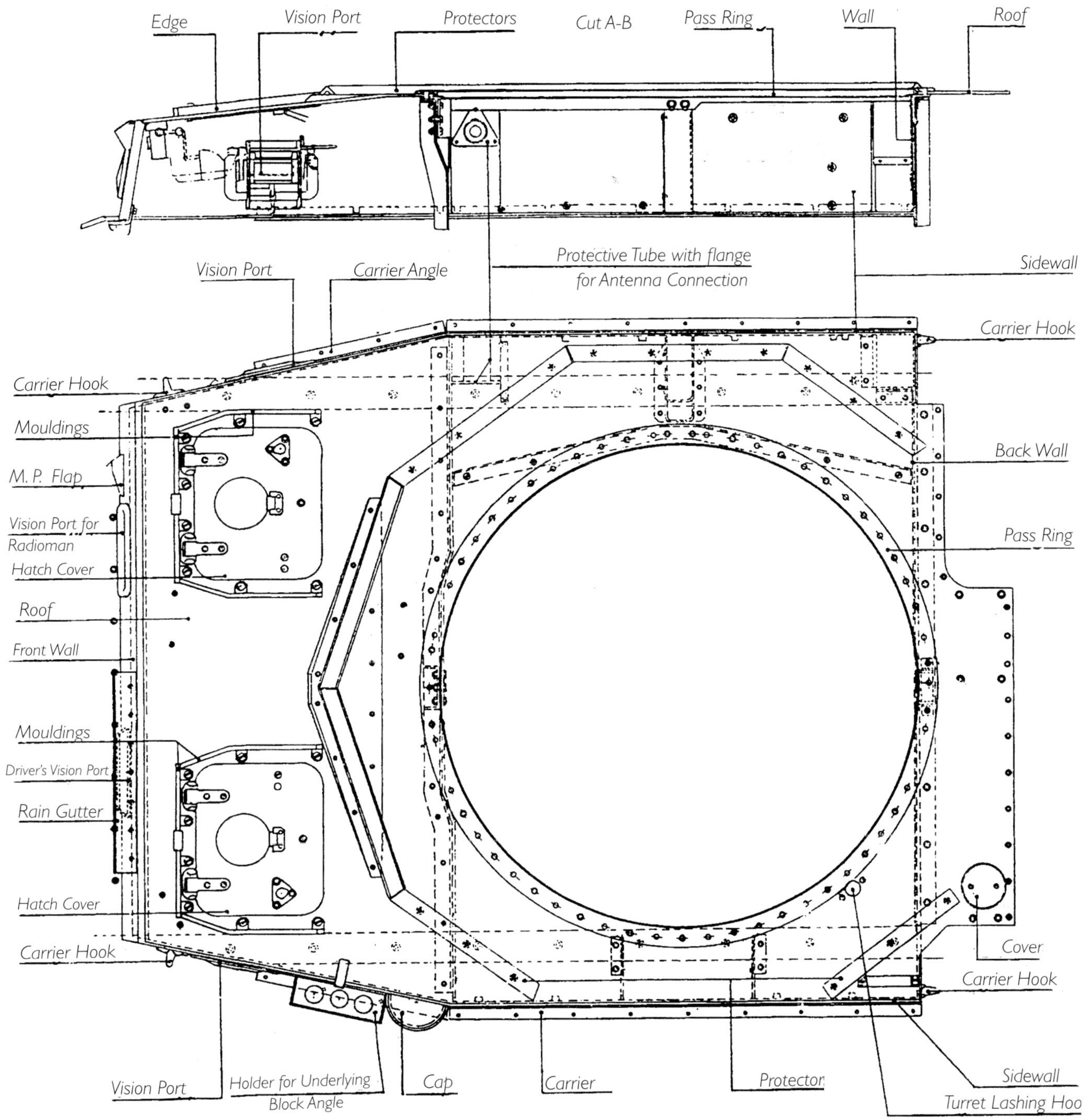

Bow armor (Drawing No. 021 A 8786 U1) for Pz.Kpfw.IV Type C With rain gutter over driver's vision port.

On May 27 Krupp answered:

Wegmann commander's cupola for Z.W. production again reworked and distributed to the building firms under 021St9261. It would be possible for us to make the armored parts available as of the 31st one (3./B.W.).

On 3 June it was agreed in a discussion in Berlin:

Testing 6/IId agrees to use new Commander's Cupola 021B9261 for Series 3./B.W. as soon as possible.

6.4.2 Rain Gutter over Driver's Vision Port

Krupp reported on December 15, 1938, that the Wegmann driver's vision port is no longer watertight after making the armor cutout narrower. The rain gutter developed by Daimler-Benz will be planned on as of 3./B.W. Bow Armor No. 58. For already finished vehicles offer and form change are to be contracted with the Gruson Works for.

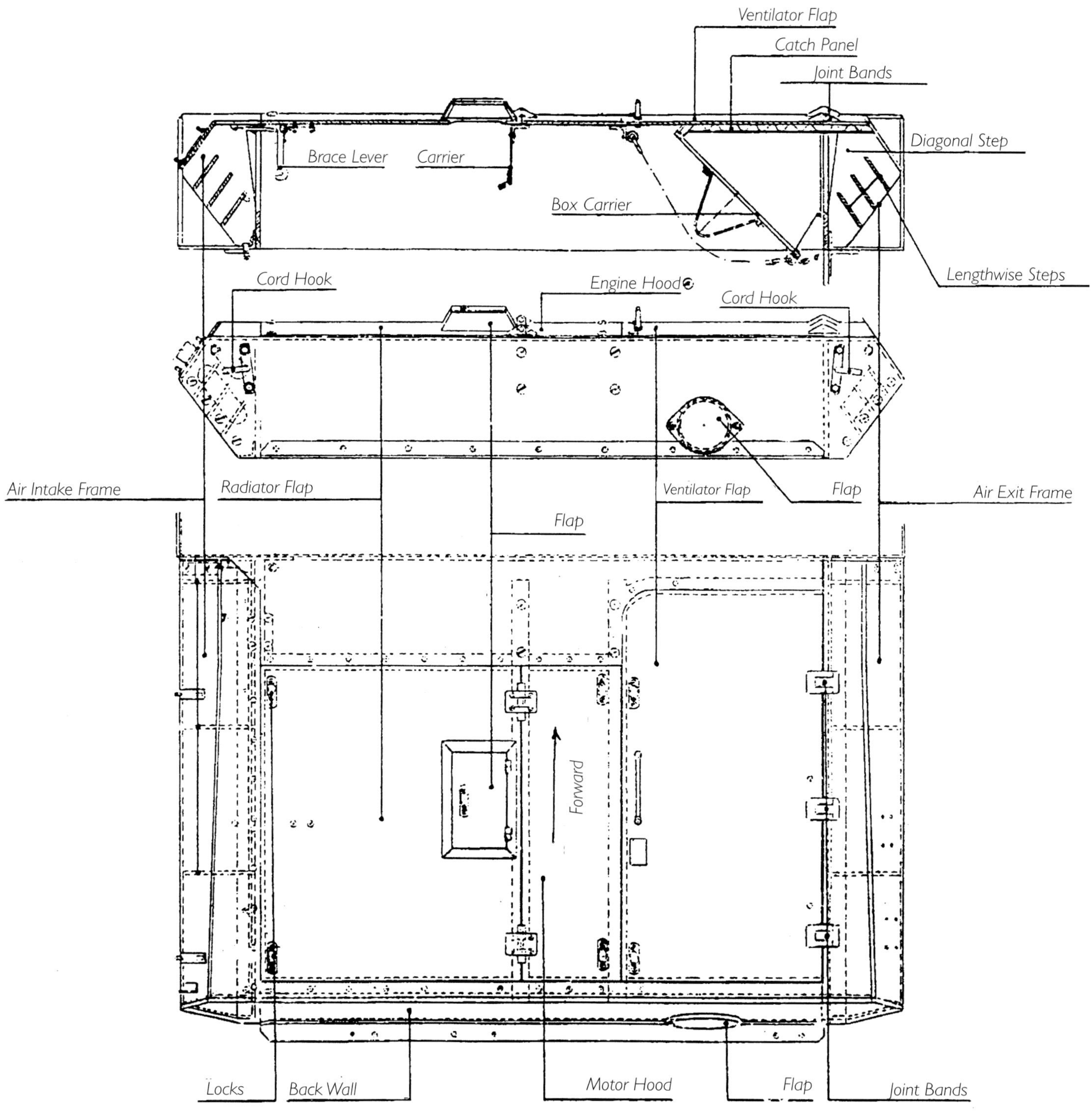

Rear Armor (Drawing No. 021 A 8787) for Pz.Kpfw.IV Type C.

6.4.3 7.5 cm Tank Gun Loading Space

On March 14, 1939, Krupp reported:

The Series 1./B.W. includes the barrels lfd. No. 1-35.
2./B.W. 36-77
3./B.W. 78-217

Until current No. 157 the barrels had the old loading area. The barrels of Series 3./B.W. Nos. 113, 115, 116, 125-157 already had the new, reworked loading space. The barrels from lfd. No. 158 on immediately received the new loading space.

By the troops. Barrels No. 8, 10, 30, 32, 41, 43, 50, 52, 54 and 61 had arrived and been sent back equipped with the new loading space.

6.4.4. Antenna Deflector

Many Pz.Kpfw.IV Type C were accepted and delivered to the troops without an antenna deflector. In the acceptance report (for chassis No. 80390, 80391 and 80393) on April 29, 1939, it was explained:

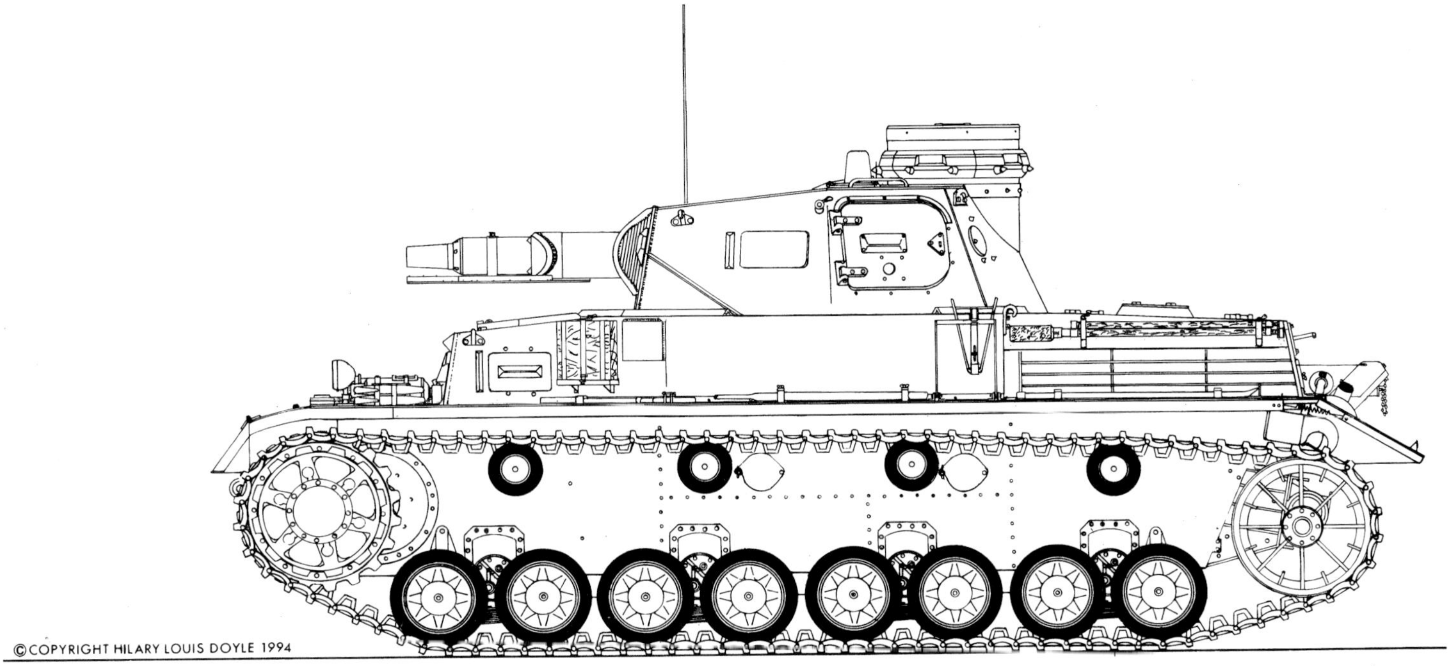
©COPYRIGHT HILARY LOUIS DOYLE 1994

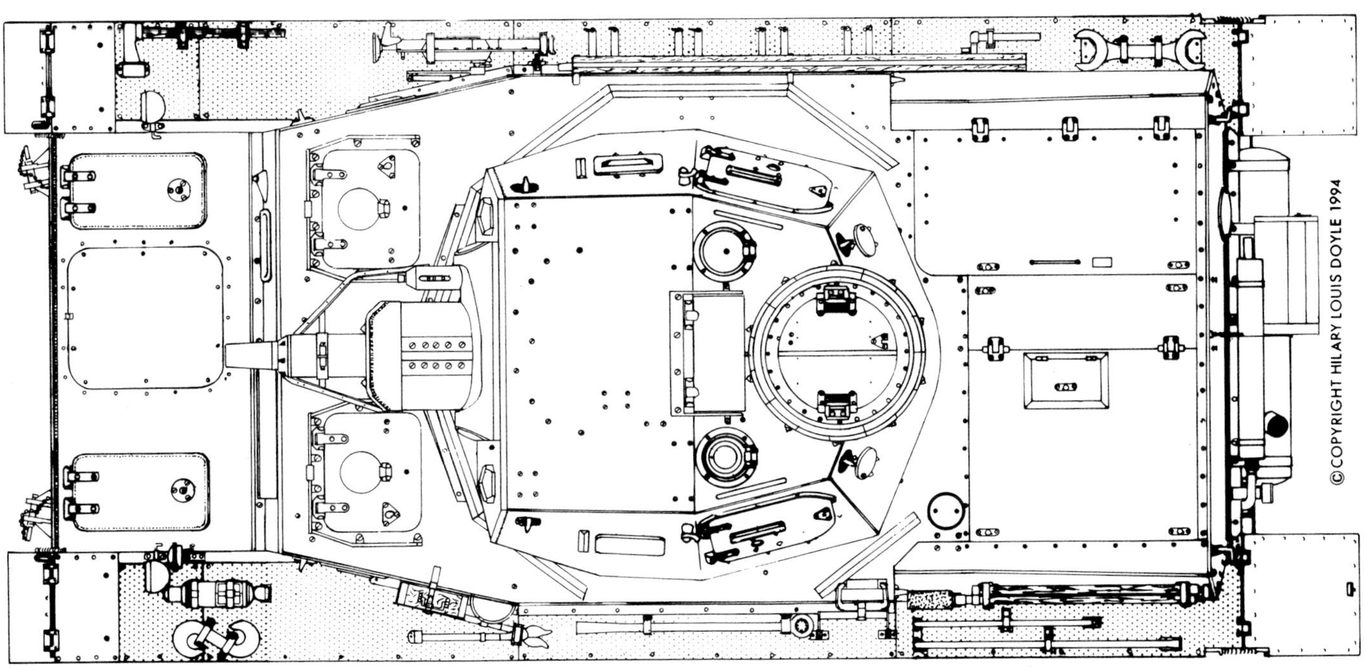
©COPYRIGHT HILARY LOUIS DOYLE 1994

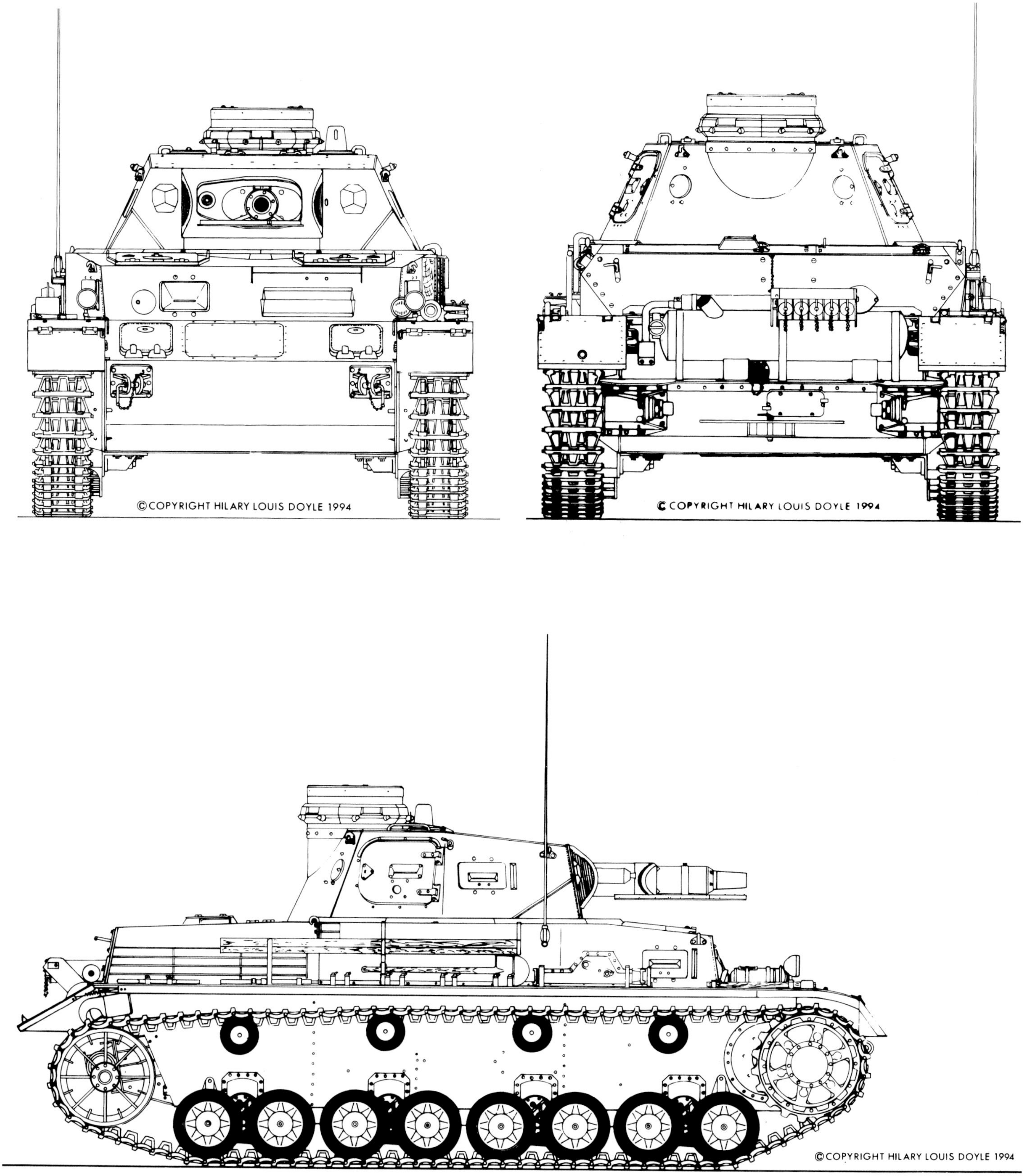

Panzerkampfwagen IV (7.5 cm), (Vs.Kfz. 622), Type C
Improved interior roller mount, armored protective shell for mounted M.G., new antenna deflector on barrel of tank gun. Improved air exit in turret roof. Modified lock in motor room cover. New attachment for tools on track apron.

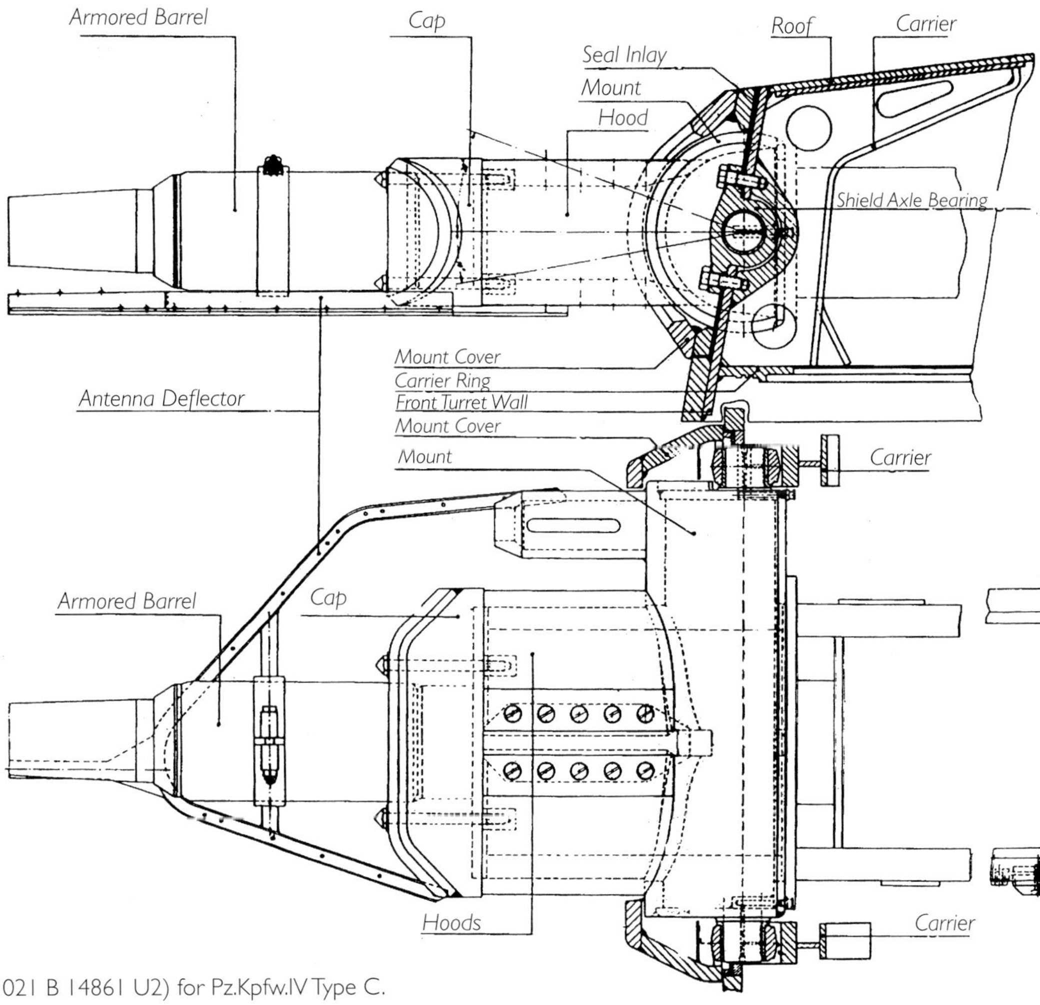

Roller Mount
(Drawing No. 021 B 14861 U2) for Pz.Kpfw.IV Type C.

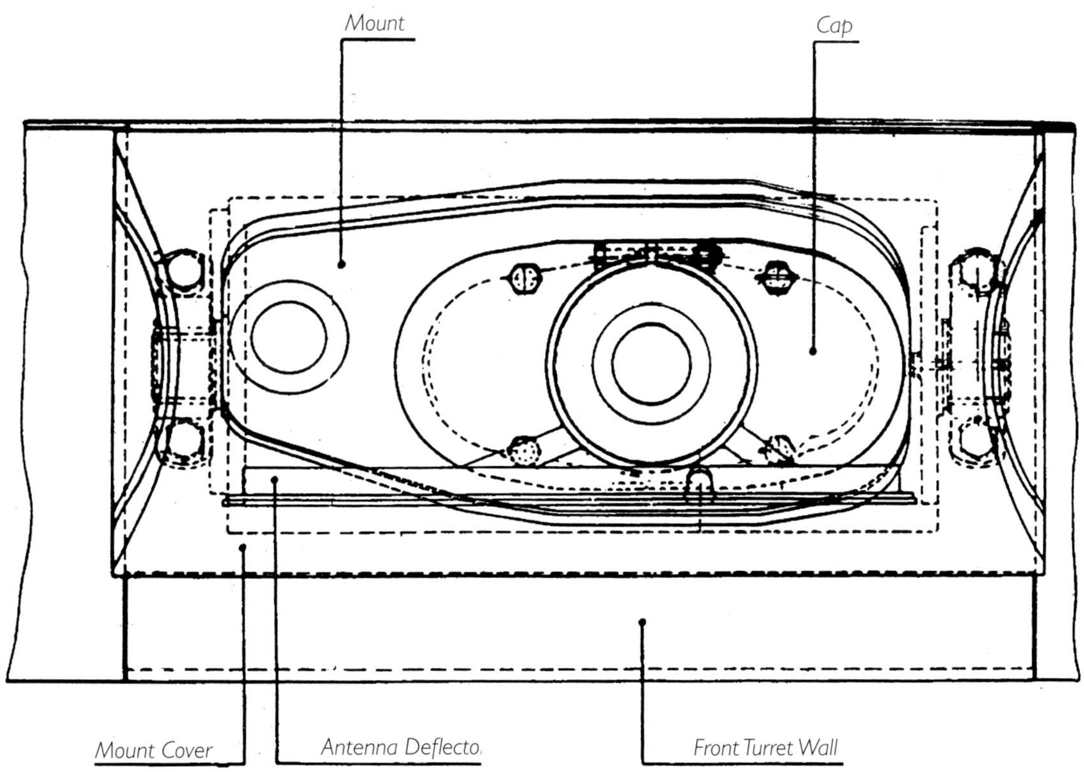

Infantry Assault Ladder Body on Pz.Kpfw.IV Chassis, Type C.

Pz.Kpfw.IV Type C, Chassis No. 80420, finished in July 1939. Panzer Regiment 11 of the 6th Panzer Division training in February 1940.

Vehicle 623, Pz.Kpfw.IV, Type C, Chassis No. 80383, photographed in April 1941 with the 6th Company, Panzer Regiment 3 (2nd Panzer Division) during the Balkan Campaign, in Greece.

B0383
623

623

Pz.Kpfw.IV Type C: White bar cross and white aircraft recognition sign on the rear roof, added at the Army Equipment Office in 1939 before delivery to the troops. This type of national emblem was used only in the Polish campaign.

One antenna deflector per vehicle is lacking; these were delivered later because of design changes (point in time indefinite).

6.4.5 Rebuilt Bridgelayer with Type E Bow Armor

On June 24, 1940, the Gruson Works, Department A.K., asked *...whether on the 3./B.W. chassis that were built for ladder structures, 6./B.W. bow armor could be used after exchanging a rim on the hull.* The answer on 25 July said: *6./B.W. bow armor could be mounted on 3./B.W. chassis by using a new rim as on Sk 3890 instead of the 021 D 33001-39 type from the hull. Apparatus Building III will deliver three rims and send welders.*

The installation of new bow armor was necessary, since on the last six 3./B.W. bow armor numbered 80435-80440, already used with bridgelayer structures, attachments for the turret were lacking. The end result was three Pz.Kpfw.IV Type C, born of an unusual mixture of 3./B.W. guns and turrets, 6./B.W. (Type E) bow armor, with 3./B.W. rear armor and rebuilt 3./B.W. bridgelayer chassis.

6.5 IMPROVEMENTS AFTER DELIVERY

A series of improvements was arranged after the delivery of Pz.Kpfw.IV Tyoe C tanks to troop units. Among others:

1) As of February 1939, 7.5 cm Kw.K. loading space enlarged.
2) As of spring 1939, adding a rain gutter over the driver's vision port.
3) As of spring 1940, attaching the "Notek" night marching device on the left track apron.
4) As of June 1940, switching from drum to belt feed for machine guns and installation of bullet-bag holders for belt feed.
5) As of February 1941, strengthening the front hull wall by welding on an additional 30 mm thick armor plate before the original 30 mm bow armor plate.
6) As of February 1941, making changes to the ventilator system and drive.
7) As of March 1941, mounting a baggage rack on the back turret wall.
8) As of June 1941, building a coupling on the back for a fuel trailer.

7. PANZERKAMPFWAGEN IV (7.5 cm) (Sd.Kfz. 161) Type D, (4. & 5./B.W.)

Chassis Number Series 80501-80748

Of the 248 ordered Pz.Kpfw.IV Type D tanks, 231 were built and accepted between October 1939 and October 1940. 16 Type D chassis were finished and delivered for a special bridgelayer test series in March and April 1940, and one Pz.Kpfw.IV, Chassis No. 80625, was built and delivered for testing purposes in June 1940.

7.1 IMPROVEMENTS AFTER TYPE C

Wa Prw 6 wanted a uniform chassis that was usable for both Pz.Kpfw.III and Pz.Kpfw.IV. But Mr. Woelfert (Department A.K., Krupp) opposed it in a note dated May 2, 1938, saying that problems with new ideas left no time to take up this major change in the production program:

Originally the Daimler-Benz uniform chassis was supposed to be used for the 2./B.W. series. Through the use of new design parts that are not sufficiently tested for tanks, there were setbacks, among others, with the steering gears built with three stages by the Zahnradfabrik, the Maybach gearbox with semi-automatic shifting, and the lubricated tracks. The experience gained with gearboxes and tracks of towing tractors could not be transferred directly to tanks. Since the first test vehicle is only coming to Kummersdorf for testing at this time, series production in a foreseeable time is not to be reckoned on. Instead of the Daimler-Benz uniform chassis, therefore large contracts for the 3rd, 4th and 5th series of B.W. are being issued.

The most important improvements, with descriptions of the devices, are listed in this chapter.

7.1.1 Armor

On March 10, 1938, suggestions for increasing the armor thickness of the 4./B.W. were discussed at Prw. 6 in Berlin:

In reference to A.K. Document No. 89812/L. of 2/21/1938, AK turns over the new Suggestion W871 to Wa Prw 6 (IId) and Wa Prw 6 (IIIg) for optional strengthening of the hull, or the hull with superstructure, or with superstructure and turret. According to a, ZK suggestion, the following thickenings were at first foreseen for better protection of ammunition, fuel and motor:

Sidewall of Hull and Bow Armor from 14.5 to 30 mm
Back Wall of Hull 14.5 to 25 mm
Bow Armor 14.5 to 30 mm
Stern Armor 14.5 to 20 mm

At Department A.K. it was decided in a discussion with Major Dr. Olbrich and Dipl.Ing. Biersack on April 12, 1938:

The armor increase for 4./B.W. is being discussed after W 874. Suggestion 1 is being carried out, but with a hull 10 instead of 8 mm thick, the bow armor with sidewalls and front bottom plate of 20 mm, and the stern armor with 20 mm sidewalls. This results in a weight gain from 1256 kg as opposed to the 2./ and 3./ B.W.

For increased safety against antitank shells, the front plate was made of surface-hardened, cemented PP494 armor steel. If softer tank shells (with enough power to penetrate thicker, homogeneous armor plates) hit this highly hardened armor plate, the tank shells would fail and break up.

The three types of armor used in building Type D were:

1) Plates up to 14.5 mm thick of homogeneous nickel-free PP694 armor steel with 153 +/-5 kg/mm^2 firmness,
2) Plates over 14.5 mm thick of homogeneous nickel-free PP694 armor steel with 125 +/-5 kg/mm^2 firmness,
3. Front plates 30 mm thick of surface-hardened cemented PP494 armor steel with 2.5 (+0.5) mm entry depth, a frontal firmness over 60 Rockwell C and a nuclear firmness of 125 +/- kg/mm^2.

7.1.2 Chassis

7.1.2.1 Armor Hull

The armor hull consisted of armor plates with the following thicknesses and inclinations (from the horizontal):

Panzerkampfwagen IV Type D

Plate	Thickness	Inclination
Front wall, upper	20 mm	18 °
Front wall, forward	30 mm	76 °
Front wall, lower	20 mm	29 °
Sidewall	20 mm	90 °
Back wall, upper	20 mm	80 °
Back wall, lower	14.5 mm	16 °
Bottom	11 mm	0 °

7.1.2.2. Tracks

On April 28, 1938, Wa Prw 6 reported that as of 4./B.W. new Kgs 6111/380/120 tracks with higher teeth were to be ordered.

7.1.2.3 Striking Block

Until Type C a contact to limit roller rising for the front and rear road wheels was attached to the vehicle's hull. As of Type D, five contact blocks were attached to the vehicle's hull, one before the attachment of each pair of road wheels, and the fifth behind the attachment of the last pair of road wheels.

7.1.3 Upper Body

In a letter to Department AK on January 18, 1938, Wa Prw 6 requested a basic change in the bow armor.

As of 4./B.W. a ball mantlet (30 mm) will again be installed for the radioman. Drawing from Daimler-Benz. Being omitted are thus the vision port and the MP opening before the radioman.

This change was extended on in a discussion with Wa Prw 6 on January 29, 1938:

When installing a ball mantlet in 4./B.W., Department AK recommends offset front walls, since lateral sight is essential for the driver. But the location of a sufficiently strong 30 mm vision flap makes difficulties because of its limited usable width. Wa Prw 6 still gives data on a turning vision flap for this purpose.

On 10 March it was decided in a discussion at Wa Prw 6:

As of 4./b/w/, Optic K.F.F.2 with simple prisms will be installed.

In reference to the simplification and uniform components used in both Pz.Kpfw.III and Pz.Kpfw.IV, it was decided on May 20, 1938:

The 30 mm Daimler vision flap 30-0 degrees to the right and left is to be taken over unchanged in 4./B.W. for supply reasons, although the Pz.Kpfw.IV Type D has only 20 mm armor. Department A.K. is also receiving through Daimler until May 23 all drawings of these flaps. Plus drawings of the driver's vision flap and commander's cupola.

On November 17, 1938, it was decided:

The new stern armor according to Sk3128 is to be foreseen as of 4./B.W. The new stern armor became simpler and more useful, built with thicker armor.

These and other decisions to improve the upper armor with accompanying components were defined in the following equipment description:

7.1.3.1 Bow Armor

The bow armor consisted of armor plates with the following thicknesses and inclinations (from the horizontal).

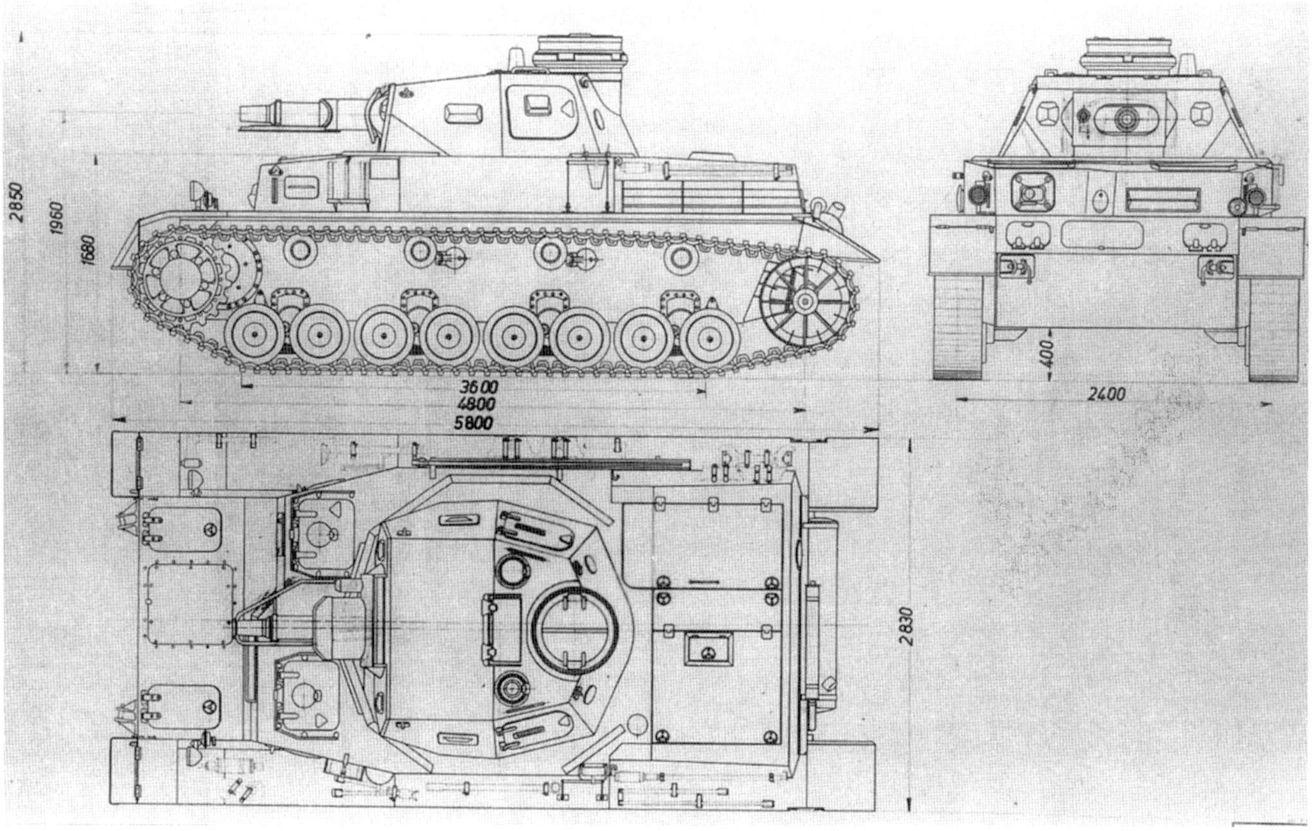

Drawings of Pz.Kpfw.IV Type D before changes to the rear armor.

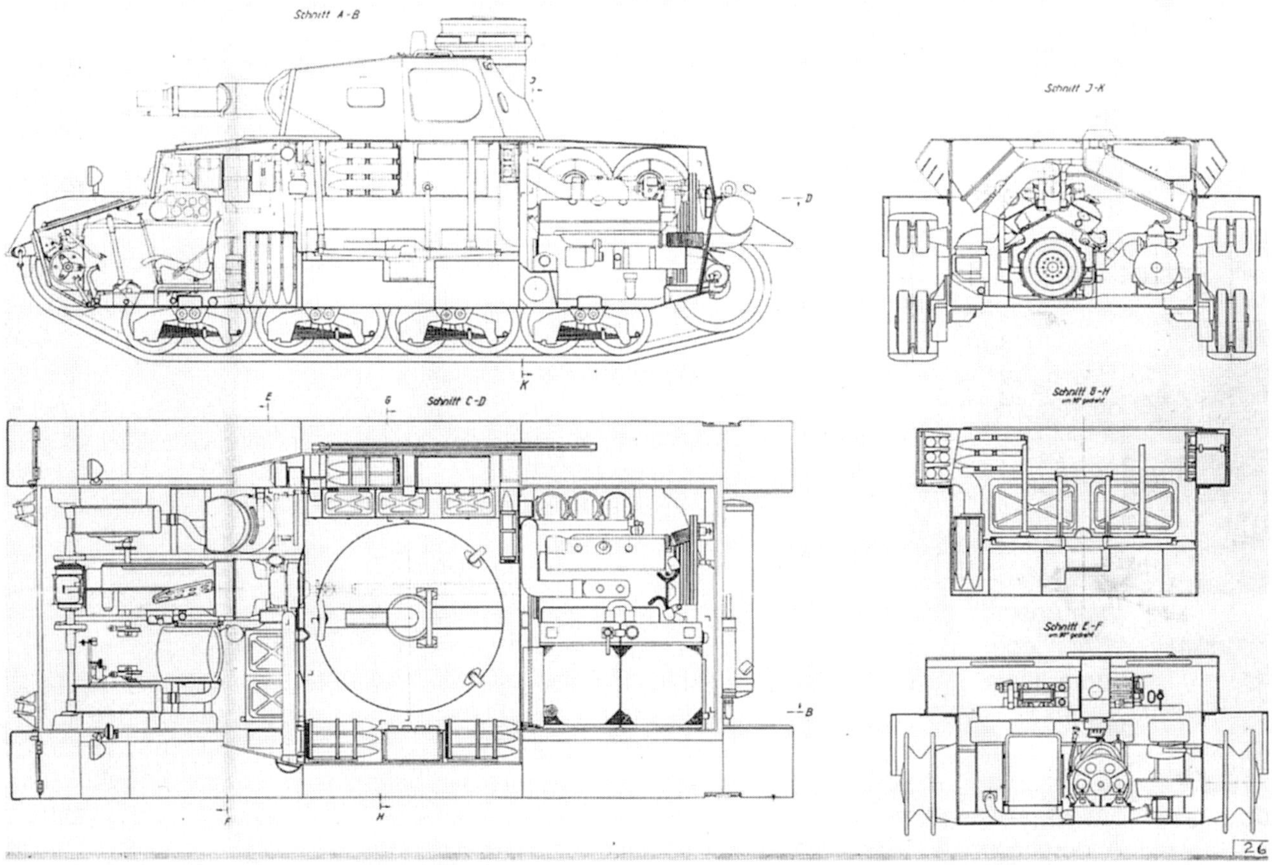

Cutaway drawings of Pz.Kpfw.IV Type D before changes to the rear armor.

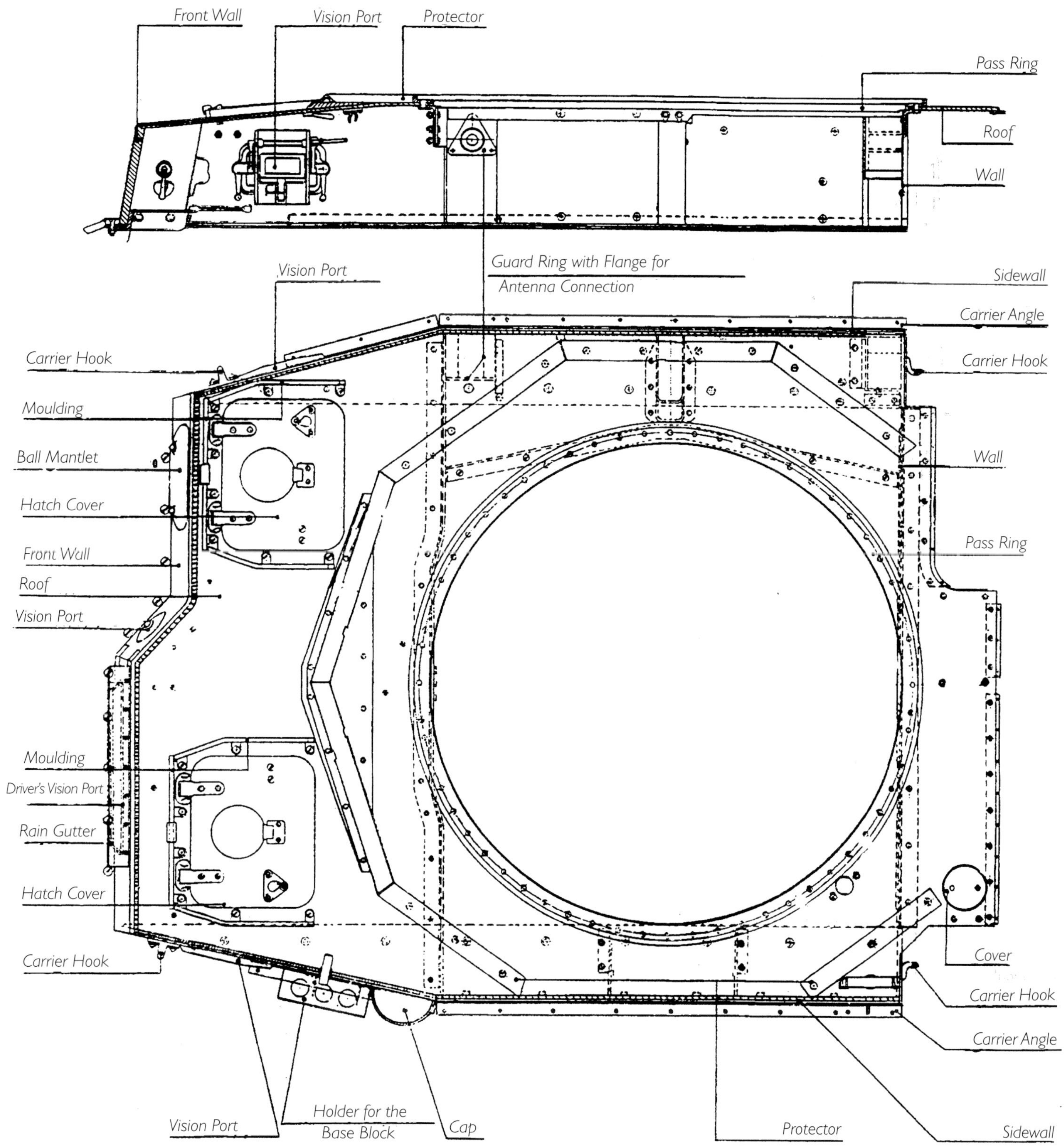

Bow Armor (Drawing No. 021 A 8886 U1) for Pz.Kpfw.IV Type D.

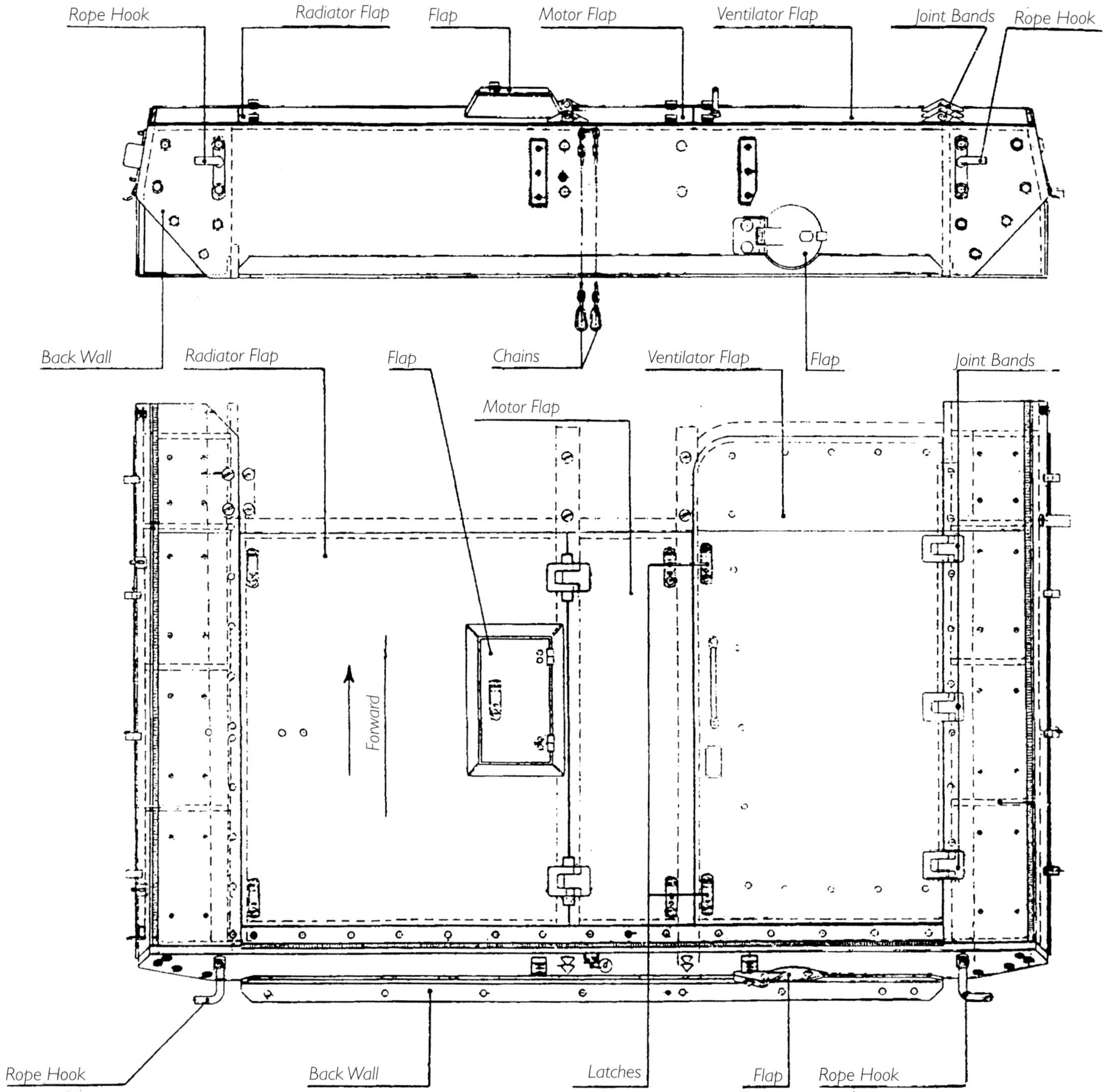

Stern Armor (Drawing No. 021 A 8887) for Pz.Kpfw.IV Type D.

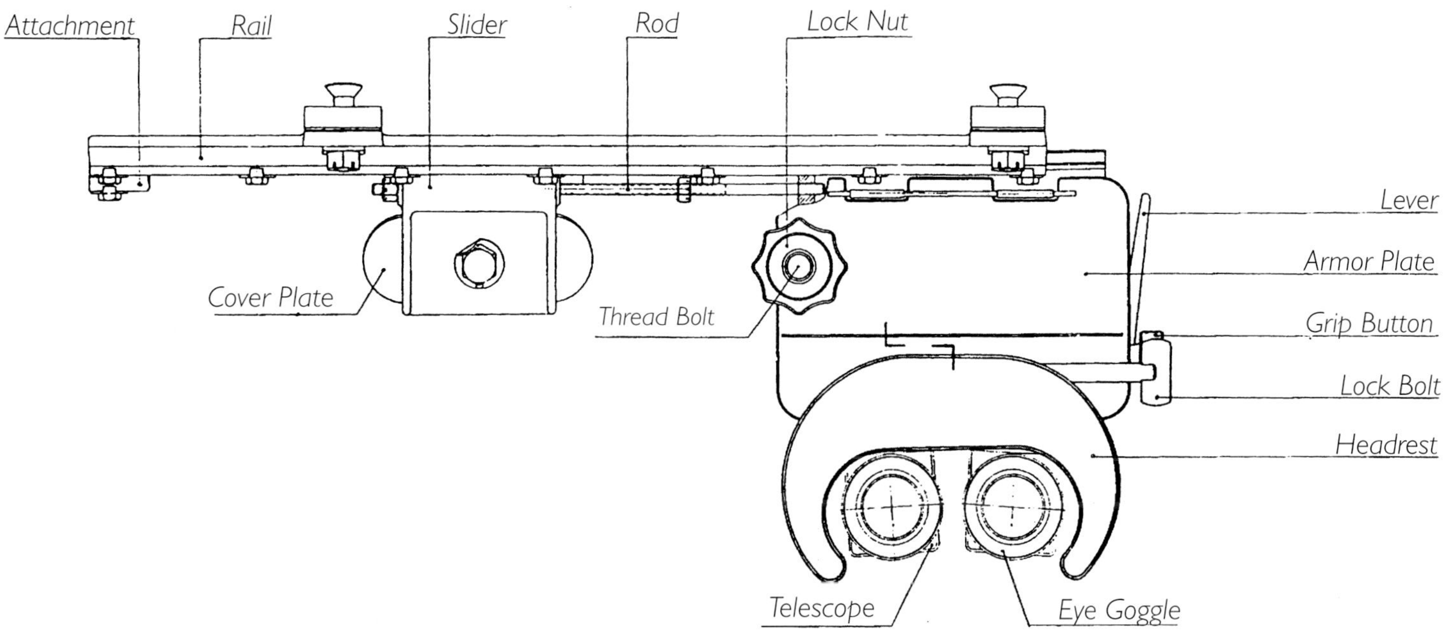

Driver's Optics (Drawing No. 021 B 8875) for Pz.Kpfw.IV Type D.

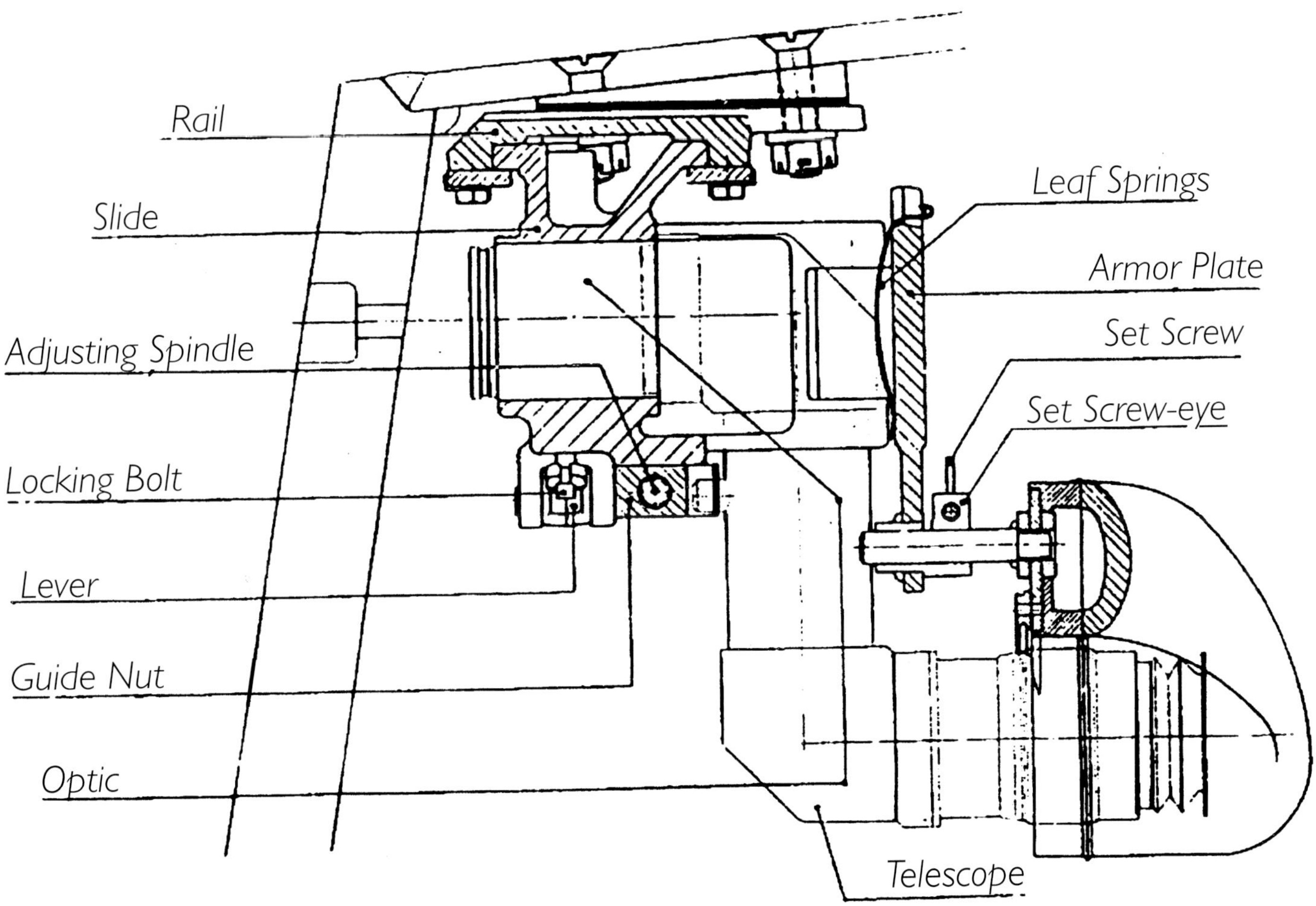

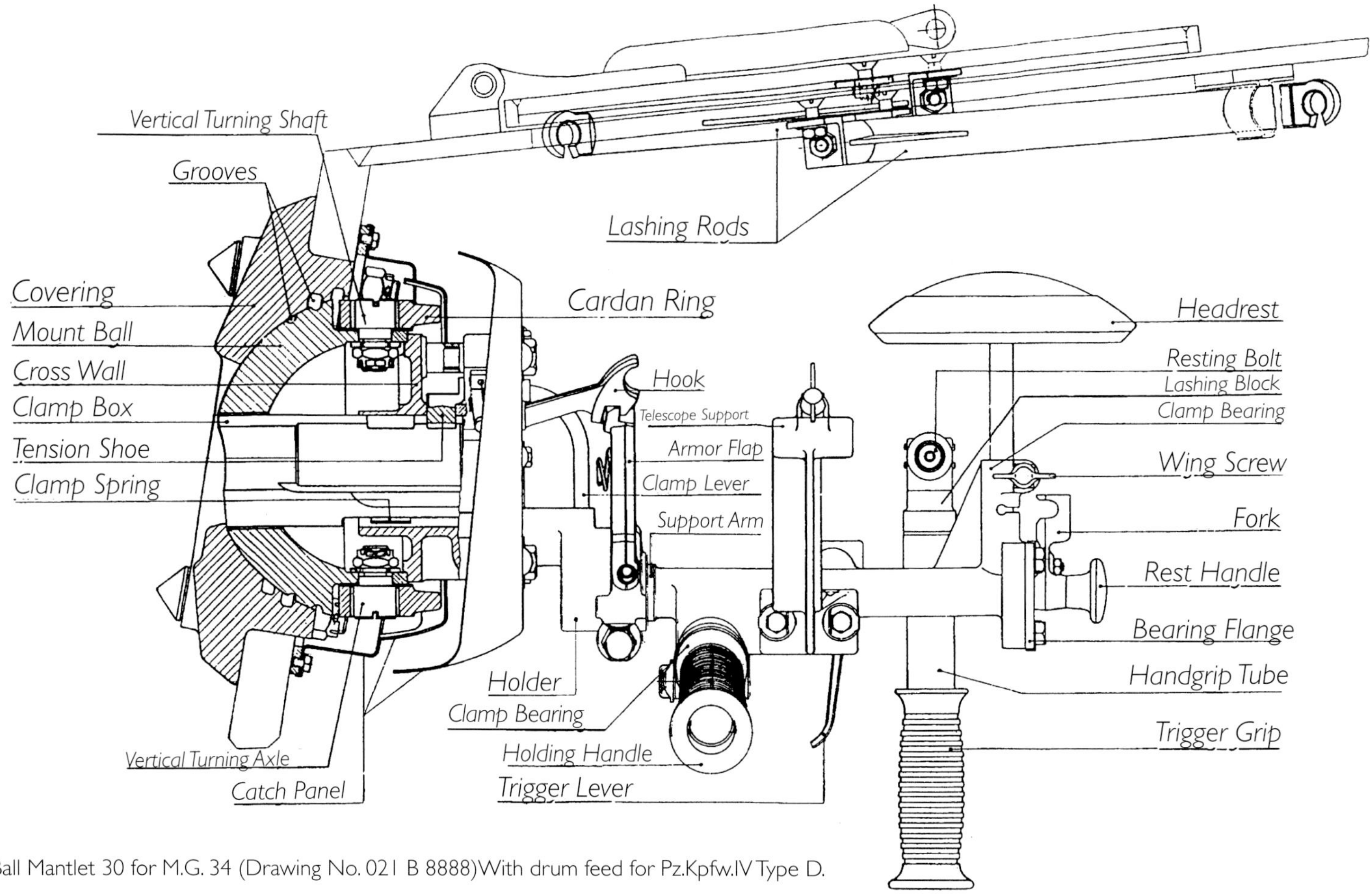

Ball Mantlet 30 for M.G. 34 (Drawing No. 021 B 8888)With drum feed for Pz.Kpfw.IV Type D.

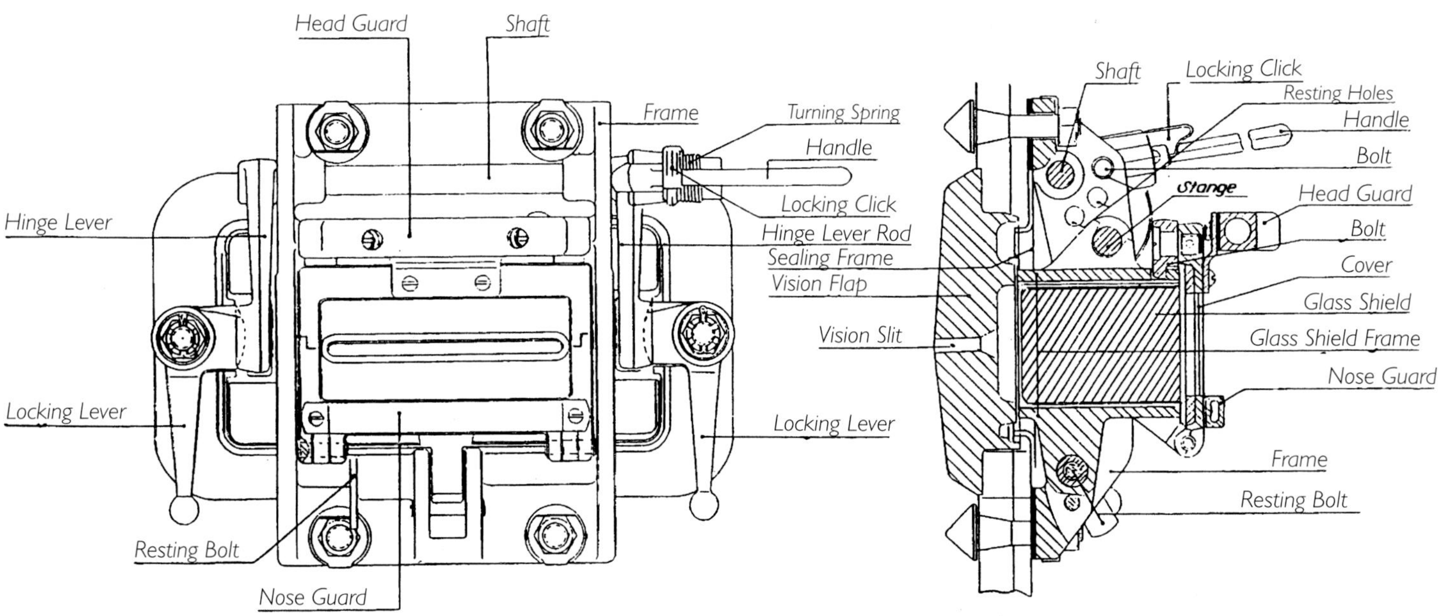

Vision Port in Sidewalls (Drawing No. 021 B 8876) For Pz.Kpfw.IV Types D and E.

Plate	Thickness	Inclination
Front Wall	30 mm	81 °
Sidewall	20 mm	90 °
Roof	11 mm	0 and 5 °
Back Wall	20 mm	90 °
Bottom Plate	8 mm	0 °

The front wall before the radioman was stepped backward in Type D (and Type E). The driver could look and shoot to the right through the vision port (also M.P. port) in the short sidewall of the front wall.

7.1.3.2 Driver's Optics

The driver's optics K.F.F.2 (formerly K.F.F.IV) had a 1.5 fold magnification and a 50-degree field of vision. The k.F.F.2 was angled at the bending places and fitted to the form of a simple triangular prism.

7.1.3.3 Ball Mantlet 30 for M G 34

In Pz.Kpfw.IV Type D (and Type E) the Ball Mantlet 30 was installed in the front wall of the bow armor. The ball mantlet allowed the radioman to fire the M.G. 34. It consisted of the mount ball and the covering. The covering overlaid the gap between the mount ball and the armor wall and also enclosed part of the mount ball from the front. The mount ball, movable in all directions, had bearings on its back for an M.G. 34 and a K.Z.F.2 ball targeting scope. The swinging extent of the ball was +/- 15 degrees to the side from the central position, + 20 degrees upward, and - 10 degrees against the horizontal.

The double-angled K.Z.F.2 ball targeting scope was used for targeting with the M/G. 34 installed in the Ball Mantlet 30. The optical value of the K.Z.F.2 was 1.8 fold magnification with 18-degree field of vision.

7.1.3.4 Vision Port

A vision port without a vision slit was installed in Type D (and Type E) to the driver's right in the short connector of the step-shaped offset front wall of the bow armor. It gave the driver a view to the right and could also be used for firing the machine pistol or a pistol.

The vision flap closed the vision openings in the sidewalls of the bow armor to the driver's left and the radioman's right. From Type D on, the vision flap was a rectangular plate with a stepped offset frame, with which the flap lay on the armor wall. To protect against lead spray, the flap was set a few millimeters into the armor wall. A vision slit 8 mm wide had been cut into the flap. It was shaped like a roof outside, so that lead spray and striking bullets were largely deflected from penetrating the vision slit. For eye protection against lead spray, splinters and rifle bullets that penetrated the vision slit, a 90 mm thick, easily replaced glass shield was set up.

7.1.3.5 Stern Armor

The stern armor consisted of welded armor plates with the following thicknesses and inclinations (from the horizontal):

Plate	Thickness	Inclination
Roof	10 mm	2 °
Sidewall (angled)	20 mm	55 °
Sidewall	20 mm	90 °
Back wall	20 mm	80 °

On the back wall of the stern armor (as of Type D) there were two chains, which attached towing lines.

In the cooling air entry and exit frame (as of Type D) one lengthwise bar each was welded in for protection against shots from the side, and four crossbars for rigidity and protection against shots in the forward direction.

As of Type D guide bars for air cooling were attached to the diagonal long bar and the angle iron between the vertical wall of the stern armor and the track apron; they also caught bullets and splinters. Over the lower guide bar, at the outer angle of the vertical wall, there was also a deflector edge welded on. Above, on the inside of the diagonal long bar and the inclined outer wall, were wooden pieces attached to direct the air and catch lead spray and splinters. As of Type D the entry and exit openings for the cooling air could be closed in cold weather by four flaps on each side of the stern armor.

7.1.4 Turret

7.1.4.1 Roller Mount

On May 23, 1938, Wa Prw 6 requested:

For further development, the design of the roller mount of the B.W. turret is changed.

1) *The heavy mount covering as well as the seam, which can be damaged by shots, between this and the roller mount are to be eliminated.*
2) *The roller parts can scarcely be hit by direct fire.*
3) *The shield panel can be strengthened without significant weight increase.*
4) *Installation and removal of the entire mount is quick and easy.*

Weight saving from 2./B.W., despite the 35 mm shield panel, 35 kg.

The design of the 4./B.W. mount, corresponding to the AKF 30523 and 30525 suggestions from the A.K. Department, was approved by Wa Prw 6 on June 3, 1938.

The roller mount of Type D (and Type E) consisted of the shield, upper and lower base pieces, the two sidewalls and the back wall. In front on the shield the shield barrel for the M.G. was welded. The spaces between the front turret wall and the mount at the left and right were covered by the shield trunnion bed, above and below by the shield of the mount. To increase protection from lead spray, catch panels were screwed onto the mount above and below.

7.1.4.2 Turret Housing

The turret armor consisted of welded armor plates with the following thicknesses and inclinations (from the horizontal):

Roller mount (Drawing No. 021 B 14871 U4) for Pz.Kpfw.IV Type D.

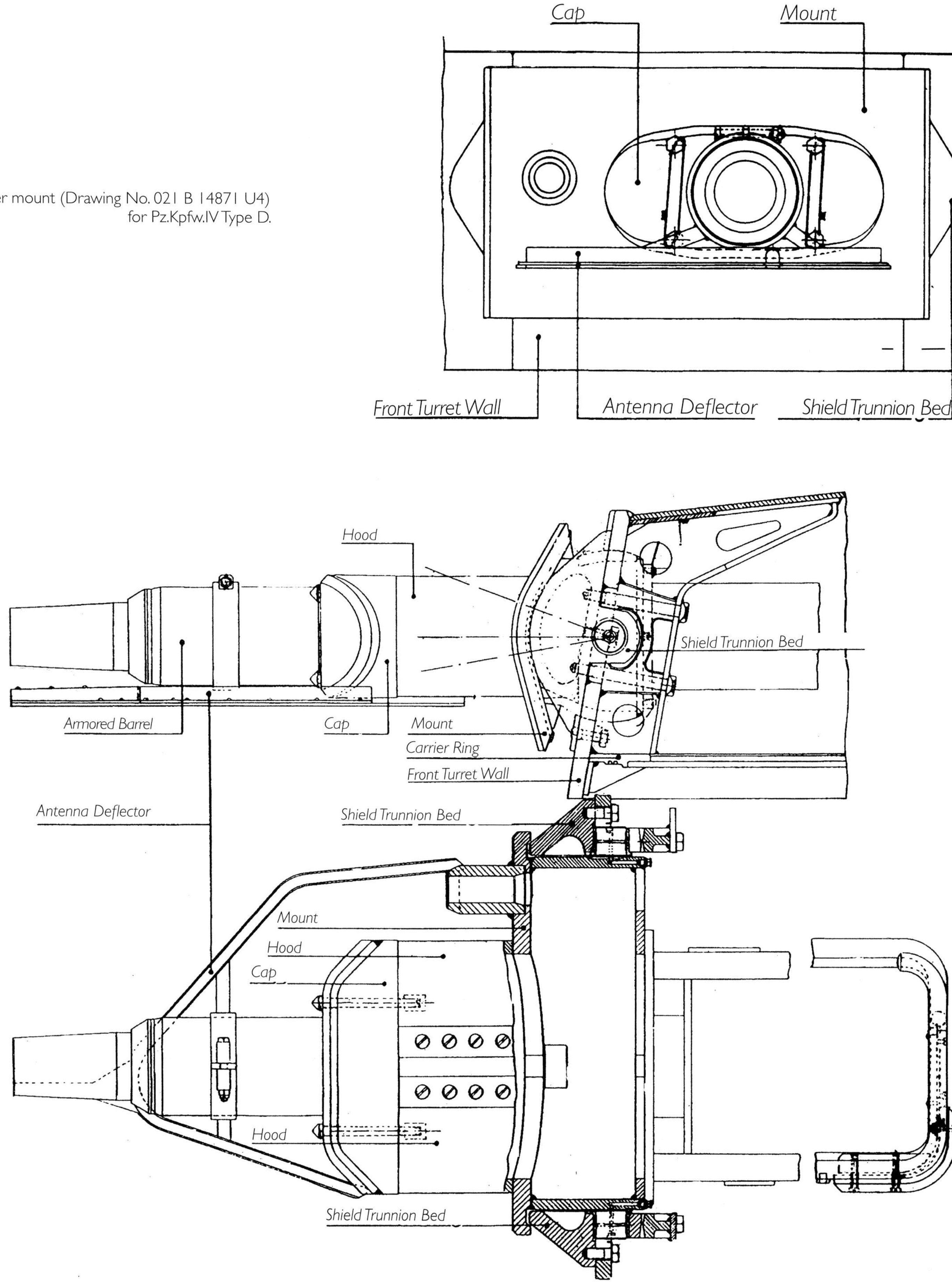

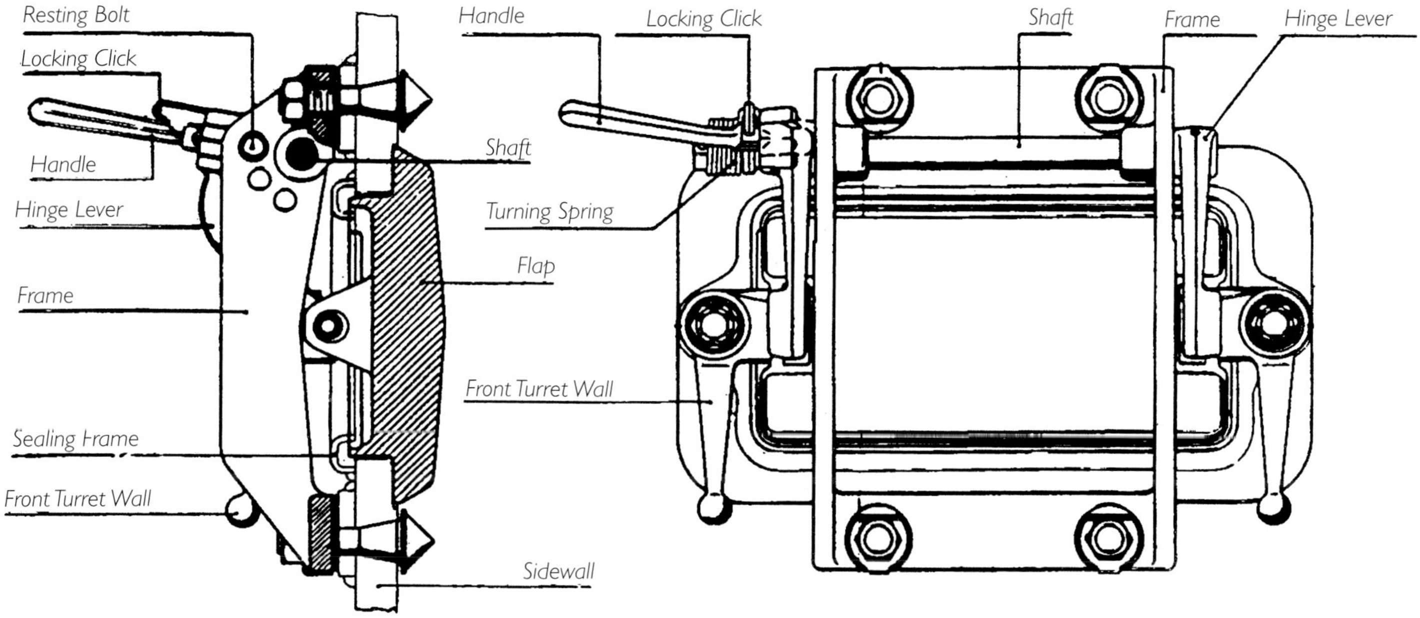

Left vision port without vision slit (Drawing No. 021 B 14871 U8) For Pz.Kpfw.IV Type D.

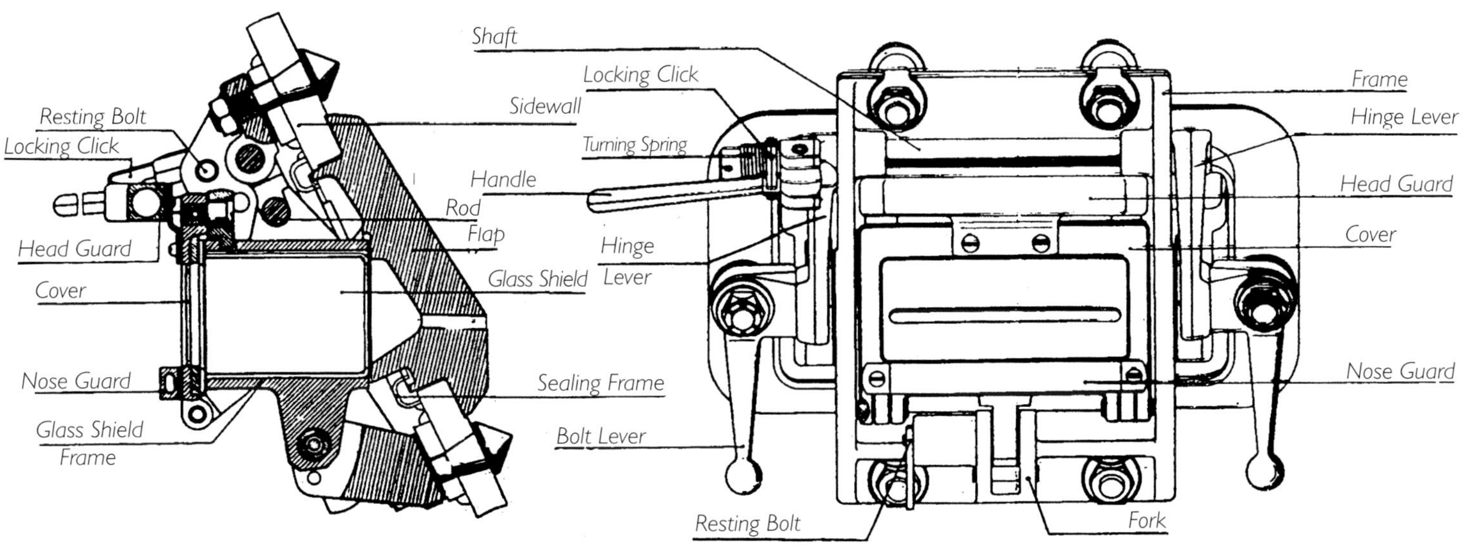

Right Vision Port without vision slit (Drawing No. 021 B 14871) for Pz.Kpfw.IV Types D and E.

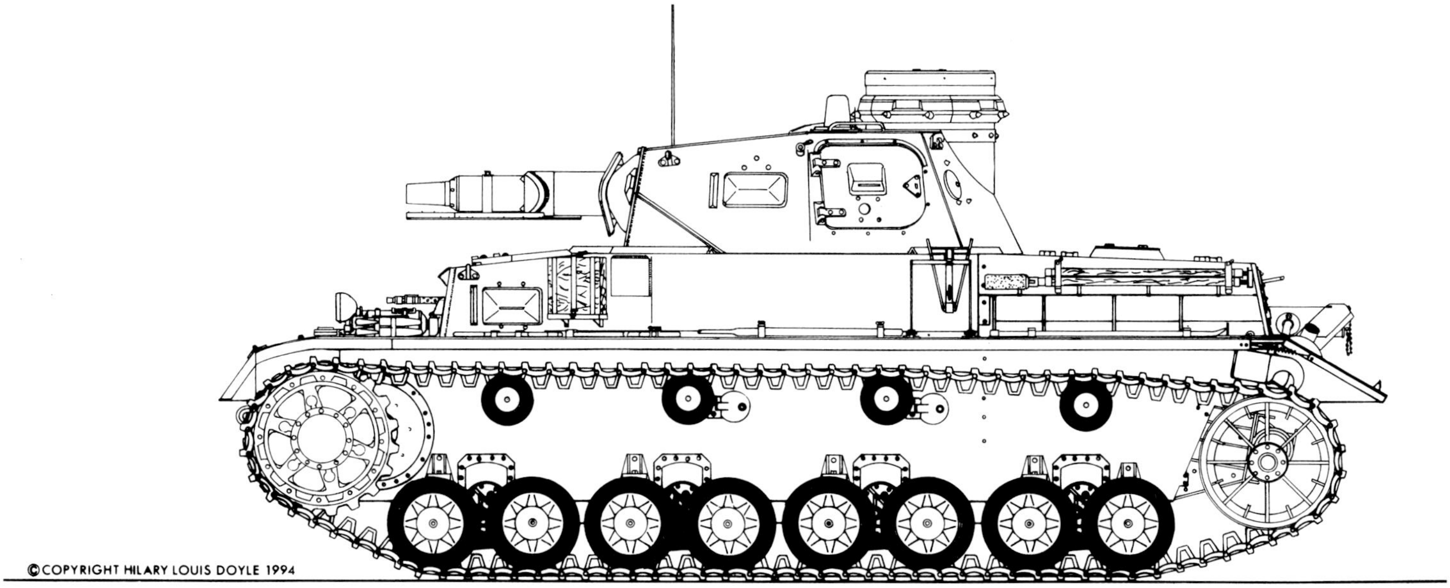
©COPYRIGHT HILARY LOUIS DOYLE 1994

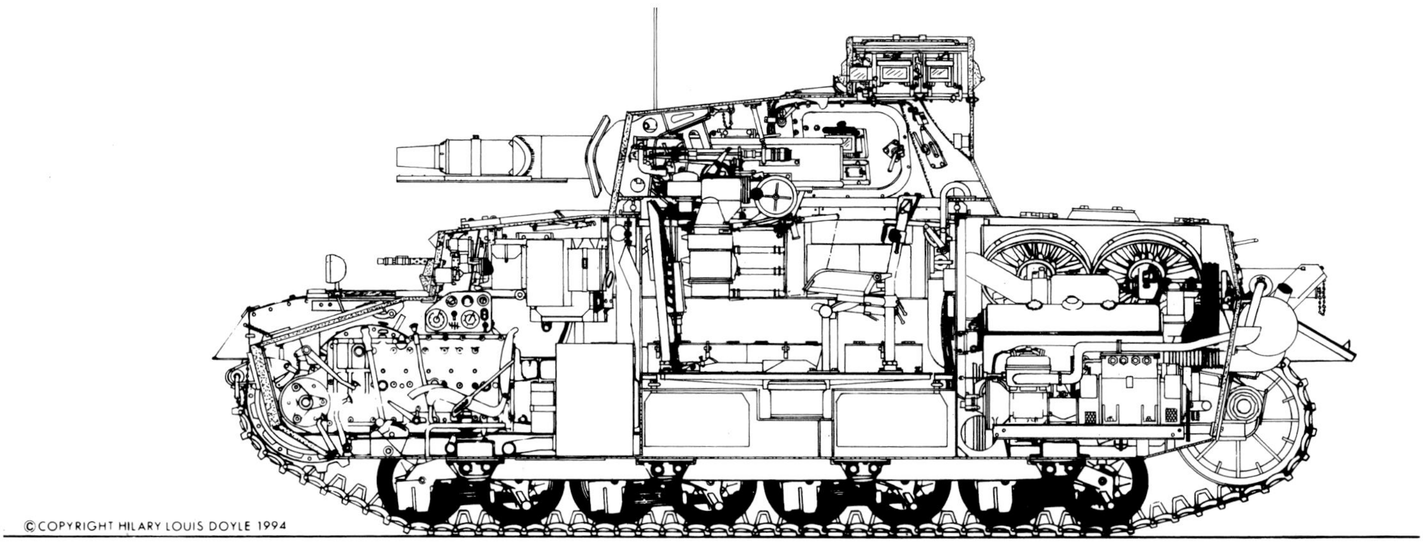
©COPYRIGHT HILARY LOUIS DOYLE 1994

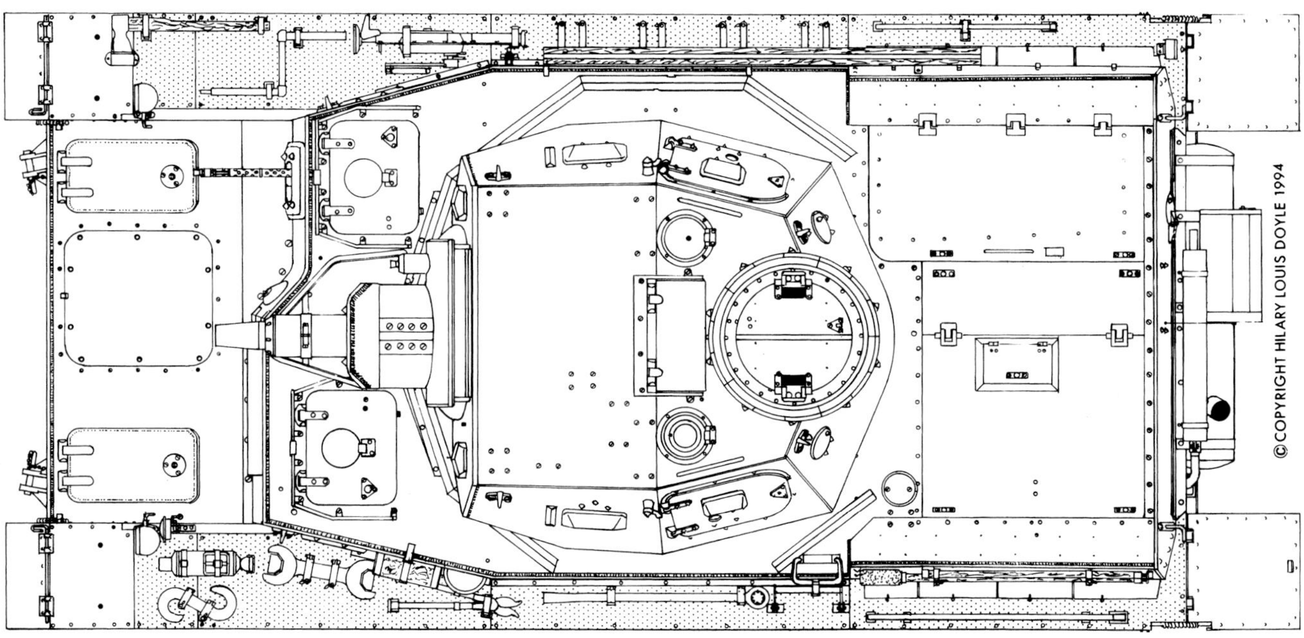
©COPYRIGHT HILARY LOUIS DOYLE 1994

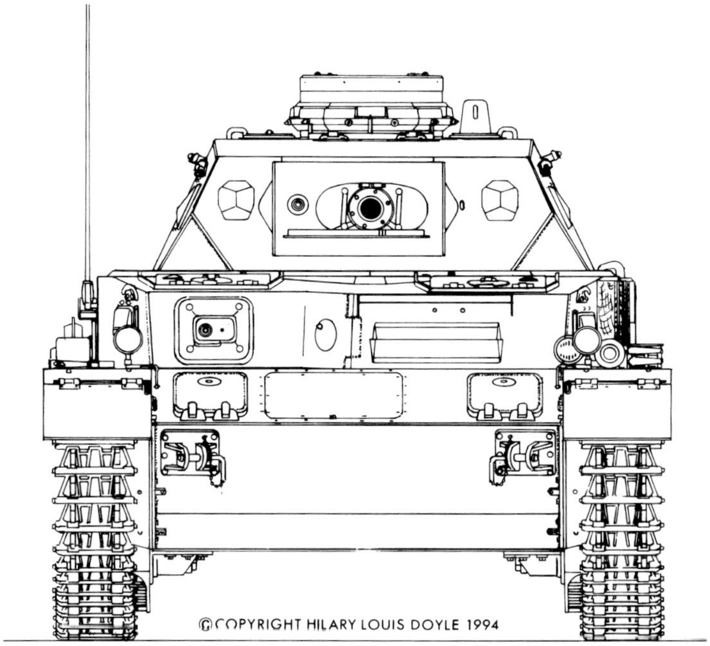

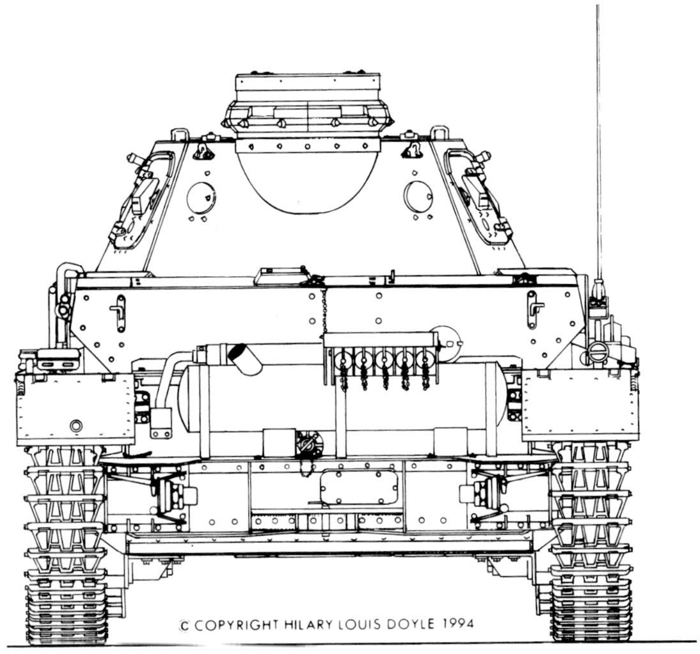

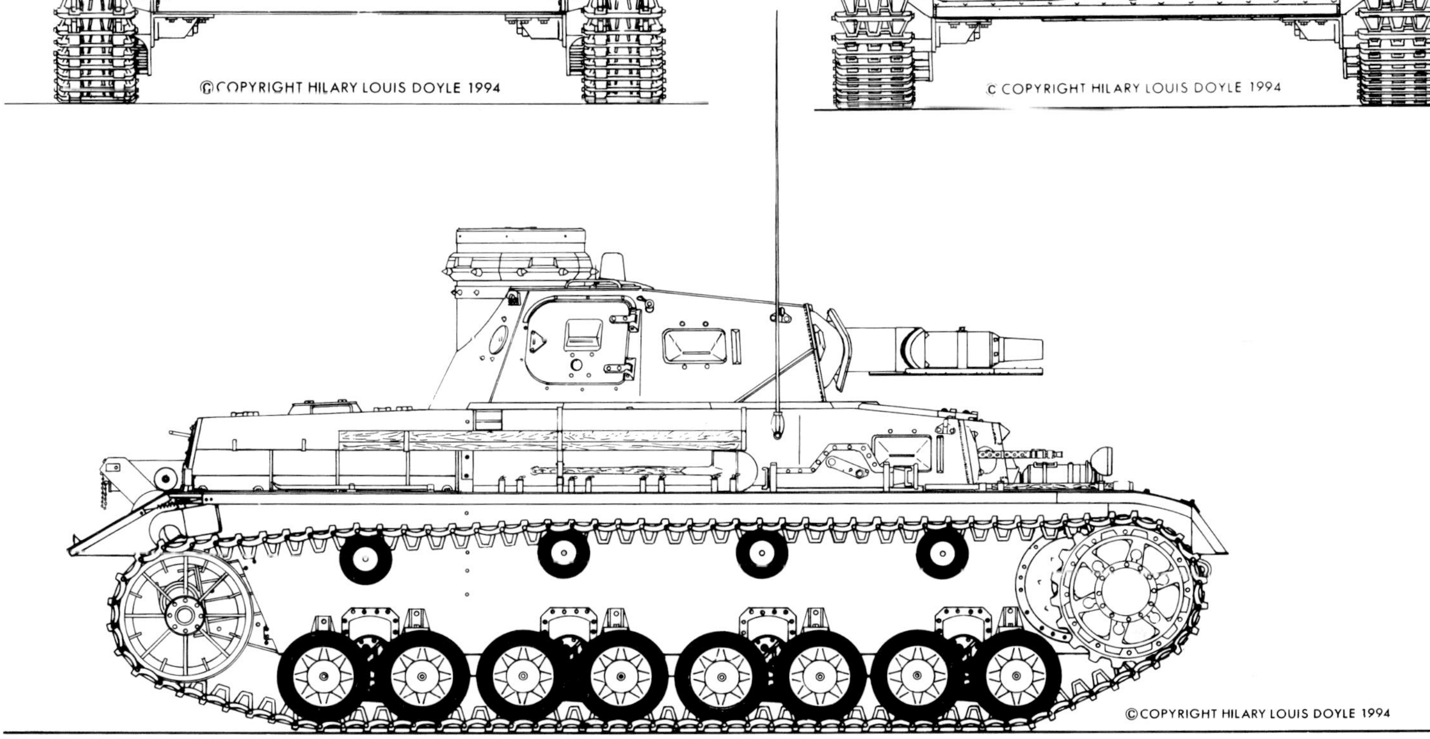

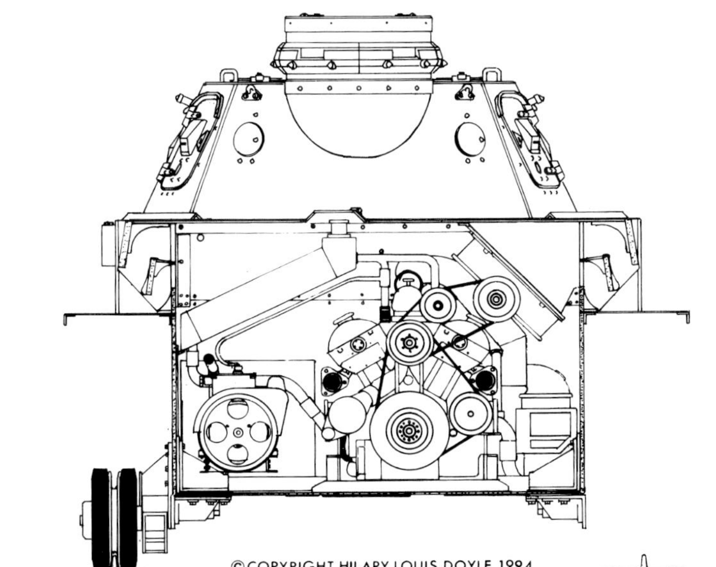

Panzerkapmfwagen IV (7.5 cm), (Sd.Kfz. 161), Type D. New hull with 30 mm bow plate (4./B.W.), 50 mm bow plate on Type 5./B.W., side armor of hull 20 mm. New body, angled layout of driver's front plate, return of Ball Mantlet 30 for radioman's M.G. and new side vision ports. New motor compartment cover, improved running gear with welded-on strike blocks before road wheels 3, 5, and 7. Type Kgs 6111/380/120 tracks with raised guide teeth. New turning turret with external roller mount and new vision slits. The cutaway drawings show the SSG 76 gearbox. Unlike the previous model, less ammunition supply, introduced as of Type B. Cuts through the engine room show the gasoline-electric machinery with DKW motor on the left side of the Maybach HL 120 motor, introduced as of Type C. The cooling air was sucked in through the radiator unit in the corner on the right side of the motor. The power for the radiator unit came via a V-belt from the left.

Commander's cupola and signal flap on turret roof of Pz.Kpfw.IV Type D (at Baunholder, April 2, 1940). Note also the commander's black padded cap. This characteristic headgear was used by the armored troops until 1941. It consisted of a passed head protector and a removable black cover.

Plate	Thickness	Inclination
Front wall	30 mm	80 °
Mount cover	35 mm	curved
Sidewall	20 mm	65 °
Back wall	20 mm	65 and 90 °
Roof	10 mm	0 and 7 °
Commander's cupola	30 mm + cast slide	

7.1.4.3 Control Apparatus for Turret-Turning Gear

Until Type C the control apparatus was mounted in a housing of sheet steel, which was closed by a cover.

[text continues on page 134]

It was attached to the horizontal flange of the housing of the turret turning apparatus. As of Type D the control apparatus was housed in a cast silumin alloy housing that was closed tightly by a cast silumin cover, flattened on one side. In the housing was a set of condensers to remove interference.

7.1.4.4 Vision Flaps

The vision flap (as of Type D) was a rectangular plate with a stepped offset rim. The plate lay with this rim on the armor wall. To protect against lead spray, the flap was set a few millimeters into the armor wall. By the vision port with the slit a vision slit, 8 mm wide, was cut. The vision flap was shaped like a roof on the outside, so that lead spray and striking bullets were mostly deflected from entering the vision slit. To protect the eyes from lead spray, splinters and rifle bullets that came through the vision

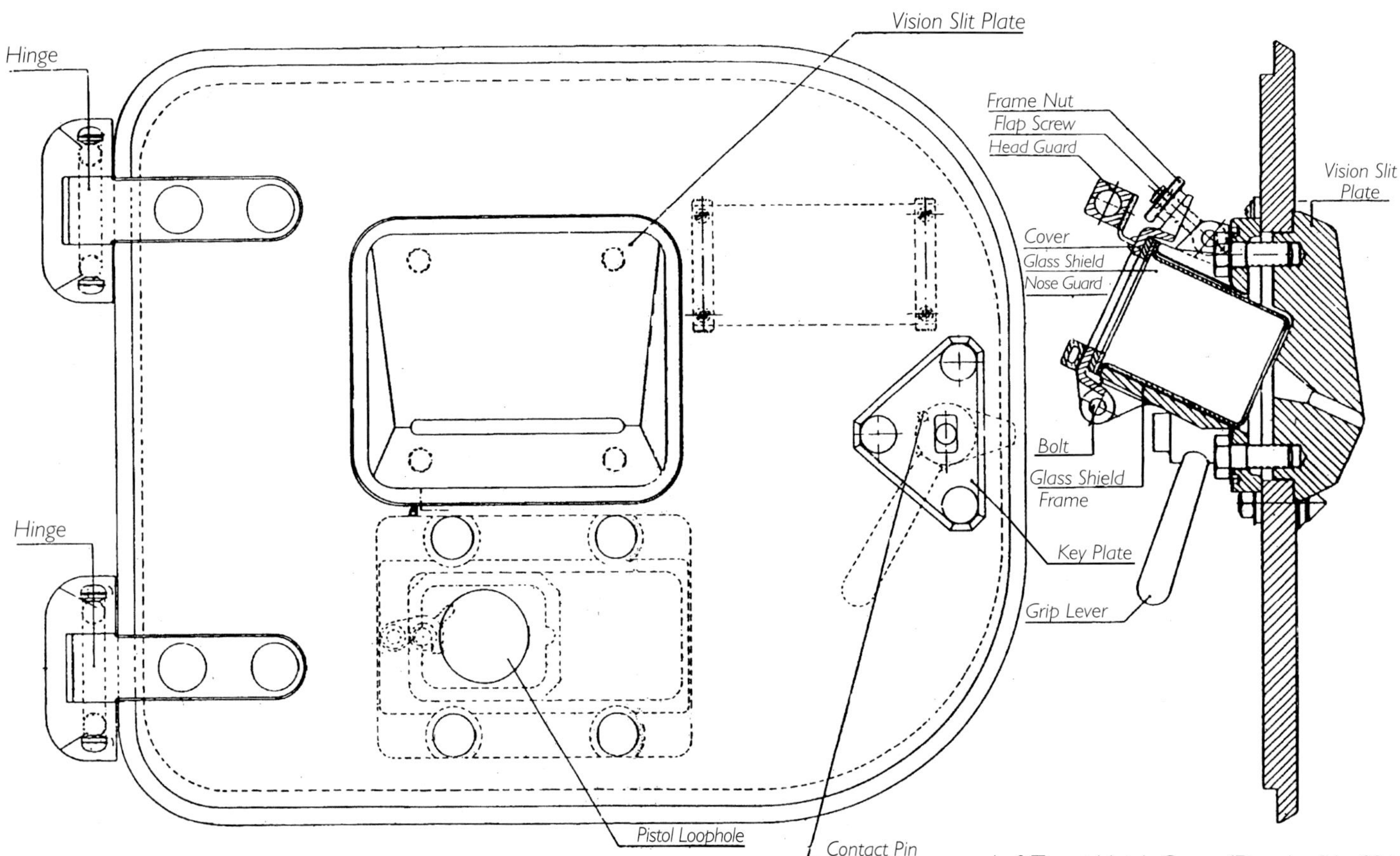

Left Turret Hatch Cover (Drawing No. 021 B 14881 U6) for Pz.Kpfw.IV Types D and E.

Pz.Kpfw.IV Type D with the 10th Armored Division.

Pz.Kpfw.IV Type D with Panzer Regiment 7 (Baumholder, April 2, 1940).

Pz.Kpfw.IV Type D, taken before March 1940.

Pz.Kpfw.IV Type D, tactical number 333, on the march somewhere in the East.

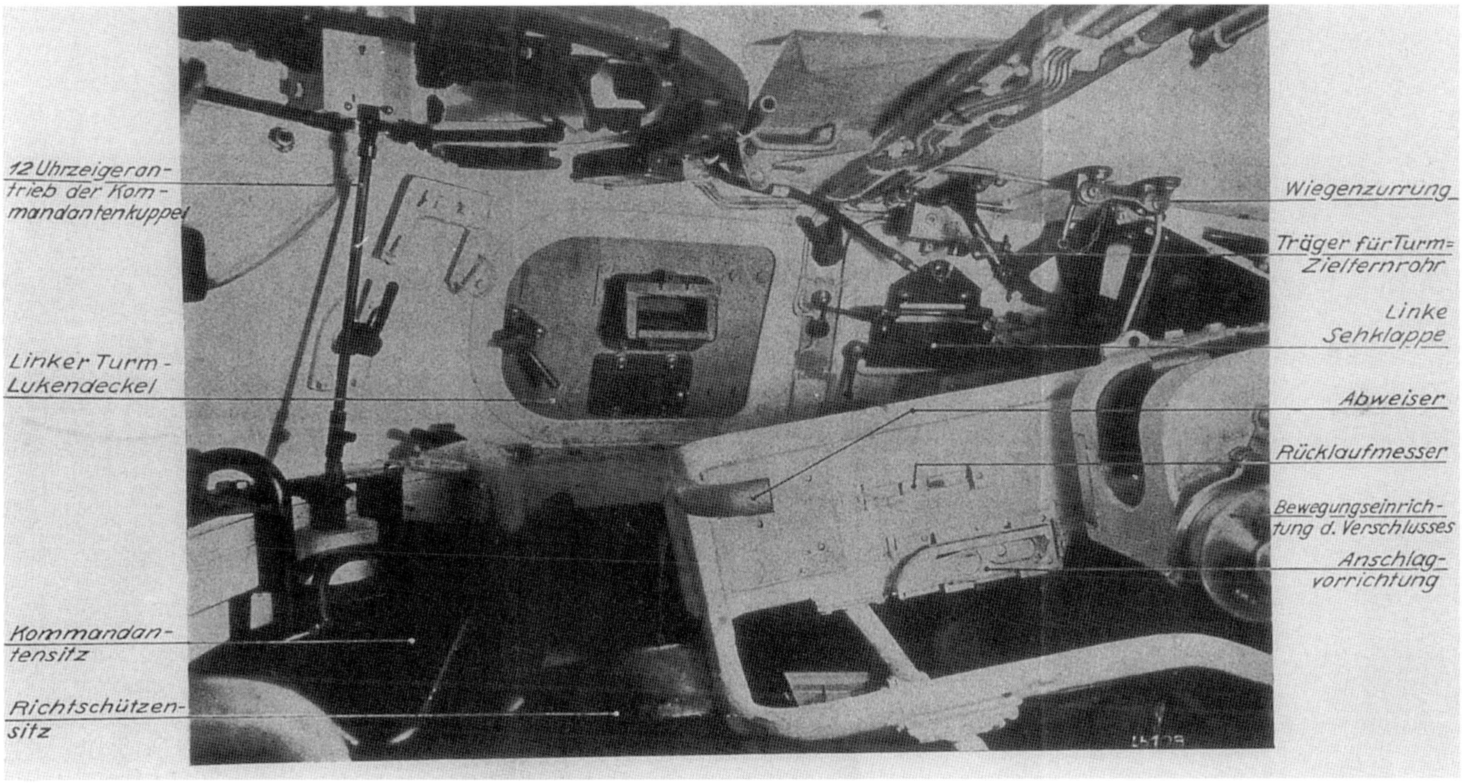

Pz.Kpfw.IV Type D, interior of the turret, Turret No. 80705, a look through the right hatch at the aiming gunner's and commander's seats.

Pz.Kpfw.IV Type D, interior of the turret, Turret No. 80705, a look through the right entrance hatch at the tank gun.

Pz./Kpfw.IV Type D, interior of the turret, Turret No. 80705, a look through the left entry hatch at the tank gun.

slit, a glass shield, 90 mm thick and easily replaced, was set up behind the vision slit. The glass was set in a frame and was held by a rubber-lined cover with a bolt.

7.1.4.5 Equipment Storage in Type D (and Type E)

In Type D (and Type E) a speaking tube was located on the left side under the roof for communication between the driver and the aiming gunner.

In the turret were the following storage places for equipment:

- At right front on the carrier ring, the holders for spare M.G. barrels and two glass shield holders.
- On the floor of the turntable, a container for bullet drums.
- On the footrest for the tank driver, two holders for gas masks.
- On the back wall to the right, a holder for a field flask, behind the commander's seat a holder for double headphones, and at left a holder for a flare pistol.
- At the left rear on the carrier ring, a container for flare ammunition and the rack for targeting rail holders.
- At the left rear brace of the turntable, a signal-flag rack and a holder for field flasks.
- At the right rear brace of the turntable, the glass shield holder with a holder for field flasks, plus a holder for gas masks and one for unloaders.

7.2 FORM CHANGES DURING PRODUCTION

7.2.1 Motor

The attachment flange of the exhaust pipe was modified on the Maybach 120 TRM motor delivered to the Gruson Works in September 1939 (with screws welded on). For assembly reasons, this change was not possible. The Gruson Works urgently requested that new drawings of the changes to attachment parts be sent.

7.2.2 Armor Strengthening

From the experiences gained in the Polish campaign, the armored troops had asked for strengthening of the armor plates as protection against antitank guns as early as September 1939. It was not at all simple to fulfill such wishes. The automotive and running-gear design of the Pz.Kpfw.IV was based on a vehicle weight of 18 tons. It had to be tested again to see whether the tank could stand the added weight of thicker armor plates. On October 24, 1938, a test was made, in which Chassis No. 80315 was driven 460 km with 20 tons of weight. For further testing, a chassis with a body without a turret, weighing 20 tons, and with heavier armor, was ready for road tests at Altengrabow in November 1939. At a discussion on 14-15 June, it was reported that the road test of 1967 km on the Autobahn with Type D was made without problems at 25 kph with a weight of 21.4 tons, with strengthened armor. On July 8, 1940, it was reported:

Test runs in Kummersdorf with the vehicle brought up to 21.4 tons gave no problems with the rubber bandages on off-road terrain; speed on the road will probably be limited to 25 kph.

On December 18, 1939, Wa Prüf 6 requested from Krupp of Essen:

Front walls of the last 68 hulls for 4th and 5th series, 50 mm thick homogeneous walls of PP794 to be used. The first 180 hulls of the 4. and 5./B.W. will be delivered with 30 mm cemented walls.

Further details as to strengthening were announced on May 15, 1940:

1) *On the 4. and 5./B.W. strengthening of the hull bow is already introduced.*
2) *Things tested were:*
   a) The strengthening of the bow front plate by applying two pieces of 30 mm surface-hardened plates.

Pz.Kpfw.IV Type D, built from July 1940 on. The driver's front plate was strengthened with two added 20 or 30 mm armor plates.

Pz.Kpfw.IV Type D, built from July 1940 on. Sidewalls are strengthened with 20 mm thick armor plates.

Pz.Kpfw.IV Type D, Chassis No. 80597, finished in April 1940, with the front wall of the hull thickened afterward. The hull sidewalls and bow armor (front and side walls) have added armor plates.

b) Strengthening the bow sides with 20 mm homogeneous plates.

c) Strengthening the hull sides with 20 mm homogeneous plates.

3) By attaching the added plates in the course of assembly, the production of vehicles will not decrease. When the Gruson Works deliver the new batch with added plates as of July, 185 vehicles of the previous production remain from the 4th series, which are to be strengthened later by the troops. A wooden model was built for later strengthening of the bow armor at the Gruson Works by May 28, 1940.

Krupp of Essen reported on 12 June that the first 100 armor plates for strengthening the front hull wall would still be delivered in cemented form, since this material was still on hand. The rest would be delivered with homogeneous armor steel without surface hardening. The armor strengthening of the driver's front of the bow of Type D consisted of two plates 240 mm apart; the sidewalls were held on by countersunk pointed-head screws. On July 5, 1940, Krupp had to notify Wa J Rue 6 that the Gruson Works would deliver the 26 B.W. vehicles to be delivered in July with strengthened armor. In this discussion it was also stated that no more repaired vehicles would leave the works without having the armor for strengthening the front hull wall of the Pz.Kpfw.IV applied.

Krupp of Essen answered inquiries from Wa Prüf (July 10, 1940) about delays in delivery of the armor-strengthened Pz.Kpfw.IV:

The punctual delivery of the Gruson Works with strengthening parts would have been possible without problems if the Department had not repeatedly postponed the decision as to the design version of the strengthening plates. Only on June 25, 1940, thus at a time at which the delivery should have begun, was the decision shared with us, and only after receiving these drawings could the production be taken up. Despite this, we have achieved through special means that today the Gruson Works could have the first set of strengthening parts delivered by truck, except the strengthening plates for the front of the hull, which will follow in a few days.

On 23 July combined statistics stated the following Pz.Kpfw. IV armor strengthening: *(See table: next column, top)*

Series	1.(A)	2.(B)	3.(C)	4. & 5.(D)		6.(E)	
	added later			later	original	later	original
I. Front Hull Wall							
Number	35	42	140	180	68	—	223
Previous thickness	15	30	30	30	50	—	50
Strengthening	30	30	30	30	—	—	—
Total thickness	45	60	60	60	50	—	50
I. Front hull wall							
Number				248		223	
Armor thickness				20 (+20)		20 (+20)	
II. Bow Sidewall							
Number				248		223	
Armor thickness				30 (+30)		30 (+30)	
II. Bow Sidewall							
Number				248		223	
Armor thickness				20 (+20)		20 (+20)	

7.2.3 "Notek" Night Marching Device

On February 1, 1940, was given the job of installing the night marching device experimentally on a Pz.Kpfw.IV. The camouflaged headlight was to be mounted in the middle of the vehicle, the stop light behind. The headlight was to be carried in a padded case in the vehicle.

On April 11, 1940, it was further decided:

A set of parts was attached to a vehicle experimentally. The backup and stop light is to be attached to the track apron for uniformity with Pz.Kpfw.III. The cat's-eye can be eliminated after the introduction of the night marching device.

7.2.4 Conversion From Drum To Belt Feed For M.G.

On June 5, 1940, Wa J Rue (WuG6) VIIa requested in a message to the Gruson Works:

It is advisable that the conversion from drum to belt feed for M.G. must already be carried out on the Series 4./B.W. and 5./B.W. vehicles still to be delivered.

You are therefore requested to carry out this conversion to the vehicles arriving for delivery in June, and in fact:

For 4./B.W. as of Chassis No. 80630, and

For 5./B.W. as of Chassis No. 80726.

The belt bag served to hold the bullet belt with 150 bullets, plus to catch the empty bullet shells and empty belt after firing. In all, 21 belt bags with 3150 rounds of M.G. ammunition were stored.

7.2.5 Type D Chassis with Type E Body

To an inquiry from the Gruson Works, Krupp of Essen answered on June 3, 1940:

The 6./B.W. bow armor can be used for 4./B.W.. New 021St9288 driver's vision ports obtained at the right time.

Then one or more of the last Pz.Kpfw.IV Type D could be built with Type E bodies. One of these Pz.Kpfw.IV with mixed Type D and Type E parts, Tactical No. 800 of 8th Company/ Panzer Regiment 5, was captured by the British Army in North Africa and sent for examination to England, to the School of Tank Technology. In the S.T.T. report, the chassis number was said to be unknown, since the chassis number shield was already removed. This Pz.Kpfw.IV was built in September-October 1940 (radiator stamped 8./40 and 9./40). It had a Type D hull No. 80673, Type E bow and stern armor No. 80801, but without added armor in front of the driver, with Type D added armor before the M.G. and on both sidewalls, and Type E turret No. 80806.

7.2.6 Rebuilt Bridgelayers

After 20 Bridgelayer IVb (16 of them on Pz.Kpfw.IV Type D chassis) were finished by May 1940, Wa Stab Ia No. 1023/40 reported:

The foreseen bridgelaying tanks are given up. Their chassis are to be utilized to further increase the supply of Pz./Kpfw.IV.

Pz.Kpfw.IV Type D, chassis with Type E body, hull No. 80673, bow and stern armor.

Pz.Kpfw.IV Type D, Chassis No. 780732, built in July 1940, rebuilt with 7.5 cm Kw.K. 40, with armor plates and side aprons added later. Presumably photographed while used in driver training by the NSKK (National Socialist Driver Corps).

General of the Armored Troops Guderian inspects a Pz.Kpfw.IV Type D (diving tank) with the 2nd Company of Panzer Regiment 18.

This provides a further possibility of increasing by 16 Pz.Kpfw.IV by the end of July 1940.

Of the original Order No. 210-3-5602/38(H) for 200 armored parts for Pz.Kpfw.IV./B.W., Krupp of Essen had finished and delivered 200 turrets and ball mantlets, but only 184 armored upper bodies. Under a special contract, 16 upper armored bodies for the 16 Bridgelayer IVb on Type D chassis were built. After receiving information about the rebuilding of the Bridgelayer IVb into normal Pz.Kpfw.IV, Krupp of Essen had written to the Wa Amt on August 30, 1940:

Of the last 16 sets of armor parts, we have already sent the turrets and ball mantlets to the Gruson Works. We will finish the 16 armored upper bodies still to be delivered and likewise send to Magdeburg.

The first three bow and rear armors, No. 80685, 80686 and 80687, finished in 1941, according to Contract No. 210/35602/38(H), were sent by Krupp of Essen to the Gruson Works A.G. on March 10, 1941.

Wa J Rue (WuG6) reported: *29 finished Pz.Kpfw.IV in May 1941, 13 of them from the 4th Series B.W. according to the report of the Army Acceptance Depot.* Some Type D chassis were assigned to the building of ammunition tractors for the Karl Device.

7.3 IMPROVEMENTS AFTER DELIVERY

A series of improvements became ordered after the delivery of the Pz.Kpfw.IV Type D to troop units. Among others:

1) From spring 1940 on, attaching the "Notek" night marching device to the left track apron.
2) From June 1940 on, conversion from drum to belt feed for machine guns and installation of bullet sacks for belt feed.
3) From July 1940 on, strengthening the front hull wall, side walls and bow armor (front and side walls) with added armor plates.
4) From July 1940 on, 48 Pz.Kpfw.IV were rebuilt as diving tanks.
5) From January 1941 on, installation of changed shapes for ventilators and radiator flaps for tropic use. (First Pz.Kpfw.IV for Panzer Regiments 5 and 8 for North Africa, followed by further use of Pz.Kpfw.IV by armored units subordinate to the Army Group South in Russia.) Three sets were already sent to the Gruson Works at the end of January 1941, and eight sets were picked up by Panzer Regiment 5 on 31 January.
6) From March 1941 on, mounting of a baggage box on the back walls of turrets.
7) From June 1941 on, installing a coupling on the stern for a fuel trailer.
8) From April 1942 on, installing flange tubes for warm air heating of fighting compartments.
9) From July 1942 on, repaired Pz.Kpfw.IV with 7.5 cm tank guns were to receive a new long 7.5 cm Kw.K. 40 with a reinforced mount, and their chassis were to be brought up to the newest state.
10) From September 1942 on, installation of a cooling-water heating device (built by Fuchs).
11) From October 1942 on, installation of connections and vents for warm-water passage.
12) As of May 1943, mounting aprons on turret and hull sides.

Pz.Kpfw.IV Type D rebuilt as a diving tank with seals for underwater use.

Pz.Kpfw.IV Type D rebuilt as a diving tank with vents and exhaust pipes for underwater use. Baggage box on the turret. Hooked to the stern armor is a trailer with fuel tanks.

8. PANZERKAMPFWAGEN IV (7.5 CM) (Sd.Kfz.161) Type E (6./B.W.)

Chassis Number Series 80801—81006

On May 10, 1939, in a discussion with Wa Prüf 6, it was decided: 6./B.W. is to be designated Type E.

Of the 206 ordered Pz.Kpfw.IV Type E, 200 were finished and delivered between October 1940 and April 1941. Four Type E chassis were finished and delivered in January 1941 for a special bridgelayer test series. The last two Type E (Chassis No. 81005 and 81006) were to have box running gear.

8.1 IMPROVEMENTS AFTER TYPE D

In the 1938-39 business year report (1 October to 30 September) of the War Materials Department of Fried.Krupp A.G. it was stated the designing and task of building B.W. chassis. Bodies and turrets with 30/20 mm armor thickness (6./B.W. Series) for the Army Weapons Department was concluded:

The B.W. has proved itself especially well in the Polish campaign and should be built further in large numbers, perhaps also by a branch firm. As a further development, designs for a B.W.40 with six-wheel running gear and modified motor room are in the works. (See Chapter 14 about test vehicles)

The 6./B.W. (Type E) had the basic form and components of Type D, with improvements including the armor, driver's vision port, mantle for NKAV, back wall of the turret, built-on ventilator and commander's cupola. The most important improvements, with equipment descriptions, are presented in this chapter.

8.1.1 Armor

Four types of armor were used in the building of Type E:

1) Plates up to 14.5 mm of homogeneous nickel-free PP694 armor steel with 153 +/- kg/mm^2 firmness.
2) Plates over 14.5 mm of homogeneous nickel-free PP694 armor steel with 125 +/- kg/mm^2 firmness.
3) Front plates 30 mm thick of surface-hardened, cemented PP494 armor steel with 2.5 +/-5 kg/mm^2 firmness.
4) Plates 50 mm thick of homogeneous PP794 armor steel.

8.1.2 Chassis

8.1.2.1 Armored Hull

The armored hull consisted of armor plates with the following thicknesses and inclinations (from the horizontal):

Plate	Thickness	Inclination
Front wall, upper	20 mm	18 °
Front wall, forward	50 mm	76 °
Front wall, lower	30 mm	29 °
Sidewall	20 mm	90 °
Sidewall (center)	20 & 20 mm	90 °
Back wall, upper	20 mm	80 °
Back wall, lower	14.5 mm	16 °
Bottom	10 mm	0 °

From Type E on, the two flaps over the steering gear for the steering and support brakes were set in at the same height as the surface of the upper front wall.

8.1.2.2 Running Gear

A new toothed arc (without side holes) for the track movement was mounted. Improved road wheels with new caps for lubrication with grease were introduced.

8.1.3 Upper Armored Body

8.1.3.1 Bow Armor

The bow armor consisted of armor plates with the following thicknesses and inclinations (from the horizontal):

Plate	Thickness	Inclination
Front wall	30 & 30 mm	81 °
Sidewall	20 & 20 mm	90 °
Roof	11 mm	0 & 5 °
Back wall	20 mm	90 °
Bottom plate	8 mm	0 °

Panzerkampfwagen IV, Type E, finished in 1941 at the Krupp-Gruson Works in Magdeburg.

8.1.3.2 Driver's Vision Hatch 30 with Turning Slide

On May 10, 1939, it was decided: *For 6./B.W., for reasons of firing, the new driver's vision port with slide is not to be planned on, but rather the 6./Z.W. (Pz.Kpfw.III Type G) turning vision flap.*

The driver's vision port for Type E consisted of a housing screwed onto the front wall of the bow armor and covered by a turning slide. Behind the slide, to protect the eyes from lead spray, splinters and rifle bullets, a glass shield, 90 mm thick, was set into the housing and could be removed to the back. To catch lead spray, there were shield plates built into the left and right rear ends of the housing between the housing sides and the glass shield.

8.1.3.3 Stern Armor

The stern armor consisted of welded armor plates with the following thicknesses and inclinations (from the horizontal):

Plate	Thickness	Inclination
Roof	10 mm	2 °
Sidewall (angled)	20 mm	55 °
Sidewall	20 mm	90 °
Back wall	20 mm	80 °

From Type E on, a foglaying launcher (NKAV) with a protective mantle was attached to the back wall at the left.

8.1.4 Turret

The Type E turret had only one signal opening, on the roof to the left, while forward on the right side a round opening for the body ventilator had been cut and covered by an armor plate. The ventilation flap and the hood of the left signal flap were not present on Type E. The arched piece on the back wall under the commander's cupola was eliminated. The back wall of the Type E turret was made of a curved plate.

The hatch covers of Type E were held open by one cover lashing each.

8.1.4.1 Turret Housing

The turret armor consisted of welded armor plates with the following thicknesses and inclinations (from the horizontal):

Plate	Thickness	Inclination
Front wall	30 mm	80 °
Mount Cover	35 mm	curved
Sidewall	20 mm	65 °
Back wall	20 mm	75 °
Roof	10 mm	0 & 7 °
Commander's Cupola	30 mm + cast slide	

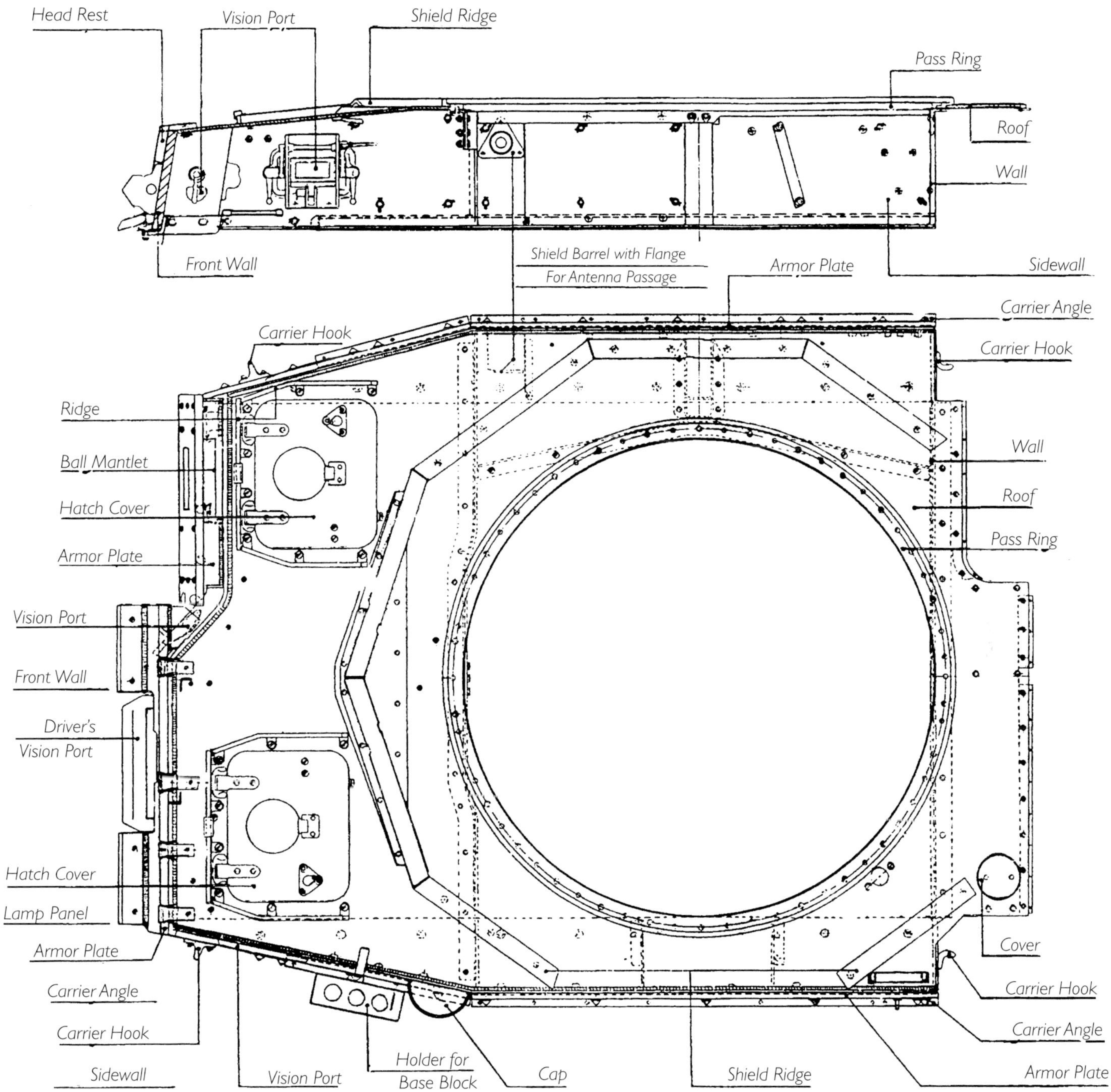

Bow armor (Drawing No. 021 A 33886) for Pz.Kpfw.IV Type E, No. 8081, and Turret No. 80806, finished in October 1940.

8.1.4.2 Commander's Cupola

On May 10, 1939, it was decided: *For 6./B.W. the 6./Z.W. (Pz. Kpfw.III Type G) commander's cupola is to be taken over.*

The commander's cupola consisted of the cylindrical mantle, the five lower and upper slides, the five side coverings, the numeral circle and the two-part hatch cover. The cylindrical mantle had five vision ports, each of which could be wholly or partly closed by two adjustable-height slides. To protect the eyes from lead spray and rifle shots, a glass shield was set behind every vision port and mounted in a shield frame.

8.1.4.3 New Motor for Gasoline-Electric Machinery

The Type E gasoline-electric machinery differed from the earlier version. It consisted of an Auto-Union DKW two-cylinder motor, Type ZW 500, and a generator. The DKW motor was newly available with a starting carburetor.

8.2 FORM CHANGES DURING PRODUCTION

8.2.1 DKW GENERATOR AGGREGATE

On October 19, 1940, during the anti-interference testing of the first 6./B.W. series vehicles by Wa Prüf 6, a big hole in the DKW

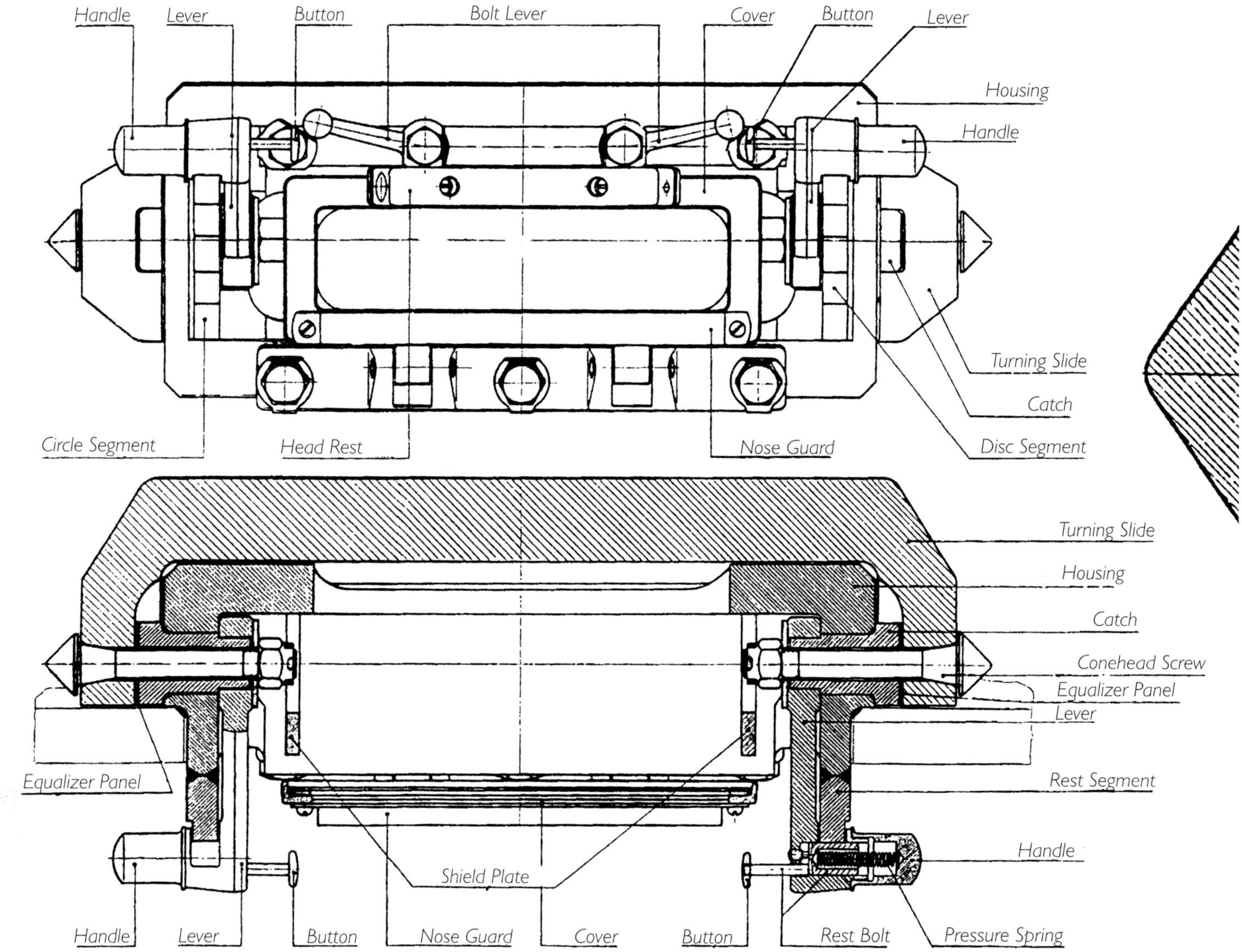

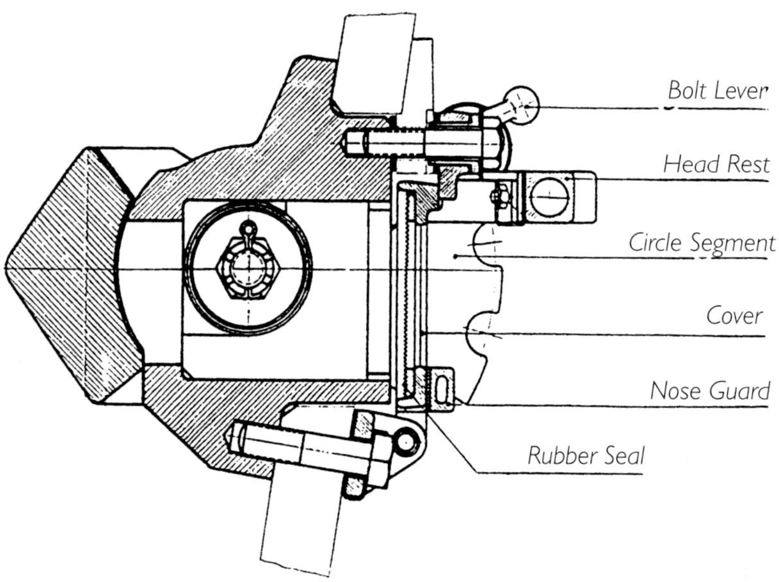

Driver's Vision Port 30 (turning flap), (Drawing No. 021 C 9288 U1) for Pz.Kpfw.IV Type E.

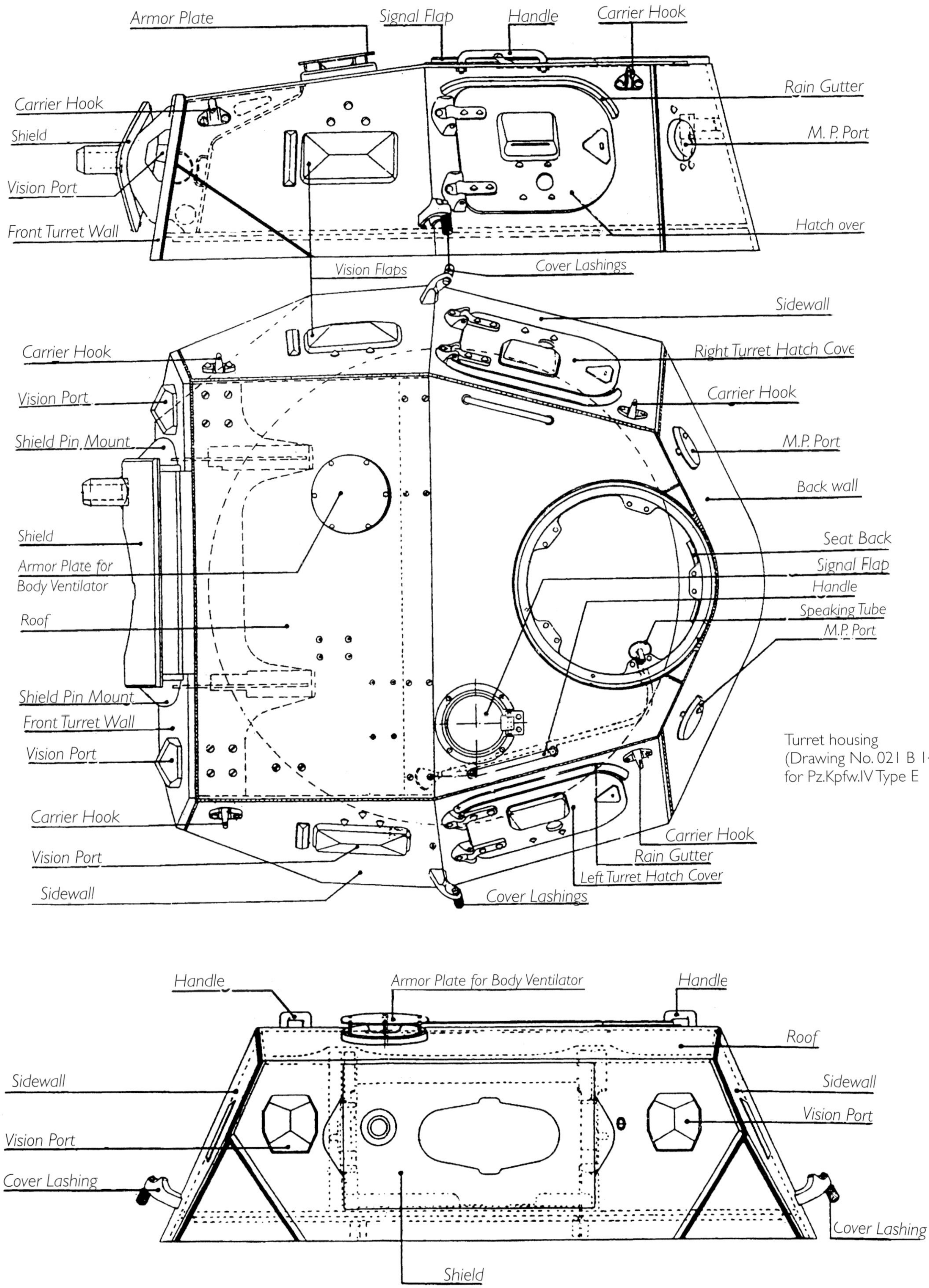

Turret housing
(Drawing No. 021 B 14881 U1)
for Pz.Kpfw.IV Type E

Commander's cupola (Drawing No. 021 B 39022) for Pz.Kpfw.IV Type E

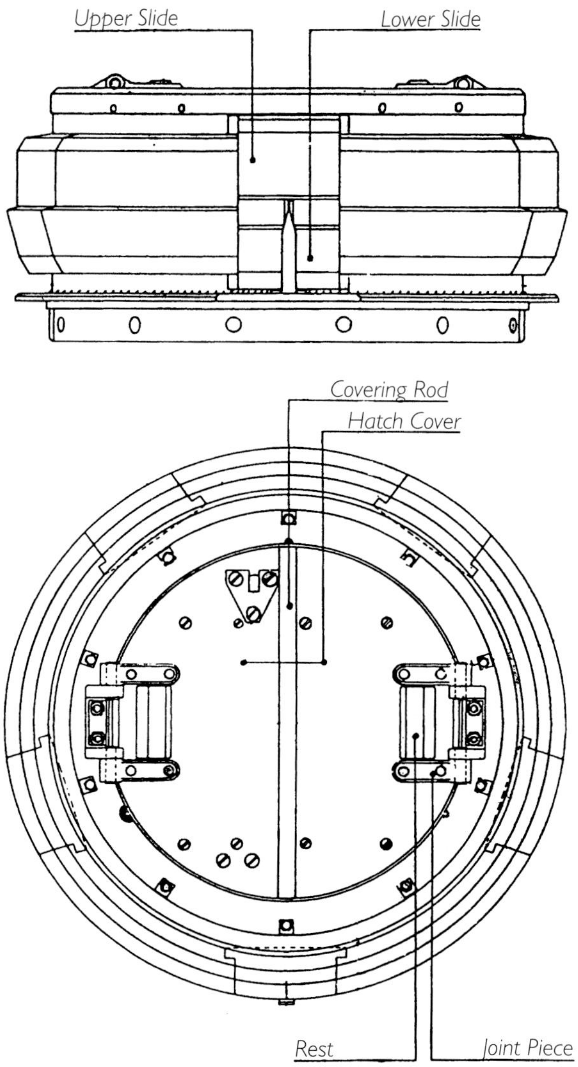

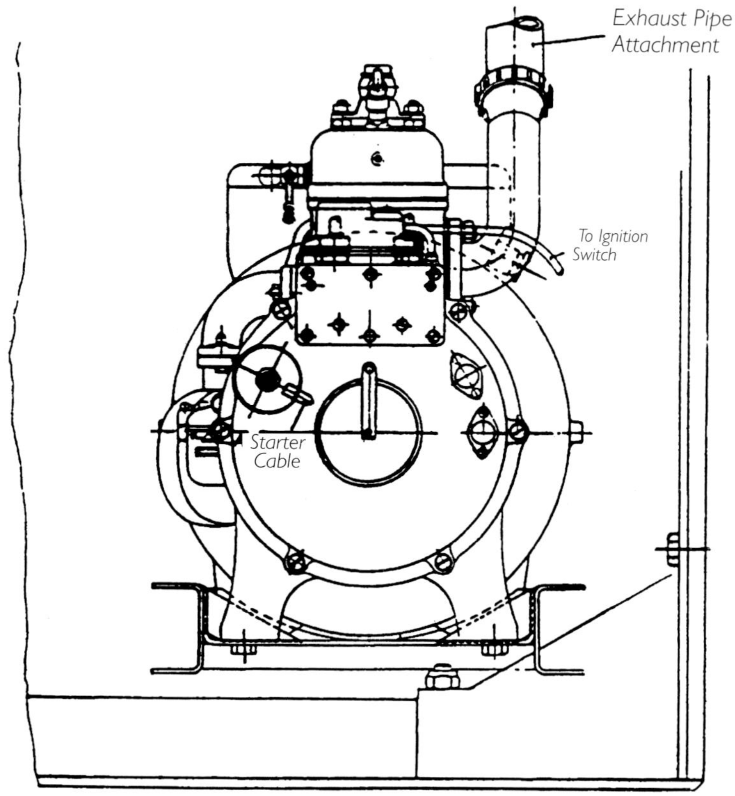

Gasoline-electric Machinery for Pz.Kpfw.IV Types E to H

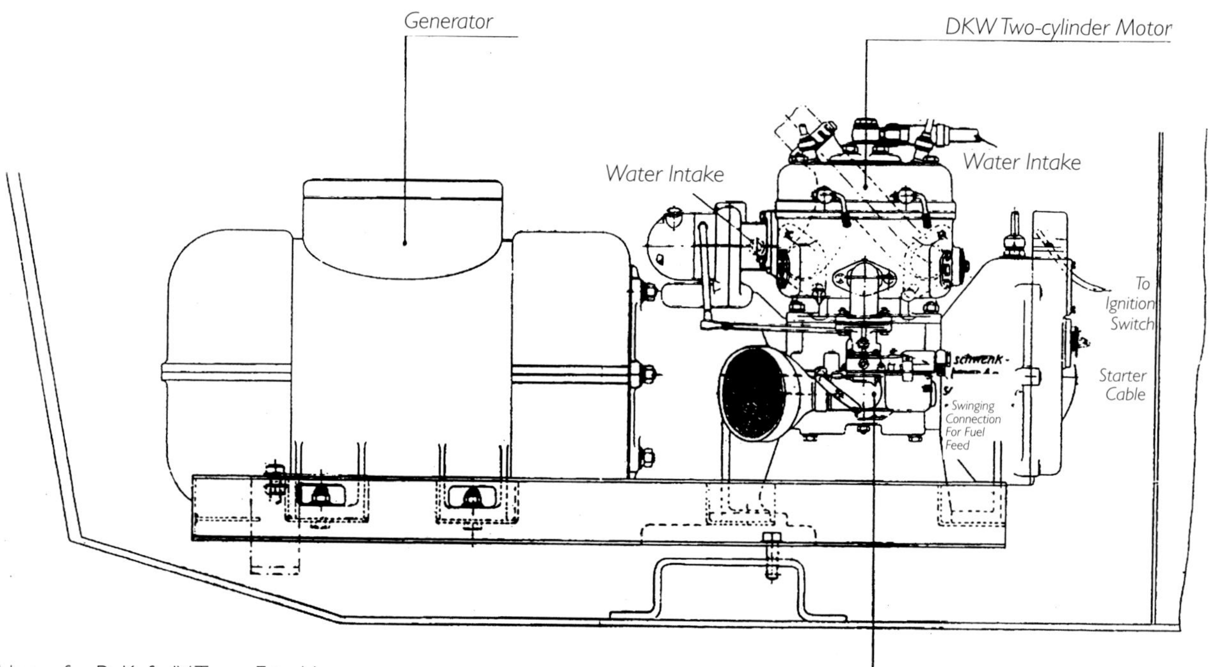

Gasoline-Electric Machinery for Pz.Kpfw.IV Types E to H

Commander's cupola of Pz.Kpfw.IV Type E. The picture was very probably taken during a drill. The commander and platoon leader, a lieutenant, waits for the order to action. The map case hangs out of the turret hatch.

generator aggregate was noticed. Thus, all DKW generators had to be removed, and none with that fault could be installed.

8.2.2. Organic Strengthening of Bow Armor to 50 mm

From the first Pz.Kpfw.IV Type E on, the armor hull received an organic 50 mm thick front wall. As to the 6./B.W. bow armor strengthening nothing was decided until June-July 1940, as was reported in a letter from Krupp of Essen to Wa Prüf 6/IId:

After the previous telephone talk, we suggest the following optionally for Type E as of Vehicle 51:

1) *Instead of subsequent strengthening, organic increase of the wall thickness to 50 mm as in the enclosed drawing Sk 3874. Angled shape remains in use for scheduling reasons, side view for driver is retained.*
   For scheduling reasons, already discontinued drivers' vision ports and ball mantlets for 30 mm walls must be retained and mounted inset 20 mm. As opposed to the subsequent attachment of 30 mm plates, in the suggestion above the added 30 mm protection of the ball mantlet and partly of the driver's vision port remain omitted. Please decide yet whether the resulting gap at the attachment place on the bow armor is to be protected with the hull by the ridge indicated in the drawing.
2) *Organic strengthening of only the driver's front according to Sk3874 to 50 mm, but adding a 30 mm strengthening plate only before the ball mantlet, according to Sk3874.*
   Side view for driver remains present, but disadvantage that this angled wall is less shot-secure.
   We ask for immediate decision by teletype.

On June 26, 1940, Wa Prüf 6 replied:

1) *Organic strengthening only on front of hull.*
2) *Bow armor strengthened by added plates as discussed for 5./B.W.*
3) *Driver's front plate 50 mm according to Sk3874 allowable only in connection with driver's vision port 021 St 33918.*

In a note on a discussion in Berlin on July 8, 1940, between Dipl.Ing. Biersack (Wa Prüf 6/IId) and Mr. Woelfert (Krupp Department A.K.) it was decided:

The bow armor strengthening was agreed on as follows: 6./B.W. soonest after 61st chassis:

Strengthening according to Sk 3874. Installation of 50 mm driver's vision port according to 021 C 33918 (assault gun). Installation of sliding driver's optics according to Daimler-Benz Drawing 021 St 39050 approximately 50 mm right of center. Installation of 50 mm ball mantlet according to Daimler-Benz Drawing 021 St 197 with somewhat smaller MG turning to right. If keeping schedule is difficult, if necessary retain 30 mm ball mantlet according to 021 St 33388, in case milling for installation on the inside is possible.

The abandoned driver's vision port according to 021 St 9288, driver's optics according to 021 St 8875 and MG ball mantlet according to 021 St 33888 are to be retained. The shortage of parts due to switching to organic version is to be mentioned to Wa Prüf 6.

On July 12, 1940, Special Branch VI stated, because of the security of the production program:

Krupp begins the conversion to organic strengthening of the 7./B.W. as of January 1941, while 6./B.W. in the present version with added armor will be delivered.

8.2.3 Ventilator and Radiator Flaps

As of February 1941, Pz.Kpfw.IV was prepared for action in hot climate (over 28 degrees) by the assembly firm with tropical equipment, including modified ventilator and radiator flaps.

8.2.4 Baggage Holder

On March 28, 1941, it was decided: *The baggage holder on the back wall of the turret is to be made as a dummy.*

8.3 LATER IMPROVEMENTS

8.3.1 Improvements after Delivery

A series of improvements was ordered after delivery of Pz.Kpfw.IV Type E to troop units. Among others:

1) As of January to March 1941: Some Pz.Kpfw.IV Type E were rebuilt as diving tanks.
2) As of February 1941: For Pz.Kpfw.IV Type E which went through repairing, mount the missing armor-strengthening parts.
3) As of February 1941, install the form change for ventilators and radiator flaps for tropical use.
4) As of March 1941, mount a baggage box on the rear turret wall.

Pz.Kpfw.IV Type E with turning Driver's Vision Port 30 without strengthened plates. Note the shot marks on the hull to the right of the radioman's machine gun.

Pz.Kpfw.IV Type E, finished in 1941 at the Krupp-Gruson Works in Magdeburg. Strengthened armor on hull and body sides as well as on bow.

Pz.Kpfw.IV Type E
with strengthened armor.

Pz.Kpfw.IV Type E rebuilt as a diving tank
with seals for underwater use.

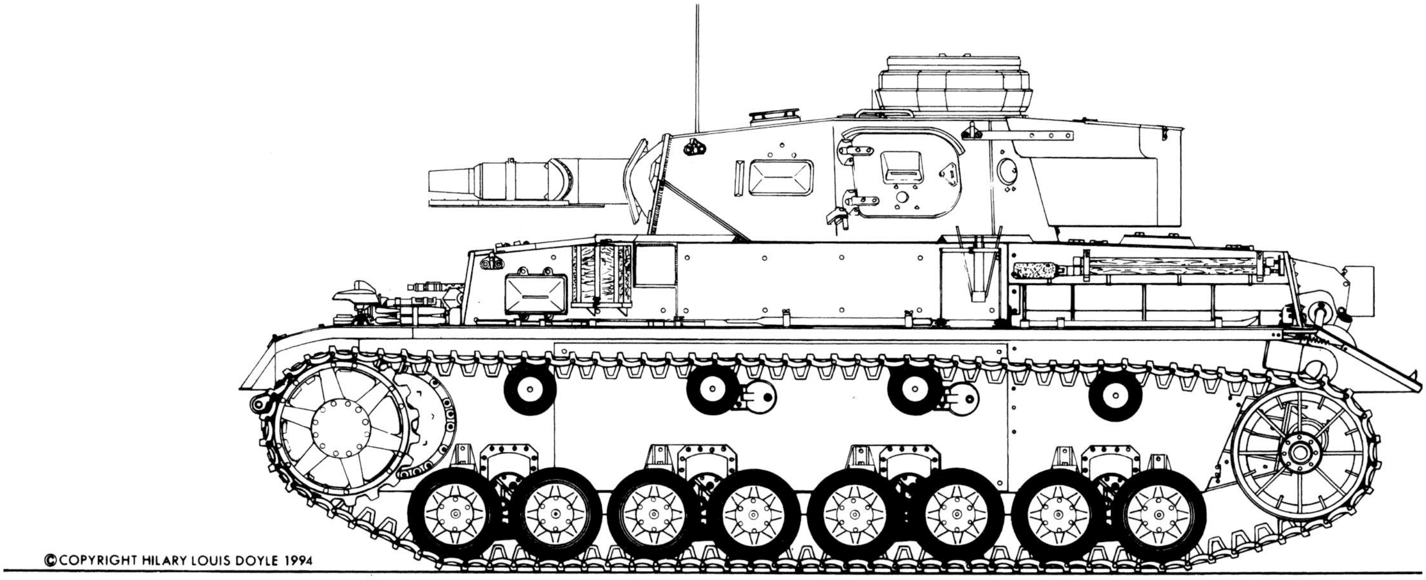
©COPYRIGHT HILARY LOUIS DOYLE 1994

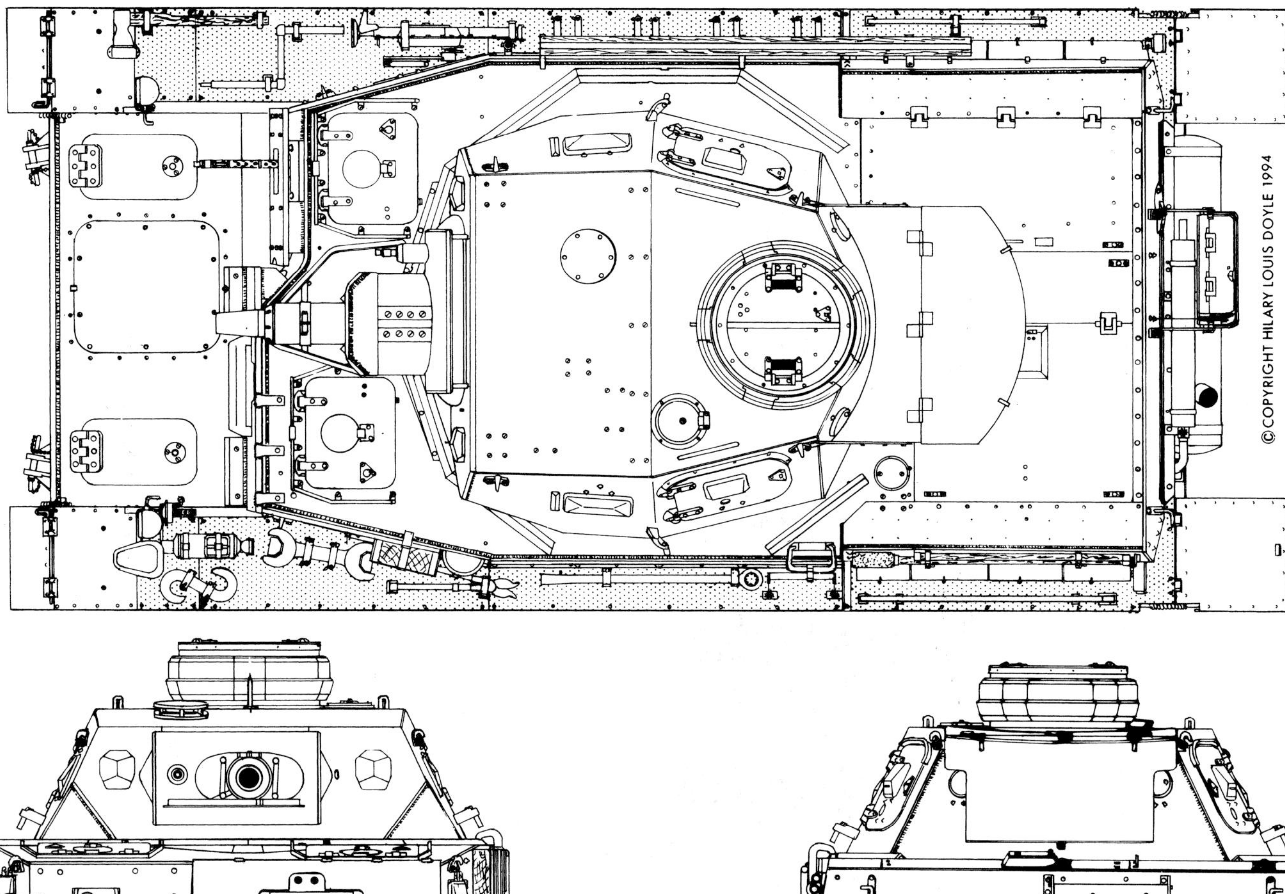
©COPYRIGHT HILARY LOUIS DOYLE 1994

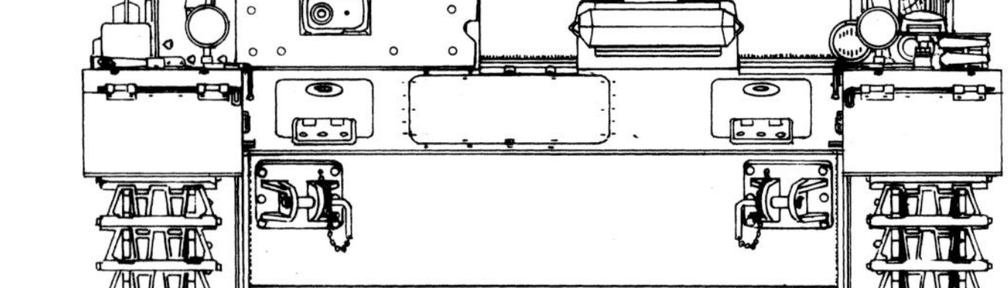
©COPYRIGHT HILARY LOUIS DOYLE 1994

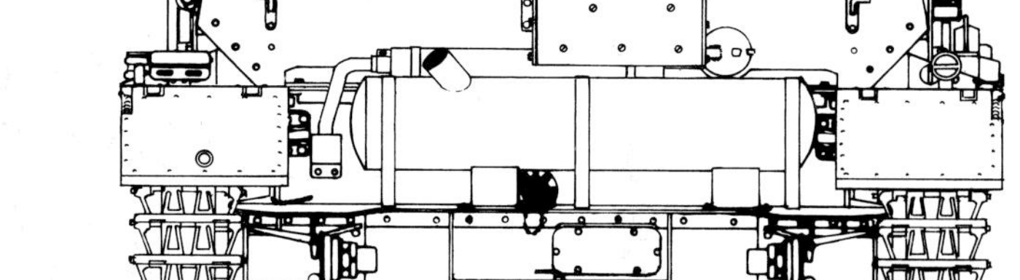
©COPYRIGHT HILARY LOUIS DOYLE 1994

5) As of June 141, install a coupling at the rear for a fuel trailer.
6) As of April 1942, install flanged ducts for warm-air heating of fighting compartment.
7) As of July 1942, repaired Pz.Kpfw.IV should receive a new long 7.5 cm Kw.K. with strengthened shield and chassis should be brought up to the latest state.
8) As of September 1942, installation of a cooling-water heater (built by Fuchs).
9) As of October 1942, installation of connections and vents for warm water passage.
10) As of May 1943, mounting of aprons on turret and hull sides.

8.3.2 Self-Made Form Changes Forbidden

From the General Army Messages of January 7, 1941, reform changes on tanks:

Counter to the often-repeated ban on self-made form changes to tanks, troops have undertaken so-called "improvements" to equipment, which is not allowed by the O.K.H.

In one special case, Panzerkampfwagen IV were strengthened on the front by an armored division by welding on 15 to 30 mm armor plates.

If such armor welding was not carried out in a tank assembly factory, then there is no guarantee of the good condition of the tank; in addition, the welding of unsuitable armor plates or even iron plates, as practical tests have shown, can lead to a reduction of shot security.

Form changes not approved by the O.K.H. are and remain forbidden; such later form changes are also not approved.

Contracts for equipment improvement resulting from particularly justified experience are, as before, to be presented to the O.K.H.

Pz.Kpfw.IV Type E with storage rack.

Opposite:
Panzerkampfwagen IV (7.5 cm), (Sd.Kfz.161), Type E:
New hull with servicing hatches for steering-brake adjustment, fitted with one-piece hinges. Body with new driver's vision port and armor protection for the foglaying apparatus.
Added screwed-on armor on the hull sides and body on several Pz.Kpfw.IV Type E. New turret with a curved back wall made of one piece. Opening for signaling on the right side of the turret roof omitted. An electric smoke-ejecting ventilator was built into the turret roof. New commander's cupola and baggage holder on back of turret. New drive wheels and cast steel caps for road-wheel hubs. The improvements carried out in the vehicle concerned, among others, the gasoline-electric machinery with DKW motor and generator. Notek night-vision lights were also introduced.

9. PANZERKAMPFWAGEN IV (7.5 CM) (Sd.Kfz.161) Type F, (7./B.W.)

Chassis Number Series 82001–82399, 82501–82564 and 82601–82613

In the 1939/40 business year report (October 1, 1939, to September 30, 1940) of the War Material Department of Fried.Krupp AG, it was stated that the design and task of building B.W. turrets with 50/30 mm armor thickness (7./B.W. series) for the Army Weapons Department had been finished. Further: *The B.W. has proved itself very well in Poland and in the West, and is being built in increasing numbers. Vomag and Steyr (Nibelungenwerk) are preparing for series production.*

Between May 1941 and February 1942, 470 Pz.Kpfw.IV Type F chassis were finished and accepted (393 by Krupp, 64 by Vomag and 13 by the Nibelungen Works). Some of them were built and accepted as ammunition tractors for the Karl Device and various testing purposes.

9.1 IMPROVEMENTS FROM TYPE E

In a discussion at the Gruson Works on June 15, 1940, it was explained: *7./B.W. (Type F) are only different from 6./B.W. in a few points (ca. 30%).*

The 7./B.W. (Type F) basically had the same form and components as Type D, with improvements that dealt mainly with thicker armor (50 mm in front, 30 mm on the sides), and among others, wider tracks, Driver's Vision Port 50 and Ball Mantlet 50. The most important improvements with equipment descriptions are given in this chapter.

9.1.1 Armor

As for the increased armor protection of the 7./B.W., Wa Prüf 6/IId decided to achieve it not only through thicker armor plate, but also through surface hardening of almost all front and side armor plates and parts. From a discussion with Building Advisor Rau (Wa Prüf 6/IId) and Mr. Korkhaus (Krupp Apparatus Building 3) on September 13, 1940, re surface hardening of the 7./B.W. series:

Of the plates that are to be surface-hardened for the 7./B.W. series, materials with too meager carbon content have been ordered and already delivered for the lower bulkhead of the hull, the front wall of the turret, the shield of the mount and the right and left corner sheets of the turret.

Herr Baurat Rau was informed that the firmness attained by surface hardening do not correspond with the requirements of his acceptance conditions.

He was asked to agree that these plates be installed with either homogeneous plates or with a surface firmness that is below his requirements. The following quantities are involved:

Lower bulkhead of the hull,	*30 mm thick*	*200 pieces*
Front wall of the turret,	*50 mm thick*	*164 pieces*
Right & left corners of the turret,	*40 mm thick*	*200 pieces each*
Shield of the mount,	*50 mm thick*	*171 pieces*

After receiving this information, Herr Baurat Rau could not at once declare his agreement that the plates would be used. He asked that of the occurring plate thicknesses, 30, 40 and 50 mm, ten plates of each be surface-hardened, and that he be informed of the attained firmness in a message via the representatives in Berlin. After the receipt of the document in Berlin, the use of the plates should be decided in a new discussion.

Further decisions were made on October 9, 1940, in a discussion with Building Advisor Rau and Mr. Mann (Wa Prüf 6) and Messrs. Lanz and Heerlein (Krupp):

...had the purpose of discussing the positions of plates and formed parts in detail that were to be surface-hardened for the 7./B.W. series. The hardness of the individual parts was discussed and can now be quickly determined in drawings, as far as they are still on hand. The parts for surface-hardening Appa 3 listed in a document of September 28 were accepted by Baurat Rau, with the exception of the housing for the driver's vision port, which will also remain homogeneous in the future. In the hull, the surface-hardening of the sidewalls shall be carried out as soon as possible. Also, hardness tests with the back wall of the turret in its present form shall be carried out, in order to determine whether in this form a trouble-free surface-hardening in the rounded central part is attainable. Also, the surfaces of the turret hatch covers are to be surface-hardened. For the turning slides the hardness degree has already been determined.

Panzerkampfwagen IV, Type F, finished in 1941 by Krupp-Gruson.

Whether individual plates from the stern still can be surface-hardened was not decided in the discussion. The individual drawings of the stern are still to be checked later.

The nickel-free PP 782 armor plates consist of a mixture of 0.4% C, 0.5% Si, 0.45% Mn, 0.02% P, <0.01% Cr and 0.2% Mo.

9.1.2 Chassis

9.1.2.1 Armor Hull

The armor hull consisted of armor plates with the following thicknesses and inclinations (from the horizontal).

Plate	Thickness	Inclination
Front wall, upper	20 mm	18 °
Front wall, forward	50 mm	76 °
Front wall, lower	30 mm	29 °
Sidewall	30 mm	90 °
Back wall, upper	20 mm	80 °
Back wall, lower	14.5 mm	16 °
Bottom	10 mm	0 °

On May 16, 1940, Wa J Rue 6 requested:

As of Type F, the side flaps for steering brakes in the upper part of the bow should remain closed and receive ventilation ducts.

9.1.2.2 Tracks and Running Gear

Because of the increased weight, Wa Prüf 6 decided that the tracks for Type F should be widened from 380 mm to 400 mm. This produced the Kgs.61/400/120 track type.

First, on May 27, 1940, the Continental firm had chosen widened tires for the road wheels, with 470 mm diameter and 95 mm width for the 7./B.W. Later, on June 13, 1940, Wa Prüf 6 decided on widening the running gear of 7./B.W.: 90 mm instead of 95 mm free bandages on road wheels. The road wheels, which had been ordered from the Continental and Vorwerk firms, had a size of 470/90-360, the return rollers had a size of 250/65-135. The front wheel was also enlarged to fit the new Kgs.61/400/120 track type. To simplify production, the new leading wheels were welded together from pieces of tubing.

9.1.2.3 Muffler

On April 11, 1940, Wa Prüf 6 decided:

For Type F the large exhaust muffler is to be shortened, and for the DKW motor a new flat muffler is planned, so that the NKAV can be fastened to the stern armor more purposefully.

The mufflers for the DKW motor were built in a compact box form instead of as a long pipe. To make room for the new DKW muffler on the back of the stern armor, a shortened exhaust muffler for the Maybach HL 120 TRM with a shorter intake was mounted on the back wall.

9.1.3 Upper Armored Body

9.1.3.1 Bow Armor

The bow armor consisted of armor plates with the following thicknesses and inclinations (from the horizontal):

Plate	Thickness	Inclination
Front wall	50 mm	81 °
Sidewall	30 mm	90 °
Roof	11 mm	0 & 15 °
Back wall	20 mm	90 °
Bottom	8 mm	0 °

As of Type F, the front wall of the bow armor consisted of one straight armor plate. Unlike Type D and Type E, the side vision on

New track type Kgs.61/400/120, factory-new Type F2f(G), with new drive wheels and road wheels for Pz.Kpfw.IV Type F.

Leading wheel of welded tubes for Pz.Kpfw.IV Type F.

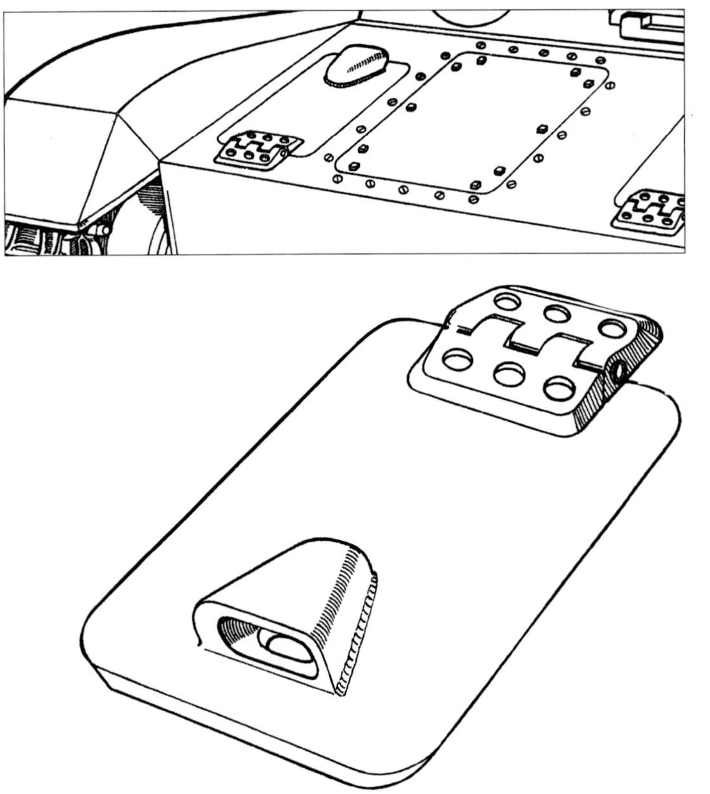

Ventilation ports on servicing hatches for the steering brakes on the upper part of the bow.

the driver's right was taken away. The two entry hatches located on the roof remained in their previous positions.

9.1.3.2 Driver's Optic K.F.F.2

On June 19, 1940, Daimler-Benz AG, Factory 40, in Berlin-Marienfelde, sent to Krupp of Essen, on order from the OKH Wa Prüf 6 IId, a Sketch 6019 of the spatial needs of the newly developed sliding driver's optic for the 50 mm front wall.

In the front wall, the two viewing holes of the driver's optic were drilled, with oval outside openings for a 65-degree lateral field of vision.

9.1.3.3 Driver's Vision Port 50

On June 14, 1940, it was announced that the sPak (heavy assault gun) driver's vision port, Type 021 St 33918 for Type F (7./B.W.) was to be changed from a screwed-in to a welded-on form.

The Driver's Vision Port 50 consisted of a housing screwed onto the front wall of the bow armor, which was covered by a turning slide. Behind the slide, to protect the eyes from lead spray, splinters and rifle bullets, a glass shield 90 mm thick was mounted in the housing, and was removable from the rear.

9.1.3.4 Ball Mantlet 50

The Ball Mantlet 50 consisted of the mantlet ball, the cover and the pass ring. The cover and the pass ring covered the gap between the mantlet ball and the armor wall and also formed the mount for the ball. The mount ball, movable in all directions in this mount, served as a mount for an M.G. 24 with armor mantle and belt feed, and a K.Z.F.2 ball targeting scope. The turning range of the ball was +/- 15 degrees from the central

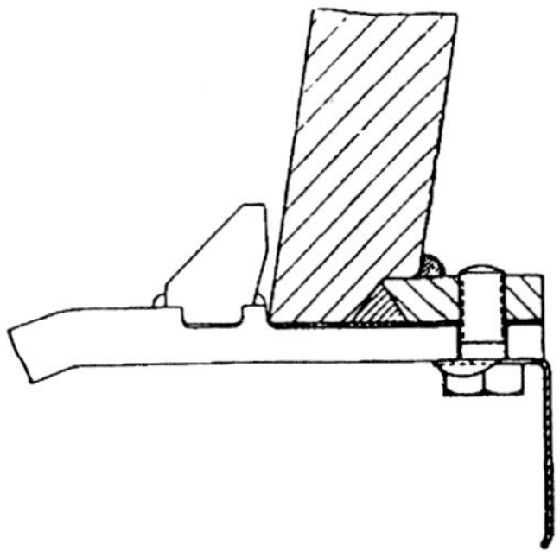

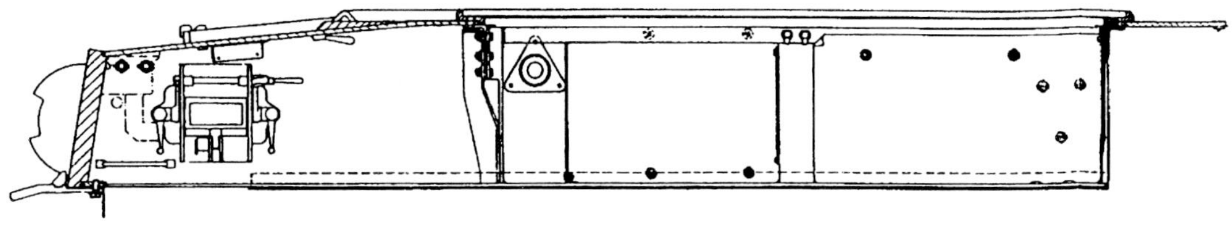

Bow armor
(Drawing No. 021 A 170 U1)
for Pz.Kpfw.IV Type F.

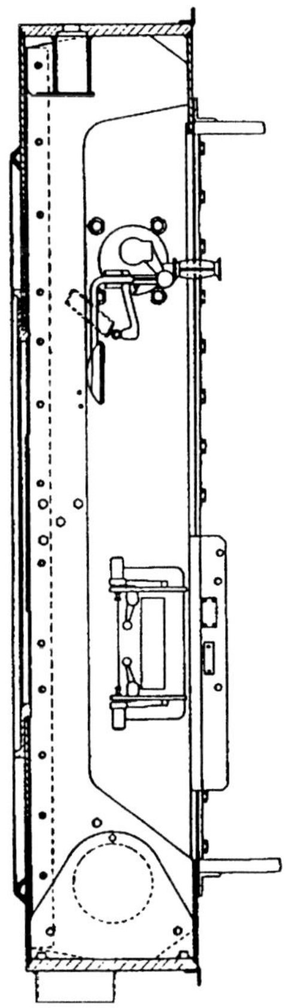

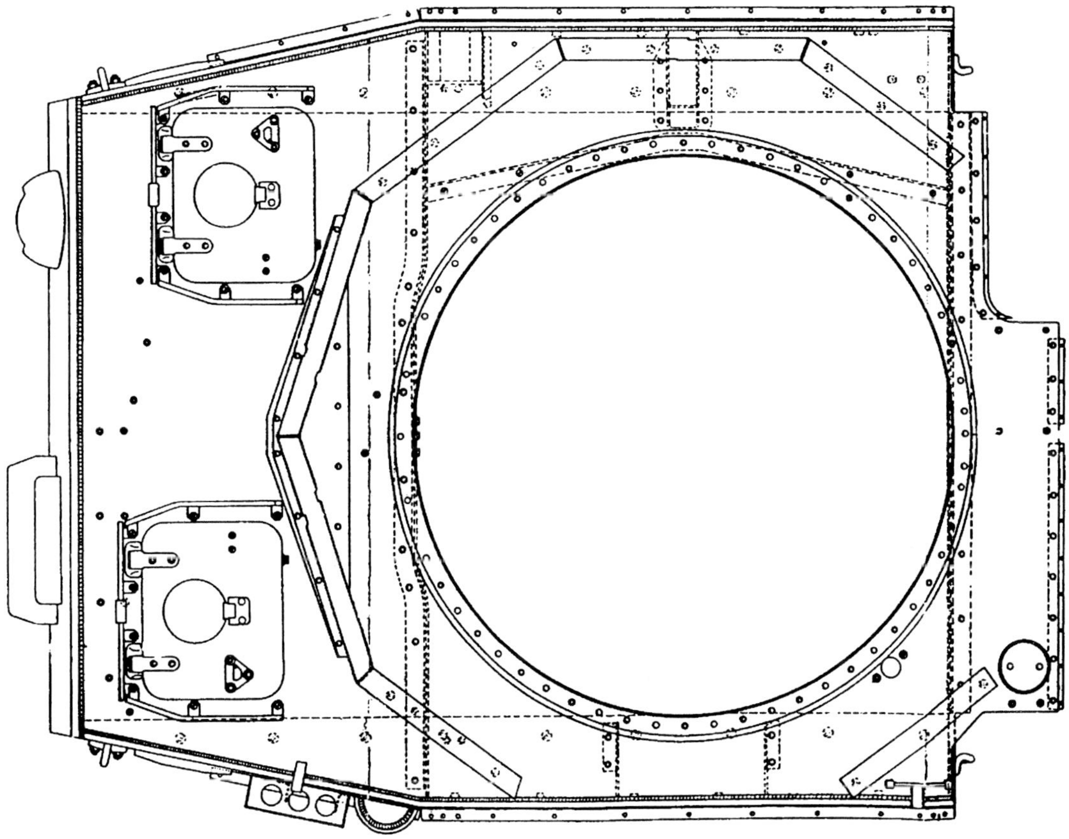

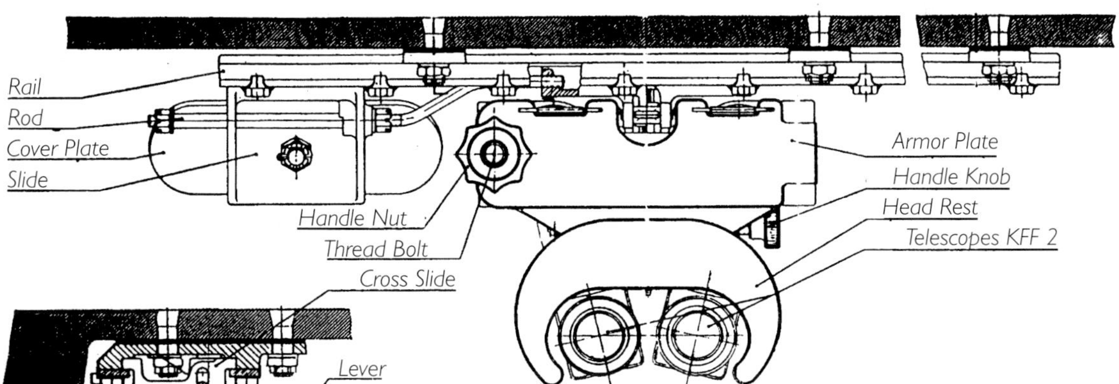

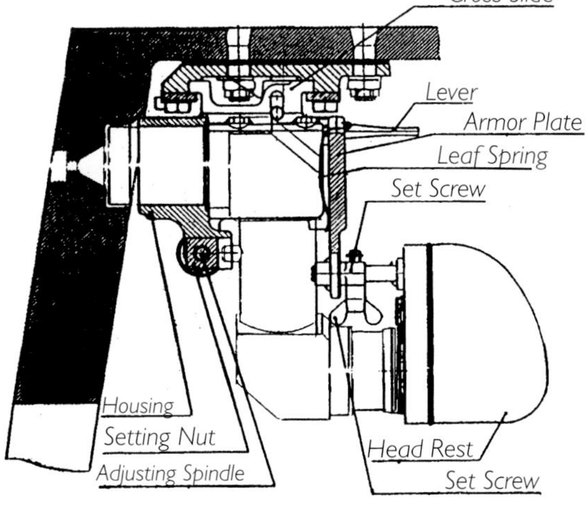

Driver's optics set for use (Drawing No. 021 B 172)
for Pz.Kpfw.IV Type F.

Pz.Kpfw.IV Type F, No. 82023, finished in June 1941, with Driver's Vision Port 50 and Ball Mantlet 50.

position to the side and +20 and -10 degrees up and down from the horizontal.

9.1.3.5 Stern Armor
The stern armor was retained almost unchanged from the earlier type.

9.1.4 Turret

9.1.4.1 Turret Housing
The turret armor consisted of welded armor plates with the following thicknesses and inclinations (from the horizontal):

Plate	Thickness	Inclination
Front wall	50 mm	80 °
Mount cover	50 mm	curved
Sidewall	30 mm	65 °
Back wall	30 mm	75 °
Roof	10 mm	0 & 7 °
Commander's cupola	30 mm + curved slide	

9.1.4.2 Machine Gun Mount (New Type)
As of Type F, a new sprung machine gun mount was mounted on the back wall of the roller mount.

9.1.4.3 Twelve O'clock Indivator (New Type)
As of Type F, a new twelve o-clock indicator with two dials was placed to the left on the turret carrier ring. Numeral Disc 1 on Dial 1 was calibrated from 0 to 100, and Numeral Disc 2 on Dial 2 from 0 to 64 and 0 to 12.

9.1.4.4 Hatch Covers
The turret hatches in the left and right sidewalls serve for entrance and exit. They were closed by two-piece covers (design of Pz.Kpfw.III turret taken over). The front flaps could be fixed firmly in the turret sidewall with a fork-shaped device on the lower bolts, so that a gap some 30 mm wide remained open. This device was intended for ventilation under limited safety from enemy fire.

On each front flap was a vision-slit plate with a slit about 8 mm wide, shaped on the outside so that lead spray and shots were mainly deflected from entering the vision slit. To protect the eyes from lead spray, splinters, etc., there was an easily replaced 90 mm thick glass shield situated behind the vision slit.

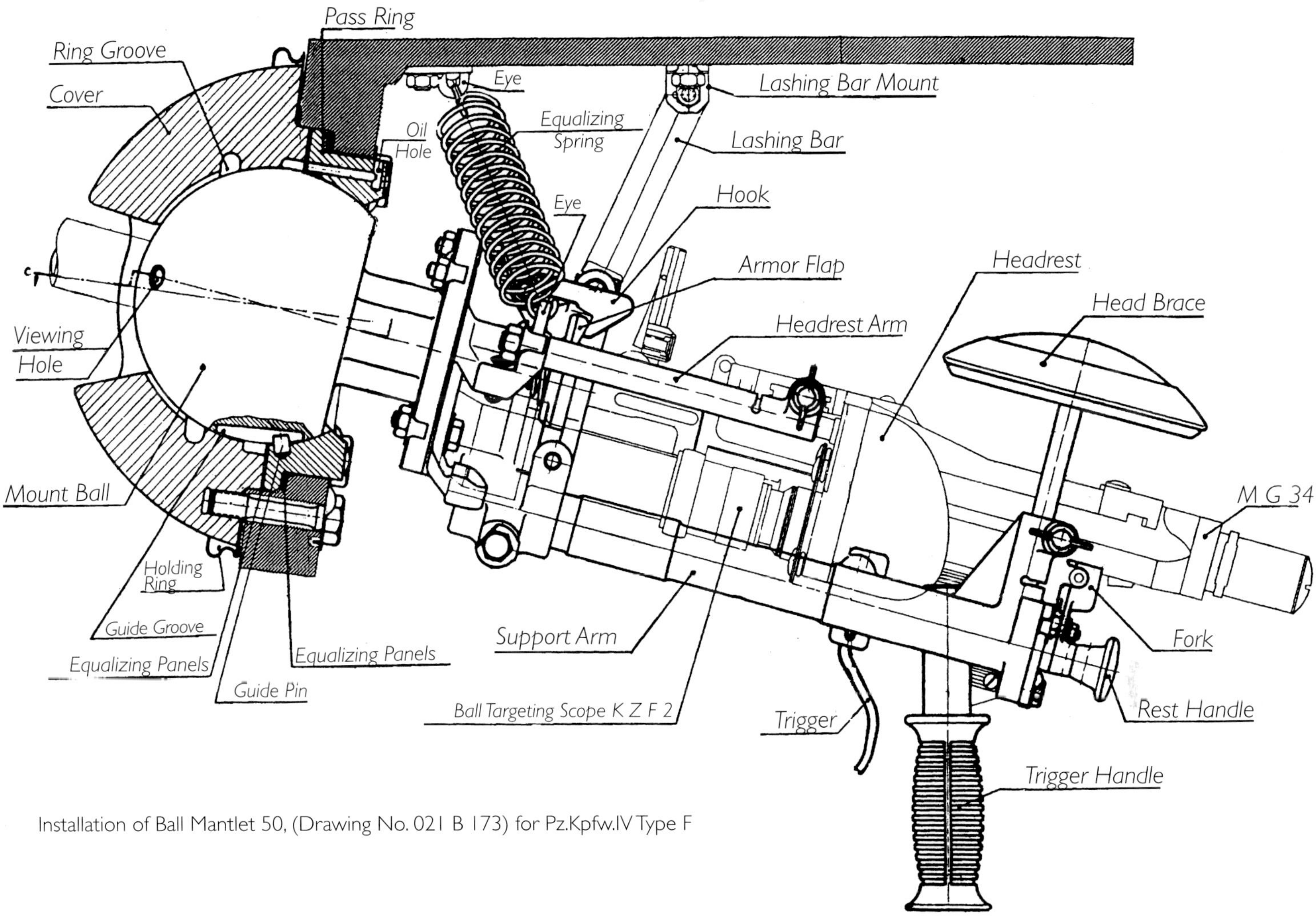

Installation of Ball Mantlet 50, (Drawing No. 021 B 173) for Pz.Kpfw.IV Type F

In each rear flap of the turret hatch cover a pistol loophole was built.

9.2 FORM CHANGES DURING PRODUCTION

During the entire building time, the Type F was produced almost without noteworthy changes. For higher outside temperatures faced in North Africa by the German Africa Corps, and by the Army Group South in the East, Pz.Kpfw.IV Type F was built with tropical equipment; among others, an added radiator intake was planned in the radiator flap and a cooling-air exit in the ventilator flap on the rear roof of the tank.

Type F, built from October 1941 on, had a rod on the right side directly before the front wall of the bow armor to brace the right track apron.

9.2.1 Fuel Trailer

On April 5, 1941, Wa Prüf 6 requested that Krupp of Essen make tank modifications as follows:

Additional carrying of fuel for tanks on trailers (Panzerkampfwagen IV Types E and F).

As a coupling between the vehicle and trailer, a suitable fork for the trailer hook by DIN KR4054 norm with a 40 mm bolt hole is to be used. The middle of the fork may be at most 700 mm over the roadway (if possible, 600 mm).

The coupling bolt must be able to be raised with the help of a line from inside the vehicle.

9.2.2 Fore-Armor

By teletype on July 7, 1941, to O.K.H., Field Marshal von Brauchitsch, from Keitel:

The Führer considers it probable, according to news that has reached him, that the British—presumably in the combat in North Africa—are aware of our armor-piercing ammunition and of the use of conical barrels to increase the penetrating power, and draws the conclusions for his armor-piercing weapons from that. The Führer therefore considers it purposeful that we strengthen our tanks additionally in future new production by fore-armor—set off from the main armor—in order to absorb the increased penetrating power of the British weapons. The weight increase and the loss of speed shall be taken in the bargain in the fuehrer's opinion. The Führer asks for opinions or suggestions.

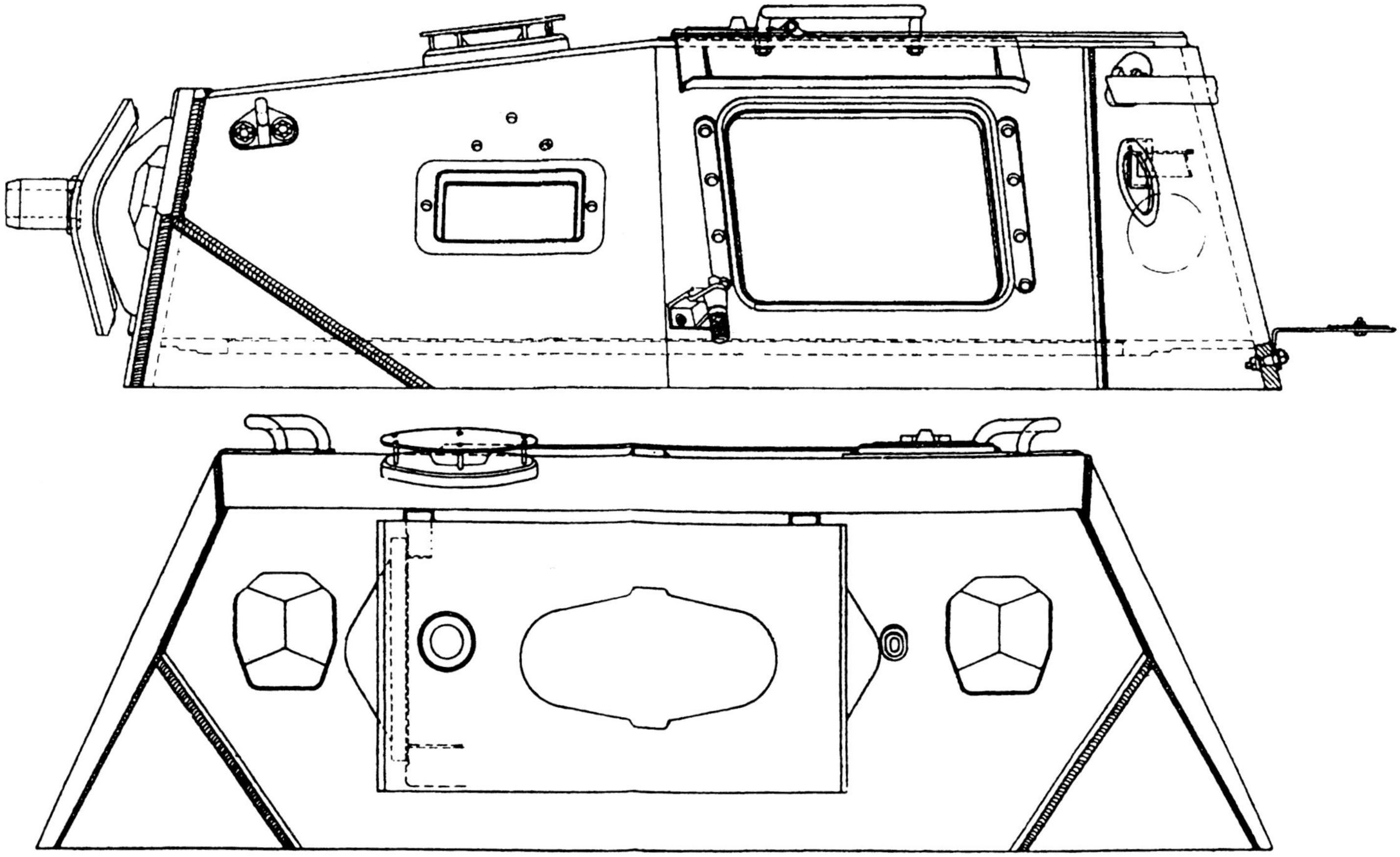

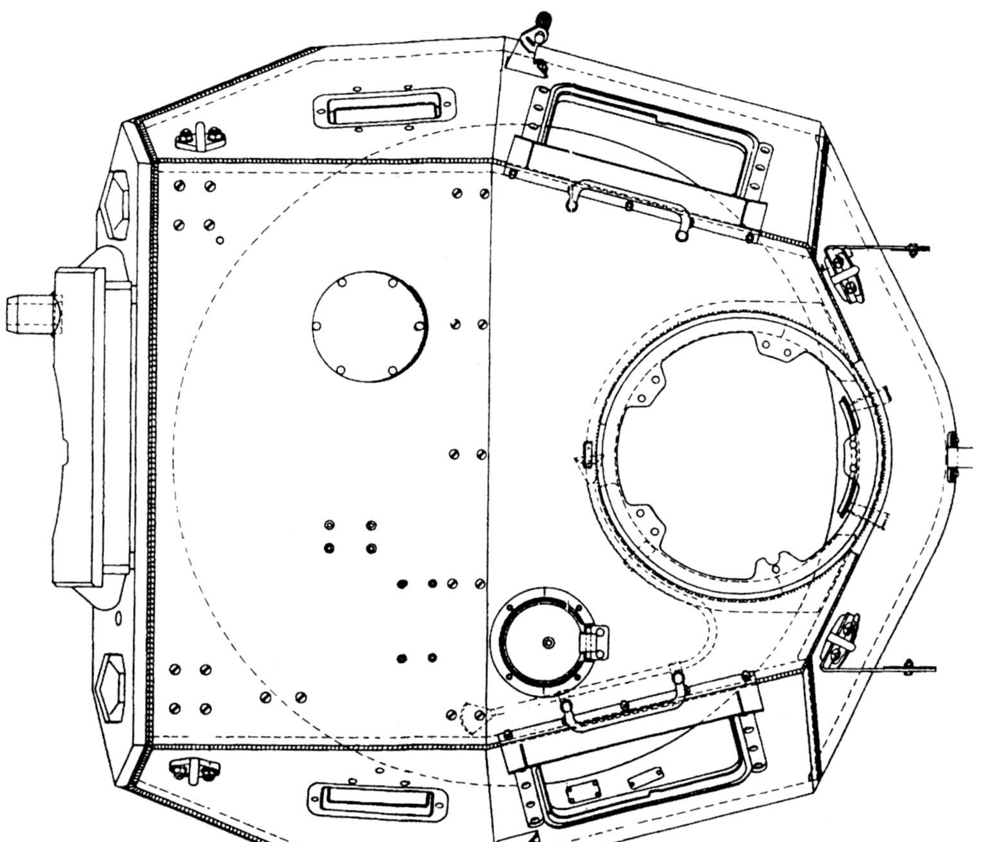

Turret (Drawing No. 021 B 181 U1) for Pz.Kpfw.IV Type F.

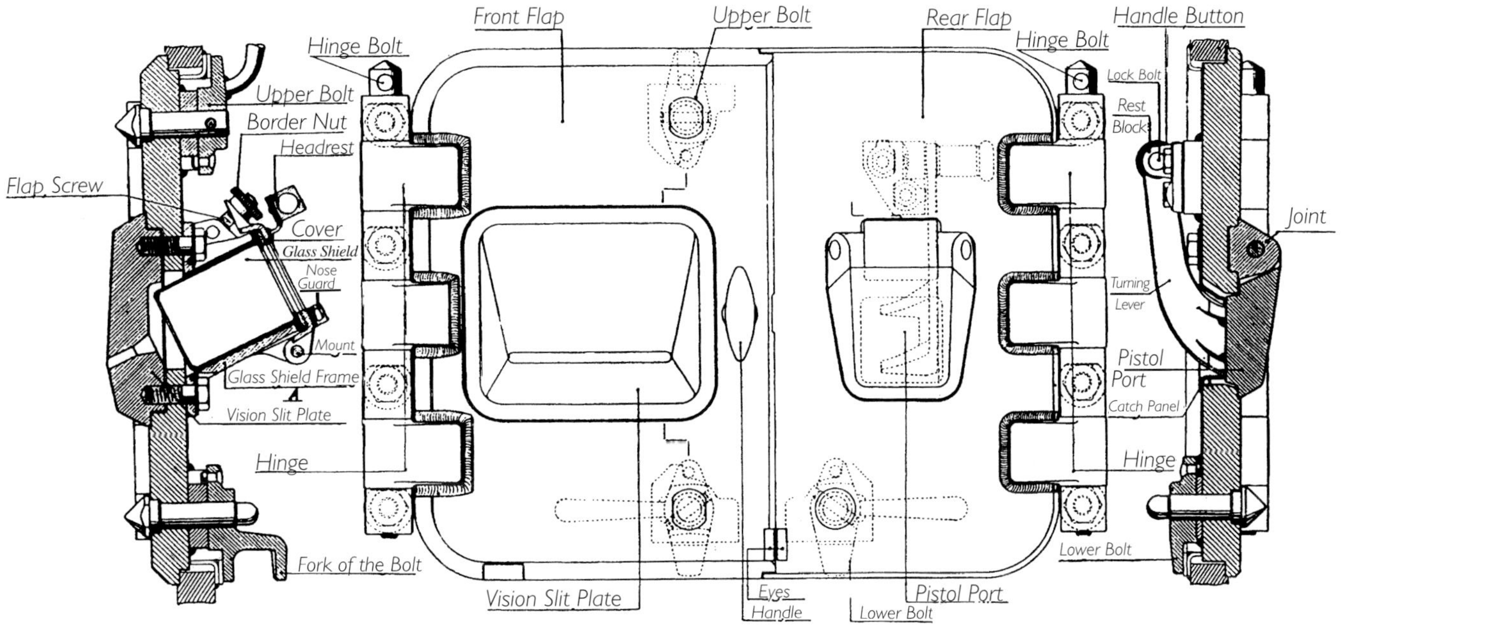

Turret hatch covers (Drawing No. 0-21 B 9293) for Pz.Kpfw.IV Type F.

M.G. Triggering (Drawing No. 021 B 185)
for Pz.Kpfw.IV Type F

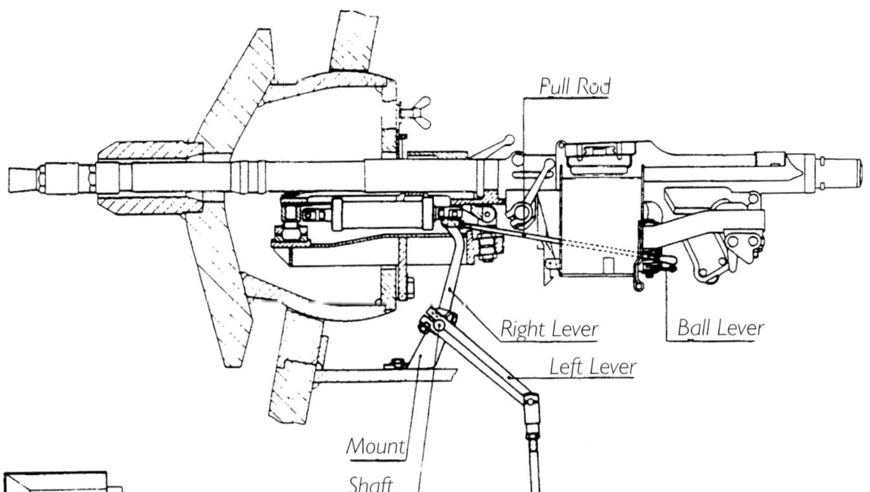

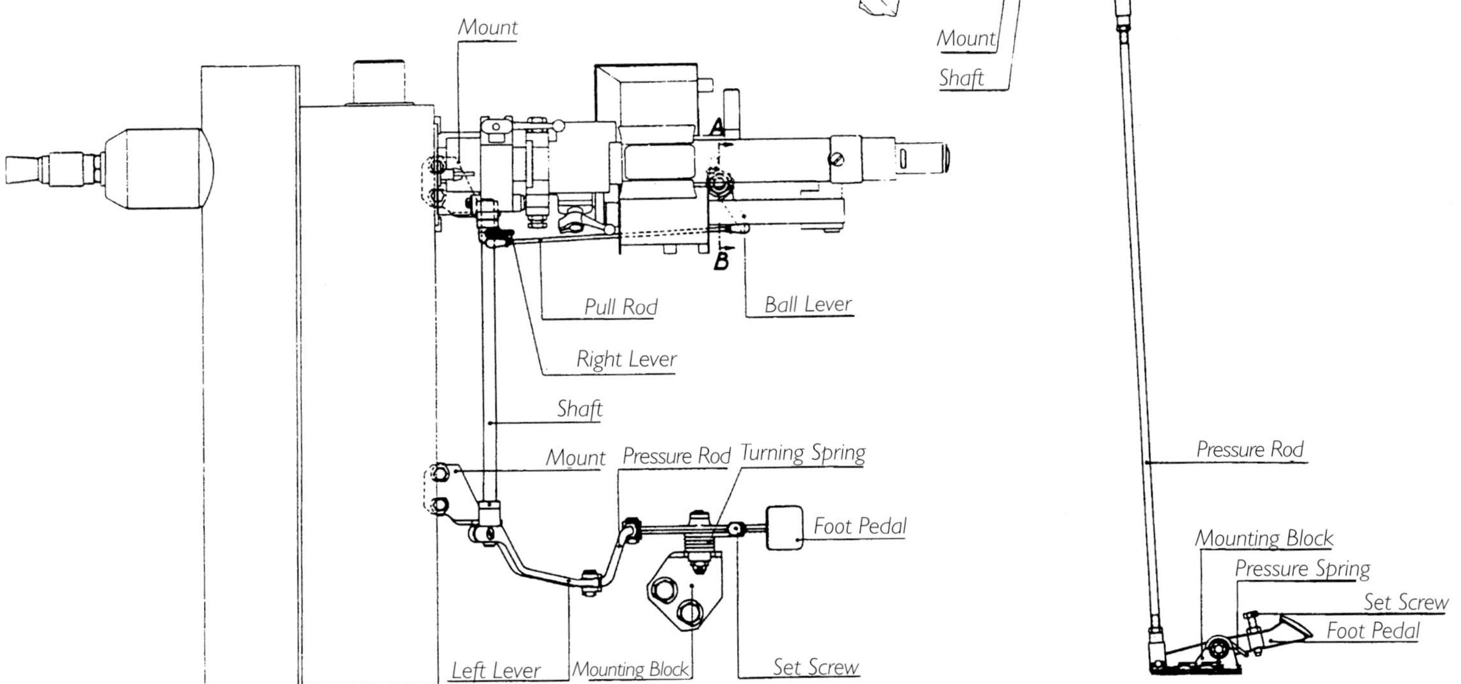

Factory-new Pz.Kpfw.IV Type F, finished in the spring of 1941.

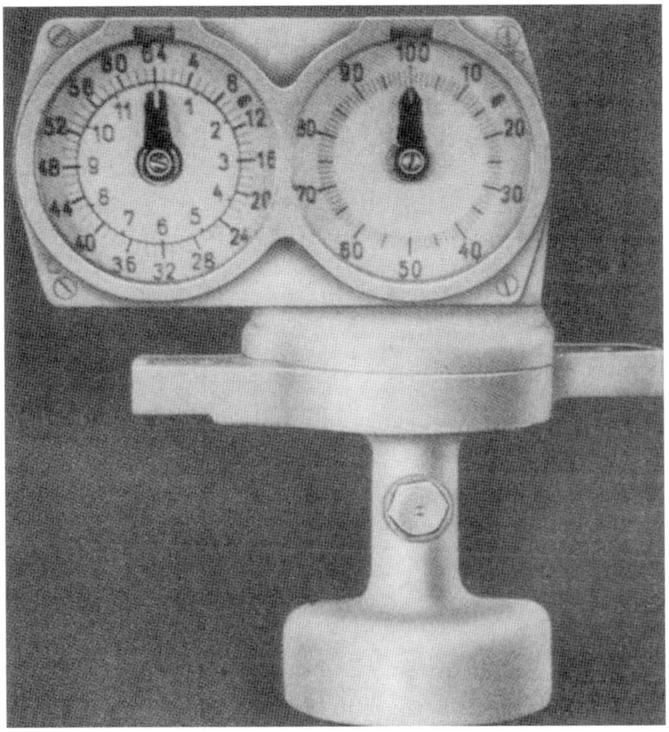

12 O'clock Indicators (new type) with two dials for Pz.Kpfw.IV Type F

Open Turret Hatch Cover.

During a discussion on November 29, 1941, about armored vehicle production and antitank defense at the Reich Chancellery, the Führer explained:

For the Antitank Shell 40, the strengthening of the armor is solved by the fore-armor. A part of the enemy antitank shells will be eliminated by the fore-armor. Not solved by the fore-armor will be protection from hollow-shell fire. At the first hit, the fore-armor will be shattered. The fore-armor is thus only a partial protection here.

The first requirement is to equip the tanks of all units with fore-armor by the spring of 1942. Where the vision slits do not allow fore-armor, the armor strength itself must be increased.

The state of the mounting of B.W. fore-armor for Series 7./B.W. was reported by Krupp of Essen to the Wa J Rue (WuG6) VII Department as follows:

In the discussion in Berlin on October 27, 1941 it was determined that the armor shell of the Pz.Kpfw.IV as of the beginning of January 1942 should be delivered with fore-armor. As already reported by our Mr. Anhenn re the discussion in Berlin on December 15, this delivery guarantee cannot be met, since it was not possible for us to obtain the material for the fore-armor at the right time. For this reason we are basically only able to deliver fore-armor for the first time as of February 1, 1942.

Also, the question of whether the turret should be delivered with or without fore-armor, has not yet been decided. Thus we are pointing out that we will also not be able to deliver turret fore-armor at the beginning of February, in case we do not immediately receive the decision on this point. The beginning of bow fore-armor delivery is planned for the beginning of February.

In the discussion in Berlin on 27 October, it was agreed with the gentlemen from the Böhler firm that… *in addition to the fore-armor for the Pz.Kpfw.IV delivery by the Krupp firm, the Pz.Kpfw. IV production of the Böhler firm of Essen should also be supplied with fore-armor. The extent of this delivery amounts to eight sets of fore-armor. We point out that the delivery of this fore-armor, as well as the delivery of the fore-armor for our own production, cannot take place in January 1942. The Böhler firm was accordingly notified by us.*

The following drawings of Pz.Kpfw.IV fore-armor were sent out by Department A.K. to Wa Prüf 6 on January 2, 1942:

St AKF 31293	Fore-armor for 4./ to 6./B.W. Turret
021 St 193	Fore-armor for 7./B.W. Turret

Wa Prüf 6/Pz. was not satisfied, and wrote to the Fried.Krupp A.G. firm, Department A.K., on January 17, 1942:

According to vocal information from your Berlin representatives, the last drawings for 7./B.W. fore-armor should be delivered to the German Pipe Works. Written confirmation of this information is requested.

It was also not reported to justify the non-keeping of the long-expired delivery schedule that the appropriate design bureau has been so burdened by work on the turret with the 7.5 cm Kw.K.44 that the announced schedule for the fore-armor drawings could not be met. Wa Prüf 6 has received information on this state of things oonly on the basis of the reminder of January 9, 1942.

Wa A, Wa Prüf 6 must make clear that such a procedure is not bearable. Wa Prüf must demand that it be notified at once when the probability is made known to the firm that a notified schedule cannot be kept, for whatever reasons.

On 26 January Department A.K. answered:

To the above we inform you: The production drawings for fore-armor 4. to 6./B.W. were demanded as of November 5, 1941. They were delivered by special messengers with AKH documents on November 13. The group drawing and several completions followed on November 15, 1941.

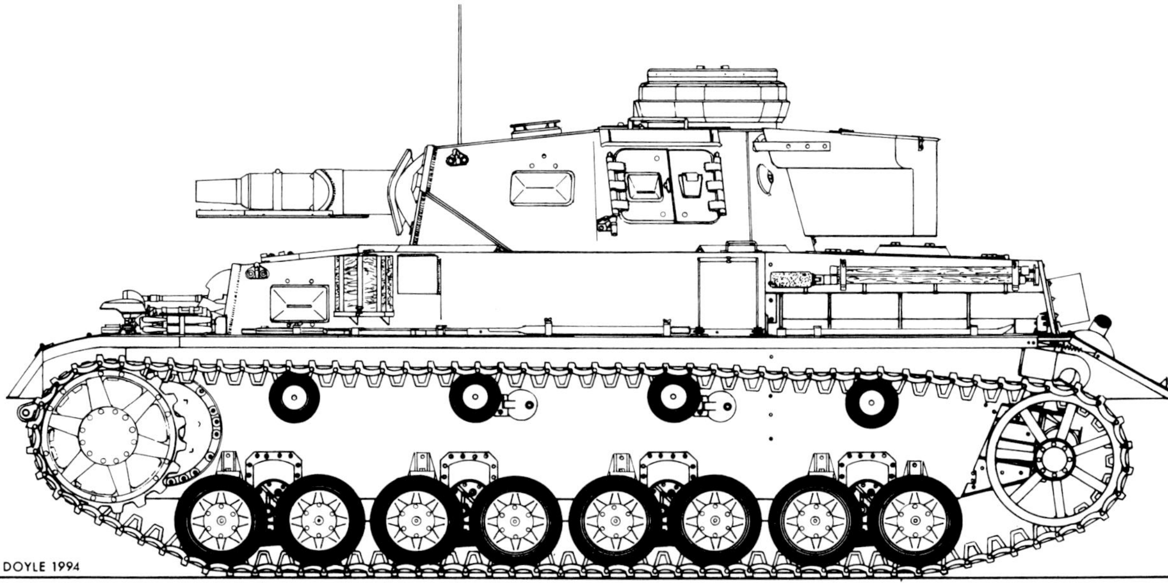

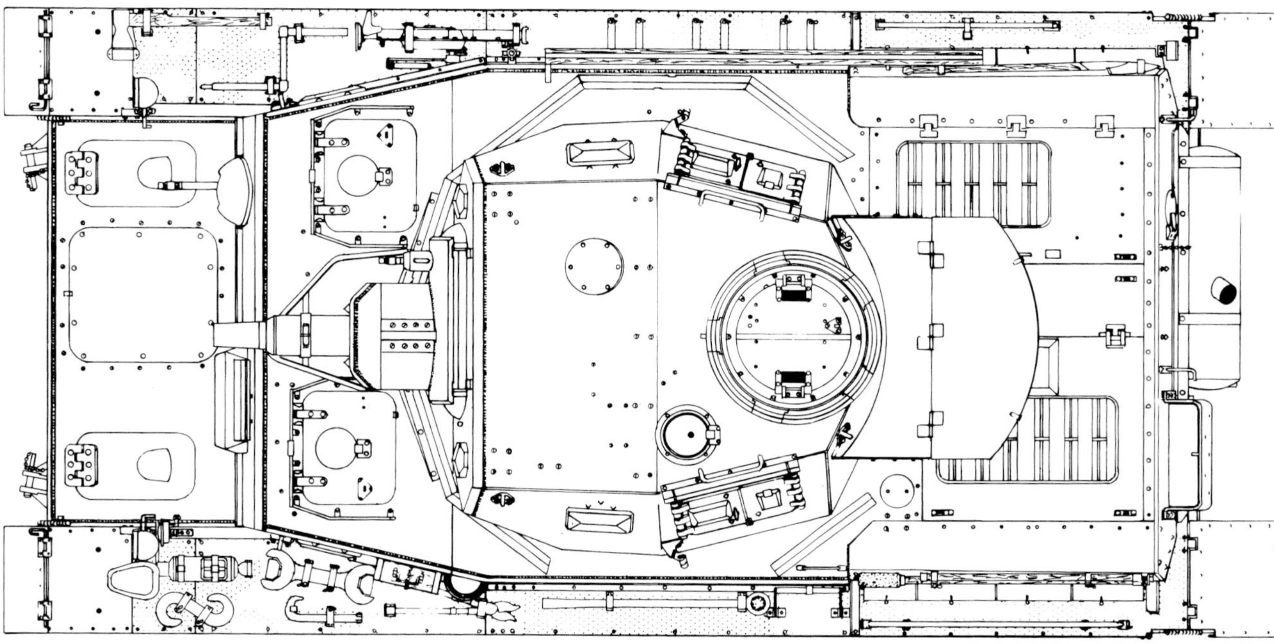

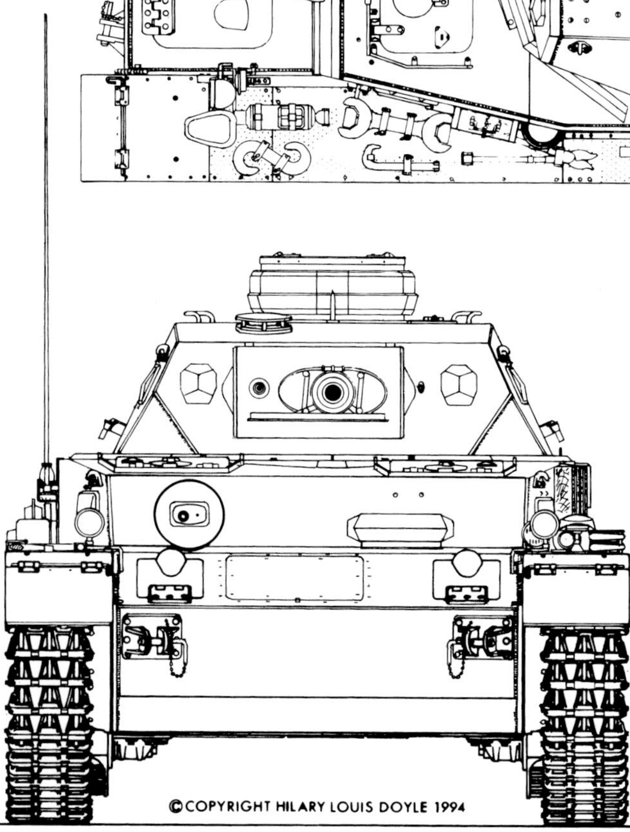

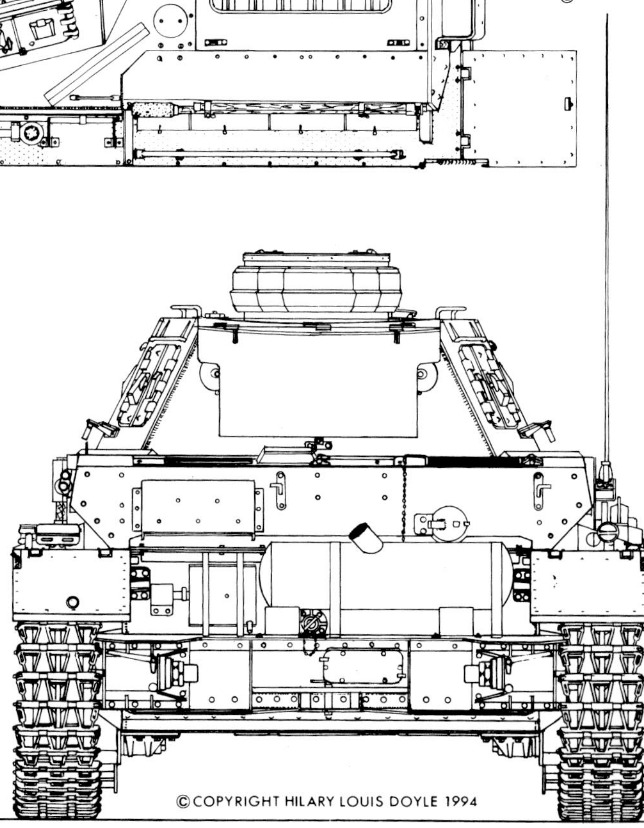

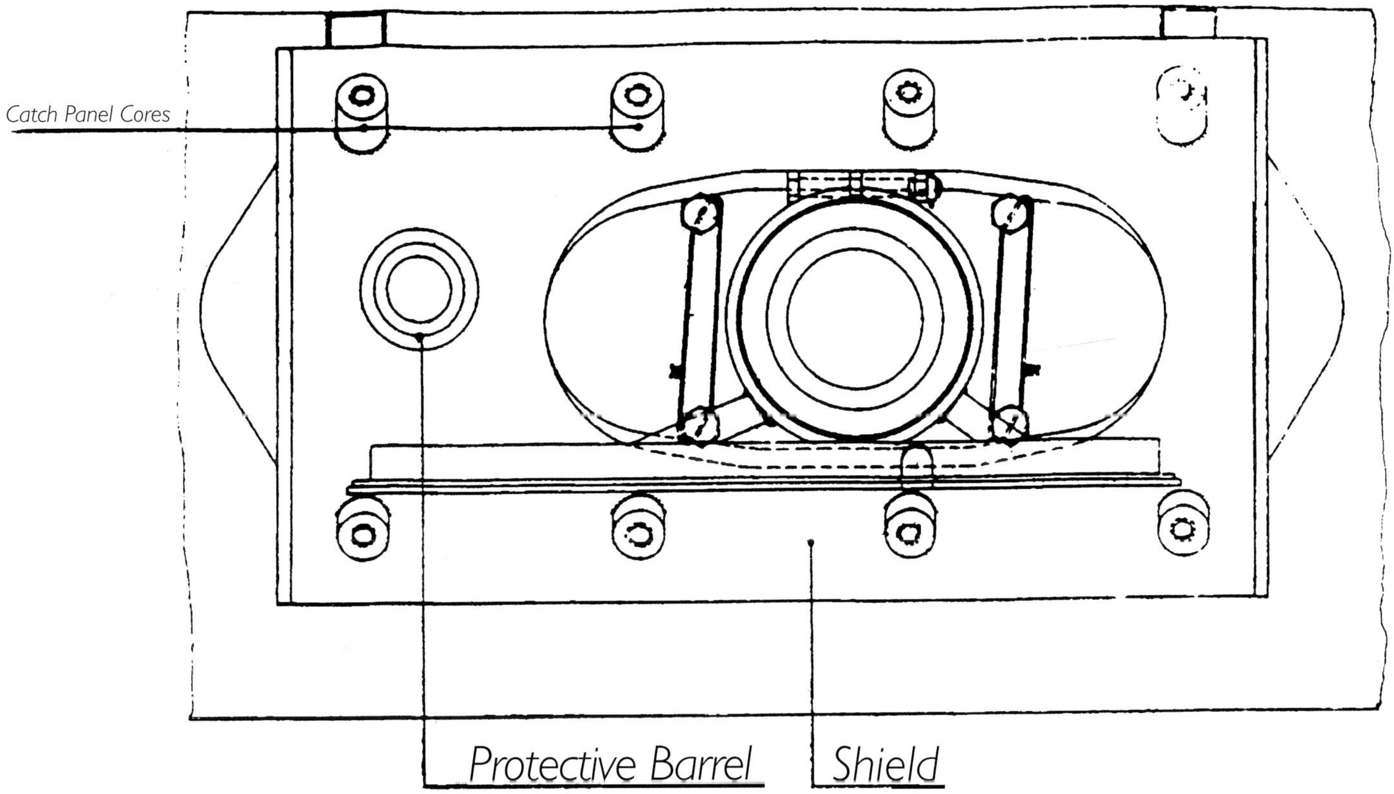

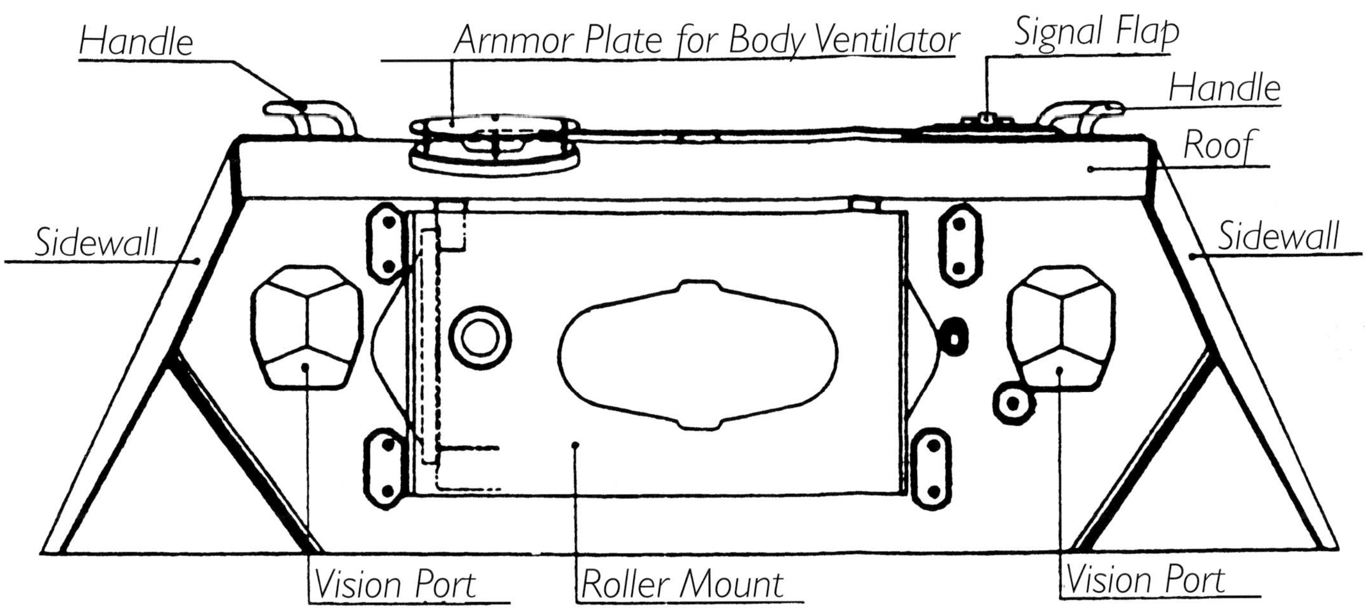

Turret and roller mount (Drawing No. 021 B 181) for Pz.Kpfw.IV Type F, with attachment parts for fore-armor.

Opposite:
Panzerkampfwagen IV (7.5 cm), (Sd.Kfz.161) Type F. As of August-September 1941 (Chassis No. 82175)—made by Krupp, new hull with 30 mm thick side plates and armored air intakes on the servicing hatches for the steering brakes. Ball-catching device on roof panel of the body. New, 400 mm wide tracks of Type Kgs 61/400/120 with wider road wheels and 470/90-360 tires. New leading wheels. New body with straight 50 mm thick front plate, K.F.F.2 optics for driver. Radioman's MG in Ball Mantlet 50. Added ventilation gratings in the motor hood in some vehicles of this type. (Some vehicles of earlier types, intended for action in tropical lands, show similar added ventilation.) Foglaying device mounted directly on the back of the motor compartment. New exhaust muffler for Maybach HL 120 and DKW motors. New turret with 50 mm thick front plate, side entrances on turret with divided hatch covers and new rear pistol loophole.

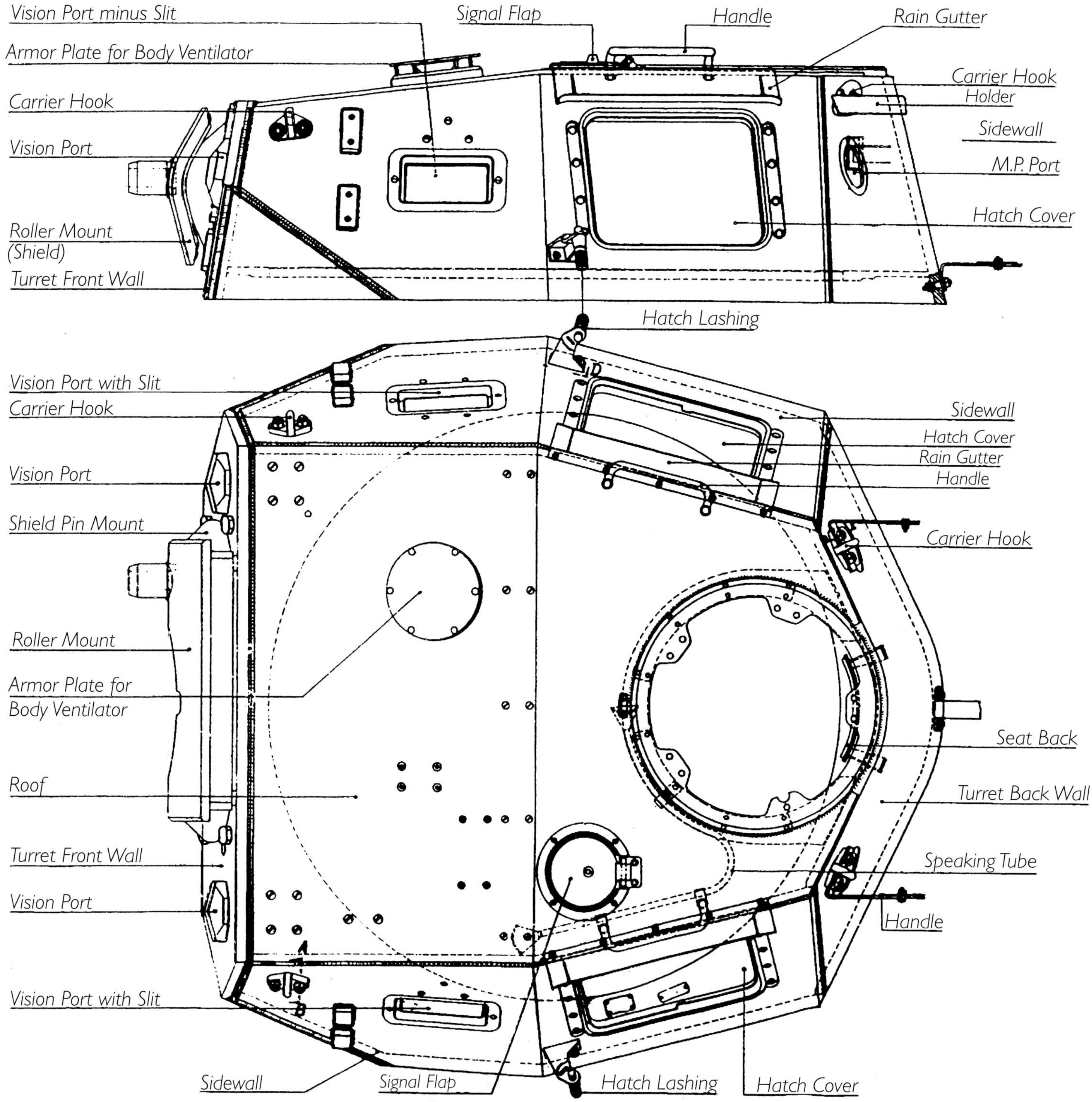

Turret and roller mount (Drawing No. 021 B 181) for Pz.Kpfw.IV Type F, with attachment parts for fore-armor. *(Continued)*

Pz.Kpfw.IV Type F: Turret with fore-armor.

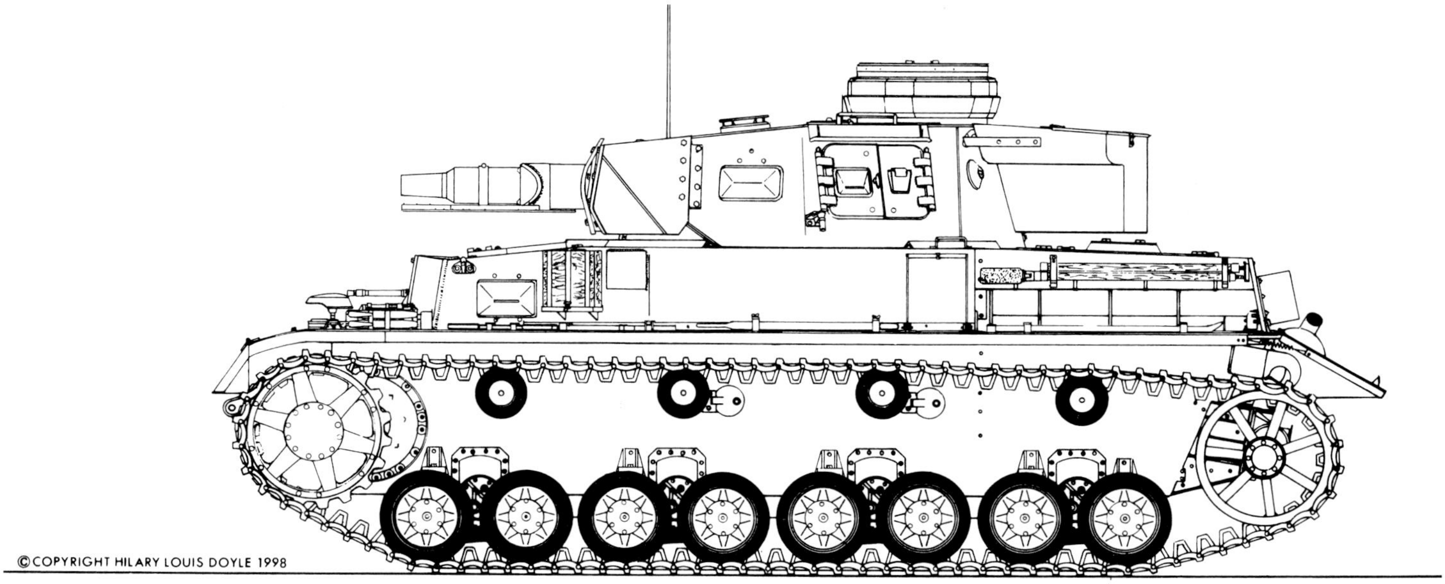

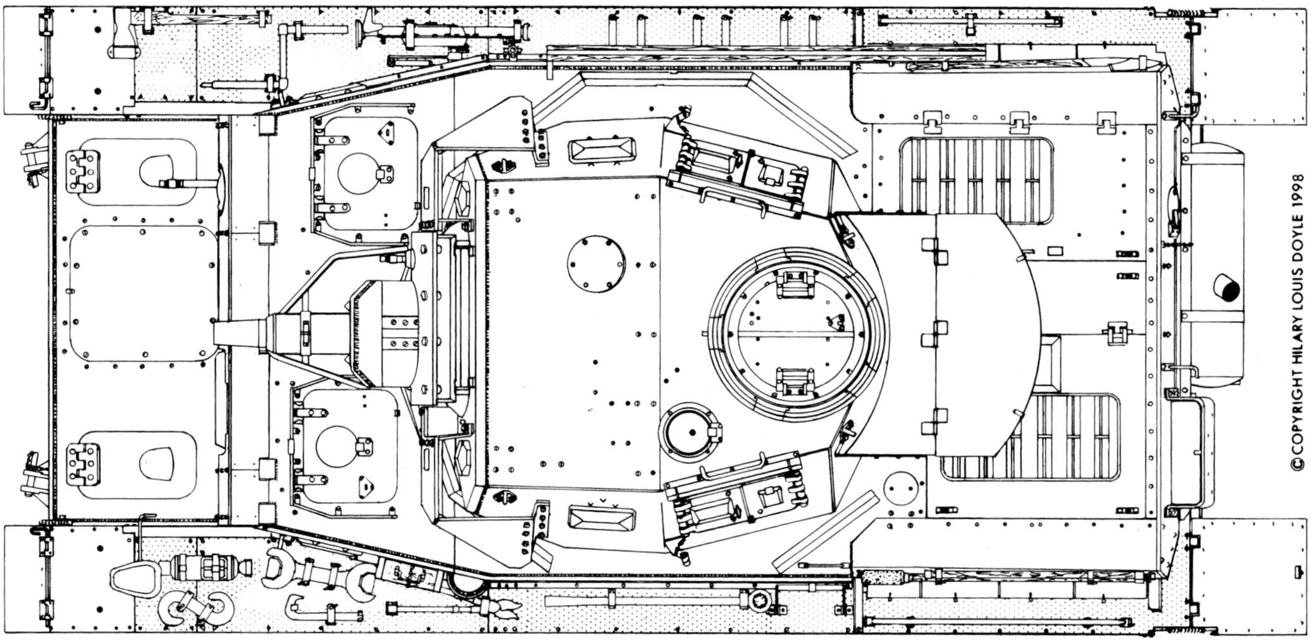

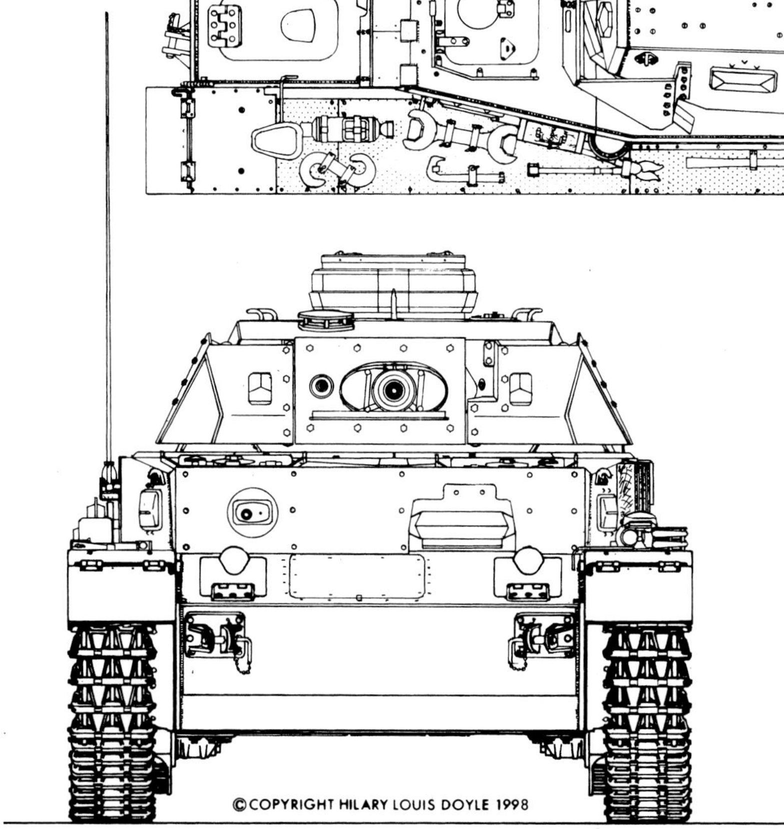

Panzerkampfwagen IV, Type F, with fore-armor. Added 20 mm thick armor plates are set in front of the turret and body armor.

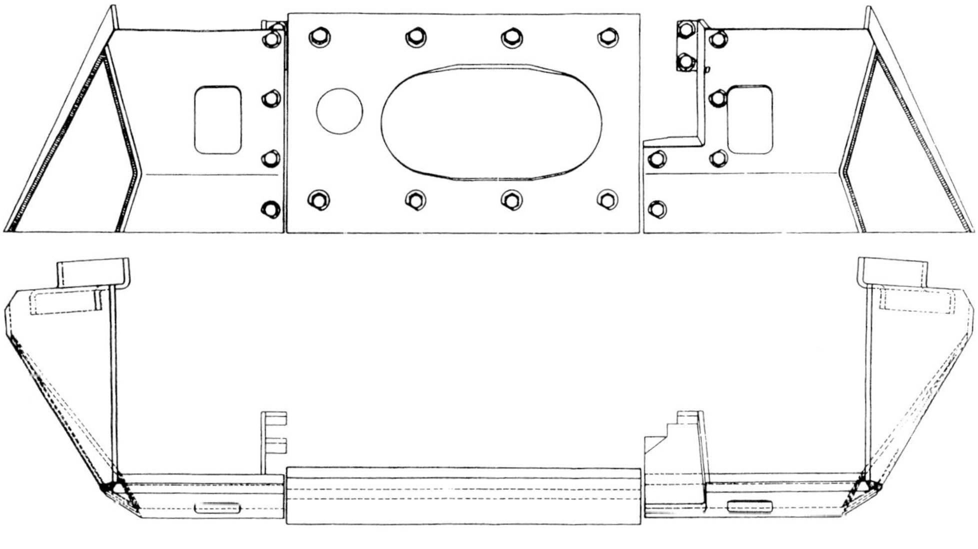

Fore-armor (Drawing No. 021 B 183) for Pz.Kpfw.IV Type F.

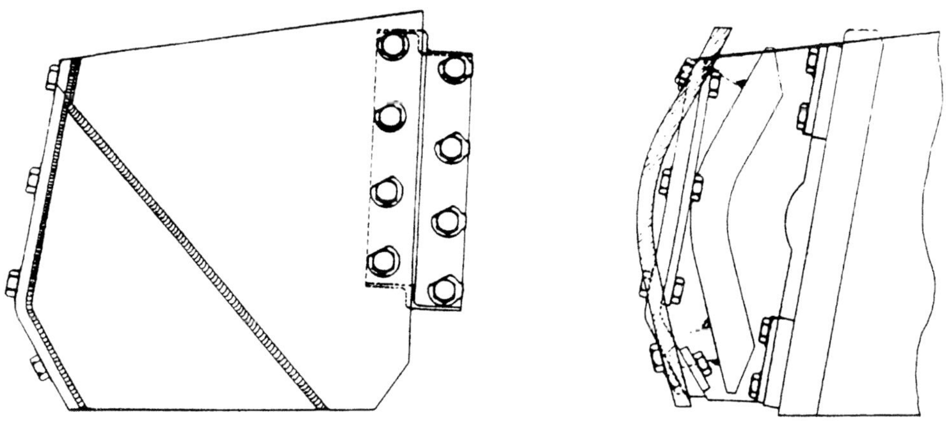

A company of Panzerkampfwagen IV with fore-armor in France in 1942.

Pz.Kpfw.IV Type F. Chassis No. 82175, built in September 1941.

The drawings for fore-armor 7./B.W. Series, Group 021 St 193, were not deliverable to the German Pipe Works. After the discussion at Prüf 6 on November 12 took place, a special attachment of the fore-armor for the already delivered 7./B.W. vehicles was supposed to be developed because of their surface-hardened layer. They were sent to the German Pipe Works on January 16, 1942. The fitting was, as far as we were told, originally planned for only three vehicles. The making of the drawings was therefore postponed somewhat in favor of the urgent installation of the 7.5 cm Kw.K.44.

9.2.3 Foglaying Launchers

Wa J Rue (WuG6) VIIId complained on February 1, 1942:

On the basis of the experiences to date it has been shown that the foglaying launcher devices (N.K.A.V.) attached to the stern of the tanks do not meet the requirements. In the future, therefore, N.K.A.V. are no longer to be delivered in the former form. New contracts with the producers of the N.K.A.V. are no longer to be issued. As far as possible, current contracts are to be canceled. Available N.K.A.V. and those still being produced by the manufacturers can be used up. In any case scrapping of already prepared material is to be avoided.

As long as new-type N.K.A.V. are not yet available and the old type are used up, the tanks can be delivered without N.K.A.V.

Krupp of Essen answered an inquiry from Wa J Rue (WuG6) VIIId that to date 452 sets of armor mantles for N.K.A.V. were delivered, and in fact for 7./B.W. The delivery was made with 359 pieces to the Gruson Works. From the 453rd device on, the armor parts were no longer delivered by Krupp of Essen.

9.3 IMPROVEMENTS AFTER DELIVERY

In the 25th Armored Division, ten Pz.Kpfw.IV Type F with 7.5 cm Kw.K. L/24 were rebuilt with 7.5 cm Kw.K.40 L/48 in the period from March to July 1943. Because of the chassis numbers (among others, 83552, 82612, 82613, 82389, 82391) of these last completed Type F., it was formerly assumed that Type F vehicles, even before they had left the factory, had been rebuilt with the 7.5 cm Kw.K. 40.

A series of improvements was assigned to the troop units after the delivery of Pz.Kpfw.IV Type F. Among others:

1) As of June 1941, installation of a coupling on the rear for a fuel trailer.
2) As of April 1942, installation of flanged tubes for warm air heating of the fighting compartment.
3) As of July 1942, repaired Pz.Kpfw.IV with 7.5 cm Kw.K. should receive a new long 7.5 cm Kw.K.40 with a strengthened mount and were to have their chassis upgraded to the newest state.
4) As of September 1942, installation of a cooling-water heating device (built by Fuchs).
5) As of October 1942, installation of the connections and vents for warm water passage.
6) As of November 1942, winter track links and snow grippers for tracks were delivered to the troops for winter equipping.
7) As of May 1943, mounting of aprons on turret and hull sides.

10. DEVELOPMENT OF THE TANK GUN FOR PZ.KPFW.IV

10.1 7.5 CM TANK GUN WITH INCREASED PENETRATING POWER

The first mention of a 7.5 cm gun with increased penetrating power is found in the Weapons Department's *Offensive Abwehr von Panzerwagen* of October 30, 1935.

The Escort Wagon (B.W.)

The 7,5 cm gun of this tank has, with an L/24 barrel length, an initial velocity of 430 m/sec. With this performance the 7.5 cm Pz.Gr. breaks through plates of a thickness up to 43 mm at a striking angle of 60 degrees at a range of 700 meters. The B.W. is thus, in terms of pure penetrating effect, fully equal to the new French tanks (until the end of 1935, about 1000 French tanks were built with armor strengthened up to 40 mm).

To be sure, it is possible, after theoretical considerations that are to be borne out by practical tests at Putlos, that with the lesser speed of the flight path caused by the small initial velocity, the hopes of striking become worse.

All the same, it remains to be determined that the B.W. is completely equal in effective technology to all tanks to be expected, with the exception of the heaviest Char 2 C, 3 C and D.

An increase in the initial velocity of the 7.5 cm gun would have to extend, in view of sufficient speed and, what could be achieved in this situation, to the effectual possibility against

Pz.Kpfw.IV Type F with a wooden dummy of a 5 cm KwK. L/60 with muzzle brake.

the Char 2 C, 3 C and D, to about 650 m/sec. Such an increase requires a completely new tank to be designed, which according to convincing calculations, with an armor up to 20 mm thick (thus not fully 2 cm-safe) would reach a weight of at least 30 tons. The Herr Ob.d.H. has recently spoken against such a tank.

10.2 5 CM TANK GUN L/60

In order to create a tank with a weapon of higher penetrating performance, Hitler ordered on February 19, 1941, the immediate assembly of one Pz.Kpfw.III and IV each with long 5 cm guns, and after their finishing, their immediate display before him.

As to the 5 cm Kw.K. L/60 for Pz.Kpfw. IV, Krupp had decided by March 13, 1941:

1) *Barrel with bottom and breechblock is to be given up at once with Pak 38 interior arrangement. The interior arrangement is to be requested from Wa Prüf 1.*
2) *A complete Pz.Kpfw.IV is to be sent to Essen at once and set up systematically for a showing by April 15.*
3) *As the date of the first test piece (turret and wagon), July 15 is to be striven for internally.*

On March 19, 1941, a director of Krupp reported on a discussion with Wa Prüf 6 in Berlin:

I asked that a completely equipped B.W. vehicle be sent to Essen, so as to hasten the experimental installation of a 5 cm L/60 weapon with ammunition storage as well as additional fuel tanks and added ammunition. Oberbaurat Kniepkamp immediately asks M.A.N. to send us the B.W. vehicle that was made available to that firm by Wa Prüf 6 for several weeks to study the installation of a 7.5 cm turret.

On March 31, 1941, M.A.N. of Nürnberg sent the following telegram to Krupp of Essen:

To your shooting range turntable B.W. vehicle No. 80668 by train.

On October 4, 1941, it was reported:

The rearmed Pz.Kpfw.IV with 5 cm Kw.K. L/60 should be delivered to Wa Prüf 6 on October 6, 1941. The vehicle was inspected at the workshop with new weapon and ammunition storage (for 149 rounds).

10.3 7.5 CM TANK GUN L/33

On March 12, 1941, at a discussion in Department A.K. of Krupp in Essen about the development of the 7.5 cm Kw.K. for Pz.Kpfw. IV it was decided:

A) *The present 7.5 cm Kw.K. L/24 has the following performance:*

Length:	*1766.5 mm*
Shot weight:	*6.8 kg*
Vo:	*398 m/sec*
Vz:	*378 m/sec*
Penetration:	*39 mm (400 m 60 degrees)*

B) *The performance of the 7.5 cm lg.s Pak L/40 (for assault gun) has:*

Length:	*3023 mm*
Shot weight:	*6.8 kg*
Vo:	*670 m/sec*
Vz:	*634 m/sec*
Penetration:	*70 mm (400 m 60 degrees)*

Pz.Kpfw.IV Type D, Chassis No. 80668, with 5 cm KwK. L/60 installed for testing purposes.

Pz.Kpfw.IV with 7.5 cm KwK, L/33, development vehicle on the basis of Type F, chassis with Type E turret.

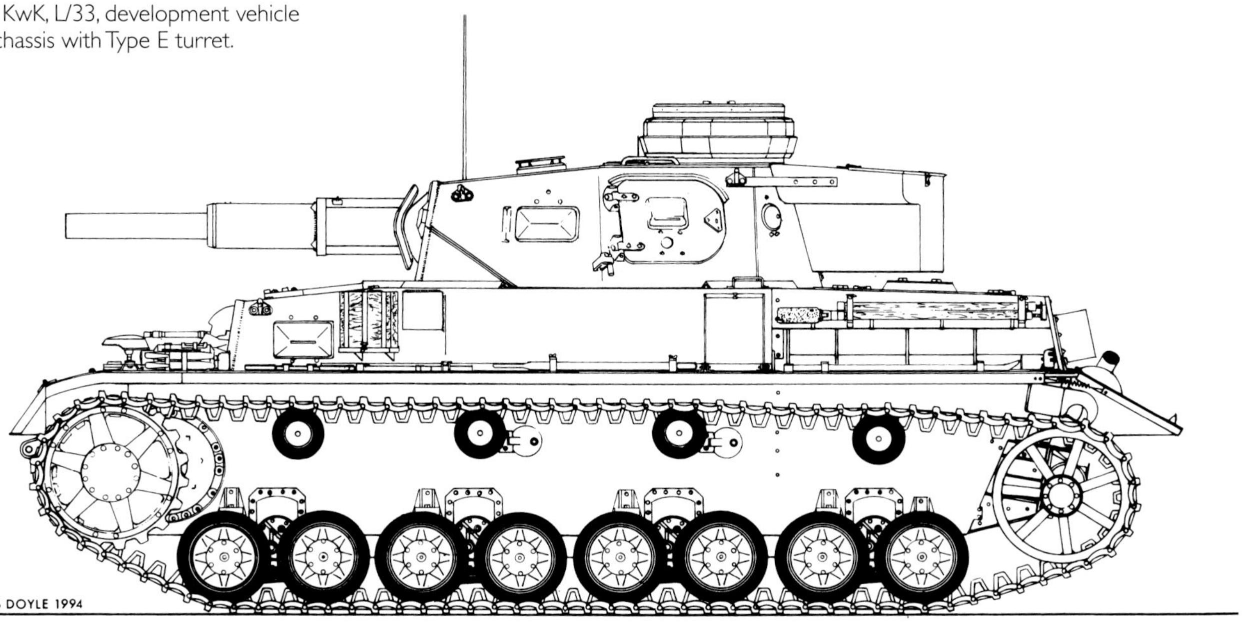

C) If the barrel is to stay secure behind the breaking angle, thus only one 7.5 cm Kw.K. L/33 with a long barrel allowed. The barrel then remains ca. 50 mm behind the angle. The performance of the gun is:

Length:	*2470 mm*
Shot weight:	*6.8 kg*
Vo:	*580 m/sec*
Vz:	*548 m/sec*
Penetration:	*59 mm (400 m 60 degrees)*
Shell length:	*660 mm*

This barrel is to be developed and all turret conditions for it are to be checked.

For the cylindrical L/33 barrel, HK and StK propellant mirror charges are to be developed. The HL gives the following performance:

Shot weight:	*2.7 + 0.3 kg*
Vo:	*875 m/sec*
Vz:	*820 m/sec*
Penetration:	*86 mm (400 m 60 degrees)*

D) As with Pz.Kpfw.III, here too an inserted conical barrel 7.5/5.5 cm Kw.K. L/45 is to be developed with the following performance:

Length:	*2475 mm*
Shot weight:	*2.8 kg (HK)*
Vo:	*700 m/sec*
Vz:	*650 m/sec*
Penetration:	*68 mm (400 m 60 degrees)*

For Wa Prüf 4 on December 28-30, 1941, a function firing of the 7.5 cm Kw.K. L/34.5 was carried out. On January 13, 1942, the shooting was continued after installing the Kw.K. 44-Rhm. Barrel.

As to the testing of the muzzle brake for Pz.Kpfw.IV it was reported:

For the muzzle brakes on the L/34.5 barrel fired in Essen on December 30, 1941, from the 5th on the effective values and performance figures were given as follows:

Muzzle Brake	Effective Value	Performance Figure
V1	*66.2*	*1.20*
V4	*58.5*	*1.02*
V5	*56.5*	*0.98*
V6	*48.7*	*0.69*
K Gr red Pz	*6.8 kg (blind) were fired.*	

Muzzle Brake

V1:	*G = 25 kg, two-chamber muzzle brake, acorn shape, with 4 lobes*
V4:	*21.5 kg lightened two-chamber muzzle brake, acorn shape without Lobes*
V5:	*23 kg firm as No. V4 =*
V6:	*15 kg one-chamber muzzle brake, ball form with two big windows*

	V1	V4	V5	V6	Without
Recoil measured kg	*1027*	*1023.5*	*1025*	*1017*	*1002*
Recoil measured m/sec	*4.08*	*4.48*	*4.58*	*5.35*	*7.15*
Mass reduced m/sec	*4.03*	*4.48*	*4.57*	*5.35*	*7.10*
Recoil mass Vo = 689 kg.sec	*211*	*234*	*239*	*277*	*363*
Shot Vo = 680kg/sec	*236*	*236*	*236*	*236*	*236*
Effective value	*66.2*	*58.5*	*56.5*	*41.7*	*—*
Performance figure	*1.20*	*1.02*	*0.98*	*0.69*	*—*

On January 19, 1942, Krupp reported to Wa Prüf 4 as to the 7.5 cm Kw.K. L/33 for Pz.Kpfw.IV:

Four full barrels prepared and confirmed for the above contract by Dortmund Hörder Hüttenverein AG, Dortmund, we have used two for the production of the test barrels RV1 n.A. and RV2 n.A. For the other two barrels there is no use, since 7.5 cm L/33 barrels are not being made.

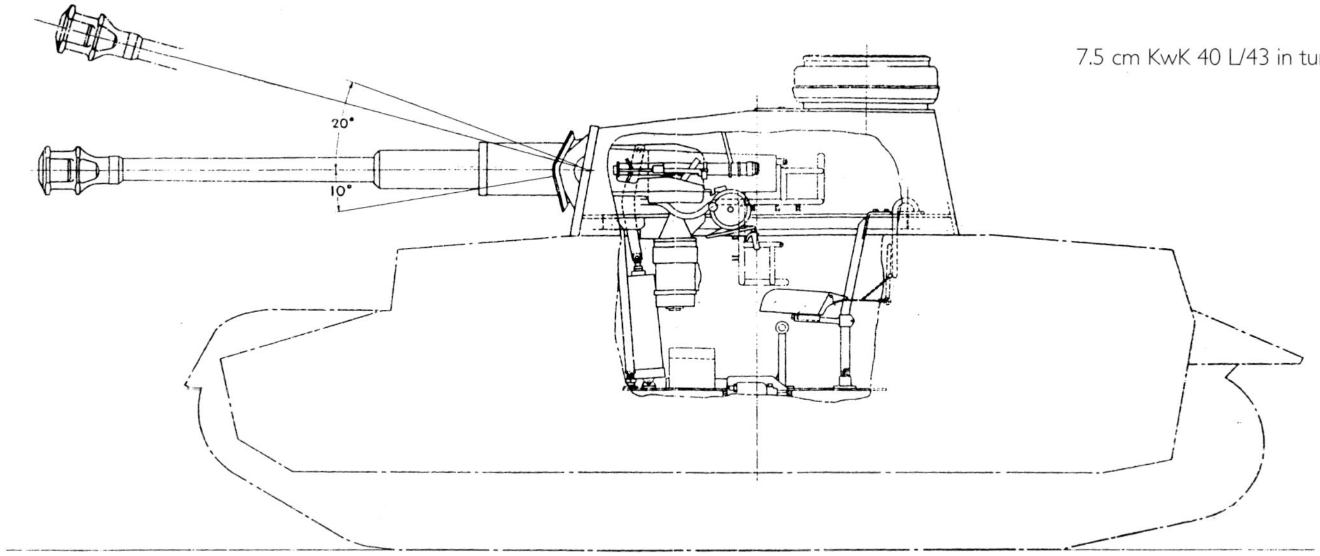

7.5 cm KwK 40 L/43 in turret.

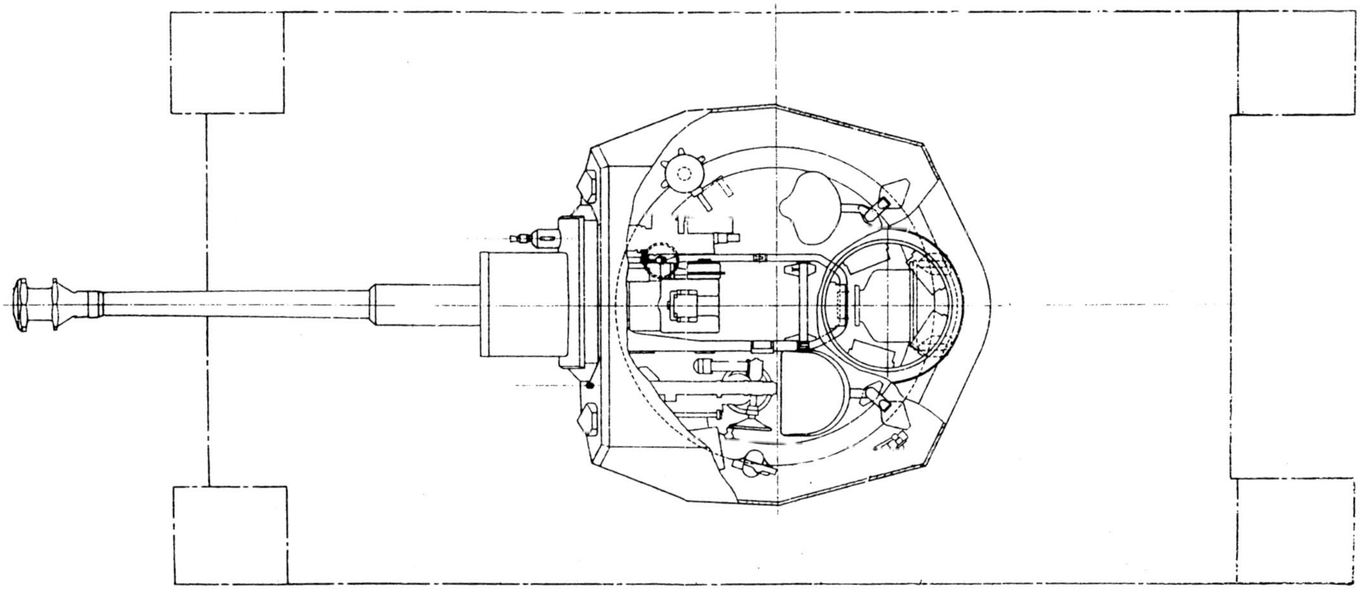

On April 23, 1942, Krupp decided: *Install the 7.5 cm KwK l?38 (34.5) in Turret No. 80979 and install this turret (in exchange for Turret No. 82127) in the Pz.Kpfw.IV Type F Chassis No. 82091sent to us by the Alkett firm.*

10.4 7.5 CM TANK GUN 40

The development of a 7.5 cm Kw.K.40 (originally called 7.5 cm Kw.K.44) with the same performance as the 7.5 cm Pak 40 L/46 (Rheinmetall) was requested by Wa Prüf 4 on November 18, 1941. Wa Prüf 4 assigned tasks to the Fried.Krupp A.G. (in cooperation with Rheinmetall-Borsig A.G.) for the development of the gun. Rheinmetall-Borsig was responsible for the interior ballistics and Krupp for the design. A penetration of 80 mm should be attained at 1000 meters at a 60-degree angle.

The barrel of the 7.5 cm Pak 40 (originally called Pak 44) with breechlock could not be installed directly into the Pz.Kpfw.IV turret because of the recoil length and the length of the shells. The rifled part (2470.5 mm long) of the 7.5 cm Pak 40 barrel was taken over unchanged for the 7.5 cm Kw.K. L/43. For simple loading of the tank gun, and in order to be able to store more shells in the tank, new short and thick shells and the necessary loading room were designed by Krupp.

Difference in size and performance between the 7.5 cm Kw.K.40 L.43 and L/48 and the 7.5 cm Pak 40 were:

Dimension		KwK 40	KwK 40	Pak 40
Barrel length caliber		L/43	L/48	L/46
Barrel length without muzzle brake	mm	3233	3615	3450
Length of rifling	mm	2470.5	2852.5	2470.5
Length of loading room	mm	507.5	507.5	729.5
Loading room content	liters	3.17	3.17	3.5
Recoil, normal	mm	485	485	900
Firing pause	mm	505	505	915
Shell casing length	mm	495.1	495.1	716
Shell casing diameter	mm	102.2	102.2	89
7.5 cm Pzgr.Patr .39, length	mm	748.3	748.3	969

The first mention that the 7.5 cm Kw.K. L/24 was to be replaced by a new tank gun appeared in the Army Weapons Department Preparation Preview Report of December 2, 1941:

Month	Dec 41	Jan 42	Feb 42	Mar 42	Apr 42	May 42
7.5 cm Kw.K. L/24	55	55	60	20	35	—
7.5 cm Kw.K.44 L/46	—	—	—	30	70	90

Monthly Army Weapons Department reports on finished and accepted weapons have shown the following production figures:

Month	Feb 42	Mar 42	Apr 42	May 42	Jun 42	Jul 42
7.5 cm Kw.K. L/24	54	29	25	15	—	—
7.5 cm Kw.K.40 L/43	—	18	104	56	87	114

Mr. Theermann (of Krupp) reported on a discussion concerning muzzle brakes for the 7.5 cm Kw.K. L/43 with Wa Prüf 4/VI in Berlin on January 16, 1942:

The barrel of the Pz.Kpfw.IV L/43 receives a rounded muzzle brake with two large exhaust openings for the first series; in later series it will be replaced by a two-chamber muzzle brake with acorn shape. To my question of why, instead of the more strongly effective acorn shape, only the one-chamber round form suffices, I was not given a satisfactory answer. I believe that some internal competence dispute is going on at the Department.

The 7.5 cm Kw.K.40 L/43 barrel had an uneven rifling angle of 6 degrees at the beginning and 9 degrees at the end. To simplify production, this L/43 barrel was replaced by a lengthened 7.5 cm Kw.K. L/48 barrel with am even rifling angle of 7 degrees. With the same powder charge in the shell, the longer barrel gave higher performance, as was measured at a test firing in Meppen on March 19, 1942:

Barrel L/43 (No. F12) antitank shell Vo = 790 m/sec
Barrel L/48 (No. F13) antitank shell Vo = 820 m/sec

The decision by Wa Prüf 4, as to when the 7.5 cm Kw.K.40 L/48 was to be built in series, originated on August 15, 1942. The 7.5 cm Kw.K.40 L/43 was still installed in Pz.Kpfw.IV until April 1943 before it was replaced by the 7.5 cm Kw.K.40 L/48 at the beginning of April 1943.

10.5 7.5 CM TANK GUN L/70

To the question from a Prüf 6 of whether the 7.5 cm Kw.K. L/70 (from the Panther) could be installed in a Pz.Kpfw.IV turret, Krupp answered on September 4, 1943:

Installation of the 7.5 cm Kw.K. L/70 in turret of Pz.Kpfw.IV is impossible for the following reasons:

1) Turret too low
2) Turret front too low
3) Weapon too heavy for turret-swinging mount
4) Turning machine too weak

Whether it is possible to house the gun equalizers and the most necessary space for loading in the available operating space in the vehicle appears unlikely.

Pz.Kpfw.IV Type F with wooden dummy of 7.5 cm KwK. L/70.

11. PANZERKAMPFWAGEN IV (7.5 cm) (Sd.Kfz. 161/1 & 2) Type G (8./B.W.)

Chassis number Series: 82396-82500, 82565-82600 and 82614-84400

Between March 1942 and June 1943, 1930 Pz.Kpfw.IV Types F2 and G were finished and accepted (907 from Krupp, 430 from Vomag and 587 from Nibelungenwerk).

Type G was the official designation for all 7./B.W. and 8./B.W. series Pz.Kpfw.IV with the 7.5 cm Kw.K. 40. The F2 designation was an interim solution that lasted only three months and was then replaced by Type G. The abandoned F2 designation is known to this day by Service Order 653/7 of April 1, 1942, and the temporary K-Device document of June 1942. The drawings for Pz.Kpfw.IV with the longer tank gun had developed as follows:

Jan. 13, 1942: 7./B.W. rebuilding
Mar. 3, 1942: Pz.Kpfw.IV Type F rebuilding
Mar. 21, 1942: Instead of Type F:
Type F1 (7./B.W.) (with 7.5 cm Kw.K. L/24)
Type F2
(7./B.W. rebuilding) (with 7.5 cm Kw.K. L/43)
May 22, 1942: Pz.Kpfw.IV (7.5 cm Kw.K.40) (Sd.Kfz.161), 7./B.W., Type F2
June 5, 1942: By document from Wa J Rue (WuG6) VII d Bd.No. 4299/42g of 6/5/1942, Pz.Kpfw.IV with Kw.K.40 was called "8./B.W. (Type G)"
July 1, 1942: By discussion with men from Wa Prüf 6 II, Wa Prüf 6 IId, Wa J Rue (WuG6/VIII) and Wa Prüf 6 IIIg, hereafter to be named:
The Pz.Kpfw.IV with 7.5 cm Kw K with 7./B.W. instead of 7./B.W.
F1, and
The Pz.Kpfw.IV with 7.5 cm Kw K 40 with 8./B.W. instead of 7./B.W. F2.

11.1 IMPROVEMENTS FROM TYPE F

Every improvement from Type F resulted from the installation of the new long gun,7.5 cm Kw.K. 40 L/43 instead of the shorter 7.5

Display of a Pz.Kpfw.IV Type F2 (G) before Hitler in April 1942.

Panzerkampfwagen IV, Type G (formerly called Type F2), Chassis No. 82574, made by Vomag in April 1942. The vision port on the right side of the turret was eliminated.

cm Kw.K. L/24. With this new tank gun came a modified hood before the roller mount, new form and position under the turret roof for the gun lashing, and a new gun cradle with elevation machine.

Because of the heavier gun, an auxiliary turning machine was installed at the right on the turret carrying ring. The aiming gunner could be assisted by the loading gunner when turning the turret by hand. One turn of the handwheel = 1.89 degrees of traversing change. As in earlier types, the smallest turning angle with electric turning was 0.08 degrees per second, the greatest angle with electric turning of the turret was 16 degrees per second. For the longer gun with higher performance, a new turret targeting scope, T.Z.F.5f, was installed. Like the earlier T.Z.F.5b, the T.Z.F.5f was a bending telescope on which, at the bending spot, an armor plate was placed to protect the aiming gunner. The T.Z.F.5f had a 2.5 fold magnification and a field of vision of 25 degrees (444 meters at 1000 meters). The turning graduated plate bore small circles for firing the Explosive Shell 34, of which the first and every second hereafter were numbered (to 3300 meters), for firing the Antitank Shell 39, smaller circles, of which the first and every second thereafter were numbered (to 2500 meters), as well as small circles for firing the Antitank Shell 40 special ammunition, of which the second and every second subsequent circle were numbered (to 1500 meters).

The highest targeted shot range for the 7.5 cm Kw.K.40 was 3000 meters.

For the larger shells of the 7.5 cm Kw.K.40, the ammunition storage was changed and more shells (87 instead of 80 in the earlier Types b through F) were stored. The location of tank-gun ammunition was done with 32 in the side space and forward in the bow armor, 55 in the chassis, for a total of 87 rounds of 7.5 cm Kw.K.40 shells.

They were divided as follows: In the left side space of the bow armor were two ammunition cases with nine shells each, in the right side space one case with six shells and one case with two shells, and behind the driver's seat three cases with two shells each. In the chassis there were three cases at right for eight shells each, behind the driver's seat one case with 23 shells, on the turntable one case for four shells, and on the back wall of the bow armor two ammunition cases with two shells each.

Factory-new Pz.Kpfw.IV Type F2(G), built in April 1942.

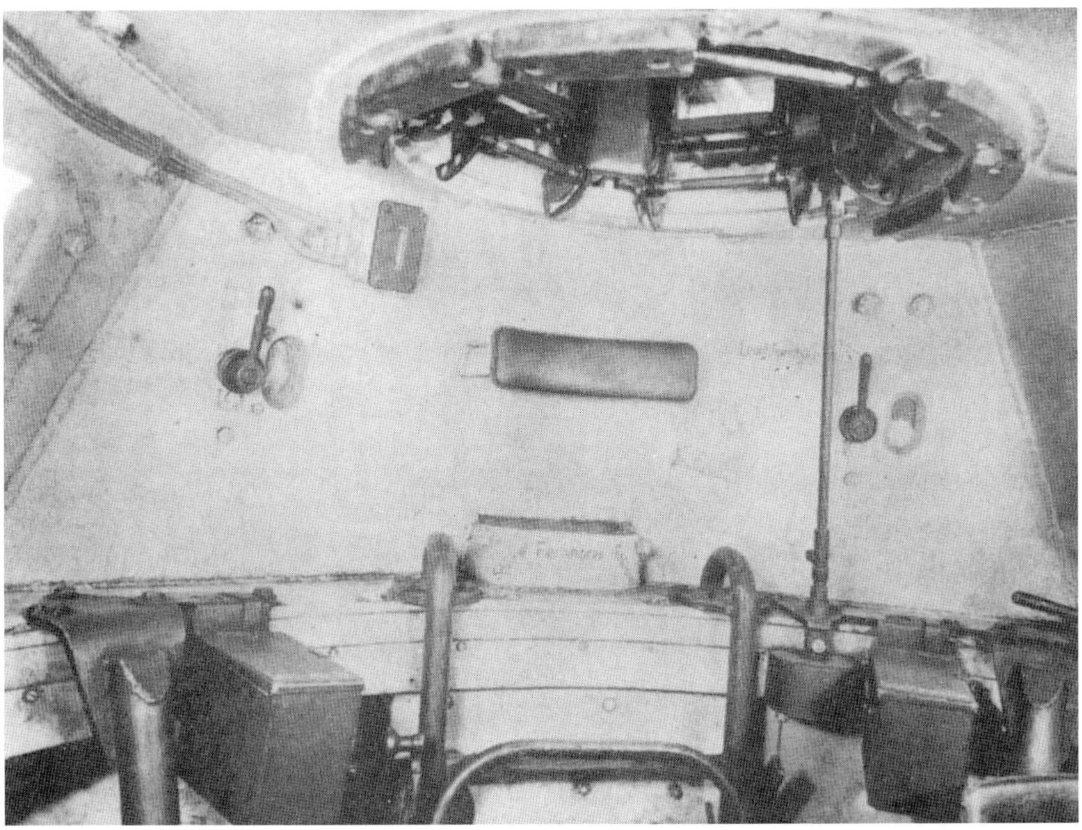

The back wall of the turret with the commander's seat and cupola.

Calibrated Plates for Turret Targeting Scope T.Z.F.5f.

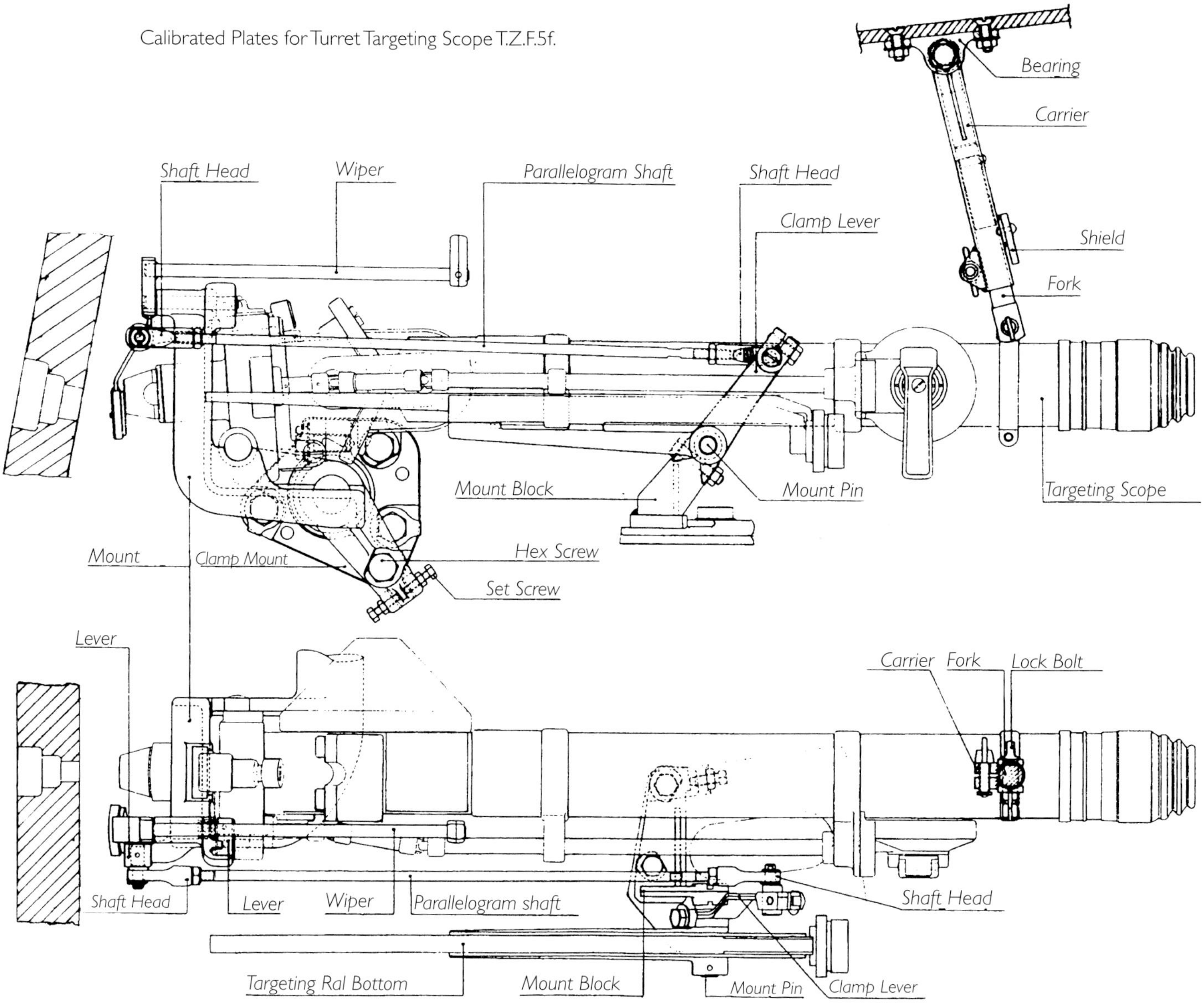

The fighting compartment of Pz.Kpfw.IV Type G with ammunition storage.

The fighting compartment of Pz.Kpfw. IV Type G with ammunition storage.

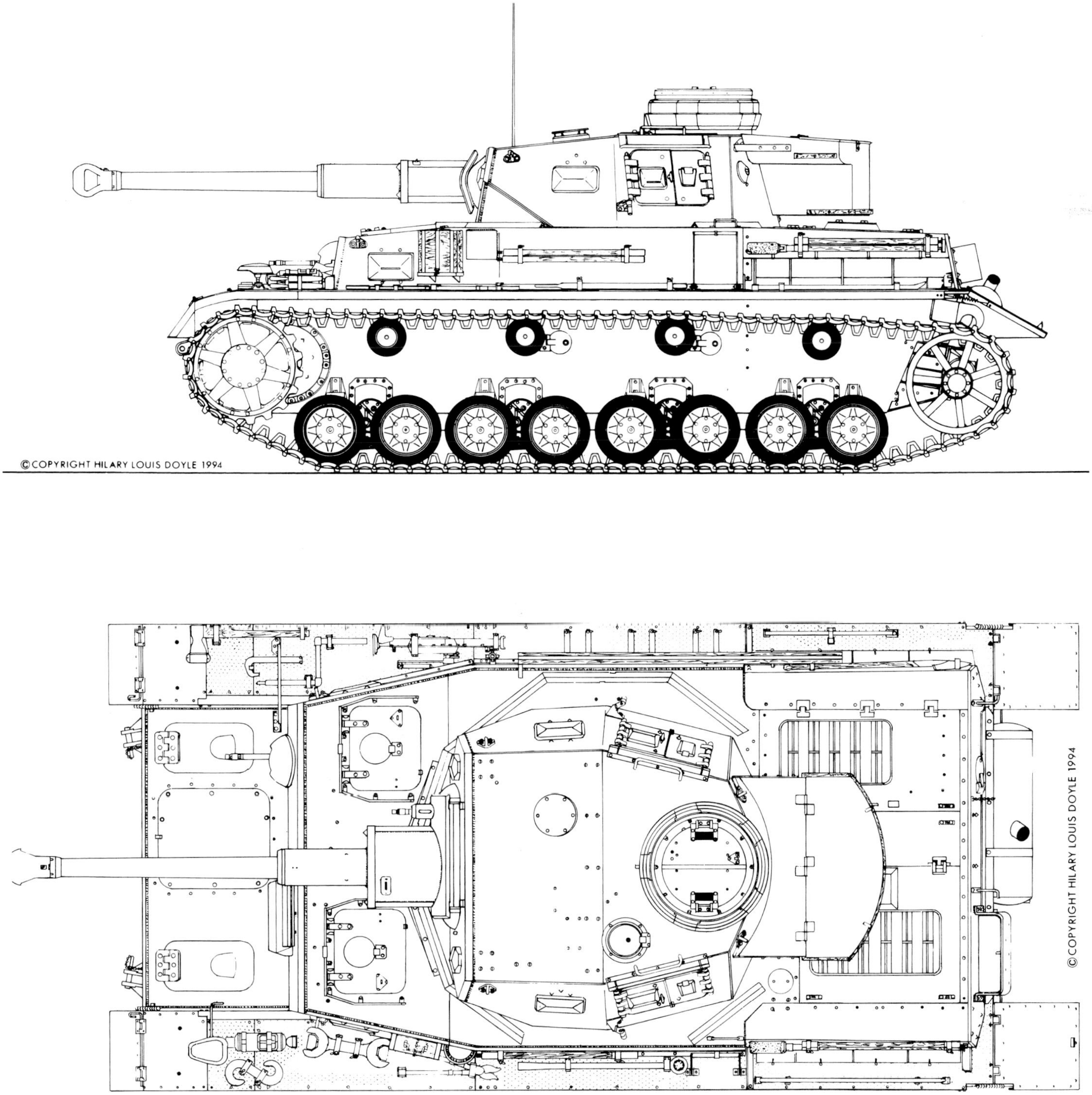

Panzerkampfwagen IV, Type G (formerly called Type F2), April 1942: One-chamber muzzle brake for 7.5 cm KwK. 40 (L/43), antenna deflectors omitted from some vehicles, others were changed. The new layout of the front plate avoided the use of a shot catcher.

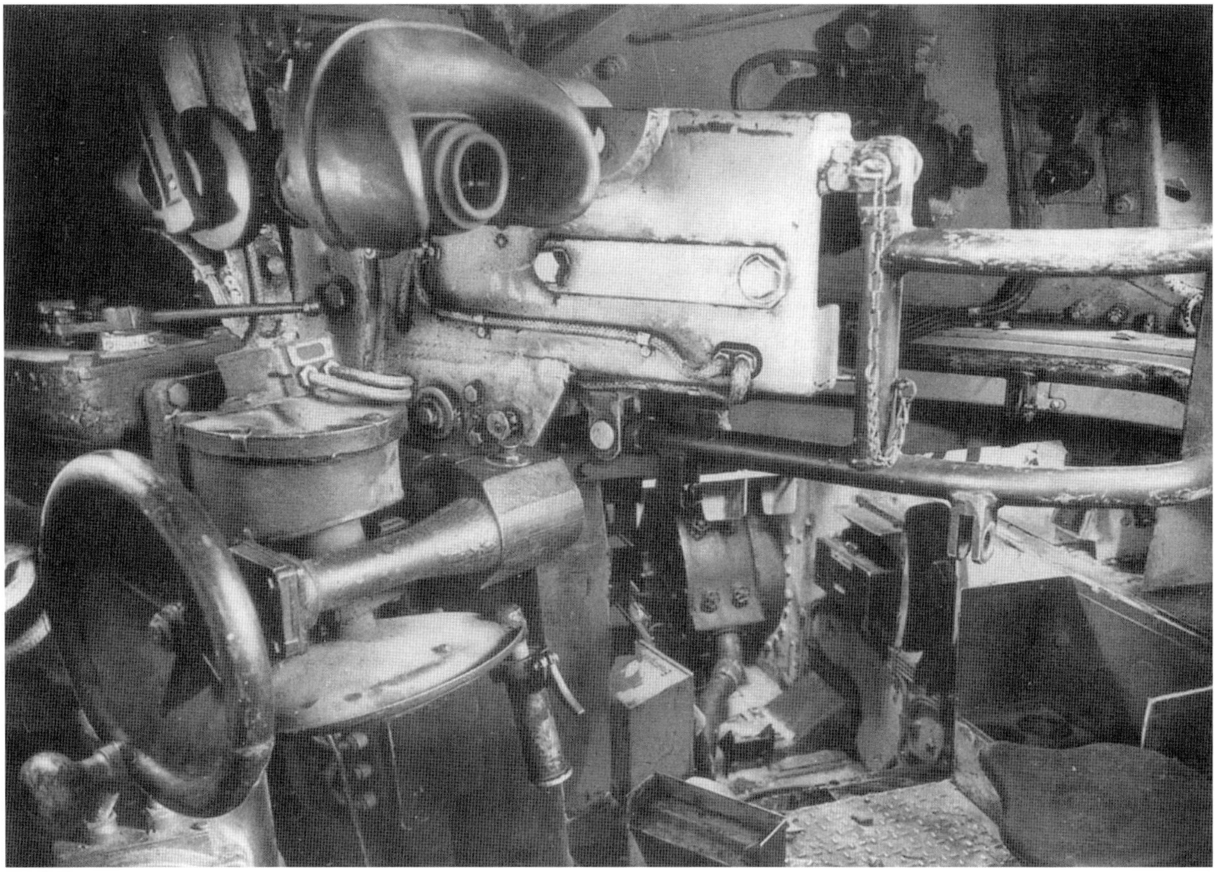

A look through the left entry hatch at the 7.5 cm KwK. 40 and turret targeting scope.

A look through the right entry hatch at the weapon layout.

Pz.Kpfw.IV Type G, Chassis No. 83072, built by Vomag in September 1942. Test muzzle brake, still vision ports on right turret front and turret sidewalls.

Pz.Kpfw.IV Type G, Chassis No. 82924, built by Krupp in October 1942. Note the two-chamber muzzle brake.

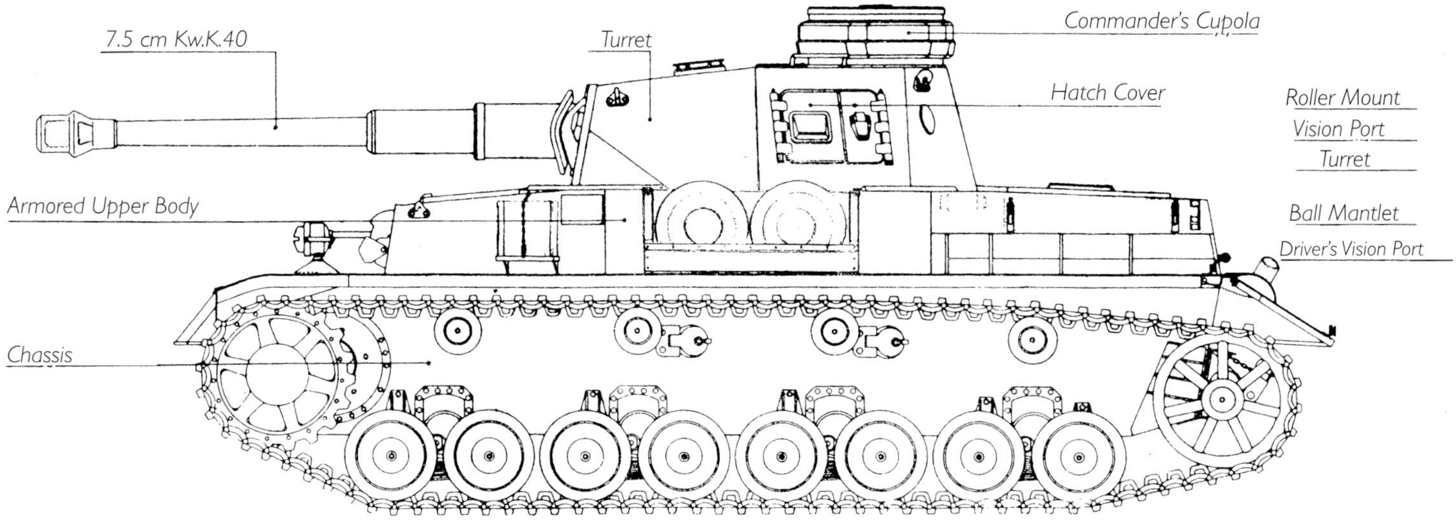

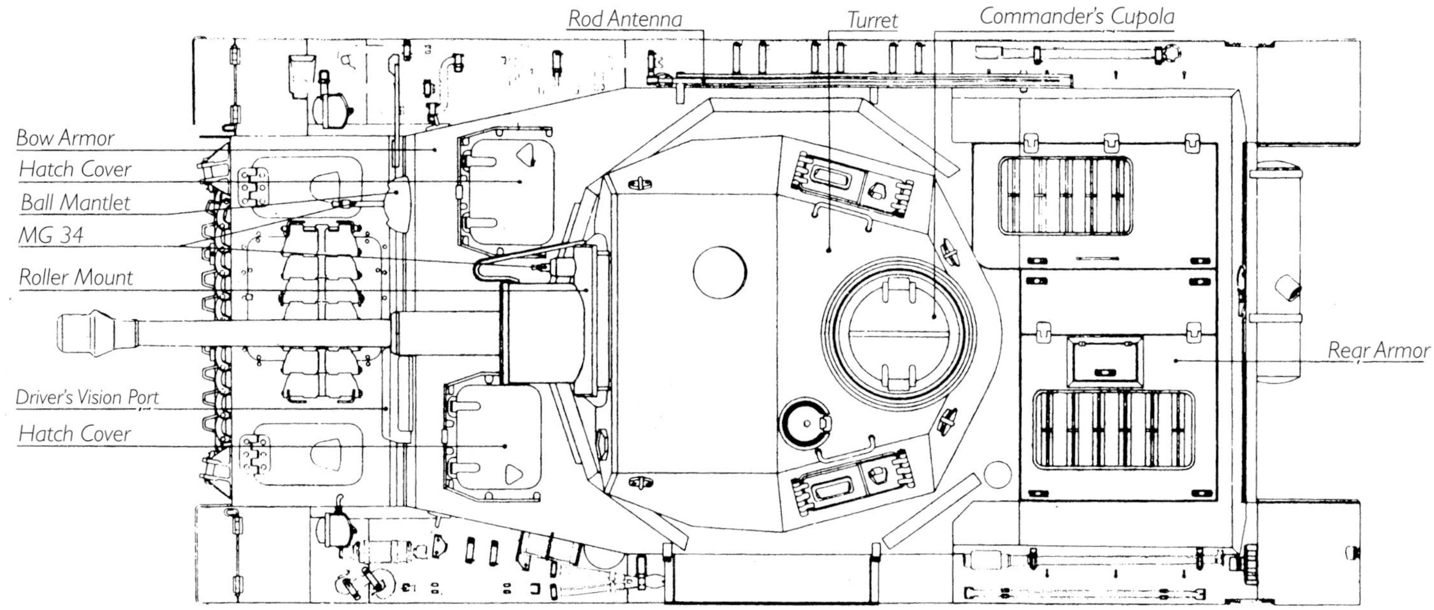

Pz.Kpfw.IV Type G (Drawing No. 021 B 48100 Bl.2 of October 8, 1942), shows, among others, the following changes: Vision ports on right turret front and turret sides plus signal flaps on driver's and radioman's hatch covers are eliminated. Two-chamber muzzle brake, antenna deflector for turret M.G., racks for 7 spare track links on upper front wall, pressure-fast headlights with shields on both track aprons, holder for two spare road wheels on left track apron, cooling air intake and exit grids on roof of rear armor plus cover for cooling water heater (built by Fuchs) on rear wall added.

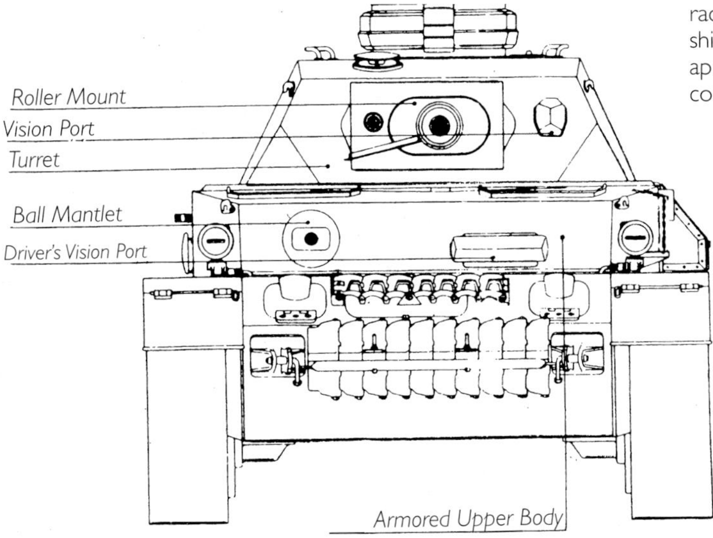

Rear View

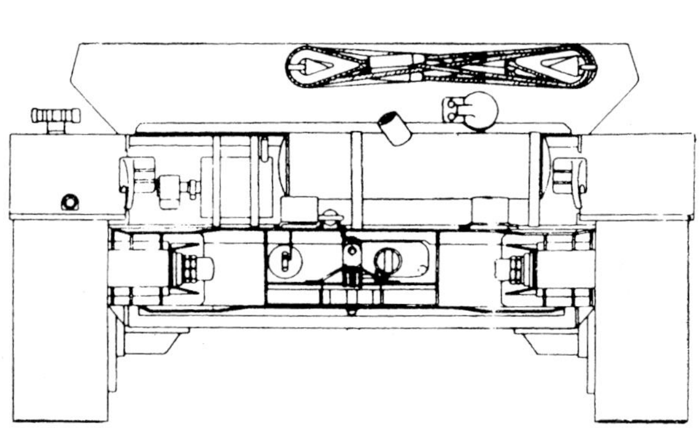

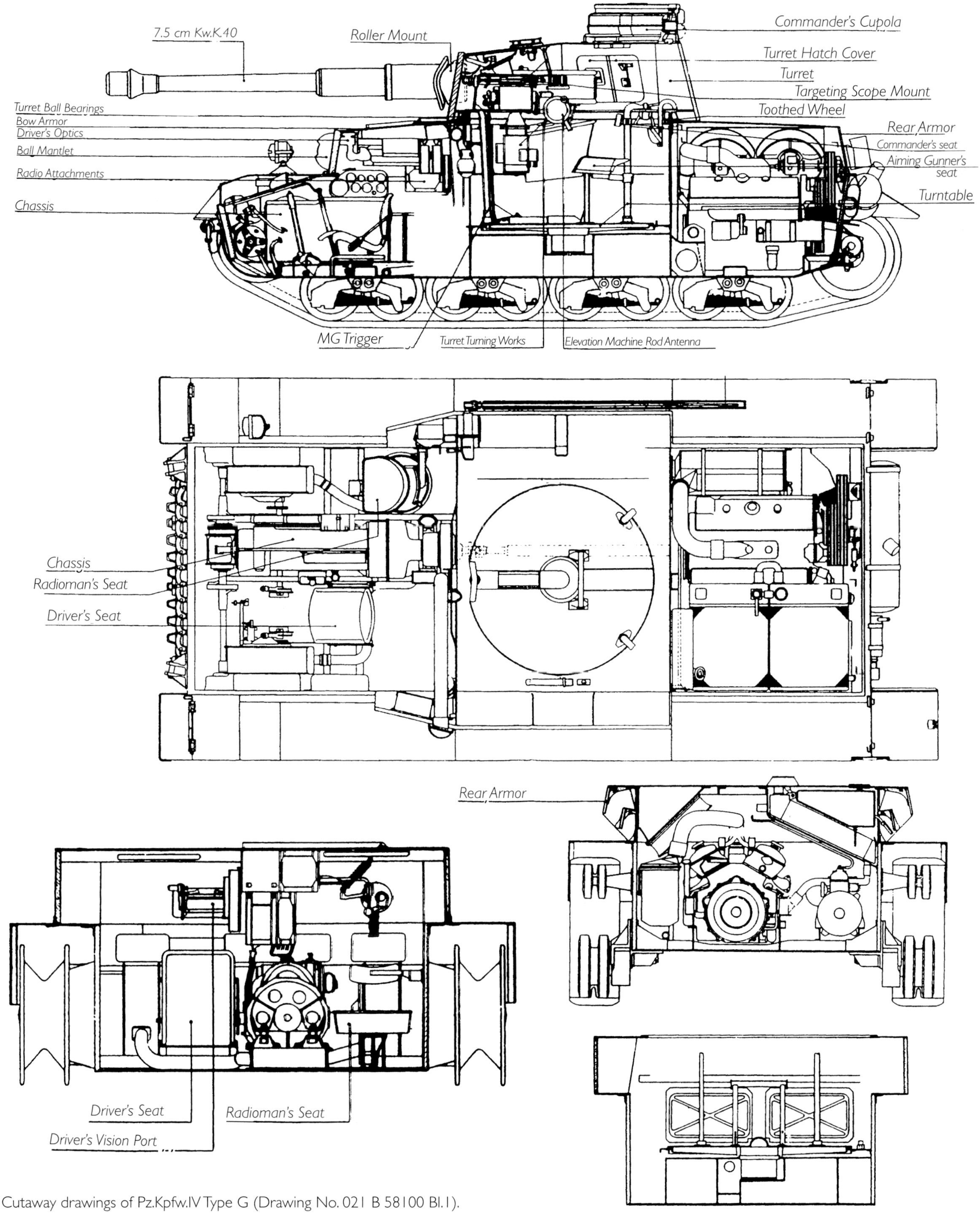

Cutaway drawings of Pz.Kpfw.IV Type G (Drawing No. 021 B 58100 Bl. 1).

11.2 FORM CHANGES DURING PRODUCTION

11.2.1 Conical Drive

As of Chassis No. 83200 (as of November 1942 at Krupp Gruson Works), strengthened cone wheels (larger module) were installed in the conical drive; the guide-seat pump was eliminated.

11.2.2 Equipment

In the realm of equipment and storage, the following improvements and changes were introduced:

For Chassis No. 82493 to 82500 (May 1942), and as of Chassis No. 82628 (as of June 1942), holders for two spare road wheels were mounted on the left track apron.

As of June 1942, racks for 7 spare track links were welded to the upper front wall.

As of August 1942, in the realm of equipment and storage, uniform winches, crowbars and mounts without leather coverings were added on the track aprons.

As of September 1942, pressure-fast headlights with shields were mounted (at first partially) for the electric system in front on both sides of the track aprons. This new type of light replaced the Notek camouflage lights. As of September 1942, the signal flaps were left off the driver's and radioman's hatch covers to simplify manufacturing. As of March 1943, the signal flap on the turret roof was eliminated to simplify production.

11.2.3 Ventilation Ports on Rear Armor

For the higher outdoor temperatures, Pz.Kpfw.IV Type G tanks were built with tropical equipment, among others, with an added cooling air intake in the radiator flap and a cooling air exit in the ventilator flap on the roof of the rear armor. As of June 1942, every Type G tank was equipped with cooling air intake and exit grids on the radiator and ventilator flaps on the roof of the rear armor.

11.2.4 Simplified Production if the B.W. Body

On January 2, 1942, Wa Prüf 6 required:

To simplify production, the right and left lateral forward vision ports of the B.W. turret and the right vision port for the loader in the front turret wall are to be eliminated at once. The Krupp firm is asked to provide the necessary drawings immediately.

Department A.K. of Krupp replied on January 6:

We shall according to precedent eliminate the two front vision ports on the turret sidewalls and the right vision port in the turret front in the set of drawings. The firms are notified by us. We shall retain the targeting rail behind the left front vision port when modifying the targeting rail block, so that the planned installation of a periscope in the turret deck is eliminated.

Already finished turrets with built-in vision ports and armored parts for turrets with holes that were already cut had to be used at first until this change was carried out. Finished Pz.Kpfw.IV tanks without this vision port were delivered partially as of April 1942 and completely as of October 1942.

11.2.5 Added Armor

For improved protection from the British and Russian tank and antitank guns, additional 30 mm thick plates were welded to the front walls of the bow armor and hull. For a special action, eight Pz.Kpfw.IV (7.5 cm Kw.K.40) were already fitted with this extra armor in May 1942. Another 16 tanks were finished at the Gruson Works in June 1942, and in fact with plans from May 1942, so that as of July 1942 every Pz.Kpfw.IV was built with added armor.

On June 30, 1942, Wa Prüf 6 wrote to Department A.K:

As ordered by the Reich Minister for Armament and Ammunition, only 16 vehicles per month are to be fitted with added armor. The delivery of tank bodies to the Krupp firm's Gruson Works is to be carried out accordingly. A supply of extra armor plates at the Upper Danube Iron Works and Böhler Brothers is being put at your disposal.

In a discussion on July 11, 1942, it was reported:

As for the added armor, Colonel Fichtner wants to suggest at the Armor Commission meeting on July 14 that the request to fit 16 B.W. bodies per month with armor also be allowed to lapse. A decision will be shared with the firms immediately after the conclusion of the Armor Commission meeting.

Pz.Kpfw.IV Type G (formerly Type F2), built by Krupp in May or June 1942. One of the first vehicles with added armor on the body and hull bow.

Pz.Kpfw.IV Type G, Chassis No. 84274, is seen crossing a low spot in the terrain.
Built in the Nibelungen Works in May 1943, Type G was given added welded-on armor (without a cutout for the driver's optics).

Further, on 16 July:

The production of armor strengthening for Pz.Kpfw.IV (16 per month) ordered on Document 76g 10/14 No. 5051/42 Wu G 6/VIII of 6/5/1942 is to be stopped at once. For armor strengthening from now on, only that material is to be used that is no longer usable for other purposes under any conditions. It is requested that you state at once how many armor strengthenings were delivered by today and how many will still come to delivery on the basis of the former order.

Krupp of Essen answered on 20 July:

We have, on the basis of the teletype, stopped the application of strengthening at once. To date, 61 sets of armor strengthenings have been delivered by us. In addition, there are on hand here:

28 front walls SKC 4991	*burned and hardened,*
84 front plates SKD 4994-1	*burned,*
248 front plates SKD 4994-2	*burned, and*
320 ridges SKE 4994-3	*burned.*

We ask for authorization that the 28 sets of which the material is completely burned and partly already hardened can be finished, and for directions as to what is to happen to the remaining material, since other use is not possible. The Gruson Works would have to equip not, as previously planned, 71 tanks, but rather 89, with the strengthenings in accordance with the above.

Pz.Kpfw.IV Type G, Chassis No. 83884, in southern Italy.
Built with added screwed-on armor at Krupp in May 1943.

Wa J Rue decided: *The 28 sets or for the most part finished fore-armor are to be delivered to the Gruson Works. The remaining material is to be kept ready until further decision is made.*

From Wa J Rue files it is known that the Krupp Gruson Works fitted 16 Pz.Kpfw.IV with 30 mm additional armor in July 1942,

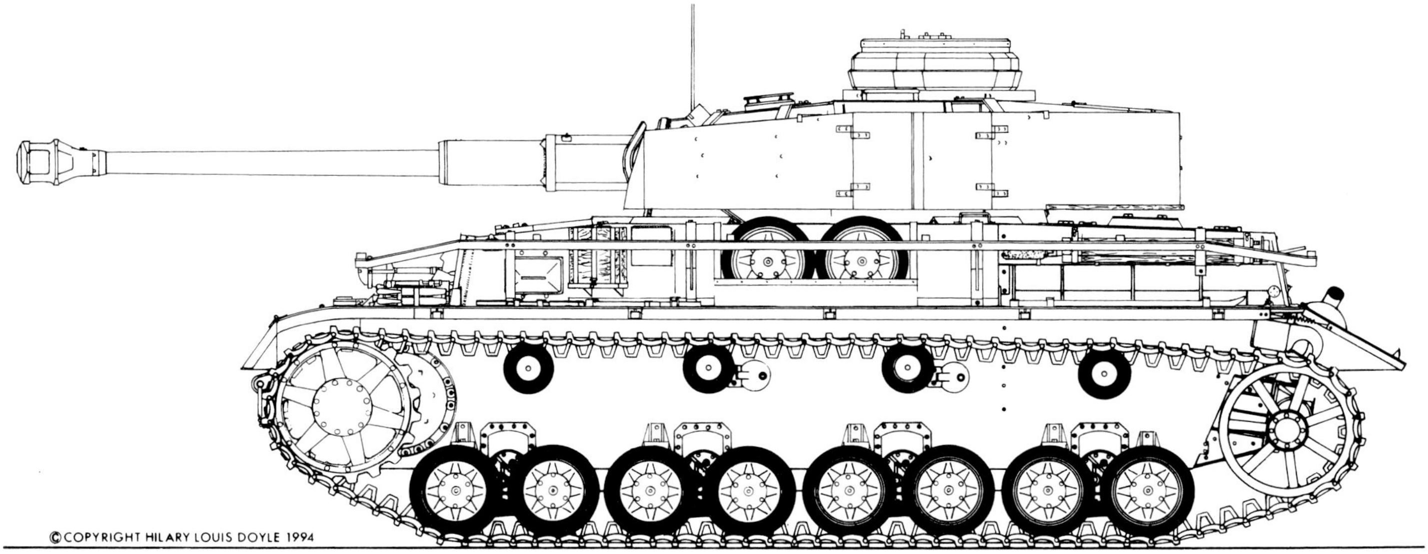

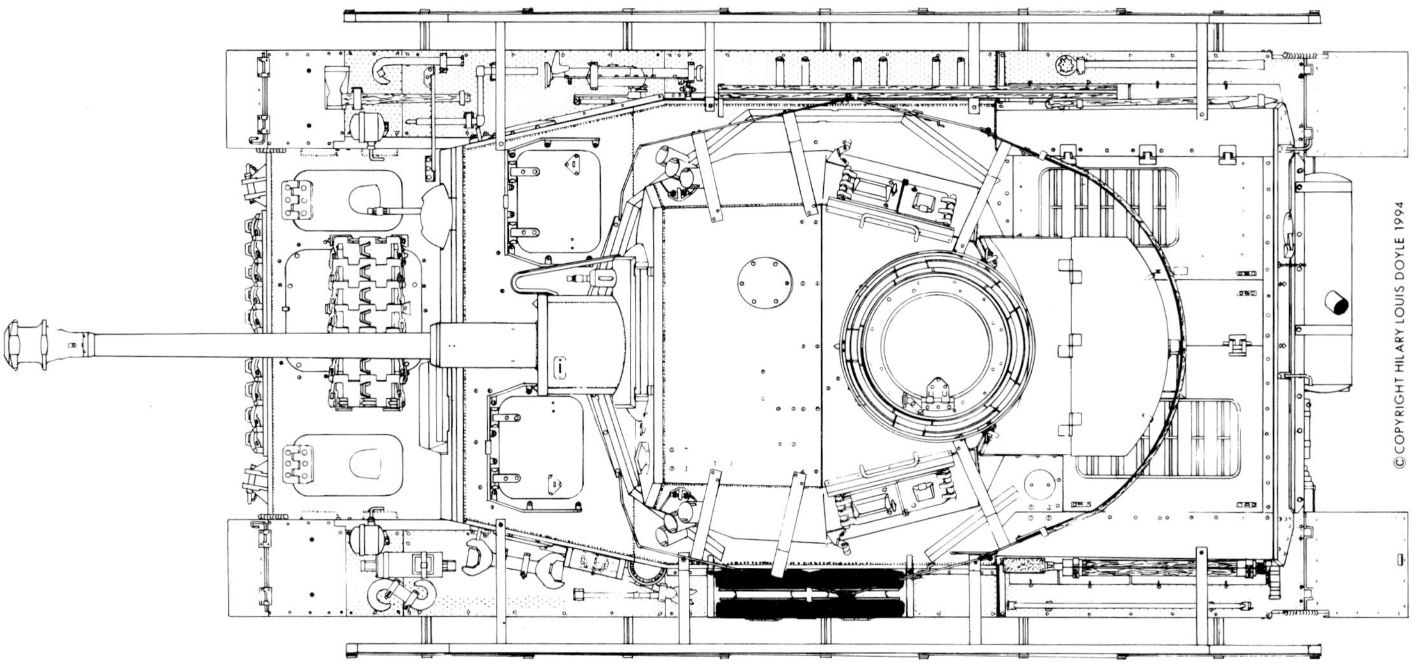

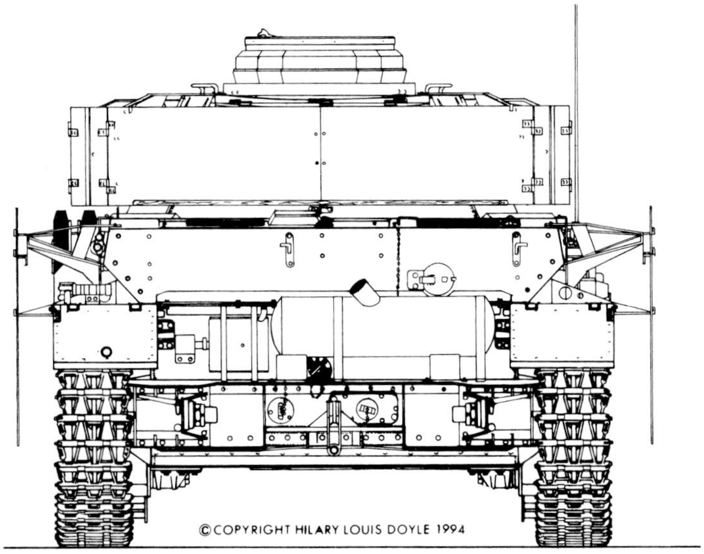

Panzerkampfwagen IV (7.5 cm), (Sd.Kfz.161), Type G, with welded added armor, finished April 1943.
7.5 cm KwK 40 L/48 with double-chamber muzzle brake, new enlarged commander's cupola, side turret periscope and vision ports at right front eliminated. Foglaying launcher attached to the front of the turret sidewalls. Openings for signaling in driver's and radioman's hatches eliminated. Added 30 mm thick armor plates welded to front of hull and turret. New stowing of spare track links on the front bow plates. Bosch lighting replaced the Notek night driving equipment. Container with two spare road wheels on the left track apron. Armor skirts on turret. Wooden cover in turret apron to protect antenna. Hangers for aprons on both sides of the chassis, openings for the Fuchs cooling-water exchange device in rear hull panel.

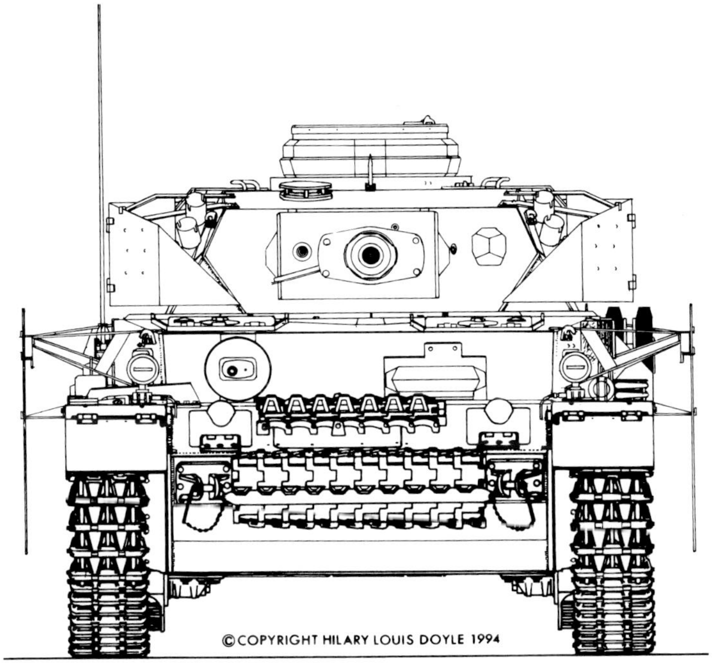

Pz.Kpfw.IV Type G with welded-on extra armor. Built as of April 1943. The K.F.F.2 Driver's Sight Device was dropped as of February 1943; the openings for it were replaced by plugs.

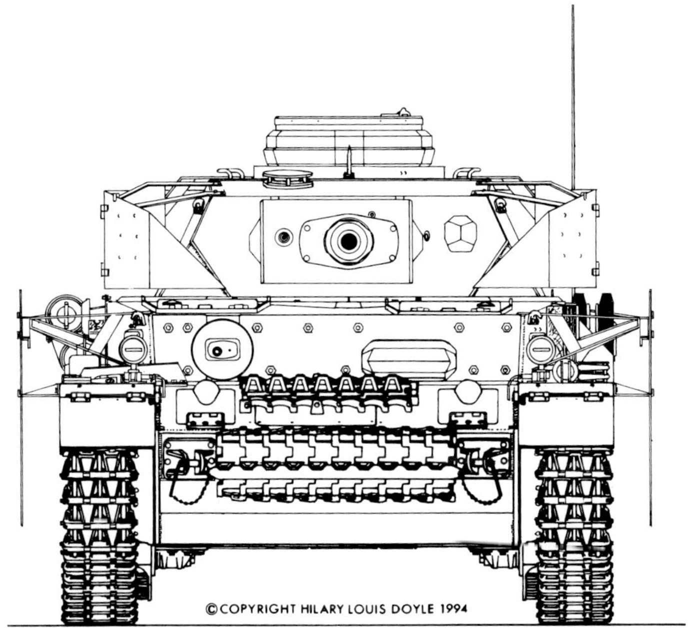

Pz.Kpfw.IV Type G with screwed-on extra armor. Built as of June 1943. The foglaying device was dropped. Dry filters were added on the right track apron. The radio antenna was now attached to the left rear motor cover by a new mount.

16 in August, and 14 in September. In a conference on October 13, 1942, it was decided:

The Führer is of the opinion that with the steadily increasing armor thickness of the enemy tanks, at least a portion of the Panzerkampfwagen IV should retain fore-armor of 80 mm as extraordinarily valuable.

On 8 November: *According to the available troop reports, the fore-armor of the Pz.Kpfw.IV has proved itself despite the related driving-technical hindrances. Since the new British six-pound antitank gun still penetrates the front of the present Pz.Kpfw.IV at 1000 meters, it is determined that 50% of the production should be fitted with fore-armor.*

And again on January 3, 1943: *The Führer decides on the basis of the test reports that for a short time the Pz.Kpfw.IV will go on with a strengthened angled bow front, on the basis of the suggestion. Until this date, all Pz.Kpfw.VI are now to be fore-armored to 80 mm. The 47 Pz.Kpfw.IV standing ready for delivery in Magdeburg should thus be fitted with fore-armor at once, since they are intended for a special action.*

Further decisions about extra armor are found in a note on the discussion on February 19, 1943, in Magdeburg. Wa Prüf 6 took the following position:

The transition from only welded-on armor thickening to screwed-on must be undertaken as soon as is at all possible. The May date appears extraordinarily late. Reference is thus made to the teletype of Dr. Rohland's office (Special Branch VI Tanks) of 2/15/1943, by which only hulls with screwed-on added armor shall be delivered by the tank firms as of March 20. Ways and means must be found to make screwing on the extra plates possible, also for the tank hulls already located at the assembly firms.

Krupp of Essen had assured WuG6 on 25 February: *Fore-armor in screwed-on form will be delivered by us as requested as of April 1943.*

On February 26, 1943, it was explained that *the entire April production is to be delivered in screwed-on form. But since the Bauer & Schaute firm already delivered the screws on March 23, efforts should be made to deliver as many of the March products as possible in screwed-on form, and to add armor afterward, if possible, by the assembly firms.*

11.2.6 Antenna Deflector

With the new longer tank gun, Pz.Kpfw.IV was delivered at first without an antenna deflector or with the older type designed for the shorter L/24 gun. As of June 1942 (to July) a new antenna deflector, a lengthened form of the older type, was mounted low on the tank gun.

In a discussion at the Gruson Works on July 14-15, 1942, it was stated: *On instructions from WaPrüf 7, the antenna deflector can be omitted. A simplified protective bow is simply to be attached in front of the M.G., so that the M.G. does not shoot the antenna off. An automatic antenna-moving device is omitted.* This decision was extended on by Reg.Baurat Biersack (Wa Prüf 6) in a discussion at the Krupp firm's Gruson Works on 30 July: *To the vehicles of the current series production, an antenna deflector is to be attached to the turret mount only so that the antenna is not shot off by the vehicle's machine gun. The antenna deflector is to be isolated by wooden blocks, so that the antenna receives no massive damage between the 3:00 and 9:00 turret settings.*

As of November 1942 (until May 1943) this new type of antenna deflector was mounted on the right side of the roller mount.

11.2.7 Winter Equipment

On July 11, 1940, the installation of the crankshaft gasoline starter (MBA) for the HL 120 TRM was discussed:

Pz.Kpfw.IV Type F2(G) with lengthened antenna deflector.

Pz.Kpfw.IV with wedge plow (built by Schmidt) mounted in front.

In the rear wall a flap of some 150 x 350 is planned for the placement of the belt. A hole with 40-60 mm diameter must be bored and is to be closed watertight by an armor cover (15 mm thick for Types A to C and 20 mm thick as of Type D). As a guide for the KBA, welded-on massive and short guiding bolts on the cover. For further tests with the KBA, besides the B.W.I test vehicle, which does not represent the newest state, a normal series vehicle was in Kummersdorf.

As of April 1942 warm-air heating for the fighting compartment was installed as follows:

In the bulkhead between the fighting compartment and motor room a hole 160 mm wide was burned, likewise in the sidewall of the ventilator box opposite the bulkhead. On the hole in the ventilator box the flange collar was attached by screws. Over the flange collar Duct No. 2 was inserted with the flange duct inserted over it. The flange duct was then attached to the bulkhead. If too much warm air was blown into the crew area, the rear ventilator flap could be opened on the outside of the vehicle.

On July 6, 1942, Wa Prüf 6 had requested of Krupp in Essen:

The following changes to the series of Pz.Kpfw. chassis produced by you should be considered as soon as possible:

1) *Warm-air heating corresponding to the drawing for Pz.Kpfw. IV sent by Alkett*
2) *Warm-water circulation heating corresponding to the Alkett drawing for Pz.Kpfw.IV*
3) *Installation of the Fuchs warming device corresponding to the pictures sent by Daimler-Benz.*

On July 28, 1942, OKH Wa J Rue, WuG 6, VIIId wrote to the Krupp Gruson Works, Vomag and the Nibelungen Works:

OKH asks that, immediately if possible, the Pz.Kpfw.IV be equipped with the aids and devices required for winter operation. These include:

1) *Warm water passage device (developed by Alkett), consisting of*
   - a) *Built-in pipes (for connection with cooling circuit)*
   - b) *ents for this,*
   - c) *Closing vents (Arado system)*
   - d) *Connection hose*
2) *Firmly installed cooling-water heating device (built by Fuchs) developed by Krupp in Essen*
   Soldering lamps for this can be ordered from the Naebel firm in Herdersdorf, Erzgebirge.
3) *Installation of starter fuel injection system consisting of*
   - a) *SUM injection pump*
   - b) *Korcher nozzles*
   - c) *Pile connections*

   Efforts must be made to equip vehicles finished after September 1 with these parts.

Pz.Kpfw.IV Type C, Chassis No. 80353, with cooling-water lines.

Pz.Kpfw.IV Type G, Chassis No. 83142, made by the Nibelungen Works in November 1942, with a round cover for the cooling-water heater (built by Fuchs).

Pz.Kpfw.IV Type G with winter track links.

Pz.Kpfw.IV Type G, Chassis No. 83672, made by the Nibelungen Works in March 1943, with commander's cupola (Drawing No. 021 B 2481) and a foglaying device on each side of the turret.

Further, on July 31, 1942, Baurat Kniepkamp (Wa Prüf 6/IIIa) explained:

The vehicles in the new series, thus not retroactively, will be set up for the crankshaft-gasoline starter by the troops. For this the motors receive an appropriate coupling shell. The holding bolts are to be attached to the vehicles.

The vehicles will be equipped with a starting fuel injection system, with retroactive effect on the vehicles with the troops.

On August 5, 1942, in a discussion in Magdeburg with Wa Prüf 6/IIIg, it was decided:

After a heating test with the Fuchs device (installed by the Gruson Works in Vehicle No. 80625), which went satisfactorily in all parts, it was decided by Wa Prüf to carry it out as a form change by the Gruson Works.

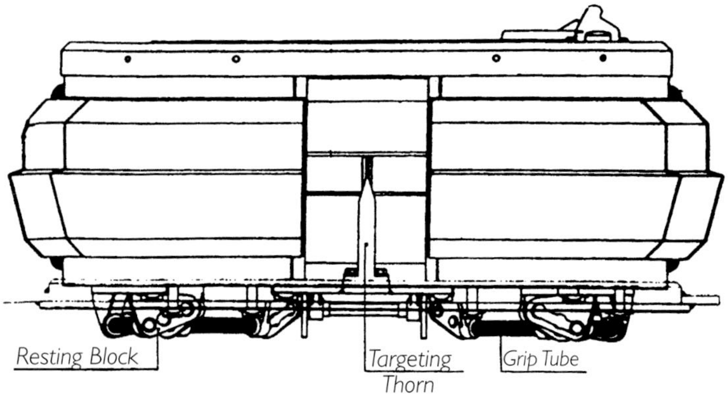

In a display of the winter equipment at the barracks of the "Grossdeutschland" Regiment on August 19, 1942, there was displayed:

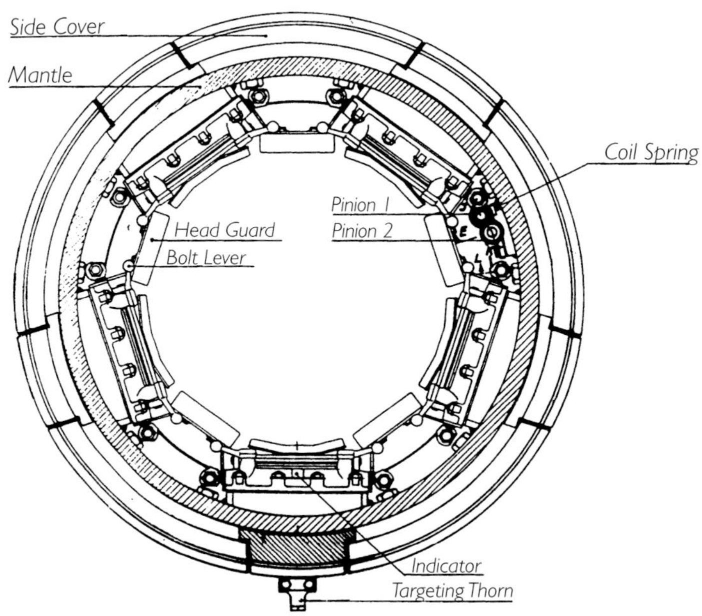

Commander's cupola (Drawing No. 021 B 2481) for Pz.Kpfw.IV Types G to J.

Factory-new Pz.Kpfw.IV Type G with 7.5 cm Kw.K.40 L/48, finished with aprons as of April 1943.

Pz.Kpfw.IV Type G, finished with aprons as of April 1943.

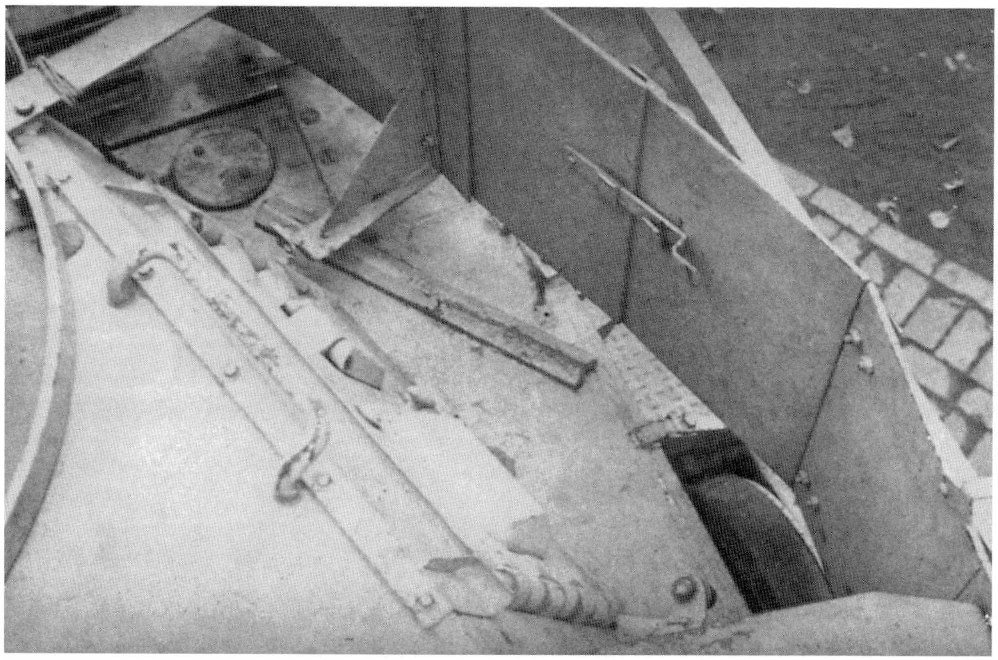

To be able to open the turret hatch cover, appropriate covers were planned for the added turret aprons.

Pz.Kpfw.IV Chassis No. 80353 with
- Cooling water connections
- Fighting compartment heating
- Cranking-power starter warming
- Starting by means of crankshaft gasoline starter
- or motorcycle crankshaft starter.

Pz.Kpfw.IV with front-mounted wedge plow (built by Schmidt) to increase snow driveability.

At a discussion in Essen on July 30, 1942, it was decided:

Wa Prüf 6 will see that reported attachments for cooling-water passage devices from Alkett to A.K. will be hastened. The hold in the back wall of the hull for attaching the Fuchs device for cooling-water heating is to be planned for the series at once. Change series drawings. Form-change drawings for repaired vehicles to be finished by end of August.

As requested, these devices for winter equipping were installed in series vehicles by the assembly firms as of September 1942.

11.2.8 Foglaying Device

From February to May 1943, the Pz.Kpfw.IV was equipped with two foglaying devices, which were attached on either side of the turret. Each device had three sprung launching containers, which were activated by the turret crew. The Army Weapons Department ordered that the equipping of Pz.Kpfw. with these devices be halted after field units reported that enemy small-caliber bullets passed through the launchers, set off the fog cartridges, and thus blinded the tank's crew.

11.2.9 Commander's Cupola

The new commander's cupola with one-piece cover and thicker armor protection was requested by Colonel Fichtner (Wa Prüf 6) on July 11, 1942, as follows:

Only the new commander's cupola is to be taken over in the building of the B.W. armored body most quickly. It should be determined by which date on this new commander's cupola can be finished. A date at the end of September's production of tanks is desired by the tank firms from the Weapons Department.

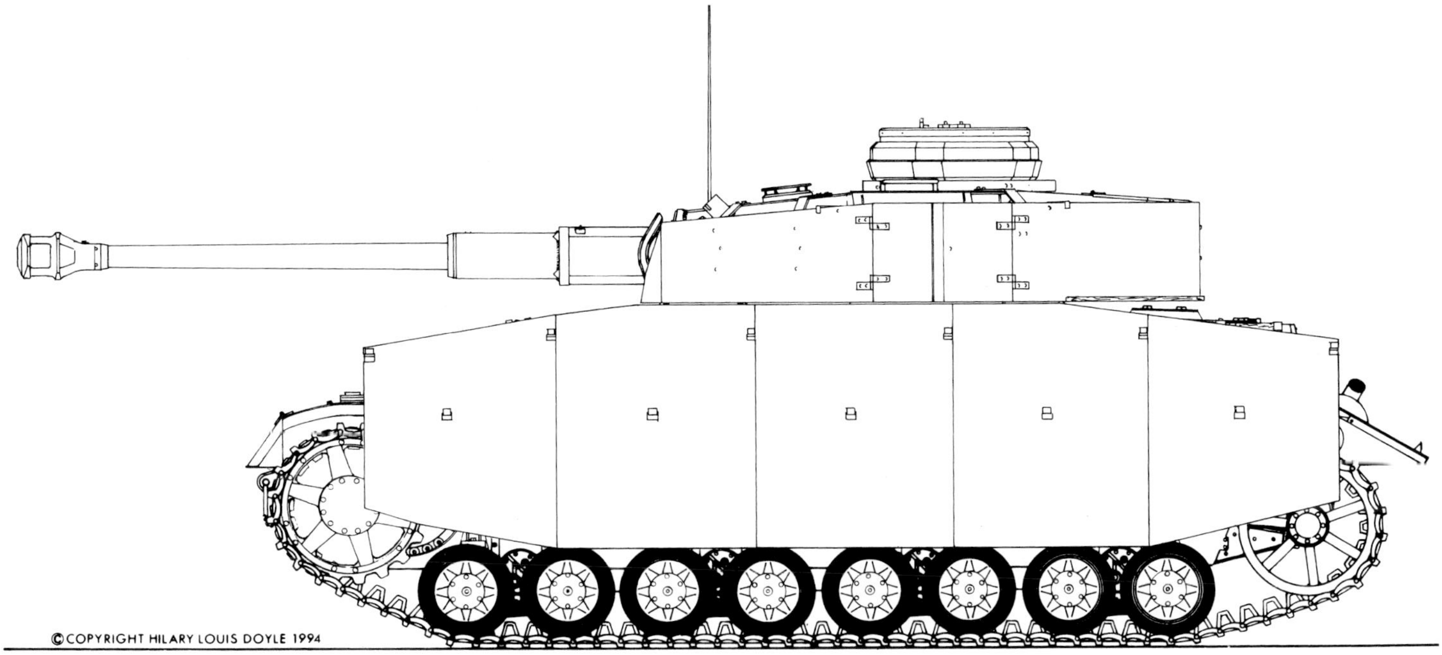

Pz.Kpfw.IV Type G with aprons, as of April 1943.

This new commander's cupola was introduced on the Type G series vehicles as of February 1943.

11.2.10 Dropping the Driver's Optics

On February 19, 1943, it was explained by Wa Prüf 6 that the driver's optic was to be dropped and: *Wa Prüf 6 agrees to the closing of the no longer needed holes of the driver's optics with a pointed-head bolt instead of the desired wedge-head screws.* As of February 1943 the K.F.F.2 Driver's Optic was dropped. At first the holes in the front bow armor plate, as requested, were closed by pointed-head bolts, then covered by the 30 mm added armor, and finally bow armor without bored holes went into production.

11.2.11 7.5 cm Kw.K.40 L/48

For unknown reasons, the introduction of the 7.5 cm Kw.K.40 L/48 was delayed. The 7.5 cm Kw.K.40 L/43 was still installed in the Pz.Kpfw.IV Type G until April 1943. The same L/43 gun arrel had already been replaced by the longer, simpler, higher-performance L/48 in assault gun cars in July 1942. In the Pz.Kpfw.IV the L/43 was replaced by the 7.5 cm Kw.K.40 L/48 by all three assembly firms as of the beginning of April 1943.

As of September 1942, the ball-shaped muzzle brakes were replaced by the two-chamber acorn-shaped muzzle brake without lobes. As of April 1943 this second type was replaced by a new two-chamber acorn form with front-lobed muzzle brakes.

11.2.12 Aprons Against Antitank-Gun Shells

On February 17, 1943:

The Armor Commission suggests to the Führer that protective armor be planned for the sides and back. The test version by Daimler-Benz with 5 mm plates before the body and 10 mm plates before the turret (total weight some 600 kg) is to be considered. The test version will be fired on along with short road tests on February 20, 1943 at the Kummersdorf Test Center and then displayed to the Führer.

On February 20, 1943, firing tests at the protective cover were made. Firing tests with 14.5 mm Russian antitank guns at a range of 100 meters (90 degrees) showed no cracks or penetrations through 30 mm side armor, which was protected by either wire mesh or plates. Firing with the 7.5 cm A.Z. explosive shell from a field gun with 2nd charge showed no damage to the hull side when protected by mesh or plate. Wire mesh and armor plate were perforated or torn away, to be sure, but still remained usable. The decision to use armor instead of mesh (though both had proved themselves and wire mesh was lighter) was based on the fact that the wire mesh required a new attachment, still to be designed, which took time to develop.

The aprons were not tested against firing with hollow-charge shells, as they were not seen as protection against that type of ammunition.

After the Führer's conference on March 6, 1943, it was reported:

The Führer was informed of the course, favorable to date, of the firing tests with aprons. Besides the current equipment of all assault guns, Pz.Kpfw.IV and Panthers of new production with aprons, in the shortest possible time a full supplying of all vehicles of this type in action or being repaired must be assured. Schedule planning for the full equipping is to be done quickly.

As of April 1943, aprons were mounted by the assembly firms on the Type G vehicles.

11.2.13 Felt Bellows Filters

On August 5, 1942, Wa Prüf 6/IIIg already decided:

For the connection of the felt bellows filter the Gruson Works had prepared two versions based on SKB 5-32 and SKB 5081. Wa Prüf agrees to the version after SKB 5032 with added armor protection. For this, form-change drawings are to be prepared.

This felt-bellows connected air filter was installed by the assembly firms, as of May 1943, on the track apron on the right side of the body.

Pz.Kpfw.IV Type G, finished in June 1943, with rubber base for the antenna at the left rear on the stern armor.

Pz.Kpfw.IV Type G, Chassis No. 83920, finished by Krupp in May 1943, rebuilt as an armored command car, with antenna on the turret roof and star antenna on the stern armor.

11.2.14 Rubber-Mounted Antenna on the Rear

In a discussion on March 4, 1943, at Department A.K., Colonel von Wilcke (Wa Prüf 6/IId) explained: *The protective armor hinders the antenna. It is decided to attach a rubber-mounted antenna on the rear of the tank. The antenna deflector for the MG shall be dropped in the future.*

As of May 1943, the antenna was mounted on a rubber base at the left rear of the stern armor during series production.

11.3 IMPROVEMENTS AFTER DELIVERY

A series of improvements was ordered after the delivery of the Pz.Kpfw.IV Type G to the troop units. Among others:

1) As of April 1942, installation of flange ducts for warm air heating of the fighting compartment.
2) As of September 1942, installation of a cooling-water heating device (built by Fuchs).
3) As of October 1942, installation of connections and vents for warm water flow.
4) As of November 1942, winter track links and snow-grippers for tracks were delivered to the troop units.
5) As of May 1943, mounting of aprons on turret and hull sides.
6) As of January 1944, Zimmerit protective paint could be applied to the armor plates by the troops and shaped into waves.
7) As of February 1944, felt-bellows connected air filters complete with air ducting could be removed up to the oil filter and the air-duct openings be closed by welding on plates.
8) As of February 1944, repaired Pz.Kpfw.IV could be rebuilt as Pz.Bef.Wg.IV command cars.
9) As of May 1944, a widened track link, called "Ostkette" (Eastern Track), was delivered to the troop units on the eastern front.
10) As of July 1944, bollards for the 2-ton emergency crane were welded to turret roofs.

Pz.Kpfw.IV Type G, Chassis No. 83920, finished by Krupp in May 1943, rebuilt into an armored command car.

12. PANZERKAMPFWAGEN IV (7.5 cm) (Sd.Kfz.161/2) Type H, (9./B.W.)

Chassis Number Series 84401—86393 and 86601—89540

Between May 1943 and February 1944, 2322 Pz.Kpfw.IV Type H were finished and accepted (379 from Krupp, 693 from Vomag and 1250 from the Nibelungen Works).

12.1 IMPROVEMENTS FROM TYPE G

On July 11, 1942, Colonel Fichtner (Wa Prüf 6) decided:

For Type H (9./B.W.) the diving capability and the B.W. 40 turret (hydraulic drive and new optics) are dropped. The strengthening of the cone drive of the side shafts and planetary drive, though, is to be carried out quickly.

The Type H showed a mounting and fastening of the drive wheel that deviated from the previous version. The entire suspension was located here in an extended neck of the cover of the side gearings. In Type H too, the end-running ring for the shaft seal of the rear sealing ring, as opposed to the former version, was not held on the running-gear lever by a wedge, but only linked with it by a press fit.

It was determined that the new turret with the cast steel front was no longer to be made, because the production of B.W. tank bodies would presumably run out at the end of 1943. The turret form of Type G was still used, but the turret roof was strengthened for better protection from low-flying aircraft attacks. The thickness of the front turret roof plate was increased from 10 to 16 mm, and that of the rear plate from 10 to 25 mm.

Krupp asserted on July 30, 1942: *For strengthened cone drive, side shafts and planetary drive, the number series 48400 and the designation of Type H (9./B.W.) are preferred.*

12.2 FORM CHANGES DURING PRODUCTION

12.2.1 Planetary Drive

On February 24, 1943, Wa Prüf 6 explained:

The planetary drive cover (No. 48440-1) is in itself, since only 30 mm thick and equipped with a 14.5 mm ring cover, a weak point. When at this time, by strengthening the front armor and attaching fore-armor aprons, the armor of the Pz.Kpfw.IV is, among other things, also strengthened against the Russian antitank guns, then it seems unbearable to make the planetary drive cover vulnerable by finishing in Cast Steel 52.81 also for fire from SmK infantry bullets. From the side of the Wa Prüf 6, it can therefore not approve the production of this cover of Cast Steel 52.81.

In a discussion with the Gruson Works in Berlin, WuG6 and Wa Prüf 6 on March 1, 1943:

Since for 369 vehicles the covers in the cast armor steel form cannot be covered, it was decided to make the missing covers of 52.81 Cast Steel. Since the expiration date for the production of the vehicles is already too soon for these cast steel lids, it was agreed that the Vomag 60 gearbox of the older type from a spare-parts order be built into the vehicles of the series.

12.2.2 80 mm Front Armor

As recorded from the discussion on February 19, 1943, in Magdeburg, Wa Prüf 6 took its position as follows:

The transition to 80 mm organic armor remains to be striven for as soon as possible. Hereby a transitional solution will probably not let itself be avoided, by which the bow front will be formed of organic 80 mm and the driver's front by screwed-on 30 mm plates, as long as to the driver's vision port and the ball mantlet an organic solution from Krupp is available.

On February 27, 1943, Mr. Lanz (of Krupp) wrote to Wa Prüf 6/II:

As was agreed in the discussion with WuG3 on February 19 in Magdeburg, we have for the sake of haste sent to the firms of Böhler, Oberdonau, and our works on February 26 the drawings for 80 mm thick plates for the hull and driver's front. After the decision of the Armor Commission on February 17, we will later remove the surface hardening.

The three armor firms prefer the organic version to the screwed-on version, because the burning out of the second plates, the boring and countersinking of 21 holes, the hardening of the plates, and other things take up too much time.

Kerupp explained on 1 March for the armor hulls, that … *as of May production a homogeneous version can be delivered.* According to a message from WaPrüf 6 (March 9, 1943), the organic version of the front plates (80 mm) was introduced.

On June 15, 1943, the production of the armored body for Pz.Kpfw.IV in its organic version was to take place.

On 4 March, there came out in a discussion with Department A.K. and Colonel von Wilcke (Wa Prüf 6/IId):

The Krupp firm presents two drawings for modifying the vision port and ball mantlet. The vision port includes a hand lever and segment lengthening, though at first through welding. Ball mantlet receives a different forged piece and is lengthened by 32 mm. The Krupp firm is concerned about the fast production and suggests, in the interest of untroubled production, that in line with the earlier suggestion they leave out the lowering of the 80 mm plate.

It is nevertheless decided to carry out the change above; condition: MG turning range no more than 3 degrees of loss to both sides. In the vision port, vision limiting of ca. 2.5 degrees taken in the bargain. There must be no schedule shortening!

Further form changes in the front strengthening were shared by Krupp with Wa Prüf 6 on March 10, 1943:

Since in the driver's vision port a makeshift change of the resting segments and lever is difficult, new lowerings for them should be created quickly. Until this point plates must be screwed to the driver's front as before. According to our suggestion of March 4, the organically strengthened driver's front could be introduced at once while keeping the previous segments and lever, if the front wall is inset 30 mm deeper.

In introducing the organically strengthened driver's front, the side field of vision is decreased, with the ball mantlet screwed on from outside, resulting from hindrance from the optic housing, from 20 to 12 degrees. Beyond that, four screw holes on the back of the front wall and a surface of some 130 sq.cm. in the area of the optic must be sunk 30 mm deep. Instead of that, we urgently request understanding for our suggestion to lower the attachment surface for the driver's vision port and ball mantlet 30 mm from outside. In this case maintaining the side aiming field , essential performance upgrading and greater safety.

After inquiry, the three armor firms prefer organic plates with corresponding working. Prompt decision necessary, so that no schedule delay occurs.

On March 11, 1943, Wa Prüf 6 decided:

Begin at once with organic version (insetting). Take up the changes to the vision port and ball mantlet decided on in the discussion with Colonel von Wilcke of March 4 at the same time and finish by March 31. From this point on, the organic version must be guaranteed with modified vision port and ball mantlet.

Panzerkampfwagen IV Type H, Chassis No. 84599, finished by Krupp in September 1943, with Zimmerit coating and modified holders for side aprons.

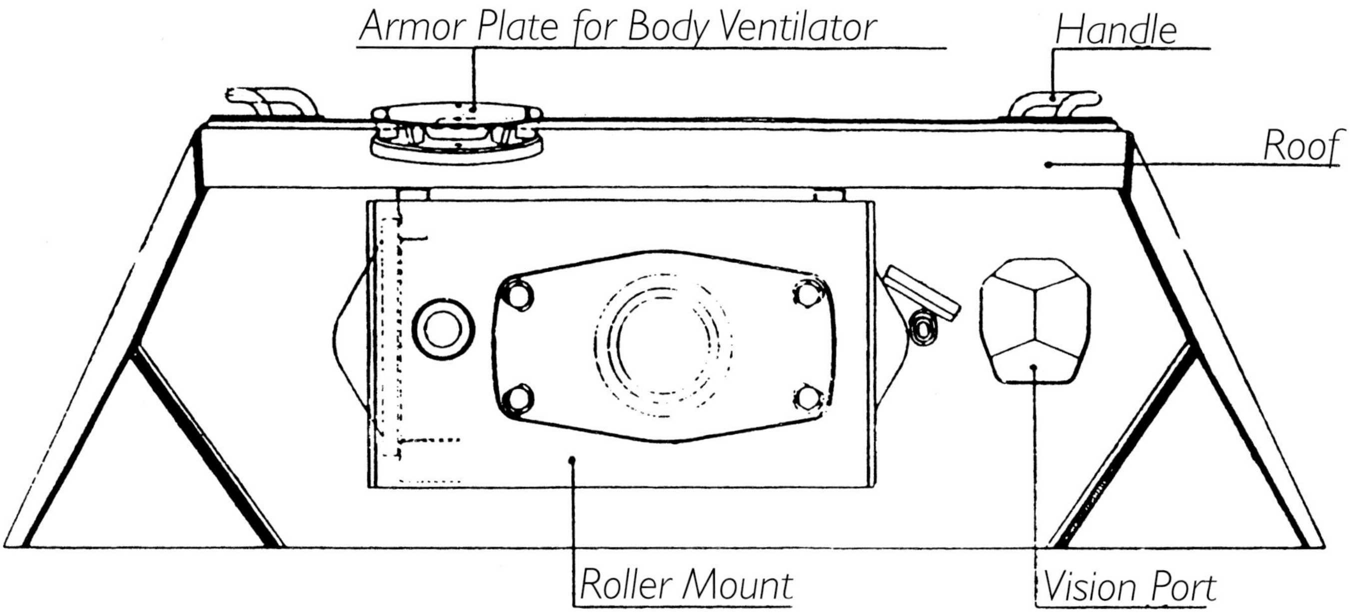

12.2.3 Side Vision Port on the Bow Armor

On March 4, 1943, it was already made known in a discussion between Department A.K. and Colonel von Wilcke (Wa Prüf 6/IId):

Protective armor covers side vision port. Whether these vision ports will still be attached in the future or not shall be decided by troop testing.

These side vision ports were installed in Pz.Kpfw.IV as of June 1943, partially dropped, and dropped completely as of September 1943.

12.2.4 Warm-Air Heating

On March 2, 1943, the Gruson Works were contracted to carry out tests with the installation of a warm-air heater for the driver in a Pz.Kpfw.IV from current series production. This warm-air heating was installed in Pzx.Kpfw.IV before the winter of 1943-44.

12.2.5 Zimmerit

In a discussion on March 1, 1943, in Berlin with the Gruson Works, WuG6 and Wa Prüf 6, it was decided:

The instruction advanced by Wa Prüf 6 to apply a coat of concrete to all vehicles is withdrawn. The department is already conducting experiments in coating vehicles with an artificial resin. The covering with this artificial resin is to be done to the whole vehicle, including the surfaces.

A teletype of June 30, 1943, about protective materials against magnetic adhesion refers to a display at Kummersdorf on June 5 and 8, 1943:

For testing the durability of protective materials against magnetic charges and fire, a troop test with a protective coating is ordered at Army Group South, 7th Armored Division, and at Army Group Center, 4th Armored Division.

As of the end of September 1943, a protective coat of Zimmerit was applied to Pz.Kpfw.IV in order to prevent the application of magnetic adhesion charges.

12.2.6 Simplification of Finishing

To simplify finishing, cast attachment blocks for running-gear levers were replaced by welded ones as of September 1943.

As of September 1943, also to simplify finishing, cast wheel hubs for road wheels were replaced by pressed ones.

As of October 1943, the rubber tires return rollers were partly replaced by those with steel surfaces. The rubberless return rollers showed differences from older versions in terms of their attachment to the axle hub of the bearing block.

To simplify finishing further, as of October 1943 a cast leading wheel was introduced (not 100%).

12.2.7 Holders for Aprons

Numerous complaints from field units reported that the aprons did indeed protect from antitank shells, but that the attachments were fully insufficient and thus the aprons were often lost.

To hinder aprons from tearing off, as of October 1943 modified side ridges with triangular holders were introduced.

12.2.8 Stilt Running Gear

On September 7, 1943, Colonel Holzhaeuer (Wa Prüf 6) reported:

A test run of a Pz.Kpfw.IV with stilt running gear has gone trouble-free for 2000 km. The installation of the middle blocks is possible among the troops everywhere, but takes a considerable lot of work.

Through these built-in middle blocks, the ground clearance of the vehicle was raised from 400 to 500 mm for better mobility in mud and deep snow.

After serious disturbances of production in November 1943, this modification was dropped because of the greater work time and delayed delivery of important parts.

12.2.9 Meshed Hull Front

To strengthen the hull front, as of December 1943 the sides of the 80 mm front plate were meshed with the side plates.

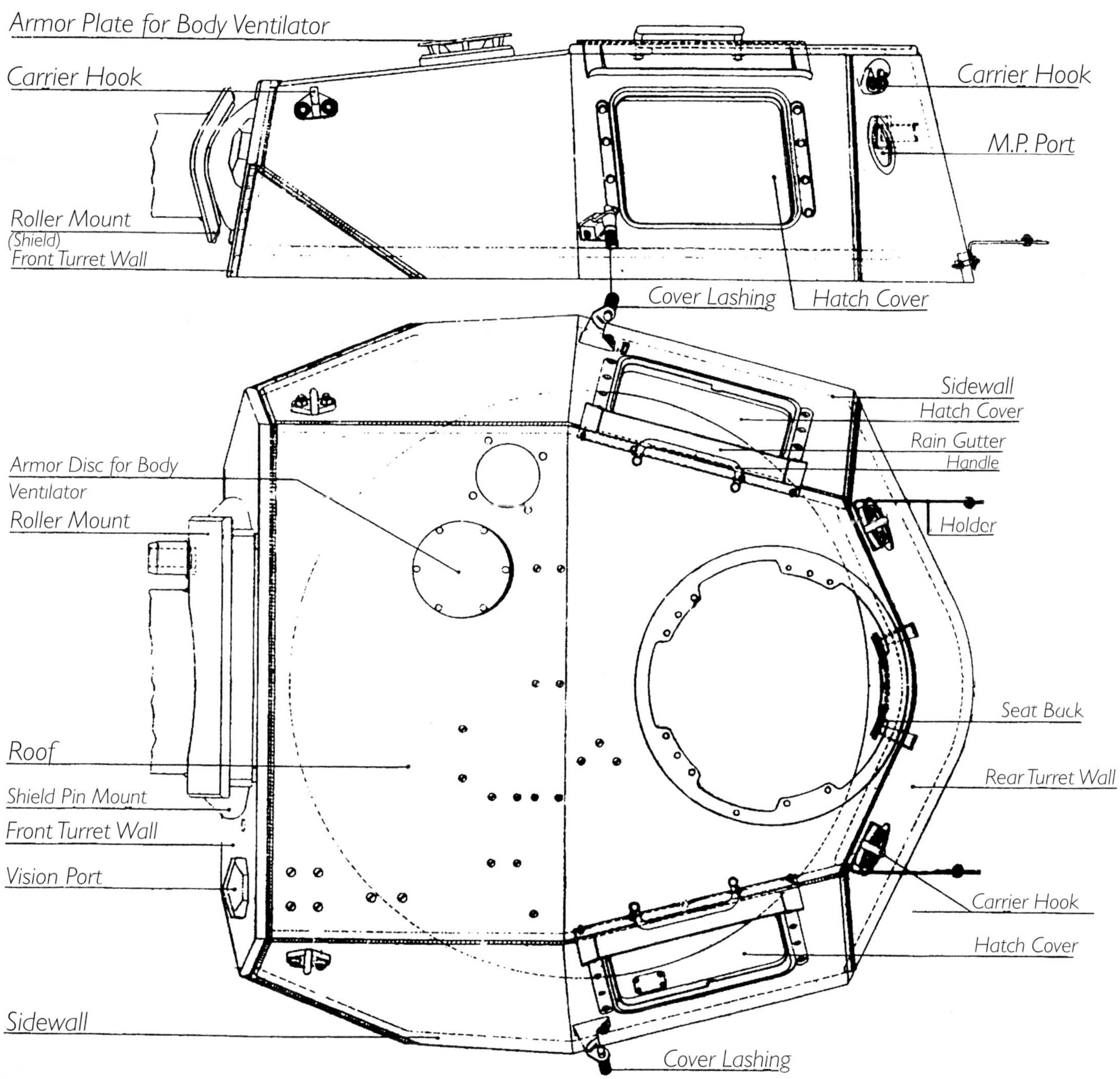

Turret (Drawing No. 021 B 48193) for Pz.Kpfw.IV Type H with strengthened turret roof.

Pz.Kpfw.IV Type H, Chassis No. 85384, with grenadiers on board, finished by the Nibelungen Works in June 1943 with screwed-on extra armor.

Pz.Kpfw.IV Type H, Chassis No. 85074, finished by Vomag in July 1943, with screwed-on extra armor on the body and hull bow.

Pz.Kpfw.IV Type H, finished by the Nibelungen Works in July 1943, with screwed-on extra armor on the body and 80 mm thick homogeneous plate on the hull bow.

Pz.Kpfw.IV, Type H, Chassis No. 85096, finished by Vomag in July 1943, with homogeneous 80 mm thick bow plate.

Pz.Kpfw.IV, Type H, Chassis No. 84599, finished by Krupp in September 1943 for testing the Zimmerit protective coating.

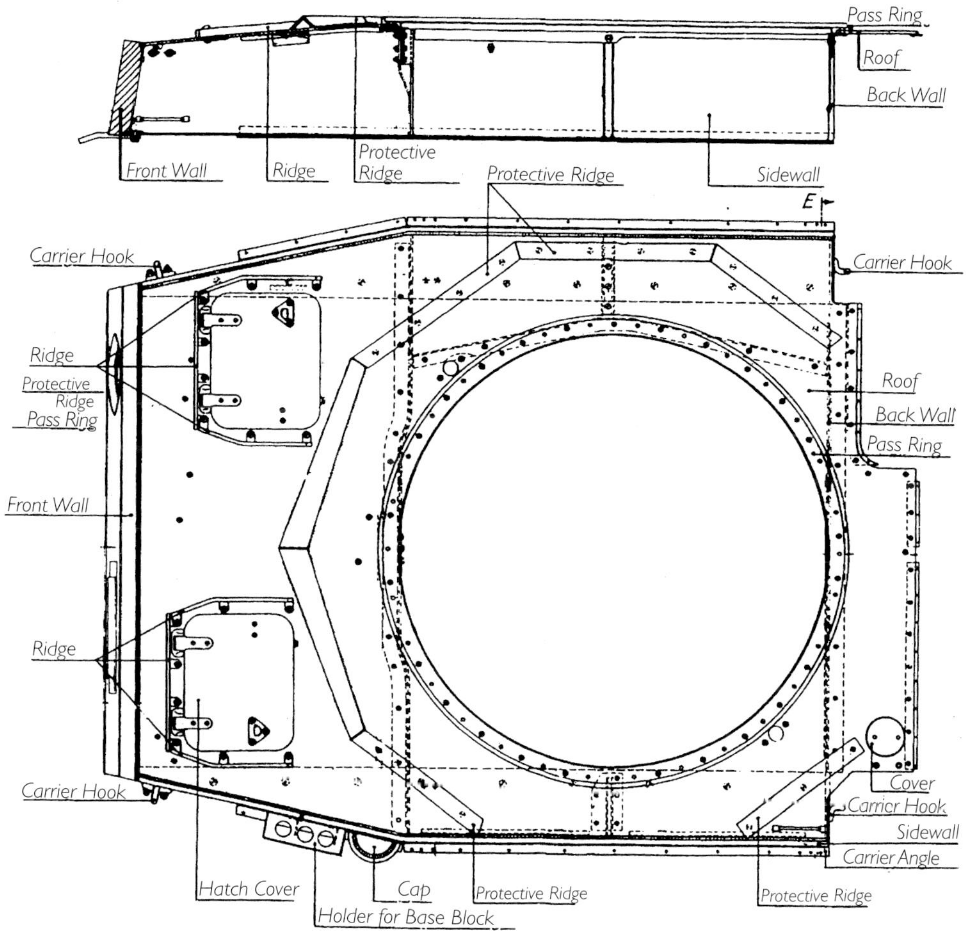

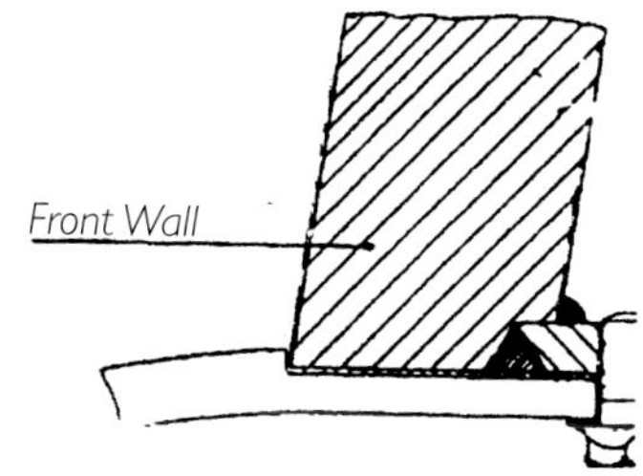

Bow armor
(Drawing No. 021 A 48450)
for Pz.Kpfw.IV Type H.

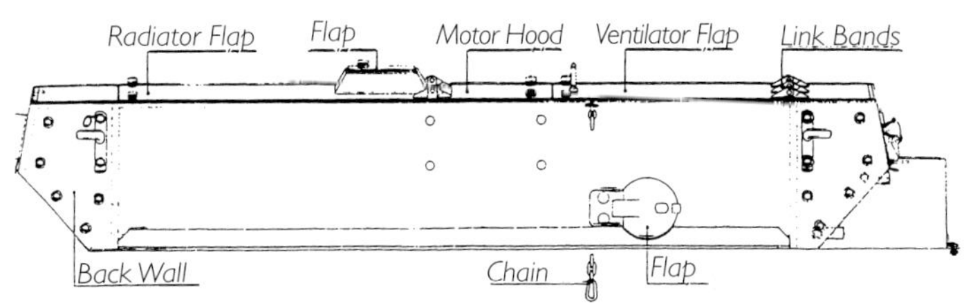

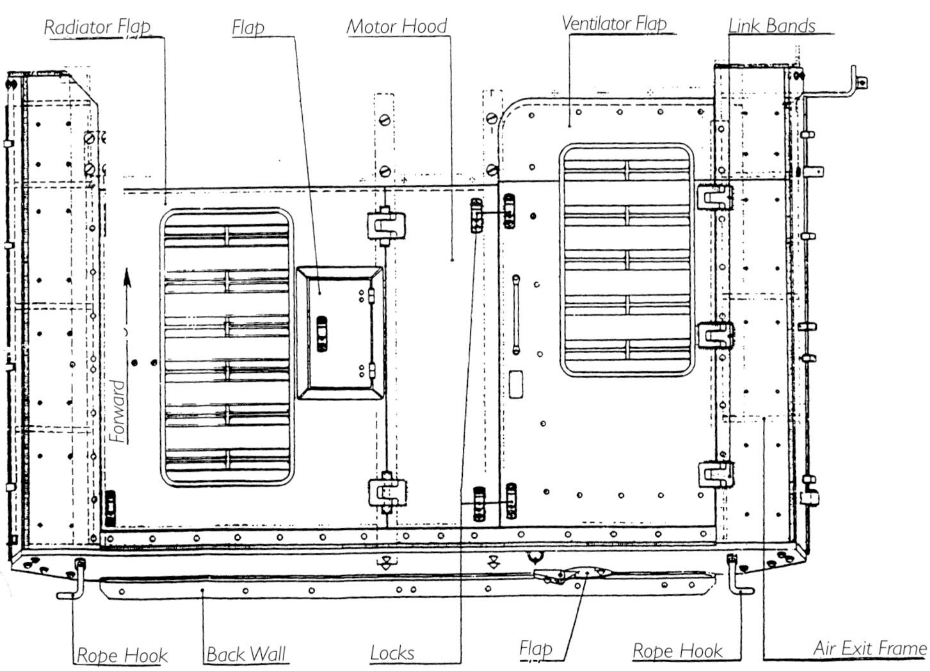

Rear armor (Drawing No. 021 B 171) for Pz.Kpfw.IV Type H with armored shield on right side for felt-bellows air filter.

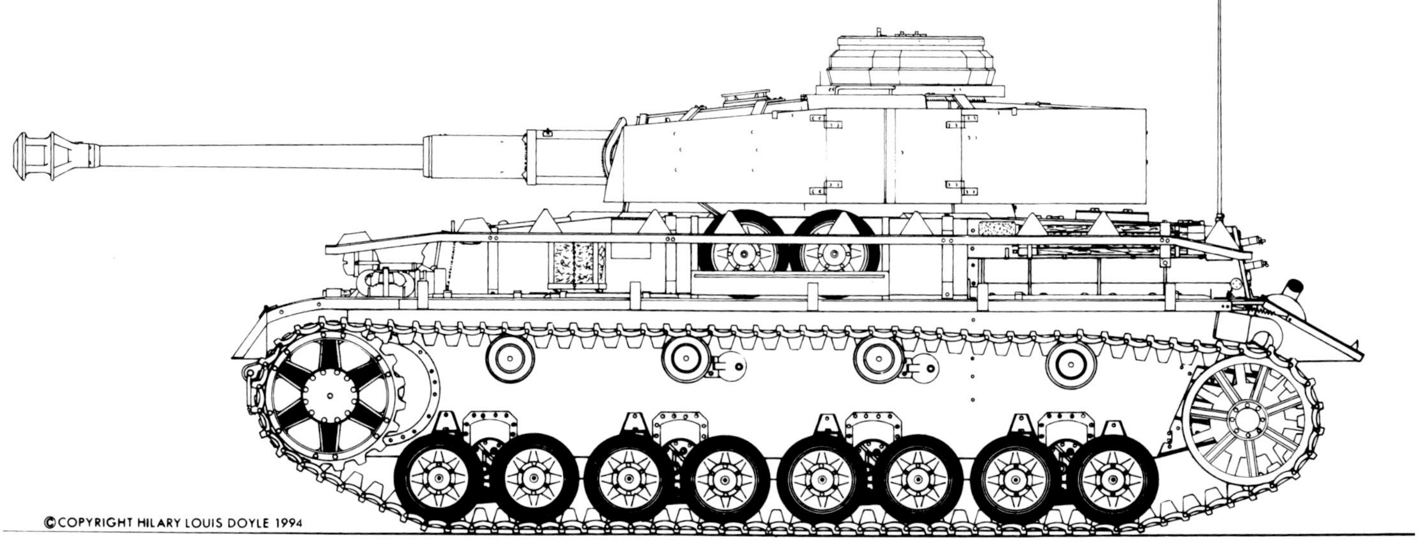
©COPYRIGHT HILARY LOUIS DOYLE 1994

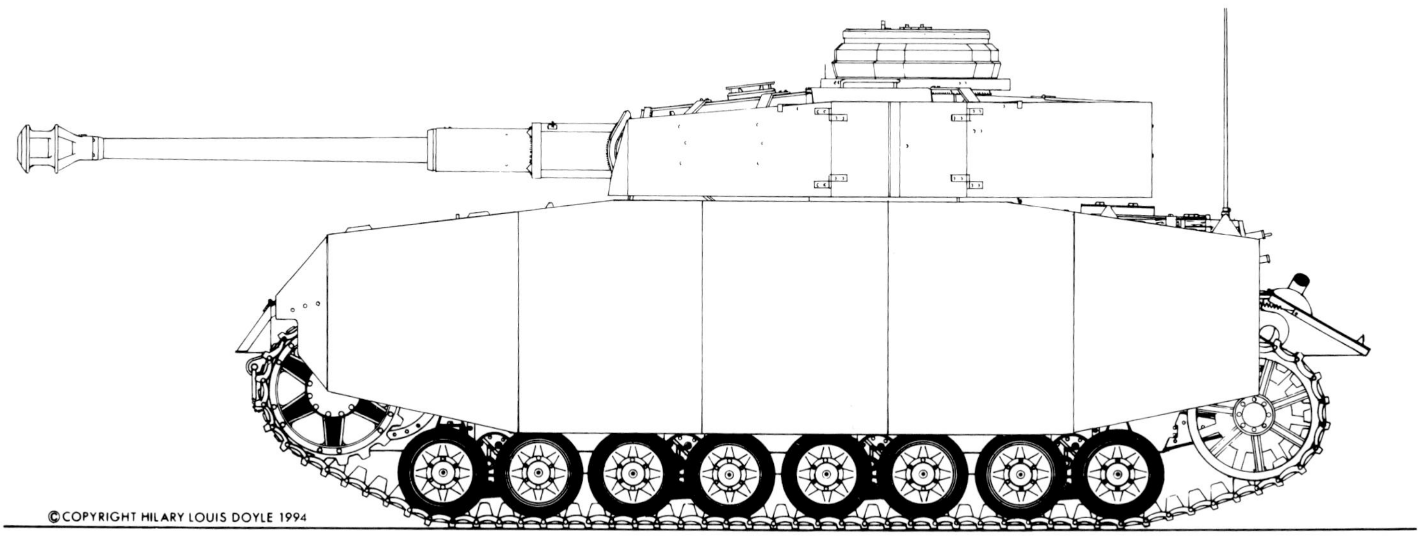
©COPYRIGHT HILARY LOUIS DOYLE 1994

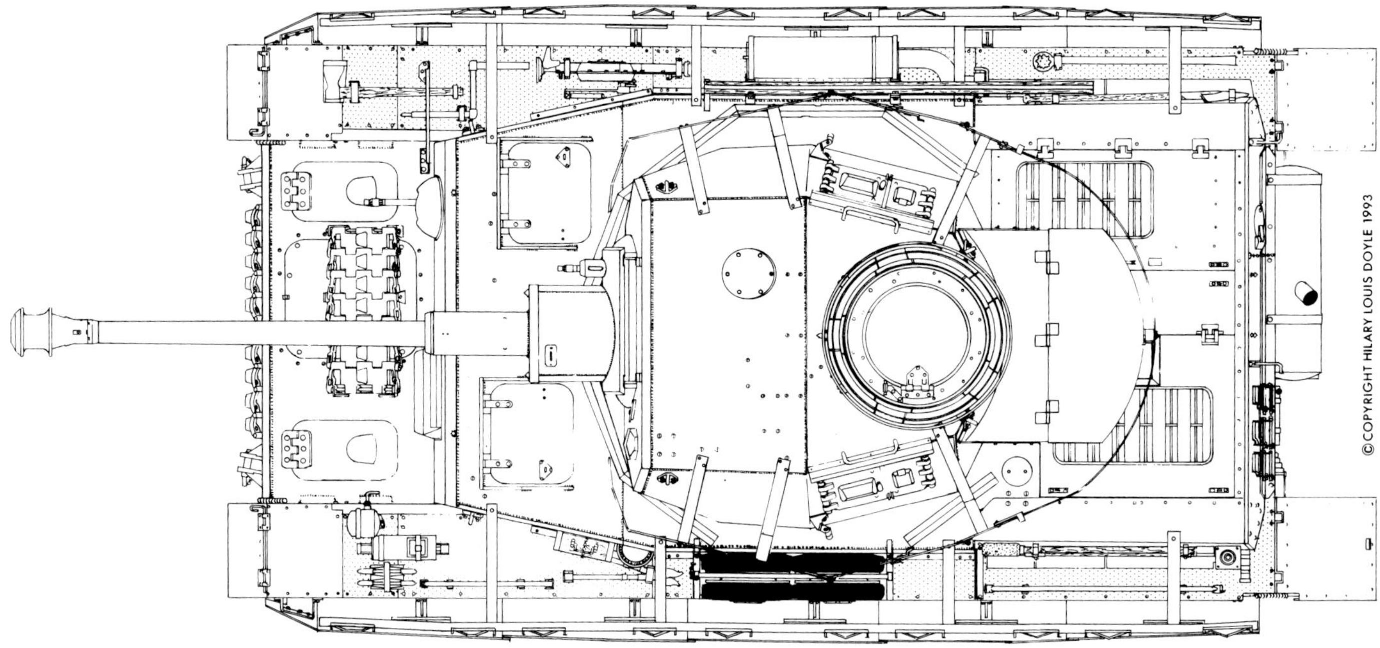
©COPYRIGHT HILARY LOUIS DOYLE 1993

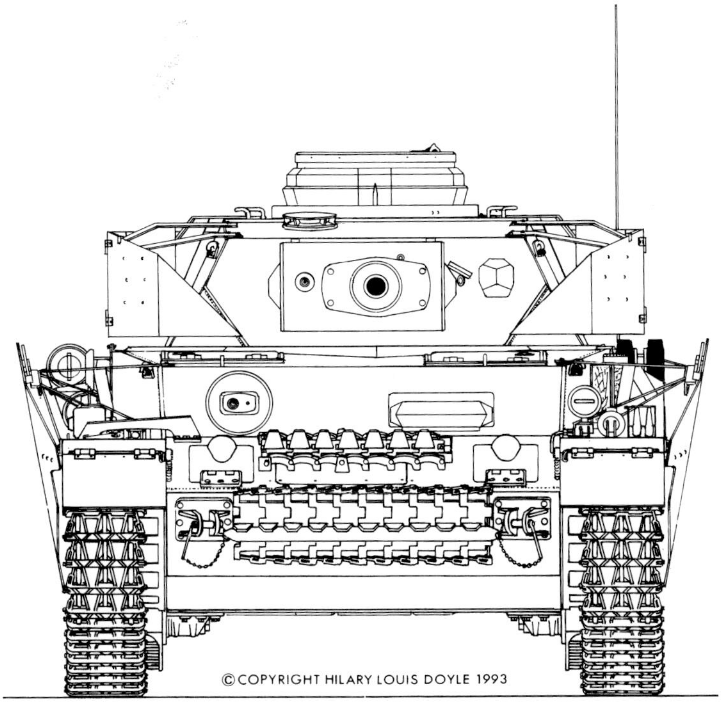

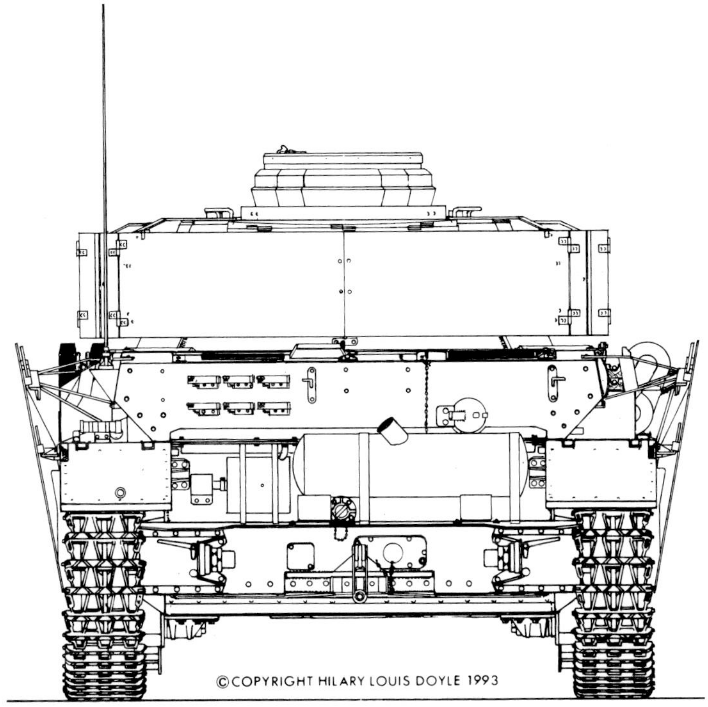

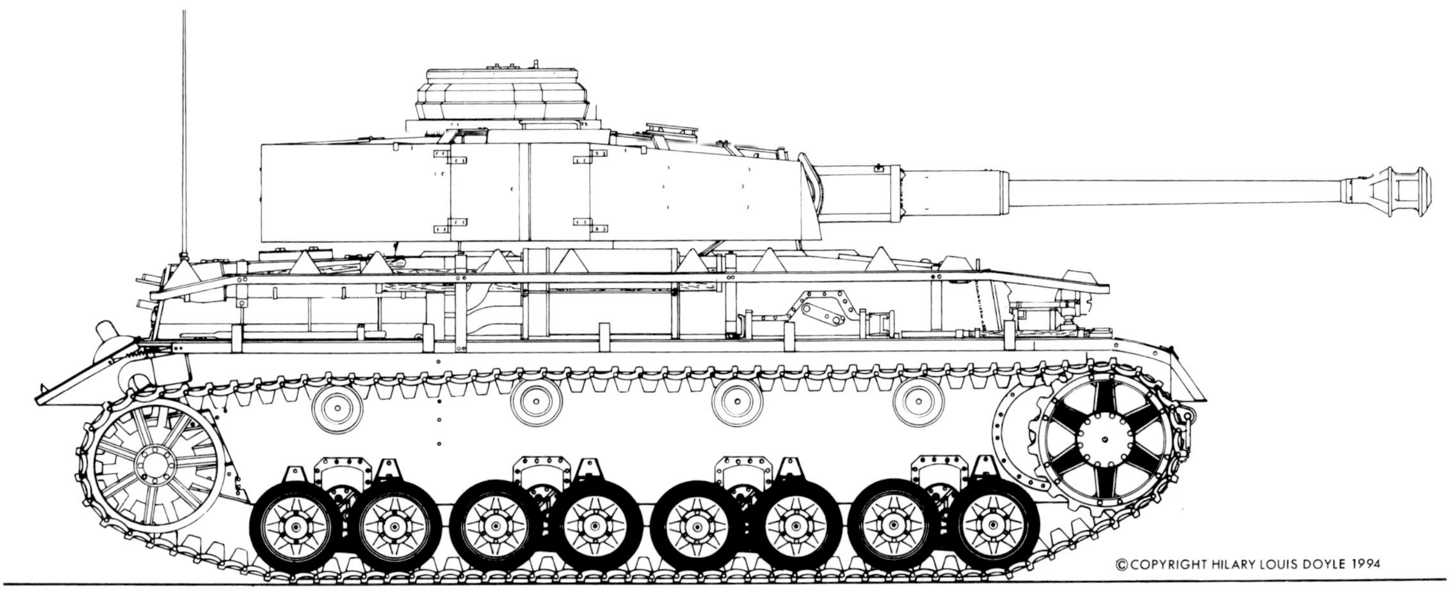

Panzerkampfwagen IV (7.5 cm) (Sd.Kfz.161/2) Type H:

Production from December 1943. Muzzle brake without modification for weight decrease. 80 mm thick front armor in body and hull. Side vision ports on body dropped. Simplified shot deflector on driver's and radioman's entry hatches. Felt-bellows filter in the middle of the right track apron. Radio antenna on left rear part of motor hood, wooden isolator rod for antenna on the turret apron dropped. Wooden antenna holder on right side retained for storing the spare antenna 2 m. Three spare track links behind by motor compartment. New rails attachments to hang aprons in angled form. New drive wheels, cast leading wheels, end attachments and pressed covers for road-wheel bearings.

Pz.KpfwIV Type H, driver's area

Pz.Kpfw.IV Type H, driver's area, just to right is radio frame.

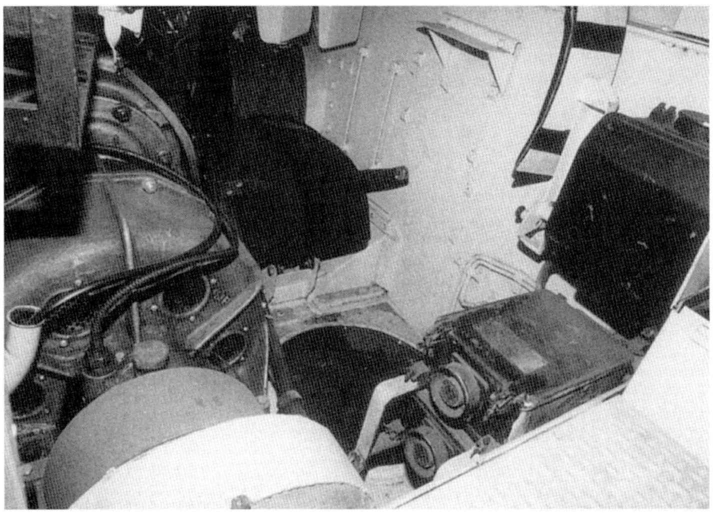

Pz/Kpfw.IV Type H, radioman's area, his seat tipped forward, with emergency exit hatch below.

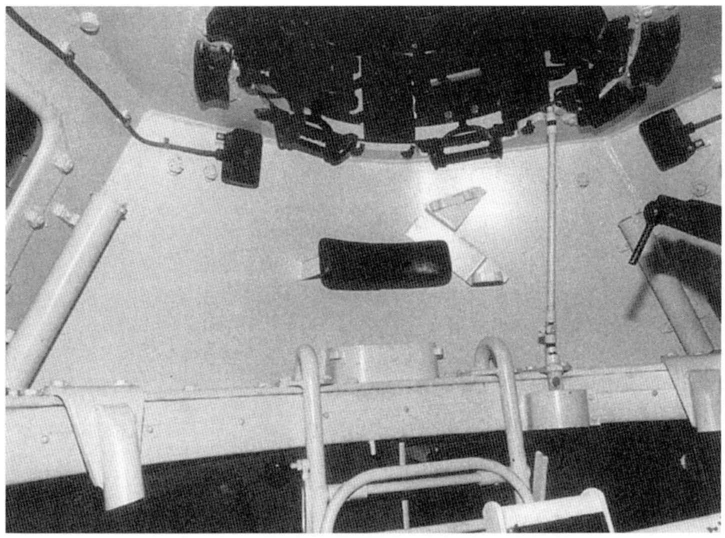

Pz.Kpfw.IV Type H, commander's area.

Pz/Kpfw.IV Type H. ahead of aiming gunner's seat.

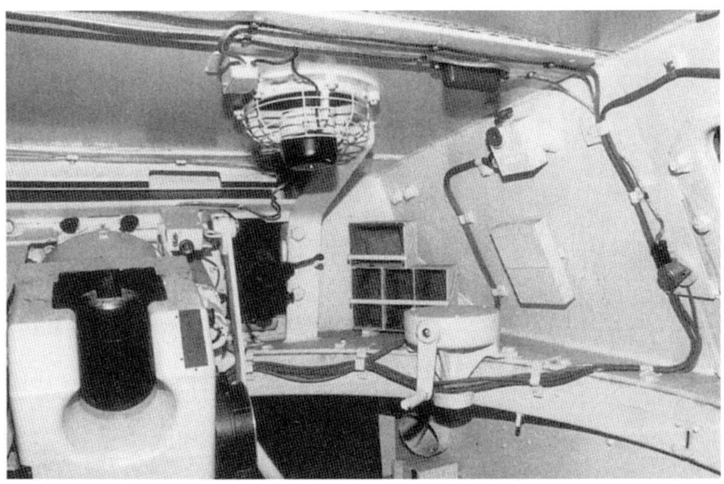

Pz.Kpfw.IV Type H, loading gunner's area with drive for auxiliary turret-turning works.

Cast wheel cap for road wheel hub.

Welded attachment blocks for road-wheel lever and pressed wheel cap for road-wheel hub.

Return roller with Continental 250/65-135 rubber tire.

Rubberless return roller.

Leading wheel of welded tubes.

Leading wheel of cast central pieces and welded curved rims.

12.2.10 Close Combat Weapon

As of January 1944, a close combat weapon was to be mounted on the right side of the turret roof. Because of production shortages, most Pz.Kpfw.IV left the factory without this weapon until July 1944. The opening in the turret roof planned for it was closed by a screwed-on cover.

12.2.11 Felt-Bellows Connected Air Filter

The felt-bellows connected air filter of the Pz.Kpfw.IV (stored on the track apron) had resulted in motor damage because of excessive dust deposits in connection with the armor aprons. The damage was caused by the dust stirred up by the tracks in increased masses being left on the felt-bag air filters and stuffing them full in a short time.

As of February 1944 the felt-bellows air filter with the air ducting to the oil-bath filter was dropped and the air openings were closed by welded-on covers. The air expulsion from the motor compartment thus was done only by the oil-bath filter.

12.3 IMPROVEMENTS AFTER DELIVERY

A series of improvements took place after delivery of the Pz.Kpfw. IV Type H to the troop units. Among others:

1) As winter equipment, winter track links and snow grippers for tracks were delivered to troop units.
2) As of May 1943, aprons were mounted on turret and hull sides.
3) As of January 1944, "Zimmerit protective coatings" could be applied to armor plate by the troops and formed in waves.
4) As of February 1944, felt-bellows air filters along with the air ducting to the oil-bath filter were removed and the air openings were closed by welded-on covers.
5) As of February 1944, repaired Pz.Kpfw.IV were rebuilt as Pz.Bef.Wg.IV command cars.
6) As of May 1944, widened tracks called "Ostkette" were delivered to troop units in action on the eastern front.
7) As of July 1944, three bollards for the 2-ton emergency crane were welded onto turret roofs.
8) As of the end of December 1944, a new towing coupling with a horizontal bolt was mounted in the middle of the rear hull.
9) As of January 1945 the cover of the commander's cupola was modified to allow 170-degree turning.

Pz.Kpfw.IV Type H, Chassis N. 89435, built by the Nibelungen Works in January 1944 (at right, Tactical No. 802) from the rear.

Pz.Kpfw.IV Type H, Chassis No. 89435, built by the Nibelungen Works in January 1944 (Tactical No. 802).

Pz.Kpfw.IV Type H, Chassis No. 89272, built by the Nibelungen Works in January 1944, with close combat weapon in turret roof.

Pz.Kpfw.IV Type H, Chassis No. 86984, built by the Nibelungen Works in November 1943 (Tactical No. 821)

13. PANZERKAMPFWAGEN IV (7.5 cm) (Sd.Kfz.161/2) Type J, (10./B.W.)

Chassis Number Series: ~86394—86573, ~89541—93250 and from 96751

Between February 1944 and April 1945, 3150 Pz.Kpfw.IV Type J were finished and accepted (180 made by Vomag and 2970 by the Nibelungen Works).

13.1 IMPROVEMENTS FROM TYPE H

From Type J of Pz.Kpfw.IV on, the electric turret turning works were dropped. Instead the tank received an additional fuel tank of 200-liter capacity in the motor compartment. The traversing machine was simultaneously modified and fitted with a second transmission, which made hand-operation easier when the tank was tilted. The switch lever for "motor" and "hand" then switched on "hand" setting to direct hand-operation and to the second transmission in the other position. Through the dropping of the electric turning works, the heavy-duty cable and the switch box with it on the forward turntable brace were gone. The inductive emergency firing, which was formerly attached to the turning motor, was now mounted on the traversing machine on a brace.

According to information from Wa Prüf 6, as of February 1 (for chassis) or February 15, 1944 (complete vehicles), the DKW aggregate with muffler was dropped. As a result, Krupp had foreseen a fuel tank in the rear part of the hull instead of the DKW motor. Thus, the group number 021St48101 was changed to 021St55001.

Panzerkampfwagen IV, Type J, built by the Nibelungen Works in August-September 1944.

Pz.Kpfw.IV, Type J, Chassis No. 89589, built by the Nibelungen Works in February 1944.

Pz.Kpfw.IV, Type J, Chassis No. 86432, built by Vomag in February 1944. Later marked with the palm tree of the German Afrika-Korps.

Pz.Kpfw.IV, Type J, without DKW aggregate and muffler.

13.2 FORM CHANGES DURING PRODUCTION

13.2.1 Flaps in the Turret Hatch Cover

To simplify production, it was decided on April 14, 1944, that the vision port (Part No. 021St9267U6) in the turret hatch cover should be dropped. Introduced as of May 1944, this change was not 100% carried out when the war ended.

13.2.2 Cover for Radiator Filler

As of May 1944, to simplify production further, the sides of the armored cover over the radiator filler pipe by the motor compartment were made rectangular. Formerly the sides had been curved.

13.2.3 Road Wheels

As of June 9, 1944, the road wheels for the Series 10./B.W. (Part No. 021Sk6267) were to be fitted with a seal ring. The seal ring running box was unhardened.

13.2.4 S-Hook Instead of C-Hook

For reasons of uniformity, only S-hooks were to be used for tanks under 30 tons. On June 10, 1944, the firms were informed by Wa Prüf 6 that the Pz.Kpfw.IV was to be equipped in the future with one S-hook instead of two C-hooks.

13.2.5 Tappets for Auxiliary Crane on Turret Roof

On June 14, 1944, Wa Prüf 6 had advised the assembly firms that three tappets for a 2-ton auxiliary crane were to be welded onto the turret roof: *It is pointed out that the equipping and arranging may be carried out only as shown in the enclosed drawings.*

13.2.6 Fuel Tank in Motor Compartment

In place of the gasoline-electric machinery for the turret turning works, from Chassis No. 91501 on a fuel tank was built in. The containers had not shown themselves to be equal to their demands. Fuel losses resulted from poor sealing, along with the danger of fire. Therefore, the vehicles with chassis numbers from 91865 to 91949 had the attachments of their installed fuel tanks sealed, and were not to be attached by the troops. From Chassis No. 91950 on, these containers were left out until the delivery of usable tanks to the assembly firms took place. In later vehicles, fuel tanks were then installed in series production. Until this point in time, replacement tanks were supplied for all vehicles from Chassis No. 91501 on.

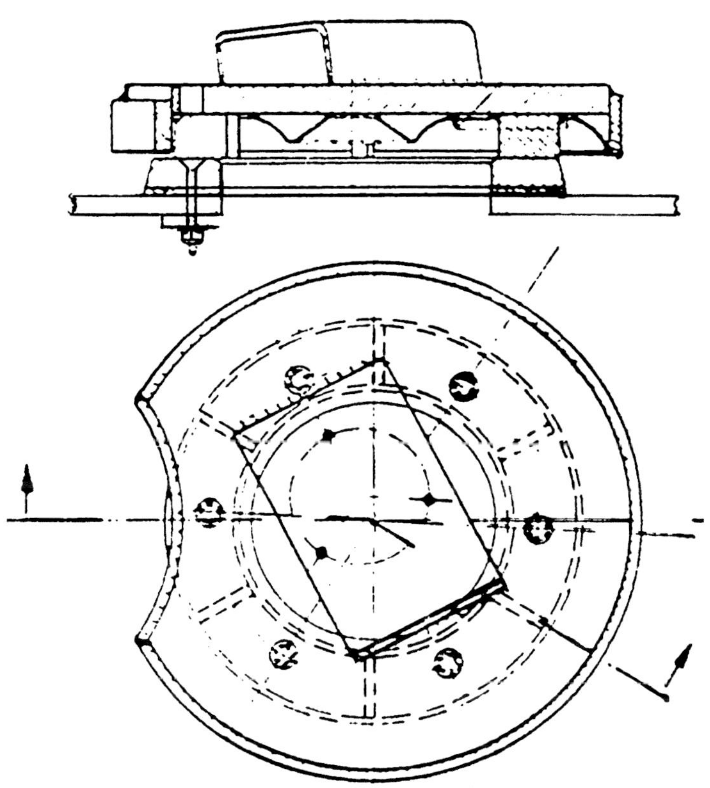

Air exit complete with standard for rubber-base antenna.

13.2.7 Prismatic Cupolas

On July 24, 1944, Wa Prüf 6 decided that *...through the new program situation, according to which in the next few months the Pz.Kpfw.IV is to be rebuilt as a tank destroyer with L/70 gun, the introduction of a prismatic cupola (similar to those of the Panther and Tiger) in the Pz.Kpfw.IV is no longer considered.*

13.2.8 Partly Homogeneous Armor Plates

On June 13, 1944, Wa Prüf 6/IIIc had decided for Krupp of Essen, because of bomb damage that endangered further manufacturing, that *...for the Pz.Kpfw.IV until 8/1/1944. all surface-hardened armor plates may be installed as homogeneous.*

Armor plates for the 80 mm front wall, the hull and body were made of Material E22, for the 30 mm sidewalls, the turret, hull, body and rear wall of the turret of Material E32.

13.2.9 Strengthened Bow Armor Roof

Because of tears in more than 50% of the roof plates, it was decided on July 13, 1944, that the work of introducing 16 mm instead of 11 mm roofs should be undertaken at once. The drawings of the bow armor with 16 mm armor were sent to the firms on July 19, 1944.

On July 19, 1944, the Oberdonau Iron Works suggested a transitional solution to ease the transition from 11 mm to 16 mm roof thickness: *The driver's and radioman's hatches will be used with 10 mm strength until parts on hand are used up. Eisenwerke Oberdonau intends to use the roof made of three combined parts.*

13.2.10 Body Ventilator in the Turret Roof

Through the new armor covering (as of July 1944) of the body ventilator, unhindered use of the close-combat weapon was not possible. As of October 1944 the cover plate was cut off on an angle and welded on.

13.2.11 Exhaust Pipe Flame-Killer

On August 8, 1944, the Nibelungen Works reported that development of the exhaust-pipe flame-killer was not far from finished. The design should prevent flames that lit up the vehicle at night. This change was already introduced before the end of 1944.

13.2.12 Hand Protection on the Deflector

Shell ejection after the shot led to injuries to the commander's hands, as he was inclined to hang onto the deflector. From Pz.Kpfw. IV Chassis No. 91851 on (August 1944), a hand protector was mounted.

13.2.13 Side Gearings

In the side gearings, damage such as cog breakage, hub breakage and shearing off of thc pass screws on the spur gear occurred. As of Pz.Kpfw.IV Chassis No. 92112 (August 1944), the following transitional solution was introduced: Strengthening of the gear mount, building up of the transition, strengthening of a hub and securing of thc pass screws with plates.

As of Pz.Kpfw.IV Chassis No. 92852 (November 1944), the following final solution was introduced: Strengthening of the gear mount at the cogs of the spur gear. The hub and disc were made of a single piece.

13.2.14 Unit Holder for Shear Telescope and Sight Rod

As of Pz.Kpfw.IV Chassis No. 92181 (September 1944), a holder was installed to hold the shear telescope or vision rod, rest holder for the shear telescope and rest holder for the vision rod.

13.2.15 Zimmerit

The assembly firms received the order in September 1944 not to apply Zimmerit anymore.

13.2.16 Wire Mesh Aprons

To save weight, offer better camouflage possibility, and cut down the dust between the apron and track cover, the wire mesh apron was introduced as of Pz.Kpfw.IV Chassis No. 92301 (September 1944).

13.2.17 Return Rollers with Plain Bearings

On May 24, 1944, it was decided:

The change to plain bearings finally takes place. The use of roller bearings is no longer considered. The return roller in new form will be made optionally of steel or tempered cast iron. All firms should find out how many rollers of the old type are still on hand. Type of the rollers as Drawing 021C55018. Plain bearing type for road wheels as Drawing 021C55014. Conversion before 9/1/1944 is not to be done.

Return-roller mounting with plain bearings was introduced as of October 1944. Only the complete plain-bearing return roller with mount was replaceable by the roller bearing with attachment.

Pz.Kpfw.IV Type J, built at the Nibelungen Works in August-September 1944, with flame-killers on the exhaust pipes.

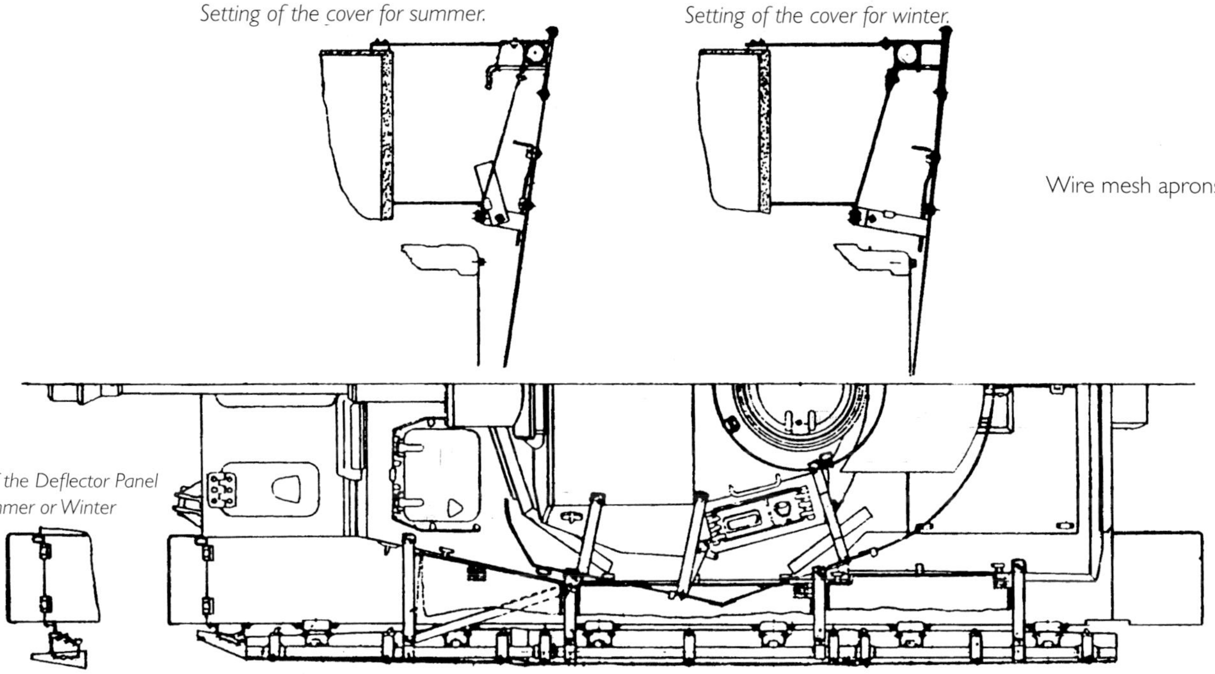

Wire mesh aprons for Pz.Kpfw.IV

Return roller with plain bearing for Pz.Kpfw.IV.

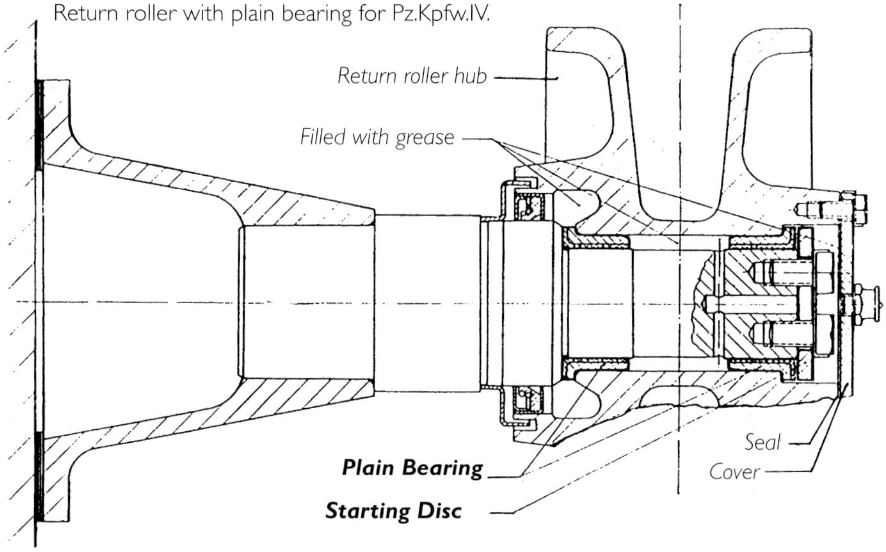

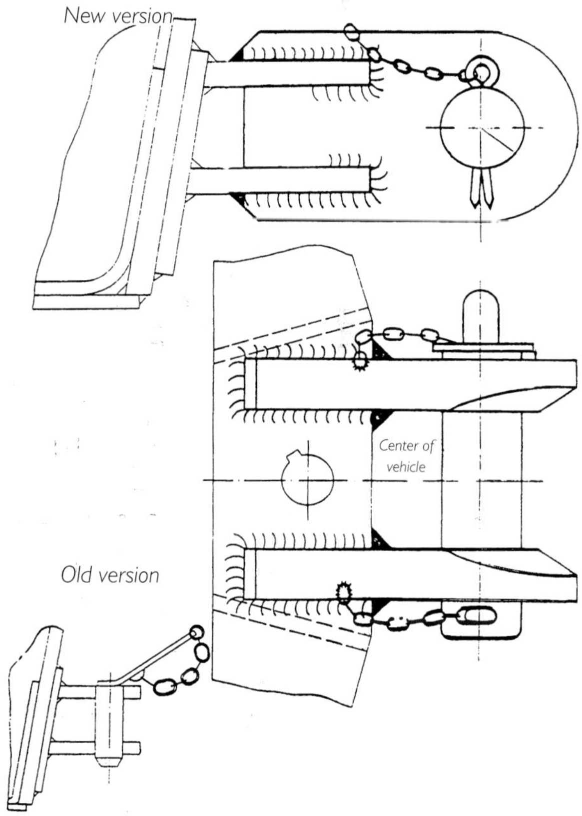

Towing coupling for Pz.Kpfw.IV.

Battery heating for Pz.Kpfw.IV.

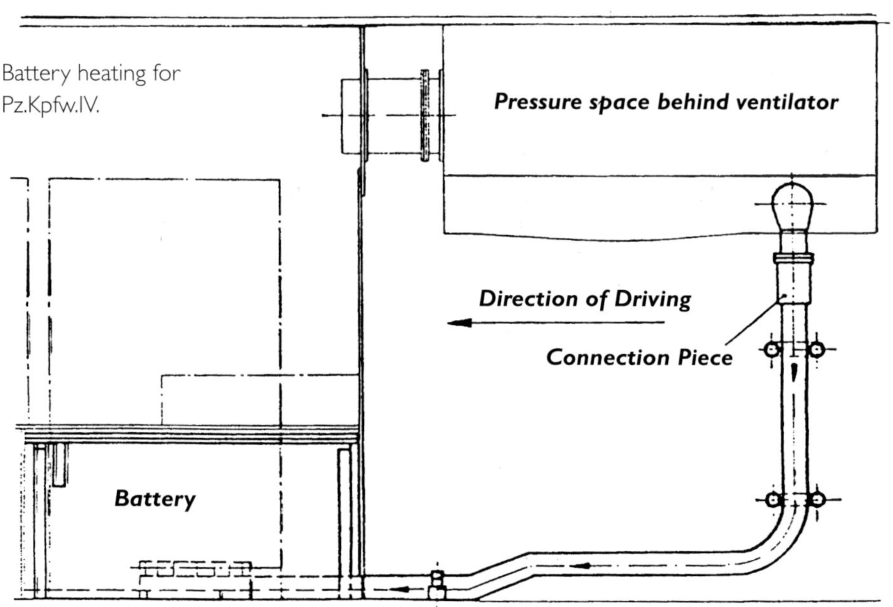

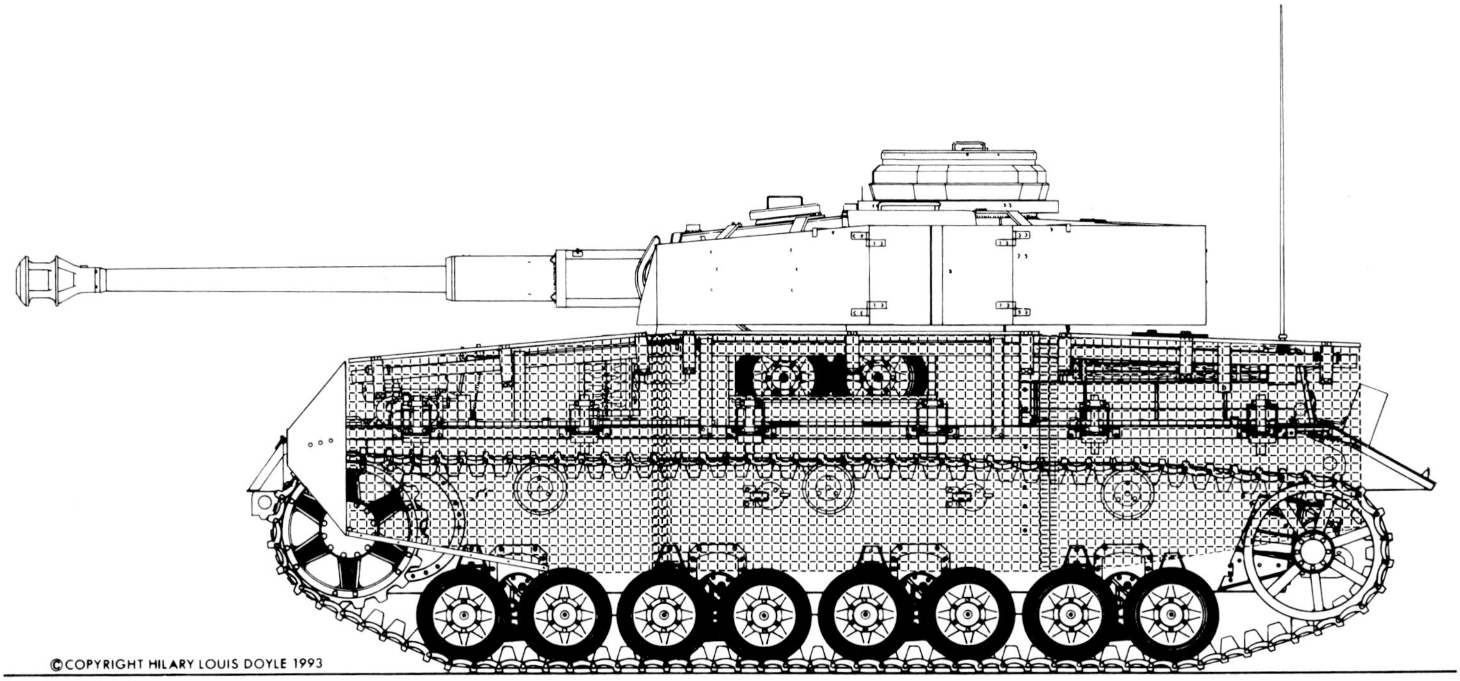
©COPYRIGHT HILARY LOUIS DOYLE 1993

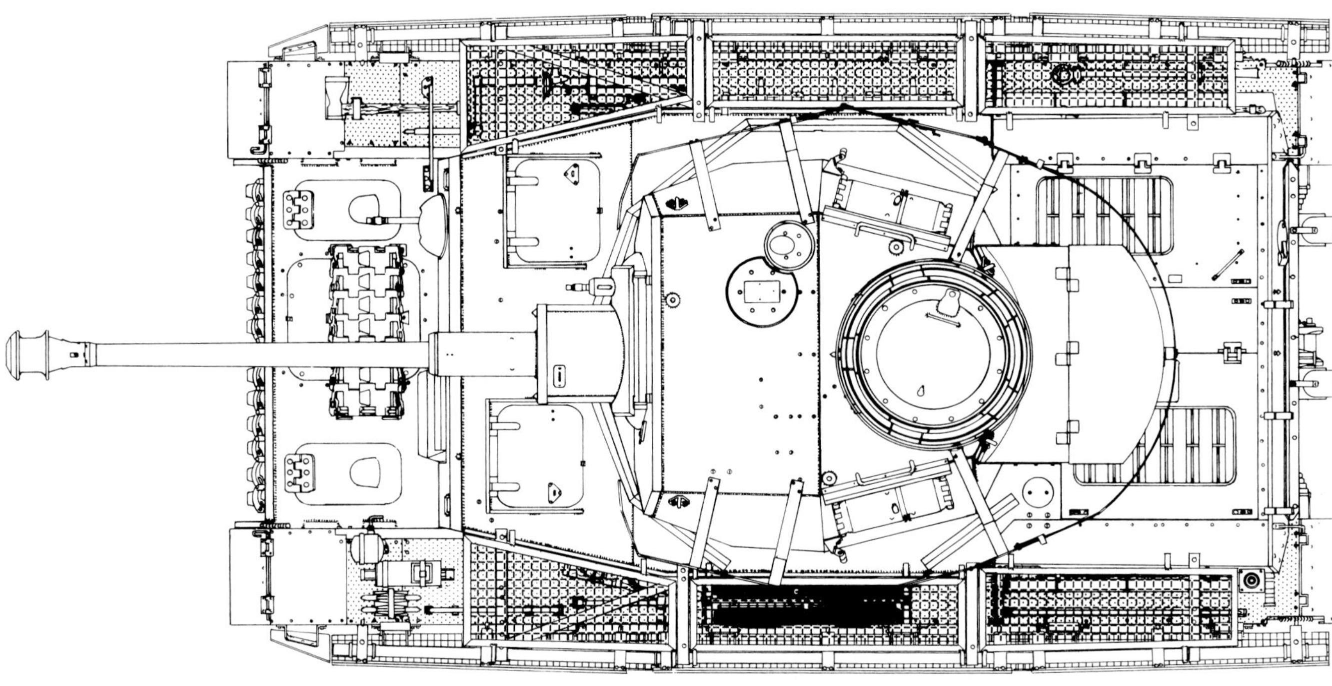
©COPYRIGHT HILARY LOUIS DOYLE 1993

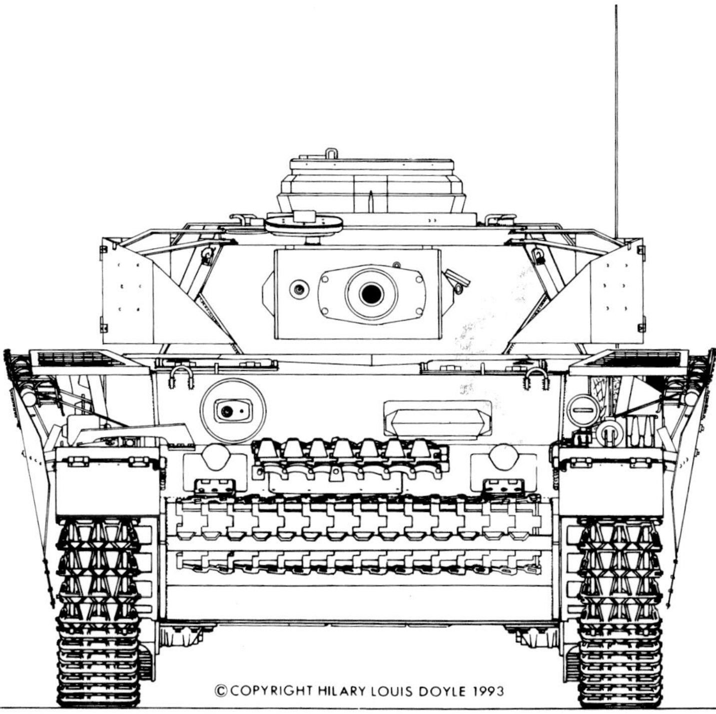

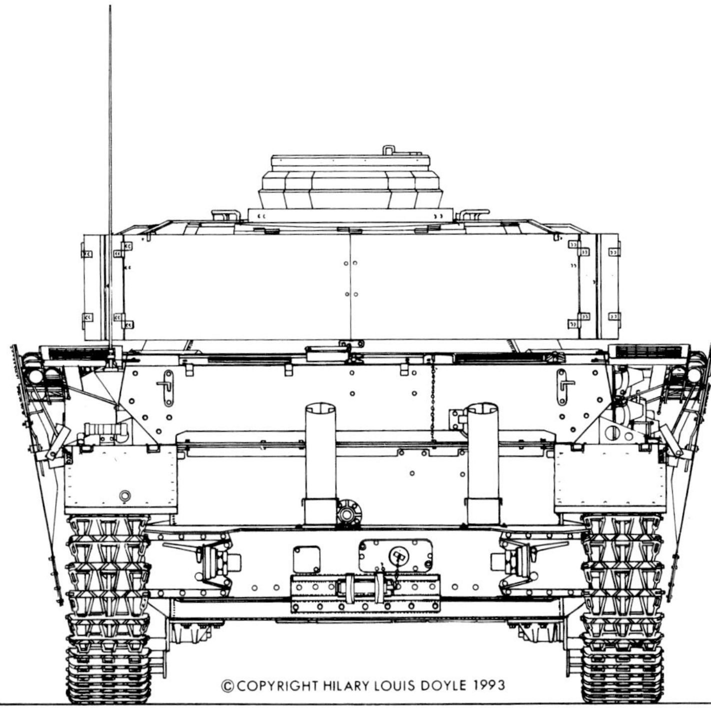

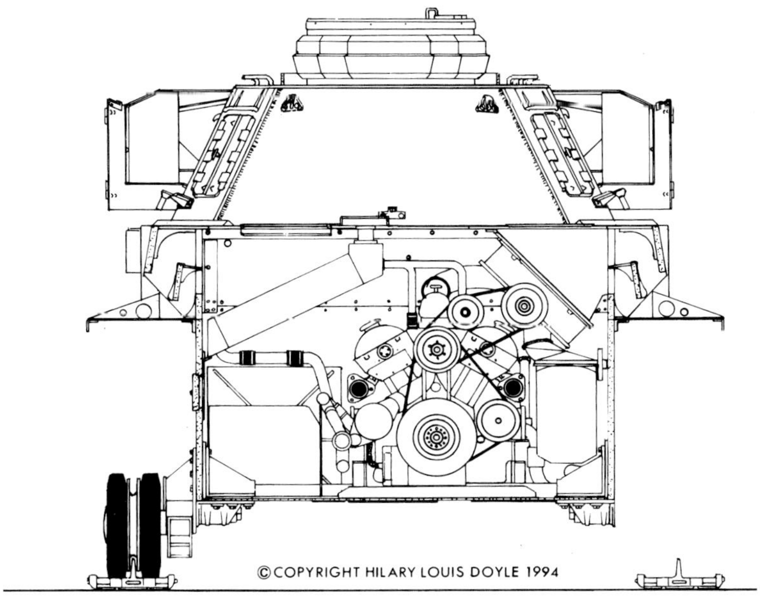

Panzerkampfwagen IV (7.5 cm) (Sd.Kfz. 161/2), Type J, as built at the Nibelungen Works in 1945. New hull with bottom panel lengthened to the rear, opening for exhaust pipe of the auxiliary motor omitted. Some armor plates of Type J with towing eyes. Lengthening for them taken from hull sidewalls. The time-consuming arrangement of the forward towing eyes was replaced later by welding U-shaped fasteners onto the upper bow plate. The normal layout of the exhaust system was replaced by the flame-killer. Heavy towing coupling in rear of vehicle. New turret with improved ventilator cover, installation of close-combat weapon in turret roof, vision port and pistol openings in side turret hatches omitted. Side aprons of wire mesh, which also covered the space between the armored body and hanging rail. The cutaway shows the fuel tank that was installed on the left side of the motor in place of the gasoline-electric system of DKW motor and generator, plus the elimination of the electric turret drive. Only three return rollers per side on some Type J vehicles, return to welded leading wheels.

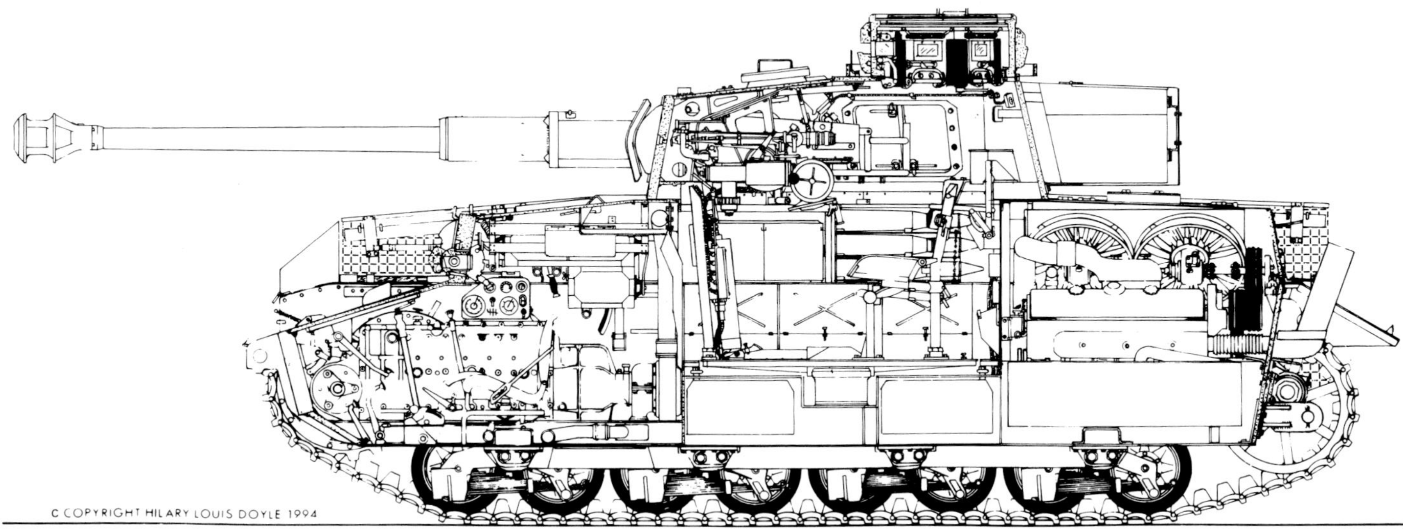

Experimentally mounted wire mesh aprons on a Pz.Kpfw.IV Type H, 1943.

Pz.Kpfw.IV Type J with three-return-roller running gear.

Tappets for attaching the auxiliary crane, close combat weapon, air exit hood and swing cover for commander's cupola.

13.2.18 Swing Cover for Commander's Cupola

On June 14, 1944, Wa Prüf 6 and Krupp of Essen discussed the production simplification of the commander's cupola (Drawing 021B41420) by changing the swinging apparatus. By August 8, 1944, the design and production had gone so far that the swing cover for the commander's cupola (021Sk41420) was to be installed in three months, so that some 400 vehicles could still be equipped with it.

This swing cover (similar to the Panther and Tiger) for the commander's cupola, as of Pz.Kpfw.IV with Chassis No. 92447 (October 1944) was not yet 100% installed by November 1944. The former cover offered a large target in ambush position.

13.2.19 Gas Protection Equipment

As of November 1944, the following new attachments for gas-protection equipment were installed:

- 5 holding clamps for gas-mask filters and T-pieces.
- 3 pieces in turret and one each by driver and radioman.
- 1 holder for gas-protection boxes in turret left near tank gun.
- 1 box for lost identity tags in niche on right side.

13.2.20 Track Circle Fastening

Because of insufficient security of track circle attachment, as of November 1944 a securer with securing plate of strip iron over two screws was introduced.

13.2.21 Battery Heating

Because the battery worked poorly when cold, as of November 1944 battery heating was installed. Warm air was conducted in through the radiator cooler and the battery isolated by wooden covering.

13.2.22 Close Combat Ammunition

As of November 1944, two cases were attached for explosive shells (close-combat ammunition), one on the turret roof, the other to the right by the ammunition cases. Later form changes were carried out only in homeland repair stops.

13.2.23 Three Return Rollers Per Side

Pz.Kpfw.IV Chassis No. 90040 was tested with six return rollers (instead of eight). On June 8, 1944, permission to introduce this procedure into the production series was sought, as the III/IV and StuG III chassis had only three return rollers per side. The Nibelungen Works expected to save 2000 bottleneck roller bearings. With the saved time, another 24 Pz.Kpfw.IV per year could be produced. This change was introduced only in December 1944.

13.2.24 Towing Coupling

As of December 1944, changes were made to the vehicles to improve their ability to be towed. The old fork hangers with bolts on the bow and stern armor were dropped; they were replaced by two eyes with bolts on the bow and stern armor (component of the hull sidewall).

Towing with tow bars had not been possible until then. As of the end of December 1944, a new towing coupling with a horizontal bolt was mounted in the middle of the rear hull.

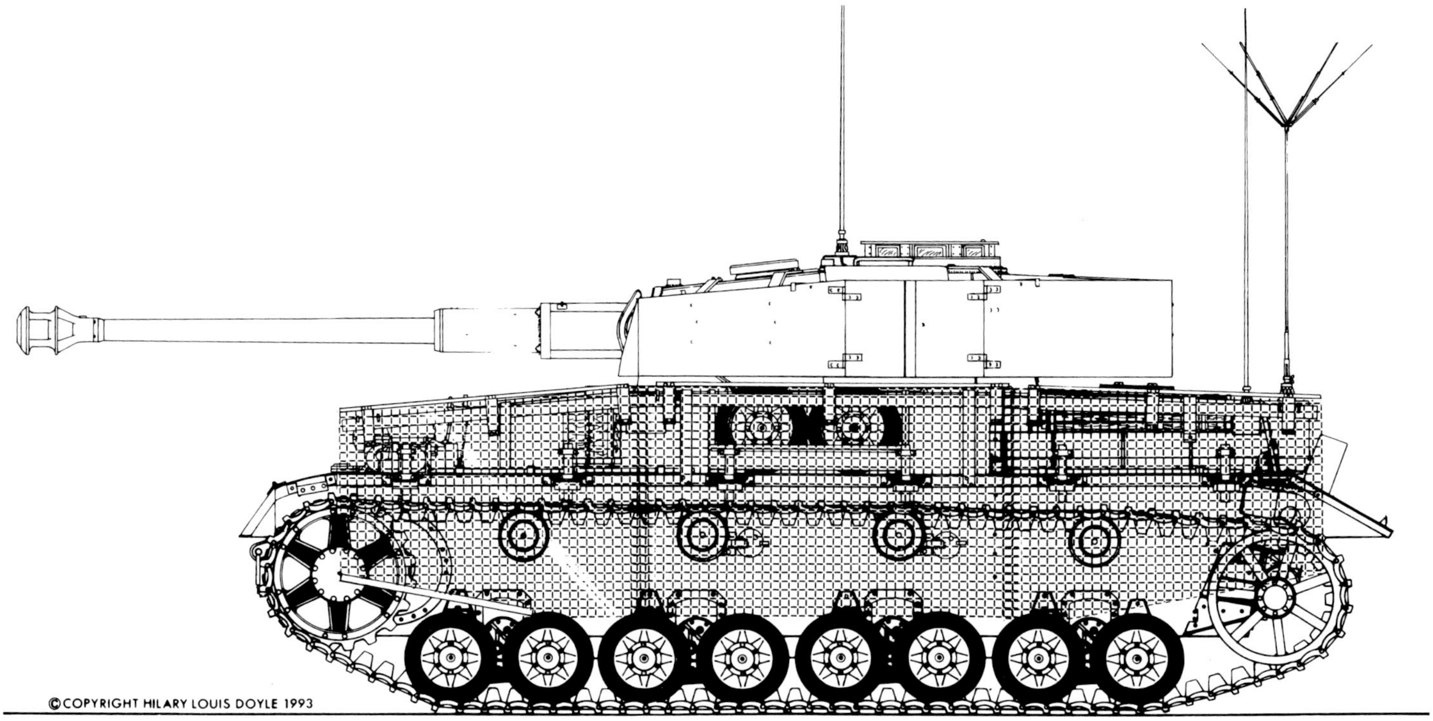

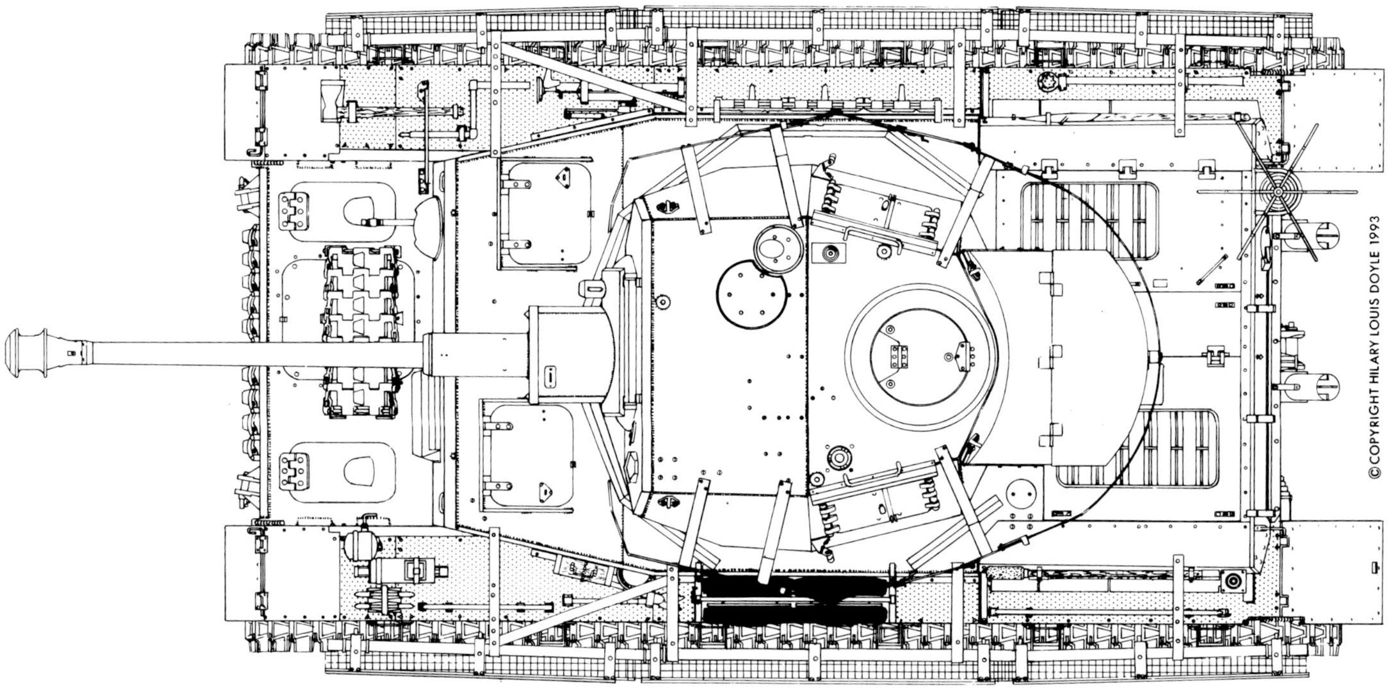

Pz.Kpfw.IV Type J rebuilt as an armored reconnaissance car with assault-gun commander's cupola, antenna on turret roof for FuG 5, Star Antenna D in a separate attachment at the back of the motor compartment for FuG8. The left rear 2-meter rod antenna was still used for the FuG. The assault-gun commander's cupola allowed the use of the 14Z periscope with closed hatches and was entrusted to the artillery. The drawing shows the vehicle with raised 560 mm wide winter tracks. The wire mesh aprons are shown in the outermost angled position. The uppermost wire mesh segments were left off for reasons of visibility.

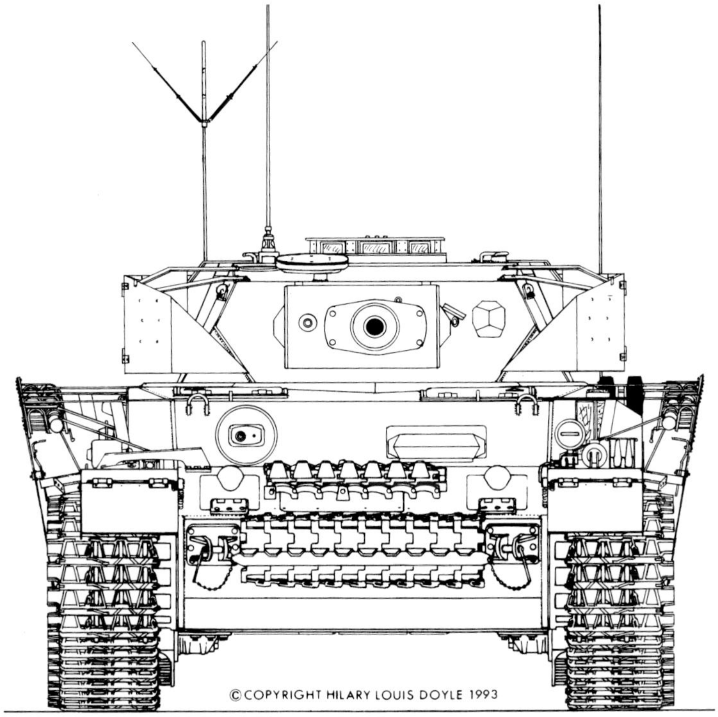

Pz.Kpfw.IV Type J, rebuilt into an armored reconnaissance car with assault-gun commander's cupola.

13.3 IMPROVEMENTS AFTER DELIVERY

13.3.1 ARMORED RECONNAISSANCE CAR I

In a discussion of production and design on September 7, 1943, it was stated:

270 Pz.Kpfw.III are released for reconnaissance purposes. The Gen.Insp.d.Pz.Tr. will cover further needs only from repaired Pz.Kpfw.IV. Pz.Kpfw.IV with the commander's cupola of the assault gun is usable as Pz.Beob.Wg.

On January 24, 1944, it was reported that at the end of January the experimental Pz.Beob.Wg. with assault-gun commander's cupola on Pz.Kpfw.IV was finished.

In rebuilding the Pz.Kpfw.IV into a Pz.Beob.Wg.IV, the turret M.G. was removed to make room for the radio set. An assault-gun commander's cupola with seven periscopes replaced the normal Pz.Kpfw.-commander's cupola with five vision ports. Three radio sets, Fu.G.8, Fu.G.4 and Fu.Spr.f, were installed for battery leaders of Wespe or Hummel batteries (K.St.N.431 of 6/1/44). In addition, a shear scope (SF 14 Z) was mounted in the turret, a vision staff (TSR1) in the turret roof and an orientation compass and turning apparatus in the upper body. Current for the radios when the motor was stopped was produced by a built-in GG 400 electric generator installed on the rear left side of the fighting compartment.

13.3.2 Armored Command Car IV

As of March 1944, Pz.Bef.Wg.IV was usually rebuilt of repaired Pz.Kpfw.IV; only 17 Pz.Bef.Wg.IV were rebuilt newly-made Pz.Kpfw.IV.

In rebuilding, Fu.G.8 and Fu.G. 5 for Pz.Bef.Wg.IV (Sd.Kfz.267) and Fu.G.7 and Fu.G.5 for Pz.Bef.Wg.IV (Sd.Kfz.268) were installed. Power for these radios when the motor was turned off was produced by a GG 400 electric generator installed in the left rear of the fighting compartment. For better vision, a shear scope (SF 14 Z) was mounted in the commander's cupola and a vision staff (TSR1) in the turret roof. To make room for the radios and generator, 15 7.5 cm shells fewer than in the normal Pz.Kpfw. IV were carried in thePz.Bef.WgIV.

13.3.3 Improvements after Delivery

A series of improvements was ordered after delivery of the Pz.Kpfw.IV Type J to troop units. Among others:

1) Ostkette for tanks (Army Technical Instruction Sheet (Ht.V.Bl.) of May 1, 1944). For armored vehicles on the Panzerkampfwagen III and IV chassis wider track links, called Ostkette (Eastern Tracks), were delivered to the troop units on the eastern front. The Ostkette was supposed to limit sinking of armored vehicles in snow and soft ground. It was intended only for the eastern front, and in case of transfer to the West, Italy or the Balkans, it was to be replaced by normal tracks before transfer.
2) As of July 1944, three attachments for a two-ton auxiliary crane were welded onto the turret roof.
3) As of the end of December 1944, a new towing coupling with a horizontal bolt was mounted on the middle of the rear hull.
4) The cover for the commander's cupola was not foldable and thus offered a large target. To make it foldable about 170 degrees, the troops had modification drawings duplicated at the Ht.V.Bl. as of January 1945.

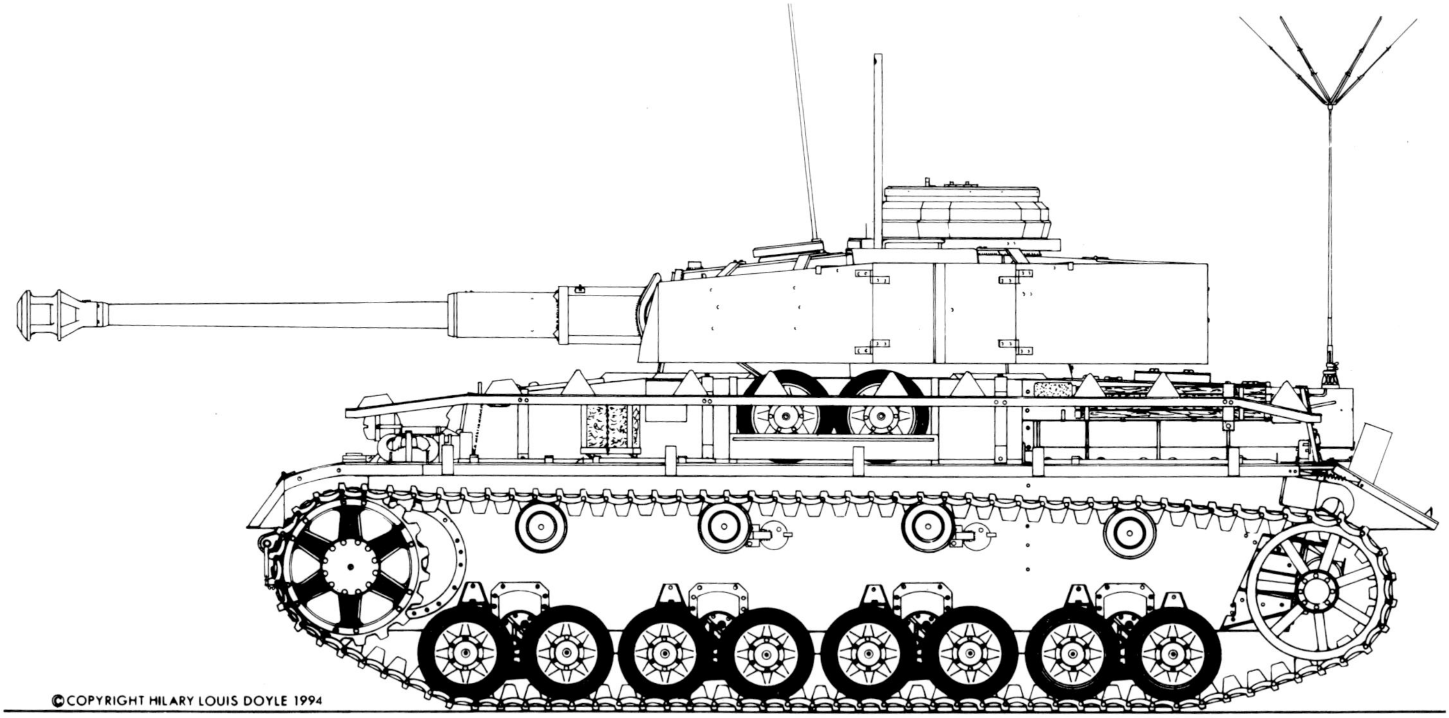

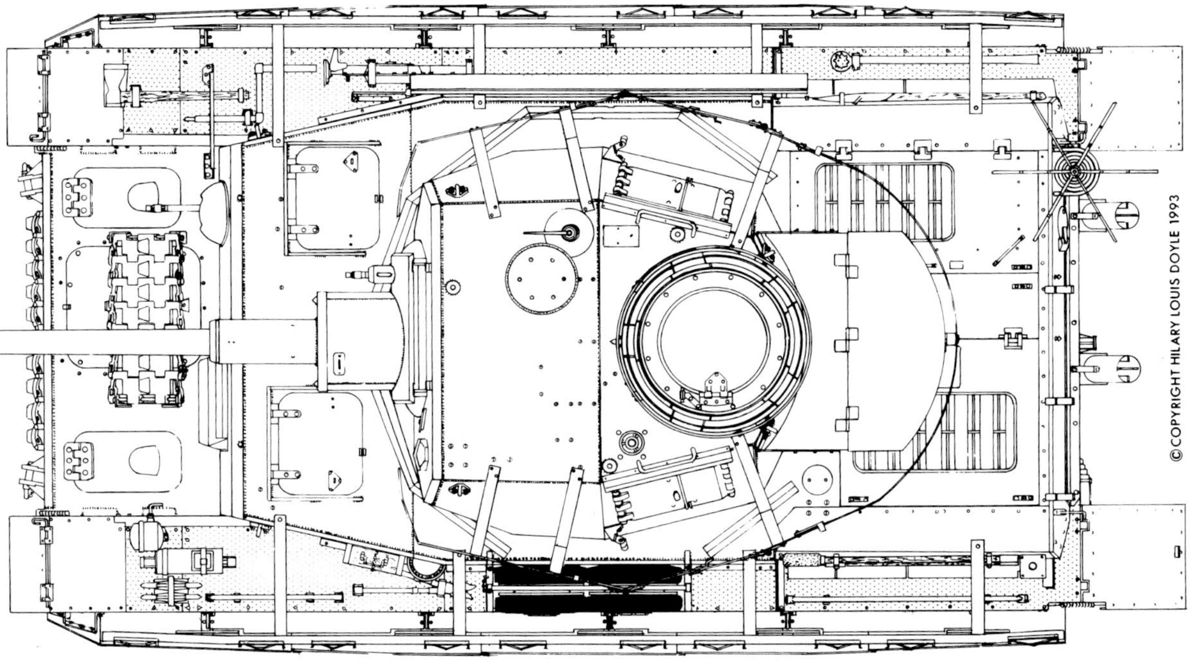

Panzerbefehlswagen IV, Type J (Chassis No. 92200), built by the Nibelungen Works in September 1944. Special attachment for the Star Antenna D for FuG 8 on the back of the hull. In early rebuildings, this antenna was usually attached to the right of the motor compartment cover. The two-meter antenna on the turret roof was intended for the FuG 5, which was located in the turret.

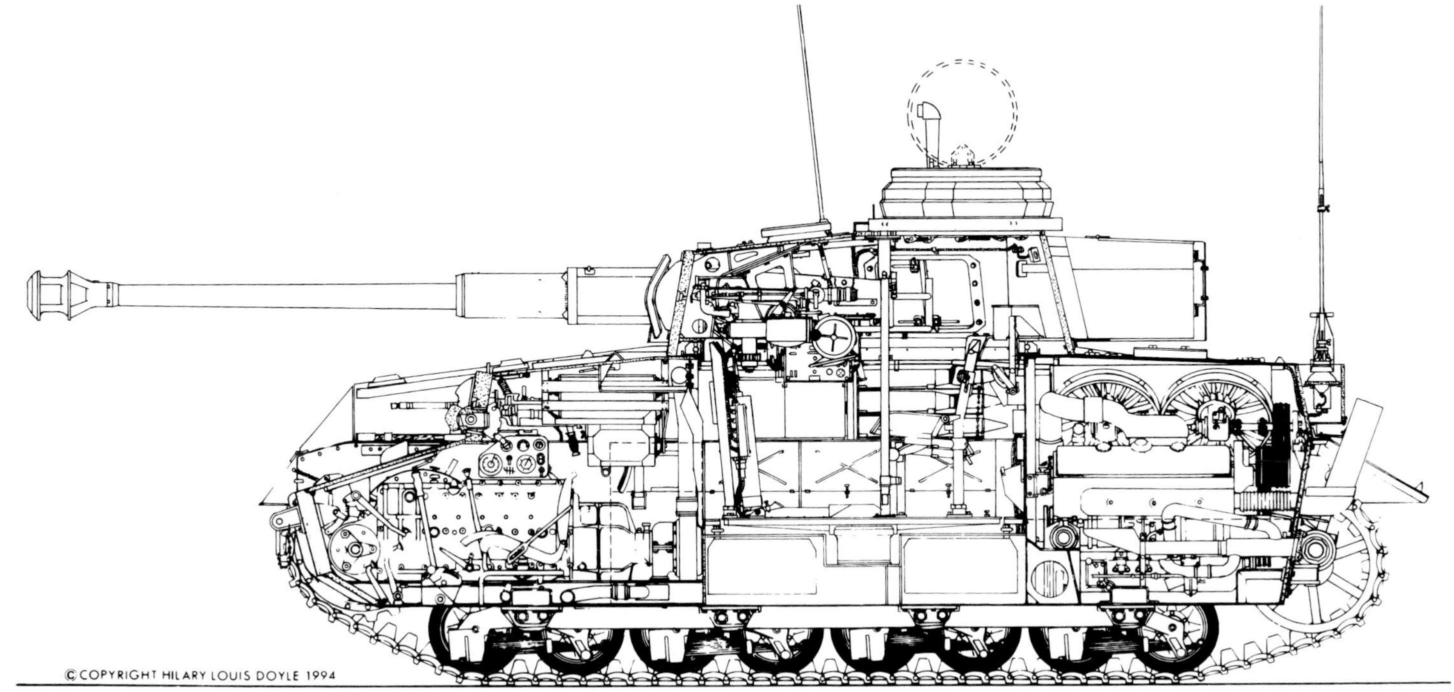

Panzerbefehlswagen IV, Type J (Chassis No. 92200), built by the Nibelungen Works in September 1944. Special attachment for the Star Antenna D for FuG 6 on the back of the hull. The two-meter antenna on the turret roof was intended for the FuG 5, which was located in the turret; the antenna at the left rear was dropped. The drawing shows the location of the FuG 5 on the right side of the turret, where the coaxial MG 34 had been. The FuG 8 is located in the hull. The star antenna can be lengthened by adding an extra rod 0.7 meter long.

Pz.Kpfw.IV Type J, rebuilt as an armored reconnaissance car with assault-gun commander's cupola, in action during the 1944 Ardennes offensive.

Pz.Bef.Wg.IV Type J, Chassis No. 92200, inside the hull, with space for the CG 400 generator at upper left.

Pz.Bef.Wg.IV Type J, Chassis No. 92200, inside in the direction of Sight Staff TSR1.

Pz.Bef.Wg.IV Type J, Chassis No. 92200, inside toward the aiming machine for elevation and traverse setting.

Pz.Bef.Wg.IV Type J, Chassis No. 92200, built by Nibelungen Works in September 1944, with three mounts for the auxiliary crane, a Sight Staff TSR1 and an antenna base on the turret roof.

Folding cover for the commander's cupola.

New Version

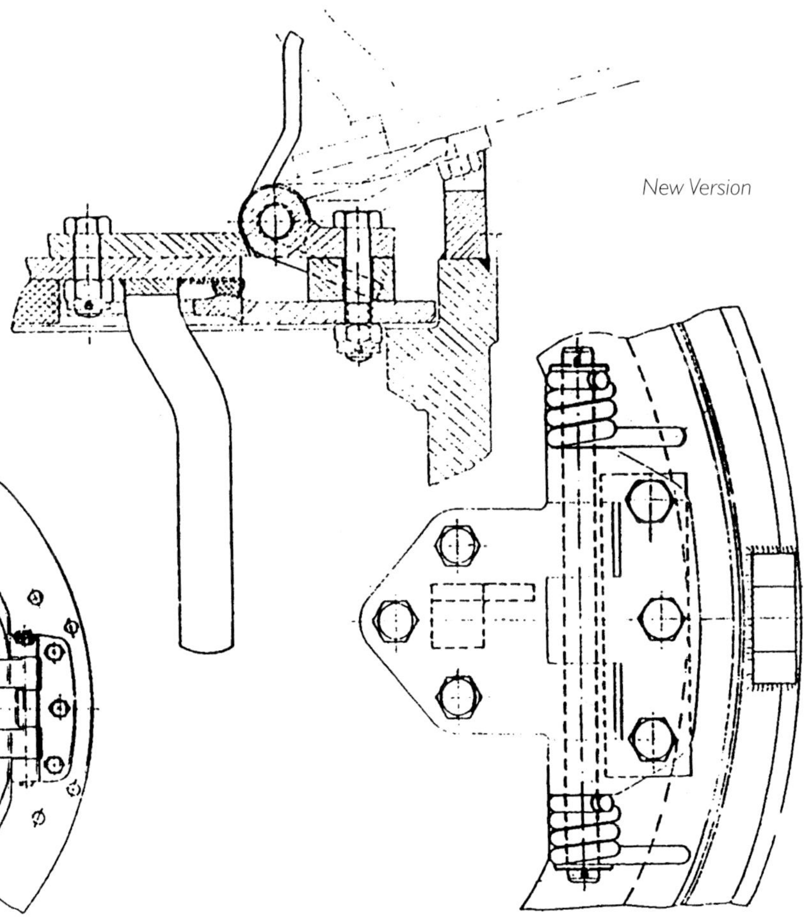

Old Version

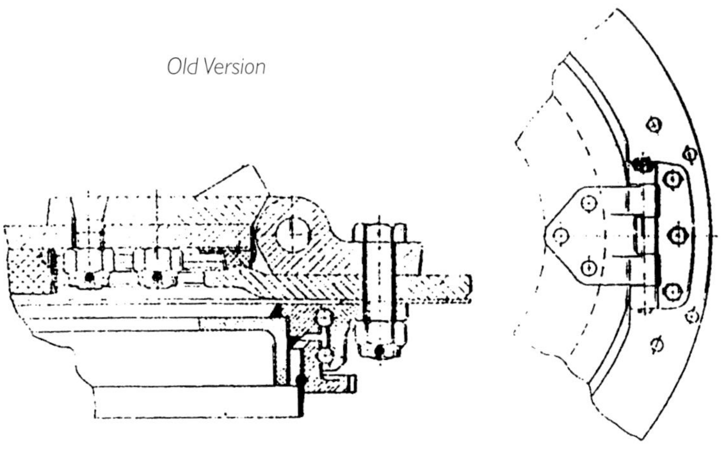

Pz.Bef.Wg.IV Type J, Chassis No. 92200.

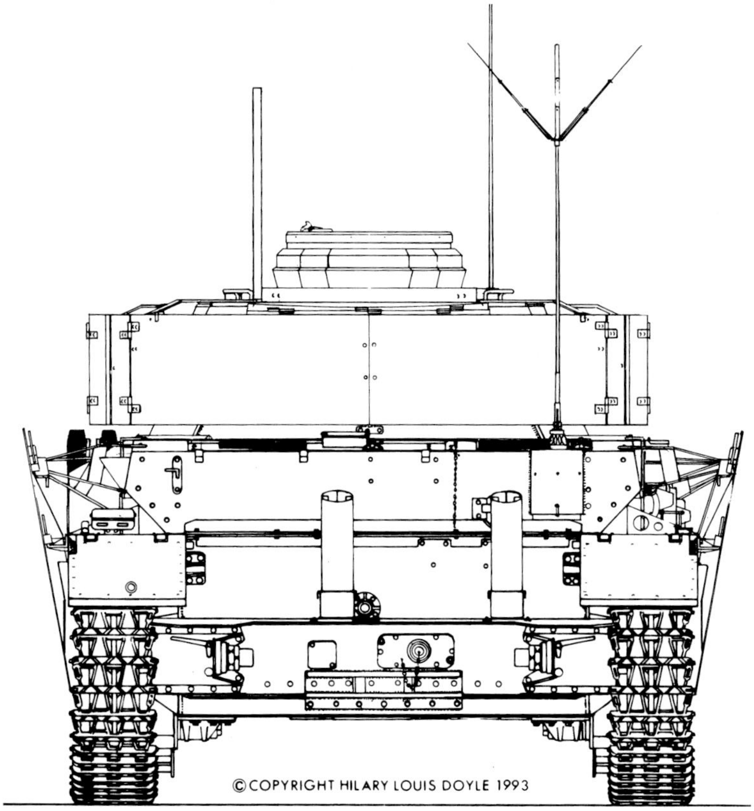

Pz.Bef.Wg.IV Type J, Chassis No. 92200, rear, with attachment for Star Antenna D at right with 0.7-meter extension rod.

Pz.Kpfw.IV Type J without Zimmerit, with side wire-mesh aprons, folding cover for commander's cupola.

Pz.Kpfw.IV Type J without Zimmerit, with swingable cover for commander's cupola. Frame with swingable attachment on the track apron for the wire-mesh aprons. To the left of the driver's vision port are pieces of the horizontal wire mesh segments of the aprons.

14. FINISHING

14.1 ESCORT VEHICLE TEST MODELS

In the War Material Department of Krupp's report on the results of the 1934-35 business year it was maintained:

Newly contracted for are two B.W. (Escort Wagons) of some 18-ton vehicle weight, which are still being designed and built. Tank turrets are contracted for by us in design and production: for their own tank designs, a gun turret for B.W. with 7.5 cm K. L/45. and a design for an M.G. turret for the B.W.

Wa Prw 6 assigned the following contracts to Krupp of Essen for designing and building of test vehicles:

Wa Prw 6/Va Contract No. 67096/36 of 7/13/1935 for one B.W. chassis "B.W.1 Kp." In cast iron (test version)

Wa Prw 6/Va Contract No. 67252/35 of 10/14/1935 for one B.W. chassis "B.W.II Kp." In cast iron (test version)

Wa Prw 6/Vb Contract No. 67941/35 of 1/3/1936 for one B.W. main turret with armored box upper part and M.G. turret in cast iron (test version)

Wa Prw 6/Vb Contract No. 68046/36 of 7/13/1936 for armor box upper body in cast iron for B.W. II Kp.

Wa Prw also issued contracts to Rheinmetall for designing and building at least one but probably two test chassis. The Rheinmetall B.W. test chassis are known from photos, but additional documentation as to their development and testing has not turned up to date.

In the "State of Tank Designing" of January 15, 1936, Wa Prw 6 reported: *The first test piece should be finished in May 1936. Contracts for an 0 series were not yet assigned.*

On a discussion with Wa Prw 6 on February 21 and 22, 1936, at Krupp in Essen, it was reported: *Major Dr. Olbrich envisioned more contracts for us in 1937. There is a larger contract for Z.W. to be expected, since the intended relationship of Z.W. to B.W. is to be 3 to 1. The ordered B.W. turret must exist on May 1, 1936; on May 2, 1936 the acceptance firing is to take place.*

On April 30, 1936, Krupp reported on Contract 41/2203-B.W. I Vehicle: *The soon-to-be introduced vehicle is today, after the end of the second shift, as promised, to be finished and ready to drive.*

On July 21, 1936, Wa Prw 6 (Va) informed the Fried.Krupp firm, Department A.K., in the following letter, that Krupp was the only firm contracted with for the designing and assembly of Pz.Kpfw.IV. The Rheinmetall firm was already closed out of the designing of the B.W. before the middle of 1936.

The O.K.H. (Wa Prw 6) confirmed that in the period from April to August 1937 a first test series of B.W. chassis was to be built at the Gruson Works.

14.2 1./B.W., PANZERKAMPFWAGEN IV TYPE A

14.2.1 ORDERS

The Fried.Krupp—Gruson Works firm of Magdeburg-Buckau had received the following contracts from Wa Prw 6 (Va) for finishing 35 complete Panzerkampfwagen IV Type A tanks.

Chassis Contract No. 67219/37 Wa Prw 6 Va (Besch.) of 12/1/1936 and Body Contract No. 67336/37 Wa Prw 6 Va (Besch.) of 12/22/1936.

The assembly firm was responsible for the assembly of the vehicles by installing the components and aggregates. While the assembly firm produced parts in its own works, most parts were made by other firms and delivered to the assembly firm ready to be assembled.

Of the 1./B.W., contracts were distributed to more than 140 sub-contractors (from the Accumulator Factory AG, Department Varta, to the Zahnradfabrik of Friedrichshafen). The following firms had produced the most important components for the Panzerkampfwagen IV Type A:

Ernst Leitz GmbH, Wetzlar	Targeting Scope T.Z.F.5b
Askania Works	Driver's Optic K.F.F.I
Auto-Union AG, Zschoppau	DKW 2-cylinder motor PZW 600
Continental, Hannover	Rubber tires
Conz Electricity GmbH, Altona	Leonard motor and generators
Daimler-Benz, Berlin-Marienfende	Machine gun ball mantlet
Deutsche Edelstahlwerke AG. Hannover-Linden	25 armored parts for turrets

Froed.Krupp AG, Essen	7.5 cm Kw.K. L/24; 30 armored hulls; 35 armored upper bodies; and 10 armored parts for turrets
Maybach Motorenbau GmbH	Motor HL 108 TR
Ritscher	Tracks
Hans Windhoff, Berlin-Schoeneberg	Radiator for motor
Zahnradfabrik Friedrichshafen	Gearbox SFG 75

14.2.2 Production Plan and Finishing

On July 21, 1936, Wa Prw 6 (Va) had notified the Fried.Krupp firm, Department A.K., as follows: *The O.K.H. (Wa Prw 6) confirms that in the period from April to August 1937 a first test series of B.W. chassis is to be built at the Gruson Works.*

After delays in design and supplying, Wa Prw 6 reported on October 8, 1936: *The delivery of the Panzerkampfwagen IV (7.5) to the troops begins in August 1937; these are the first of the 0 series, which consists of 35 tanks. After the delivery of this 0 series in January 1938, the delivery of the second test series begins in February 1938.*

After further delays for unknown reasons, the first Type A chassis (Chassis No. 80101) was finished, tested and released for adding the body on October 26, 1937. In November 1937, three more chassis, and in December 1937 six chassis, were finished and released.

The acceptance of the first two complete 1./B.W. vehicles with armored body and gun turret by Wa Prw 6 (IId) at the Gruson Works in Magdeburg-Buckau took place on November 30, 1937.

To this historic event there came the following gentlemen, who were mainly responsible for the design, development and building of the Pz.Kpfw.IV:

From Wa Prw 6:	Fichtner, Dr.Ing. Olbrich, Kutzki, Kreckel, Dr. Bruun, Dipl.Ing. Biersack
From Gruson Works:	Dr. Griessmann, C.L. von Gohlen, Scheinig, Baier, von Bergh, Janicke, Dr. Finkelnburg, Sternberg
From A.K.:	Dorn, Woelfert, Heerlein, Lanz

The further acceptance of 2 B.W. was planned for 7 December, plus 5 B.W. for December 22, 1937. But delays occurred, and in fact because of the changes that Wa Prw 6 called for on inspecting the first two vehicles.

Only on December 28, 1937, was the third vehicle (Chassis No. 80103) accepted.

Seven Pz.Kpfw.IV Type A were accepted in January 1938, but because of the acceptance of the last two only on 31 January, only five were delivered to the troops in that month. The rest of Type A was finished, accepted and delivered according to the following plan of Wa Prw 6 of February 21, 1938:

The finishing of the 1./B.W. was confirmed according to the already cited deliveries on the following dates: 6 complete vehicles in February, 8 in March, 8 in April, 3 in May and the last two in June 1938.

14.3 2./B.W., Panzerkampfwagen IV Type B

14.3.1 Contracts

The Weapons Department assigned Chassis Contract No. 66801/38 (Vorbesch.) OKH Wa Prw.6 IIc 1 (Besch.) No. 6516/37 of 10/25/1937 and Body Contract No. 67257/37 Wa Prw.6, IId (Besch.) to the Fried.Krupp-Gruson Works A.G. of Magdeburg-Buckau, for 42 complete 2./B.W. vehicles.

Further sub-contractors were drawn in, among others for main parts:

Ardelt	Tracks
Eisen- und Huettenverein, Bochum	12 armored hulls
Nordbnau (Norddeutsche Motorenbau GmbH)	20 HL 120 TR motors
Stahlwerke Harkort Eicken GmbH, Hagen	18 armored hulls

For this second series, Krupp of Essen had received contracts from Wa Prw 6 for 12 armored hulls, all 42 armored box bodies, but only two armored turret housings.

14.3.2 Production Plans

A second series was first mentioned by Wa Prw 6 on October 8, 1936: *After delivery of this 0 series, the delivery of the second test series begins in February 1938.* The earliest known contracts for components were issued by Wa Prw in April 1937. As with every new series, delays occurred. Discussions of changes of parts were still going on in March 1938.

A discussion at Wa Prw 6 in Berlin on February 21, 1938, explained the production situation of Type B as follows:

A.K. of Essen agrees that by February 28 five upper bodies will be delivered; the hatches belonging to them, though, only after March 20 or March 30, 1938. The Edelstahl firm will deliver four turrets and commanders' cupolas by the end of February, the roller mounts belonging to them only by March 10. The dates for further deliveries of turrets and commanders' cupolas were set: March 10, April 10, May 10.

Since the hull delivery for the 2./B.W. in any case is assured, the following numbers of the 3./B.W. can be given preference, which A.K. of Essen delivers as follows: 3 by March 15, 6 by March 31, 4 by April 15 and 4 by April 30, in all 17 tanks.

In addition, from May to July three each hulls are coming from the delivery of the Eisen- und Hüttenverein of Bochum. As to the hull delivery by the Harkort firm of Eicken, a message from Wa Prw 6 is still to come.

For the 2./B.W. the following dates for complete vehicles can be determined: May 3m June 4, July 8, August 9, September 9 and October 9.

14.3.3 Finishing

The finishing of complete vehicles ready for the troops was endangered by delays in the acceptance firing test at Meppen. On 22 and 23 April it was reported: *7.5 cm barrels no. 36, 37, 40 and 41 tested and fired at Meppen.* On 27 May it was reported: *Guns No. 38 and 39 are already in Magdeburg. Another six guns can be fired only at the end of June or beginning of July because of other urgent use of the firing range.*

To ease this situation, the Gruson Works requested support in the following letter of May 30, 1938:

Reckoned from the arrival of the fired guns at Magdeburg, our works need a total of three weeks for the acceptance of the guns and their installation in the tanks. To date six guns have been delivered, so that we, on the basis of our June delivery, thus urgently need one more gun. We need eight guns for July, but they will not be fired until late in June.

Please do not leave us in the lurch, and arrange for an earlier delivery, at least of the missing gun, for the June delivery. An exchange of guns of the first series for the second series is unfortunately not possible, since the tanks are already completely fitted and the guns with the tanks are displayed for acceptance. An exchange of the missing turrets is also not possible, since the designs of the individual series differ.

On June 2, 1938, the Army Acceptance Center I reported to the Fried.Krupp Gruson Works as to the 2./B.W. series: *Schedule of complete devices newly established, planned as follows:*

3 in May 1938, 4 in June, 8 in July, 14 in August and 13 in September 1938.

Chassis No. 80226 was already finished and released for mounting the body on July 29, 1938. Another 12 chassis, No. 80227 to 80239 (minus 80236), were finished and released by the end of August.

The last four chassis, No. 80236, 80240, 80241 and 80242, were finished and released by September 16, 1938.

On 10 September the Gruson Works reported: *Instead of 12, only 9 complete vehicles were finished in September.* On 7 October a further notice to Department A.K.: *Gruson Works need by 4:00 this afternoon sender and receiver attachment devices for three vehicles, otherwise the suppressor acceptance on the 11th of this month and the final acceptance cannot take place on 10/18/1938.*

As far as explanatory documents are still at hand, the Pz.Kpfw. IV Type B were finished, tested and accepted almost as in the 21 February plan: *3 in May, 3 in June, 9 in July, 9 in August, 9 in September and the last 9 in October.*

14.4 3./B.W., PANZERKAMPFWAGEN IV TYPE C

14.4.1 Contracts

The Weapons Department assigned Chassis Contract No. 66 801/38 (Vorbesch.) OKH No. 6516/37 Wa Prw 6 IIc (Besch.) of 10/25/37 and Body Contract No. 66 807/38 (Vorbesch.) OKH No. 5908/37 Wa Prw 6 IId 1 (Besch.) of 10/14/1937 to the Fried.Krupp firm's Gruson Works at Magedburg-Buckau for 140 complete 3./B.W. vehicles.

Further sub-contractors were called in for, among other things, main parts: 7.5 cm Kw.K. by Knorr Bremse, Berlin, and also by the Heidenreich Harbeck firm of Hamburg.

For this 3rd series Krupp of Essen had received contracts from Wa Prw 6 for 110 armored hulls, all 140 armored bodies, and all 140 armored turrets.

14.4.2 Production Plans

The situation and plans for the further building of Pz.Kpfw.IV were clarified in the following notes on a discussion at Krupp on May 31, 1937:

A production gap between the 1. and 2./B.W. series is not to be feared. With the groups received with A.K. documents No. 52183 of 5/8/1937 and BM No. 30085 of 5/14/1937, the material ordering can be maintained without problems. For the parts that will be completely new, Essen has provided the materials and measurements, with the exception of a few cast-steel parts, which can be ordered without problems. Essentially more difficult is the transition from the second to third series. According to word from Prw.6, two test vehicles should be finished in January. But the Weapons Department as not yet made a final decision as to this version, so that presumably the finishing date of these two test vehicles, according to information from Essen, is postponed to April 1938. Since the Gruson Works will end the finishing of the second series in August 1938, the Gruson Works would have to have the design plans for the third series available in October 1937 at the latest, in order to be able to start obtaining materials at the right time. This schedule, though, cannot be kept, according to the aforementioned situation with the designing. In case a decision is made on the designing of the third series by the Weapons Department by mid-June, two test vehicles can be delivered in April 1938, so that series production is not to be expected before June. But that would mean a production gap of eight months for the Gruson Works. The Gruson Works would thus have to receive an interim contract for B.W. vehicles (Series 2a) for these eight months.

In a discussion in the Weapons Office on June 27, 1937, Krupp was informed that in the Gruson Works, after delivery of the 2./B.W. series, further Escort Wagons were to be produced in a great numbers as possible.

How big these new contracts for the 3./B.W, were supposed to be was determined in the contract of September 21, 1937, as follows:

On September 6 of this year, Krupp of Essen received with No. 66 806/38 a preliminary notice of delivery of tank parts for armored bodies and turrets for 40 3./B.W. bodies. Meanwhile the number of 40 has been raised to 140.

The schedules for finished and released chassis to mount bodies on at the Gruson Works were set as follows on 27 October: *8 vehicles in July, 15 in August, 17 in September and 20 per month from October 1938 to February 1939.* A month after the chassis were finished, complete vehicles with turrets were to be finished, accepted and delivered. On November 12, 1937, Wa Prüf 6 ordered that the B.W. production program should be ended in April 1939.

14.4.3 Finishing

By September 30, 1938, two months after the original date, the first Type C chassis (No. 80301 to 80308) were finished and released for mounting the bodies. Then came the finishing of 12 in October, 12 in November, and 14 in December 1938. Another 88 chassis were finished between January and July 1939 and released for mounting the body and turret. According to Regulation AHA In 6 (IVd) No. 156/39, Secret Command

Material, of February 22, 1939, for special bridgelaying test series by Wa Prüf 5, the last six Type C chassis (No. 80435 to 80440) were finished and delivered in June. The first complete Pz.Kpfw.IV Type C were finished in October 1938, accepted by Army Acceptance Center II (K), and delivered to troop units. As the following table shows, the original production goal of 20 per month was never reached. Delays occurred, partly from late delivery of necessary parts. After March 1939 the monthly goal was deliberately reduced to bridge the gap in the production program until the start of finishing the Pz.Kpfw.IV Type D in September 1939.

Month	Planned	Finished	Accepted and Delivered
October 38	*	4	2
November 38	*	11	13
December 38	*	17	14
January 39	*	15	16
February 39	16	17	17
March 39	16	16	18
April 39	12	12**	9
May 39	11	11	11
June 39	11	10	11
July 39	11	11	11
August 39	11	10	11

* The finishing preview was first reported on March 1, 1939 by Wa J Rue (Wu G 6).

** The Pz.Kpfw IV Type C (Chassis No. 80392) was finished in April 1939, but not accepted or delivered to the troops. It had not passed the acceptance test, as the back wall of the hull (No. 80408 Kr) was bent outward 37 mm. This vehicle was still in the repair shop on September 12, 1939, as the Gruson Works reported, that it could be repaired by the Gruson Works themselves and need not go to Essen. The back wall went back to 10 mm in the removal of 37 mm of bending. Cracks in the armor plate were not present, so that this vehicle was made ready for the troops at the Gruson Works.

14.5. 4. AND 5./B.W., PANZERKAMPFWAGEN IV TYPE D

14.5.1 Contracts

The Fried.Krupp Gruson Works A.G. of Magdeburg-Buckau had received the following contracts from Wa Prw 6 (IIc/1) for finished assembly:

200 4./B.W. chassis, Contract No. 210.3.5002/38, of July 13, 1938
200 4./B.W. bodies, Contract No. 210.3.5602/38
48 5./B.W. chassis, Contract no. ?
48 5./B.W. bodies, Contract no. 210.5901/39

On November 5, 1938, Krupp of Essen received from Wa Prüf 6, II c/1 Contract no. 210.3.5801/39 for 48 5./B.W. armored hulls, followed on 9 November by Contract No. 210.3.5202/38 Wa Prüf 6, II c/1 for 200 4./B.W. armored hulls, and somewhat later Contract No. 210.3.5602/38 for 200 sets of armor parts for armored bodies and turrets with commanders' cupolas.

The Fried.Krupp AG firm of Essen had received, on December 27, 1938, advance notice for Contract No. 210.5401/39 Wa J Rue 6 VIII a (Army) for 48 sets of armor parts for armored bodies and turrets (30 mm commander's cupola). This last contract came from Wa J Rue VIIIa instead of as before from Wa Prüf 6, Wa J Rue (Wu G 6), as a new sub-group of the Army Weapons Department set monthly production goals and collected monthly production statistics from Army Acceptance Center inspectors to find out whether the goals were attained, and if not, why not.

14.5.2 Production Plan and Finishing

The first mention that Pz.Kpfw.IV on B.W. chassis would still be built after the 3rd series came on March 22, 1938, in a note about joint contracts for 4. and 5./B.W. from the chief of Department A.K.:

I explained to Captain Wollenhaupt that we, in order to finish promptly the contracts in view for 200 4./B.W, and 80 5./B.W. tanks, which were to be delivered about January 1940, would have to undertake an extension of our Apparatus building III. The steel allotment was already arranged through Wa B 4.

The Führer and Reich Chancellor Adolf Hitler had declared in his command for the armed parts of the SS on August 17, 1938, that for the SS troops four standards, numbered like motorized infantry regiments of the Army, should be set up. Also per standard:

1 engineer platoon
1 armored reconnaissance platoon
1 tank company
(gun vehicles) in structure like the 4th company of the Army armored units.

According to a document, Adjutant of the Führer and Reich Chancellor, of October 17, 1938, No. 13a/38, 48 Pz.Kpfw.IV should be delivered to the Führer from January 1940 on, with five delivered per month. To make this possible, an increase of capacity from 15 to 20 Pz.Kpfw.IV per month was ordered.

This series of 48 Pz.Kpfw.IV ordered by the Führer (for SS troops) was called the 5th Series B.W. The two B.W. series, 4th and 5th, were identically built, and both were thus known as Type D. The former 6th Series (for the Army) was renamed the 6th Series B.W. (Type E).

On December 20, 1938, the leader of the Fried.Krupp Gruson Works suggested to Wa Prüf 6, II d/2, to change the delivery schedule for the following reason:

We confirm the telephone talk of your Major Kreckel with our Mr. Jaenicke ask you for approval to bring first the 4./B.W. Series and then the 5./.W. Series to delivery with a monthly production of 20 as of January 1940. The reason for our request is that we do not receive the parts for the 5./B.W. Series so promptly that we could begin the 5th Series already in January

1940 parallel to the 4th Series. In administrative technology too, it requires considerably more work in production when two series of the same type and version run parallel to each other. In order to be able to deliver 20 of the 4th Series a month as of January 1940, it would be necessary to set the dates for the Reich's own deliveries so that first the Reich's own parts for the 4th Series be delivered and then those for the 5th Series. Otherwise we would have to receive official permission to use parts for the 5th Series on the 4th Series. At this time the dates are set so that the Reich's own parts for the 5th Series arrive parallel to the Reich's own parts for the 4th.

For your examination we send you the dates for the armored turrets of the 4th and 5th Series. Accordingly the 4th Series Contract 210.3.5602/38 will be delivered with 15 tanks from May 1939 to May 1940 and 5 in June 1940, while for the 5th Series the turrets will be delivered with 2 as of September 1939, 3 in October 1939, 5 from November 1939 to May 1940 and 8 in June 1940. These dates have validity if the vehicles of the 4th and 5th Series are delivered parallel from January 1940 on. But if, according to our wish, the 4th Series be finished first, the last 12 vehicles of this series would be delivered in July 1940. This would, though, allow that the delivery of turrets would take place not in June 1940, but already in April 1940. Otherwise we would have to get permission from the Department to use turrets of the 5th Series for the 4th Series.

Wa Prüf 6 did not agree to the change of the original schedule and declared that the 5./B.W. series, as scheduled, was to be turned out as of January 1940. Allowed or not, the Gruson Works had used 4./B.W. parts in building the 5./B.W. vehicles and vice versa.

Wa Prüf 6 IId tried to eliminate a problem in production on March 9, 1939 by requesting of Krupp of Essen that *...new schedule setting for delivery of 50 new type steering apparatus for the 4./B.W., since 9/15/1939 was not workable. Suggested was June 125, 1939.* Krupp answered on 1 April:

After the drawings for the new radio-desensitized steering apparatus have been sent to the workshop on March 24, 1939, it can be depended on that the first eight steering apparatus, through preferential production, can be finished by 8/1/39. The further delivery can be done at a rate of 25 per month as of September.

The Gruson Works also requested earlier delivery dates on April 6, 1939, since it needed steering apparatus as follows:

1) *8 items of July 15, 1939 (for vehicle finishing for August,)*
2) *15 per month until 11/15/1939 (for vehicle finishing in October to December),*
3) *then 20 per month until delivery.*

Krupp of Essen answered again with the same numbers and dates as on April 1. Because of this difficulty, the beginning of delivery of Type D vehicles was delayed from the original schedule to a later date of 10 Pz.Kpfw.IV in September. This goal was not attained either in the first month of the war because of delayed delivery (1 steering apparatus by express on 9/8/1939, 9 of them by rush order on 9/19/1939.

Production Statistics for Panzerkampfwagen IV Type D

Month	Projected Production	Finished Pz.Kpfw.IV		
		4./B.W.	5./B.W.	B.L. Chassis*
September 39	10	0**		
October 39	10	20		
November 39	14	11***		
December 39	16	14****		
January 40	20	15	5	
February 40	20	13	5	
March 40	20	9	5	6
April 40	20	5	5	10
May 40	20	15	5	
June 40	20	18	5	
July 40	22	21	5	
August 40	30	23	5	
September 40	30	9*****	8	
October 40		10		

* 4./B.W. chassis for Bridgelayer IVb.

** In a discussion with Wa J Rue 6 on September 29, 1939 it was explained: The production of 10 Pz.Kpfw.IV not achieved in September was made up in October. The steering apparatus had arrived in the meantime. Further delivery of steering apparatus assured. 20 B.W. would thus be delivered at the Gruson Works in October.

*** Delivery delay in favor of faster repair of damaged vehicles. From Wa J Rue 6 on November 3, 1939, in a survey of production of Pz.Kpfw.IV:

	4. Nov	6. Nov	13. Nov	18. Nov	Total
Repaired	9	7	4	5	25
Newly built	4	0	4	3	11

**** Delivery delays were caused by hastened repairs and spare-parts production.

***** Planned production was not attained because of delayed delivery of parts from the Schlemper & Co. firm.

In a discussion with Wa J Rue 6 on September 29, 1939, it was explained: *Further delivery 14 in November, 14 in December and 2o per month as of January 1940. From Krupp of Essen 30 tank bodies should certainly come per month. Thus the Gruson Works can increase their achievement to 30 as of June 1940.*

Further production increases were requested by the Army Weapons Department in a message to Krupp of Essen on May 31, 1940: *Führer orders via Reich Minister Dr. Todt increase in production of Pz.Kpfw.III and Pz.Kpfw.IV. Increase is to be put in the works at once, so that it will work out already in July, August, September, October.*

In a filed note on the session of the Subcommittee on Tank Production in the Special Branch for Tanks in Hannover on July 11, 1940, it is stated: *At Fried.Krupp AG the increase in the program at first to 30 and then to 40 B.W. is planned. It is to be determined whether Krupp can carry out an increase sooner than stated.*

Wa J Rue (WuG6) had reported the production of only 199 4./B.W. instead of 200. Data from Krupp and chassis numbers from original files confirm that all 200 4./B.W. and 48 5./B.W. were produced. The 200th vehicle had not gone through Army Acceptance Center inspectors in order, but as was reported on June 5, 1940: *By the In6 one vehicle (Chassis No. 80625) was released for testing purposes and stationed in Kummersdorf.* On 11 July Wa Prüf 6 ordered: *The series test vehicle in Kummersdorf stays there for further tests by Wa Prüf 6.*

14.6 6./B.W. PANZERKAMPFWAGEN IV TYPE E

14.6.1 Contracts

On July 17, 1939, the Fried.Krupp Gruson Works had received the following contracts from Wa J Rue for assembly of complete vehicles:

Contract No. 210-5002/39(H) for 223 chassis
Contract No. 210-5902/39(H) for 223 bodies

Krupp of Essen received contracts from Wa J Rue 6/VIIIa on May 11, 1939: Contract No. 210-5802/39 for 223 complete 6./B.W. armored hulls and Contract No. 210-5402/39 for 223 sets of 6./B.W. armor parts for upper bodies, turrets, mounts and commanders' cupolas.

On August 3, 1939, both contracts for Krupp of Essen were lowered from 223 to 203 armor parts for hulls and bodies. At this time, though, the assembly contract for the Gruson Works remained at 223. At first other firms were to be responsible for 20 sets of armor parts.

Counselor Rau (Wa Prüf 6) had decided on February 1, 1940:

The Böhler firm in Kapfenberg shall be set up for the production of B.W. hulls. Since the material question has not yet been cleared up and the setup of the mechanical workshop has not yet made so much progress that the B.W. housings can be built completely in the workshop, the following view prevails at Wa Prüf 6:

The Eisen-Huetten-Works in Bochum shall obtain the material for 7 to 10 B.W. Series 6./B.W. tanks (exact number will still be determined) and send it, worked and prepared, to the Böhler firm. At the Böhler firm only the assembly and welding work will be carried out.

Since the obtaining of forged and pressed parts, what with the small number for the Eisen-Huetten-Works firm, will cause difficulties, these parts shall be obtained from Fried.Krupp from the 6./B.W. Series.

The turrets shall not be made by Böhler. These parts shall be delivered by Fried.Krupp from the Series 4. and 5./B.W. (instead of the turrets kept out for the 16 ladder bodies).

On 29 May the Böhler Brothers & Co. A.G. wrote to Krupp:

We have noted with best thanks the contract with your letter of 5/4/40 for delivery of three complete armored bodies for Pz.Kpfw.IV.

On March 12, 1941, the Krupp Gruson Works were notified:

The number of 223 bodies assigned to us (Contract No. 210-5902/39(H) is hereby decreased by 17 pieces. Thus from this contract only 206 troop-ready bodies are to be brought to delivery.

And: *The number of 223 chassis assigned to you on 7/17/39(Contract No. 210-5002/39(H) is hereby decreased by 17 pieces. Thus from this contract 206 chassis (with numbers 80801-81006) are to be brought to delivery.*

14.6.2 Production Plan

The first mention that a 5th series was to be built came on March 22, 1938, in a notice about assigned contracts for 4./ and 5./B.W. from the chief of Department A.K.: *I explained hereby to Captain Wollenhaupt that we, in the in the contract planned from him for 200 4./B.W. tanks and 80 5./B.W. tanks...*

On July 11, 1938, In 6 (Inspector-General of the Armored Troops) assigned a continuing contract for the 1939 fiscal year for 223 Pz.Kpfw.IV (7.5 cm). The entire orders that ran out to date thus included 640 (35 in 1st series, 42 in 2nd series, 140 in 3rd series, 200 in 4th series and 223 in 5th series) Pz.Kpfw.IV. At this point in time the right to make decisions on the number of tanks in each series belonged to In 6 (checked by the budget of the Army General Staff) and also of distributing the contracts to Wa Prüf 6. Wa Prüf 6 took the contract from In 6 and divided contracts among the design, component, manufacturing and assembly firms.

After ordering a series of 48 Pz.Kpfw.IV for use by the SS troops, the earlier 5th Series was renamed the 6th Series of B.W. by the Army.

Only on April 18, 1939, were the Gruson Works finally named as the assembly firm for all 223 tanks, as Wa Prüf 6 reported: *The number of the 6th Series has been raised from 175 to 223.*

The original delivery dates from May 1939 for the turrets became 17 in July 1940, 20 a month as of August 1940 to May 1941, and the last six in June 1941.

14.6.3 Finishing of Armor Parts

Krupp-Cast Steel in Essen produced the armor parts for the 6./B.W. as follows:

Month	Hulls	Bodies	Turrets
June 40	30	16	13
July 40	30	34	17
August 40	25	30	35
September 40	28	30	40
October 40	29	31	28
November 40	48	39	42
December 40	13	23	38

After the production and delivery of the last parts, Krupp submitted a bill on February 18, 1942 for RM 2,591,295.00; in a bill for 6./B.W. Contract 210-5802/39 for 203 armored bodies for armored hulls and 203 sets of armor strengthenings for hulls; and on February 19. 1942 RM 4,548,215.00 for 6./B.W. Contract No. 210-5402/39 for 203 armored bodies, 203 armored parts for turrets, 203 ball mantlets and 203 sets of bow armor strengthening.

On 24 August, because of the 6./B.W. armored parts production by Böhler of Kapfenberg, it was announced: *Krupp Cast Steel and others are not in a position to produce, beyond the planned three armored hulls and armored bodies, to make further armor plates available in a worked condition.*

14.6.4 Production of Complete Vehicles

On September 5, 1940, Special Branch VI, Tank Production, presented the production programs for the assembly firms. The Krupp Gruson Works were to produce. Under the immediate program, 30 Pz.Kpfw.IV monthly from August to December 1940 and 35 monthly from January to April 1941. In a discussion on November 25, 1940, at Special Branch VI, Tank Production, Krupp stated that the firm could produce only 30 instead of 35 in December 1940 and 35 instead of 40 in January 1941, since only a portion of the required work force had been assigned, and these too late.

Production Statistics for Panzerkampfwagen IV Type E

Month	Planned	Finished 6./B.W.
October 1940	30	20*
November	30	30
December	30	30
January 41	30	31**
February 41	30	26***
March 41	30	28***
April 41	33	36****

* Plus 1o Type D finished in October 1940.
** Plus 4 chassis for Bridgelayer IVc.
*** Minimum delivery for lack of workers.
**** The last two Type E (Chassis No. 81005 and 81006) were to have box running gear installed.

14.7 7./B.W., PANZERKAMPFWAGEN IV, TYPE F

14.7.1 Contracts

On December 28, 1939, the Krupp Gruson Works received Contracts No. 210.5003/39(H) for the final assembly of 500 7./B.W. chassis for Pz.Kpfw.IV.

In 1940-41 contracts to other firms for final assembly of 100 7./B.W. chassis each with complete bodies were issued to Vomag and Steyr (Nibelungen Works).

Krupp of Essen was at first the only firm with a contract for the production of armor parts. On December 8, 1939, Krupp Cast Steel was issued Contract No. 210.5803/39(H) for 500 armored hulls (7./B.W.) and No. 210.5404/39(H) for 500 complete armored bodies by Wa J Rue 6, VIIIa. Further contracts were given to Krupp and other steel firms, but these were for the most part not for Type F, but for the following Type G.

14.7.2 Production Plan and Completion

On December 6, 1938, In 6 had already issued the next succeeding contract for 128 Pz.Kpfw.IV (7th Series) for the 1939 fiscal year (for the budget). By November 1939 the ordered number of the 7th Series had been raised to 500 Pz.Kpfw.IV.

Through the December 1939 contracts, delivery dates for 50 chassis per month were required in connection with the 6./B.W. Series. The original goal (set in May 1939) for producing the last PZ.Kpfw.IV in the 6./B.W. Series was June 1941. Despite various plans for raising production, the Pz.Kpfw.IV Type F were first produced by the Krupp Gruson Works only in June 1941, as originally planned.

Krupp Cast Steel was to first supposed to deliver all 500 armored parts only to the Krupp Gruson Works. When the Vomag firm was moved into the production of Pz.Kpfw.IV tanks, Krupp was, as planned in December 1940, to deliver armored parts to Vomag, as is seen in a report of February 19, 1941: *Wa J Rue 6 has knowledge that of four hull deliveries for January 1941, 13 hulls were delivered to the Gruson Works and 12 to Vomag. By February 19 16 hulls were on hand at Vomag. Wa J Rue 6 now asks that the plan be adhered to according to the discussion of December 16, 1940, whereby Vomag in February receives a total of 6 hulls.* Krupp answered: *Essen, to Vomag were sent: 9 hulls on February 9, 7 hulls on February 15 and 8 hulls on February 26 1941. The last-named 8 hulls possibly arrived at the recipient only at the beginning of March.*

With the start of Pz.Kpfw.IV production at the Nibelungen Works, armored parts were delivered from Krupp Cast Steel.

On September 5, 1940, Special Branch VI, Tank Production had presented the Production Program I for the assembly firms. In Program I (as of May 1941 to March 1942), the firms were to strive for the following monthly goal (10 B.W. from Vomag, 10 from the Nibelungen Works, and 50 from the Gruson Works. With the exception of Vomag (from November 1941 on), these monthly goals were never attained.

Production Statistics for Panzerkampfwagen IV Type F

Month	Planned Production	Finished by			
		Krupp	Vomag	Nibelungen	Total
May 41	35	16*			16
June 41	39	38			38
July 41	41	38**			38
August 41	42	42	2		44
September 41	24***	42	4		46
October 41	45	43	8		51
November 41	54	39****	12	1	52
December 41	46	45	14	2	61
January 42	57	45	12	4	61
February 42	58	45	12	6	63

* Mimimum delivery resulting from starting difficulties for 7th Series B.W. Individual Type F chassis (among others, No. 82008) were used for "Karl Device" ammunition tractors.
** Production difficulties resulting from lack of workers at Krupp Gruson Works.
*** Mimumun delivery resulting from lack of workers.
**** Minimun delivery for lack of specialist workers at Nibelungen Works and electric disturbances.

14.7.3 Conversion to Tank Gun 40

On February 11, 1942, the Krupp Gruson Works were informed:

As of March 1942, you are to install in the Panzerkampfwagen IV (7./B.W.), instead of the former 7.5 cm KwK the 7.5 cm KwK40 (L43) by Rheinmetall with T.Z.F.5f.

To the extent that it is not possible to obtain the KwK 40 at the right time, the vehicles are to be kept by you until the guns arrive.

A delivery of Pz.Kpfw.IV with old guns may not in any case take place in March.

With this simultaneous order to Krupp, Vomag and the Nibelungen Works, the production of Type F came to an end. The Krupp Gruson Works had completed 393 Type F chassis (Chassis No. 82001-82393), Vomag 64 (Chassis No. 82501-82564) and Nibelungen Works 13 (Chassis No. 82601-82613).

14.8 8./B.W., PANZERKAMPFWAGEN IV TYPE G

14.8.1 Contracts

All three factories, Grupp-Gruson Works, Vomag and Nibelungen Works, were to finish the remainders of their original contracts for Type F with the longer tank gun as Type G (F2), as follows:

- 107 with chassis numbers 82394-82500by Krupp-Gruson Works
- 36 with chassis numbers 82565-82600 by Vomag, and
- 87 with chassis numbers 82614-82700by the Nibelungen Works.

From 1941 to the spring of 1942, further contracts were issued to the assembly firms, as follows:

- 300 with chassis numbers 82701-83000 to Krupp-Gruson Works
- 100 with chassis numbers 83001-83100 to Vomag
- 100 with chassis numbers 83101-83200 to Nibelungen Works
- 200 with chassis numbers 83201-83400 to Krupp-Gruson Works
- 150 with chassis numbers 83401-83550 to Vomag
- 100 with chassis numbers 83551-83650 to Nibelungen Works
- 300 with chassis numbers 83651-83950 to Krupp-Gruson Works
- 150 with chassis numbers 83951-84100 to Vomag
- 300 with chassis numbers 84101-84400 to Nibelungen Works

To increase the monthly production of armored parts, the secondary firms of Böhler Brothers & Co. A.G. of Kapfenberg and Upper Danube Iron Works (Eisenwerk Oberdonau) of Linz became more and more involved. Contracts for armored hulls and all armored parts of B.W. bodies were assigned as follows:

- 384 to Krupp-Cast Steel in January 1941
- 116 to Böhler in February 1941
- 50 to Eisenwerke Oberdonau in June 1941
- 230 to Krupp-Cast Steel before September 1941
- 160 to Eisenwerke Oberdonau before September 1941
- 200 to Böhler in September 1941
- 200 to Eisenwerke Oberdonau before January 1942
- 175 to Krupp-Cast Steel in January 1942
- 325 to Böhler in February 1942

14.8.2 Production Plans and Assembly

One of the original plans for the Adolf Hitler Program II set the following monthly goals:

	1942									1943		
	Apr	May	Jun	Jul	Aug	Sep	Oct	Nov	Dec	Jan	Feb	Mar
Krupp	60	67	45	45	45	53	57	57	59	59	59	60
Vomag	15	17	13	13	20	22	26	31	36	39	41	41
Nibelungen	5	8	14	10	16	19	23	32	37	43	51	52
Total	80	92	72	68	81	94	106	120	132	140	151	153

The higher goals for April and May were supposed to make up for the lost production of March, caused by the change to the new 7.5 cm Kw.K. 40. These planned figures should be compared with the following list of Type G completion statistics.

Production Statistics for Panzerkampfwagen IV Type G

Month	Planned Production	Completed Type G			
		Krupp	Vomag	Nibelungen	Total
March 42			1		1
April 42	80	60	15	5	80
May 42	92	57	20	8	85
June 42	72	45	13	14	72
July 42	68	54	20	14	88
August 42	81	50	20	14	84
September 42	94	50	22	21	93
October 42	106	50	26	23	99
November 42	120	50	31	32	113
December 42	132	65	45	45	155
January 43	140	67	46	50	163
February 43	155	66	50	55	171
March 43	200	80	55	70	205
April 43	215	98	65	50	213
May 43	235	97	7	100	204
June 43		18		86	104

14.9 9./B.W., PANZERKAMPFWAGEN IV TYPE H

14.9.1 Contracts

Assembly contracts were issued for complete Pz.Kpfw.IV Type H tanks by Wa J Rue (WuG 6) as follows:

- 500 from Krupp with chassis numbers 84401-84900
- 450 from Vomag with chassis numbers 84901-85360
- 400 from Nibel. with chassis numbers 85351-85750
- 400 from Krupp with chassis numbers 86761-86150
- 450 from Vomag with chassis numbers 86151-86600
- 500 from Nibel. with chassis numbers 86601-87100
- 500 from Krupp with chassis numbers 88100-88600
- 500 from Vomag with chassis numbers 86601-89100
- 1500 from Nibel. with chassis numbers 89101-90600

Pz.Kpfw.IV Type G during assembly of chassis and inside fixtures (Nibelungen Works, April 1943).

Final assembly of Pz.Kpfw.IV Type G: After acceptance of the chassis, the turret and the tank gun were installed. (Nibelungen Works, April 1943).

After release by the Army Acceptance Center in the Nibelungen Works,
Pz.Kpfw.IV Type G were loaded onto railroad cars and transferred to the Army Arsenal in Linz (autumn 1942).

Pz.Kpfw.IV Type G with antenna deflectors for turret machine guns, completed at the Nibelungen Works (autumn 1942).

Armored parts were at first ordered, as before, from Krupp, Eisenwerke Oberdonau and Böhler, and as of mid-1943 from Eisen- und Hüttenwerk Bochum.

14.9.2 Production Plans and Assembly

In a discussion at the Gruson Works on July 14-15, 1942, it was already decided that production with strengthened cone drives, side shafts and planetary drive for Type H could start from June to July 1943. The first 9./B.W. hulls for Type H were finished and delivered by Krupp of Essen in March 1943.

One of the original plans in the Adolf Hitler Program II set up the following monthly goals:

	1943									1944		
	Apr	May	Jun	Jul	Aug	Sep	Oct	Nov	Dec	Jan	Feb	Mar
Krupp	60	60	60	70	80	80	80	80	80	80	80	80
Vomag	45	45	45	50	50	50	50	50	50	50	50	50
Nibelungen	70	70	80	95	100	105	110	115	115	115	115	115
Total	175	175	185	215	230	235	240	245	245	245	245	245

Because of the front situation, increasing production was requested, as the following monthly production goal shows:

	1943								1944		
	Apr	May	Jun	Jul	Aug	Sep	Oct	Nov	Dec	Jan	Feb
Total	215	235	255	275	295	260	290	350	300	350	310

Production increases were partly attained by Vomag, but mainly by the Nibelungen Works. Minimum deliveries from Krupp-Gruson Works were caused by difficulties with the constant program changes during 1943, as follows:

April 1943: Krupp receives orders to give up Pz.Kpfw.IV production had quickly switch to 150 Panther I or II per month.

May 1943: Krupp is informed that Panther II is not to be produced, rather only Panther I.

June 1943: Directive that from 150 anther I monthly, a number of old recovery
Tractors with cranes should be delivered.

Aug. 12, 1943: Order to stop Panther I at once and switch to 150 Pz.Kpfw.IV monthly.

Aug.20, 1943: Monthly production is limited to 100 tanks.

Dec. 1943: Immediate switch to production of 110 Assault Gun IV monthly as of February 1944.

Dec. 1943: Contract issued for armored command cars; contract withdrawn in January 1944.

The Krupp Gruson Works calculated that these and other program changes cost a total loss of 300,000 work hours, adding up to 150 vehicles. Before the switch to Assault Gun IV production in December 1943, Krupp had produced only 379 vehicles (chassis numbers 84401-84779) of their first contract for 500 Type H tanks.

The production of Type H was ended in February 1944 by switching to Type J.

Production Statistics for Panzerkampfwagen IV Type H

Month	Finished Type H			Accepted by H.WaA
	Krupp	Vomag	Nibelungen	
May 43		68		68
June 43	49	67	34	150
July 43	54	65	125	244
August 43	67	75	141	283
September 43	64	75	150	289
October 43	73	85	170+10*	338
November 43	50	68	120	238**
December 43	22	90	210+30***	354
January 44	0	100	200+50*	300
February 44		0****	100****	100****

* Chassis for assault tanks finished at Nibelungen Works.
** Production was interrupted by attempts, later given up, to increase ground clearance.
*** Chassis for Assault Gun IV finished at Nibelungen Works.
**** Estimate. According to one report that production of Pz.Kpfw.IV Type J began on February 1 or 15, 1944.

14.10 10./B.W., Panzerkampfwagen IV Type J

14.10.1 Contracts

At first, Pz.Kpfw.IV Type J was built by delivering the last of the contracts originally meant for Type H, with ca. 180 from Vomag (chassis no. 86394-86573) and 1060 from Nibelungen (chassis no. 89541-90600). As the last Pz.Kpfw.IV assembly firm, Nibelungen Works had received contracts for 1950 Pz.Kpfw.IV with chassis numbers 91300 to 93250, and then further contracts for at least 2000 Pz.Kpfw.IV with chassis numbers from 96751 onward.

Armored parts were ordered by Krupp in Essen, Eisenwerke Oberdonau in Linz, Böhler in Kapfenberg, and Eisen- und Hüttenwerke in Bochum.

14.10.2 Production Plans and Assembly

The production of Pz.Kpfw.IV was replaced by that of Jagdpanzer IV at Vomag. Every month from January to May 1944 more Jangpanzer IV and fewer Pz.Kpfw.IV were completed, so that as of June 1944 only Jagdpanzer IV were built. The Nibelungen Works was supposed to begin building Oz.Kpfw.III-IV as of November 1944, so as to replace the Pz.Kpfw.IV gradually. On July 18, 1944 Hitler's office had decided that Pz.Kpfw.IV production as tanks should also stop at the Nibelungen Works and be replaced by Jagdpanzer (at first Panzer IV/70(A), then Panzerjäger III/IV, and finally Panzer IV/70(V). At the end, the Pz.Kpfw.IV was to be replaced by the Panther Type F. With the goal of replacing Pz.Kpfw. IV by higher-performance tanks, the programs were changed very often.

Automatic lathes for gun-mount production.

Road-wheel production.

Wasmuth Program of 3/13/44:	Apr	May	Jun	Jul	Aug	Sep	Oct	Nov	Dec	Jan	Feb	Mar
Pz.Kpfw.IV	280	290	300	310	310	310	310	300	290	260	210	130
Chassis	20	20	10									
Pz.Kpfw.III/IV								10	20	50	100	180

Wasmuth Program of 4/8/44:	Apr	May	Jun	Jul	Aug	Sep	Oct	Nov	Dec	Jan	Feb	Mar
Pz.Kpfw.IV	280	290	300	310	310	310	310	310	310	310	310	308
Chassis	20	20	10									
Pz.Kpfw.III/IV												2

Wasmuth Program of 7/21/44:	Jul	Aug	Sep	Oct	Nov	Dec	Jan	Feb	Mar	Apr	May	Jun
Pz.Kpfw.IV	300	300	250	200	150	100	50					
Chassis	20	20	20	20	20							
Pz.IV/70(A)		50	100	150	200	250	300	350	348	347	345	250
Pz.Jg.III/IV									2	3	5	100

Ministry Program of 8/12/44:	Jul	Aug	Sep	Oct	Nov	Dec	Jan	Feb	Mar	Apr	May	Jun
Pz.Kpfw.IV	300	300	250	200	150	100	50					
Chassis	20	20	20	20	20							
Pz.IV/70(A)		50	100	150	150	100						
Pz.IV/70(V)					50	150	300	350	350	350	350	350

Dr. Blaicher Program of 9/7/44:	Jul	Aug	Sep	Oct	Nov	Dec	Jan	Feb	Mar	Apr	May	Jun
Pz.Kpfw.IV	300	300	250	250	250	250	200	150	100	50		
Chassis	20	20	20	20	20	20						
Pz.IV/70(A)		50	100	150	150	100						
Pz.IV/70(V)					50	150	300	350	400	450	500	500

OKH Program of 10/4/44:	Jul	Aug	Sep	Oct	Nov	Dec	Jan	Feb	Mar	Apr	May	Jun
Pz.Kpfw.IV	300	300	250	250	250	250	200	250	50			
Chassis	20	20	30	30	30	30	30	30	30			
Pz.IV/70(A)		50	100	100	100	100	100					
Pz.IV/70(V)							20	50	200	200	150	100

Teletype of 10.6 and Letter of 10.10.44 from WuG 6:	Jul	Aug	Sep	Oct	Nov	Dec	Jan	Feb	Mar	Apr	May	Jun
Pz.Kpfw.IV	300	300	250	250	250	250	250	250	200	200	150	100
Chassis	20	20	30	30	30	30	30	30	30	30	30	30
Pz.IV/70(A)		50	100	100	100	100	100					
Panther								10	20	35	60	80

Program Suggestion of Dr. Blaicher after Talk with Dr. Judtmann, 10/24/44:	Jul	Aug	Sep	Oct	Nov	Dec	Jan	Feb	Mar	Apr	May	Jun
Pz.Kpfw.IV				200	220	215	213	370	300	150		
Chassis				30	30	30	30	30	30	30		
Pz.IV/70(A)				50	100	120	157					
Pz.IV/70(V)									70	150	300	280
Panther										5	20	40

Program via Teletype with Mr. Wilhelm on 11/2/44:	Jul	Aug	Sep	Oct	Nov	Dec	Jan	Feb	Mar	Apr	May	Jun
Pz.Kpfw.IV					200	200	200	200	200	200		
Chassis					30	30	30	30	30	30		
Pz.IV/70(A)					100	100	100	100	44			
Panther										5	20	40

Preview of Tank Assembly Program, State of 1/30/1945:	Jul	Aug	Sep	Oct	Nov	Dec	Jan	Feb	Mar	Apr	May	Jun
Pz.Kpfw.IV							180	250	250	250	200	150
Pz.IV/70(A)							50	60	60	60	60	8
Panther												

Turret of Type 10./B.W.

Upper body of Type 10./B.W.

Hull of Type 10./B.W.

Production Statistics for Panzerkampfwagen IV Type J

Month	Production Preview	Completed Type H by		Accepted by H.Wa.A.
		Vomag	Nibelungen	
February 44	310	52	100	152 + Type H
March 44	270	60	250 + 10*	310
April 44	310	40	259 + 21*	299
May 44	310	28	274 + 21*	302
June 44	300		300 + 20*	300
July 44	300		300 + 20*	300
August 44	300		300 + 20*	300
September 44	250		180	180**
October 44	250		187 + 30*	187***
November 44	200		200 + 30*	200
December 44	200		195	195
January 45	170		170	170
February	200		150	150
March 45	180		55	55
April 45	200		50****	54****

* Chassis fir Sturmpanzer were completed at the Nibelungen Works.

** In-house problems.

*** Bomb damage.

**** Production in April 1945 was known from the following reports in the last days of the war:

From the Army Arsenal in Linz, 8 Pz.Kpfw.IV were shipped out on April 11, 1945 for the Army Group South (4 for 13th Armored Division and 4 for "Feldherrnhalle").

From the Army Arsenal in Linz, 13 Pz.Kpfw.IV were rolled out on April 18, 1945 for the 7th Armored Division. The next ten Pz.IV were assigned to Army Group South. Further arriving Pz.Kpfw.IV up to 10 tanks were assigned to the 7th Armored Division.

Note from Work Echelon/Abt.Org.K. to Inspector-General of the Armored Troops on April 29, 1945: *Tank assignment since 4/15 in southern area:*

5 Pz.Kpfw.IV to 6th Armored Division

5 Pz.Kpfw.IV to 2nd SS Armored Division "Das Reich"

10 Pz.Kpfw.IV to 20th Armored Division intended (for 7th Pz.Div)

4 Pz.Kpfw.IV to Army Group Center (from Raupach firm, Warnsdorf)

Further tank deliveries to end of April: about 9 Pz.Kpfw.IV

Prognosis for May: 50-100 Pz.Kpfw.IV. depending on arrival of missing parts (cast covers for gearboxes and motors).

14.11 Panzerbeobachtungswagen IV

From April 1944 on, every 25th newly built Pz.Kpfw.IV was to be rebuilt as Pz.Beob.Wg.IV.

Production Statistics for Armored Observation Car IV

	Assigned For Rebuild	Accepted by Insp.	Delivered To Arsenal	Transported to Units
March 44	0	0	0	0
April 44	10	0	0	0
May 44	22	0	0	0
June 44	13	0	0	0
July 44	13	7	6	6
August 44	25	18	15	0
September 44	25	21	15	12
October 44	25	25	30	28
November 44	20	20	23	23
December 44	25	11	14	20
January 45	0	16	14	2
February 45	25	0	2	6
March 45	0	15	5	15

14.12 Panzerbefehlswagen IV

From March 1944 9n, Pz.Bef.Wg.IV were generally rebuilt from repaired Pz.Kpfw.IV. Only 17 Pz.Bef.Wg.IV were made from newly completed Pz.Kpfw.IV.

Production Statistics for Armored Command Car IV

	Rebuilt from Repaired Pz.Kpfw.IV	Delivered to Army Arsenal		Transported to Troop Units
		from Inst.	New	
February 44	0	0		0
March 44	15	11		0
April 44	30	21		18
May 44	30		9	18
June 44	9	33		5
July 44	4	5	0	23
August 44		0	8	11
September 44		0	9	10
October 44		1	0	5
January 45		6		

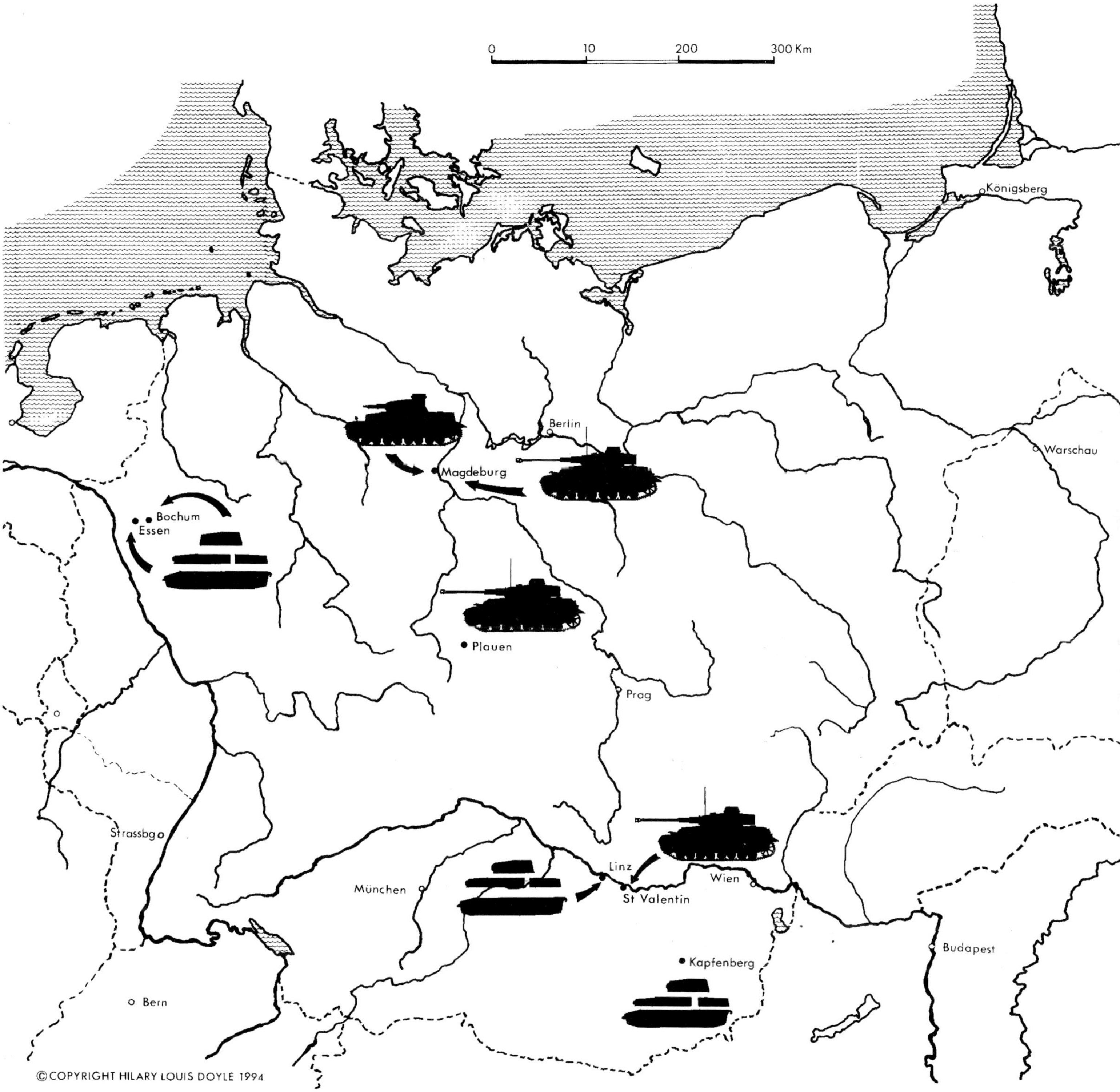

Locations of Manufacturers of Panzerkampfwagen IV.

Welding equipment for hull and turret armor at Eisenwerke Oberdonau in Linz.

15. TEST VEHICLES

15.1 SMOKE WAGON (R.W.)

The first hint that such a heavy armored vehicle was planned was discovered in a report on the results of the 1934-35 business year from the War Materials Department of Fried.Krupp A.G. in Essen:

Tank turrets are to be designed and built by us per contract for our own tank design: Preliminary drawings for a foglaying gun turret for the B.W. with 10.5 cm L/16 gun.

Further explanations were taken from the Army Weapons Department's report, *Offensive Defense of Armored Vehicles* of October 30, 1935 under *The 10.5 cm Tank (Foglaying Tank):*

This vehicle shall, to be sure, be developed chiefly for the purpose of fogging the enemy antitank weapons; in addition, it shall also fire on suitable targets with the effective field-howitzer shells (splinter shells). The wish to use the same chassis as the B.W. has led to a limitation of the performance to Vo = 230 m/sec, which suffices completely for the noted, originally considered purposes. If one would consider. Along with the foglaying and the splinter shell, a third type of shot, the antitank shell, which it not out of the question from the standpoint of carrying on board and supplying, then under the same conditions as before (700 meters, 60-degree striking angle), a penetrating power of 25 mm would result. Thus this vehicle would also be battle-capable against all French tanks with the exception of the 40 mm thickly armored parts and the heaviest tanks (2 C. 3 C and D).

In the Fried.Krupp Technical Yearbook of 1935-36, it was elucidated under Tanks:

The following contracts were assigned to us: Gun turret for ***R.W.*** *with 10.5 cm Howitzer L/16. Test model is under construction.*

A file note by Mr. Woelfert (of Krupp) on a discussion with Messrs. Olbrich and Biersack (of Wa Prüf 6) in Berlin on January 29, 1938, re R.W. and B.W. II, reveals: *The work on modifying the bow armor of B.W. II in 2.d Series manner, as well as the ensuing test version of ammunition storage for R.W. is to be hastened, so that in April-May 1938 if possible, the R.W. turret can be mounted on the B.W. chassis.*

Krupp reported on November 17, 1938:

The B.W. II vehicle with R.W. turret has arrived at Essen on 11/16. The turret is to be removed and mounted on a firing stand. The chassis with armored upper body is to be sent to the Gruson Works and made available to Wa Prüf 6 III for rebuilding.

On September 9, 1939 Krupp reported:

On September 7, 1939 an inquiry was made in a telephone call from Wa Prüf 6 III to the Gruson Works as to whether a B.W. chassis with body can be made ready for a special contract. The chassis is to hold the R.W. turret, which will be delivered from Kummersdorf on 9/8. Photos of the vehicle are to be sent to Berlin on 9/8/1939.

15.2 4./Z.W. (PZ.KPFW.III TYPE E) CHASSIS FOR 3./B.W.

On June 1, 1937, Department A.K. of Fried.Krupp A.G. was informed by Wa Prw 6 on the subject of further Pz./Kpf.w. (B.W.) chassis:

The necessity of also making the chassis of the tanks uniform compels us to give up the design of the B.W. chassis developed by you with the end of the 2./B.W. series. For 3./B.W. the chassis of the Z.W. Model 4 now in development is planned with a corresponding armored hull. As a result, there remains the beginning of a special development contract for the B.W. chassis.

As to the contract negotiations between Department A.K. of Fried.Krupp A.G. and Oberbaurat Kniepkamp (Wa Prw 6), which led to Krupp being closed out of future contracts, the directors of Krupp reported as follows on December 15, 1937:

The loss of new design contracts on the basis of the contract negotiations came about as follows:

AK No. 51548 of 2/10/37: *Contract for further development of the tank turned over by Prw6 to AK.*

AK No. 51548 of 3/12/37: *Conclusion of second half of March agreed on by AK. In numerous discussions in Berlin and Essen in the next three months, the various members of Prw6 regretted very much that the cooperation of AK with Prw6 would cease if an agreement were not attained soon. I then requested* *repeatedly in Essen that the responsible gentlemen take a position soon.*

AK No. 52340 of 4/5/1937: *AK asks for retention of the former contract.*

AK No. 52592 of 4/19/1937: *Prw6 asks for submission of the objections in writing. The contract with AK must also be concluded, since it would otherwise be untenable with the other firms. (Note: According to Kniepkamp's statement, some ten firms.) To conclude the letter from Prw6 it is noted: "In case of refusal, you close yourselves out unilaterally of a design association of the development firms."*

AK No. 52592 of 4/30/1937: *AK promises an answer soon.*

AK No. 52960 of 5/7/37: *In a discussion in Berlin, Kniepkamp repeated that the contract with some ten firms was out of the question and stated 5/15/37 as the last date for clarifying the contract, since the new tasks were urgent and he was otherwise compelled to suggest a different firm to Philipps. Thereby AK would close itself out of the further development and no longer receive the essential experiences of the troops. On agreement, AK could immediately begin the design of the 3./B.W. (formal contract already at hand). I promised to inform our responsible offices.*

AK No. 52978 of 5/8/37: *It is also urgent for Olbrich to sign the contract. In connection with these discussions. I immediately informed the responsible offices of the urgency. Even if AK could not accept without further ado the conditions of Prw6 because of the effect on other deliveries from Prw6, we would at least have to sit at the negotiation table as soon as possible. I stated: "With the personnel change in Prw6 (Fichtner and Kniepkamp instead of Phillips and Blasberg), there is the greatest danger for AK of losing rapport and hitherto good relationship to Prw6 in the realm of tanks. The 3./B.W. rebuilding is a transitional contract for us until the decision on new design contracts. For 3./B.W. the new concepts from Kniepkamp regarding steering and running gear shall be developed for the first time on a larger tank. It is thus decisive that we come into the business with him. In no case can we do without the experiences of the troops, and also for foreign contracts we need, as already because of the control numbers, the agreement of Prw6." Up to the furthest date given by Kniepkamp, our basis for negotiation was essentially more favorable, since the further development of the B.W. by AK urgently needed his new ideas for setting dates.*

AK No. 53051 of 5/14/37: *Call by Eberhard to Kniepkamp with request for postponement.*

AK No. 53023 of 5/21/37: *Confirmation of this call to Prw6.*

AK No. 53347 of 6/1/37: *Information from Prw6 that the further development of the 18-ton tank was transferred to Daimler-Benz, and that thereby the conclusion of a special development contract for B.W. with AK was superfluous.*

From then on, Kniepkamp showed little interest in further negotiations as to the contract. (Compare AK 80239 of 9/1/37). In general it is to be noted that we must cultivate the connection with the Weapons Department even more than the Berlin firms of Daimler-Benz and Rheinmetall, since they can communicate daily, personally and by telephone, with Prw 6, and also have mobile testing facilities. In addition, the chief officers Olbrich and Kniepkamp are disagreeable about our foreign business in tanks and tractors, particularly the first, who looks unwillingly on the newest developments.

From the preceding it can clearly be seen that the delay of the contract negotiations, AK was closed out of new contracts for tank chassis. In the age of motorizing this is particularly regrettable, since we have the best facilities for producing turrets and chassis and the longest experience in this field. From us come more tanks and with better success than from the other firms. These firms were included only later and have for the most part built on the La.S. drawings made available by us.

It is to be hoped that through further negotiations the rapport with Prw6 will be established again.

On May 2, 1938, Mr. Woelfert (Department A.K., Krupp) explained that new production contracts (but no design contract for the chassis) would again be assigned to Krupp:

Originally the 2./B.W. series was to be connected with the Daimler-Benz chassis. Through the use of new design parts that had not been tested sufficiently for tanks there resulted reverses, among others with the three-stage steering gear built by Z.F., the Maybach gearbox with semi-automatic shifting, and the greased tracks. The experience gained with transmissions and tracks for towing vehicles could not be transferred to tanks without further ado. Since the first test vehicle is only arriving at Kummersdorf for testing at this time, series production cannot be reckoned on in the foreseeable future. Instead of the Daimler-Benz uniform chassis. Large contracts are thus assigned for the 3rd, 4th and 5th series.

While Mr. Blasberg directed the chassis department at Wa Prüf 6, the B.W. chassis had been developed almost without collaboration from the Weapons Department. Mr. Kniepkamp too, who took the place of Blasberg, did not trouble himself about the B.W. chassis, for the reason that he was not interested in leaf-spring running gear and the former transmissions. Since the new development of the 4./Z.W. was delayed constantly, Krupp received the information on the further development of the B.W., promised by the end of 1938, only in April 1939, and then not definitely. As a result, Wa Prüf constantly had to hold back further series of earlier types, up to the 7th Series. In August 1939 the decision on the guidelines for tanks was transferred from Oberbaurat Kniepkamp to Lt. Col. Olbrich.

15.3 B.W.40

Also called VK 2001 (IV) and VK 2001 (B.W.)

In the 1938-39 business-year report of the War material Department of Fried.Krupp A.G. it was stated: *As a further development, designs for a B.W. 40 with 6-wheel running gear and modified motor compartment are underway.*

Through the discussion in Berlin on September 15, 1939, between Baurat Kniepkamp of Wa Prüf 6/III and Mr. Woelfert of the Krupp A.K. concerning VK 2001 (IV), the further development of the Pz.Kpfw.IV was to be defined:

To be used is the motor compartment of the VK 2001 (III) with HL 116 motor.

Top speed as before 42 kph. The running gear can be chosen so that with a light width of the body of 1820 mm as before, and a maximum vehicle width of 2900 to 2950 mm, a total weight of 20 tons is not exceeded. The box running gear based on S 3238 is not possible, as a maximum width of some 3040 mm and a too-high weight result. A.K. is studying the installation of the 6-wheel torsion-bar ZW 40 running gear (wheel diameter 535 mm, bandage width 100 mm) based on Drawing S 3584, which will be turn over by Wa Prüf. The wheel diameter is to be increased appropriately for the greater wheel diameter. Wa Prüf approves sending in a weight estimate including shock absorbers for Running Gear ZW 40 by Daimler-Benz and of the 021D38488 norm page for bandages by HZV. A.K. suggests 6-wheel running gear based on Drawing SK 3413, SK 3414 and SK 3431, which will be discussed. In this running gear, as opposed to a type with full-length torsion bars, the electric system and the ammunition for the 7.5 cm gun can be housed more easily, besides which the shock absorbers, most vital for torsion-bar running gear but sensitive, can be avoided. The 6-wheel running gear is built with the proved parts of the 8-wheel system, but has the following advantages: The wheel hub and the spring softness are increased considerably, the ground clearance improved, since thee bearings for wheel mounting are inside the hull, besides which the gap of the tracks from the spring tension is increased and the wheel resistance is diminished. The new tracks with higher teeth, as on Drawing H 4254, are to be installed. A.K. also enclosed leading-wheel designs SK 3428 and SK 3429 in conjunction with the Daimler-Benz leading wheels as on Drawing 432560311. Tubular braces as on SK 3429 are not possible, since apparently the braces are too strong. The layout as on SK 3428 is to be built for testing.

Since in the present situation long road tests are not possible, A.K. considers only the use of aggregates that have already proved themselves among the troops to be important, and therefore still takes a position as to the question of the gearbox and steering drive. Installation Drawing S 3649 for VK 2001 steering and shifting gears will be turned over by Wa Prüf.

At a subsequent discussion with Dipl.-Ing. Exner (Wa Prüf 6/IVc), the bandage width was discussed. A.K. had chosen an 84 mm width for the six-wheel running gear, whereby the greater burden than the bandages of the eight-wheel running gear was to be balanced by the increase of the roll diameter from 470 to 630 mm diameter. Thus it was also presumed that the greater total rubber weight increased the life expectancy. With a look back at the use of exchange materials, though, a bandage width of 100 mm is desired.

On October 28, 1939, Wa Prüf 6/III had made further decisions as to the road wheels and tracks of the VK 2001 (B.W.) (also called B.W.40):

With the six-wheel running gear planned for this, Design AK, the bandage width should be increased to 100 mm, so that even with the use of exchange materials sufficient life expectancy is maintained. With the installation of the new Kgs.61/400/120 tracks with 400 mm width, the following values result, according to the enclosed Drawing SK 3507: 83.3 kg/cm line pressure, 336 kg total rubber weight. Since with this six-wheel running gear, as opposed to the former 8-wheel type, the bandage width was increased from 75 to 100 mm, the diameter from

470 to 530 mm, and the totals rubber weight from 192 to 336, an increased weight of some 90 kg for the running gear and 130 kg for the new tracks, thus 220 kg together, cannot be avoided. Here the increased weight from increasing the track width by 50 mm is not yet considered.

On November 15, 1939, the Zahnradfabrik Friedrichshafen was ready to finish four SSG76 test gearboxes for VK 2001 (B.W.) test chassis. On December 8, 1939, Maybach-Motorenbau had submitted drawings for the installation of the HL 116 P motor in VK 2001.

By December 13, 1939, Wa Prüf 6 issued contracts to E.H.W. (Eisen- und Huettenverein A.G.) in Bochum) for two armored hulls for B.W.40 test chassis with the wall strength of 50 instead of 30 mm for the front wall. On 21 December, E.H.W. sent the red pauses back to Krupp of Essen, since E.H.W. had been released by the Weapons Department from carrying out the contract for two test hulls for B.W.40. According to the drawing list, this test chassis was made of normal Pz.Kpfw.IV chassis parts; the hull was modified only for the mounting of the new running-gear and motor components.

On January 4, 1940, Wa Prüf 6/III decided that for the B.W.40 or VK 2001 test chassis, two hulls of armor steel, one of cast iron and three upper bodies (bow and stern armor) of cast iron were to be made by Krupp. On January 16, 1940, Dipl.Ing. Biersack (Wa Prüf 6 IId) explained that three armored upper bodies for B.W.40 test vehicles were to be made of armor steel. In further discussions on February 15, 1940, between Oberbaurat Kniepkamp (Wa Prüf 6/III) and Mr. Woelfert (Krupp Dept. A.K.), it was reported as to the Krupp design for B.W.40:

A.K. submits SK 3683 for the whole arrangement, with divided two-stage steering gear by SK 3670, six-wheel leaf-spring suspension, HL 116 motor and transverse cooling. Wa Prüf basically agrees with the arrangement. But A.K. takes a position on the installation of a rear-mounted cooling system with four radiators, by Drawing 155059/8 by Maybach. Wa Prüf turns over this drawing and states that 14-15 HP are needed for starting, and that Maybach is preparing automatic setting for the belt drive. For the steering clutch envisioned by A.K., additional power must be

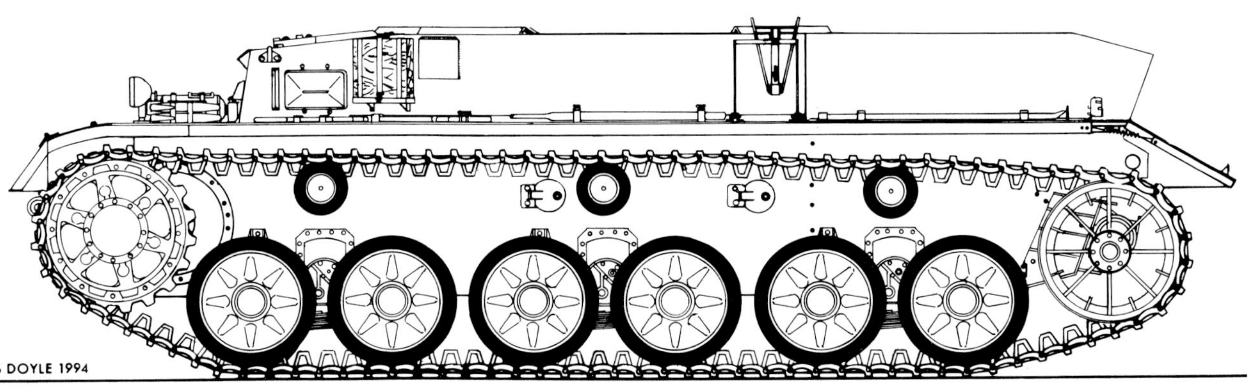

B.W. 40 with six-wheel running gear.

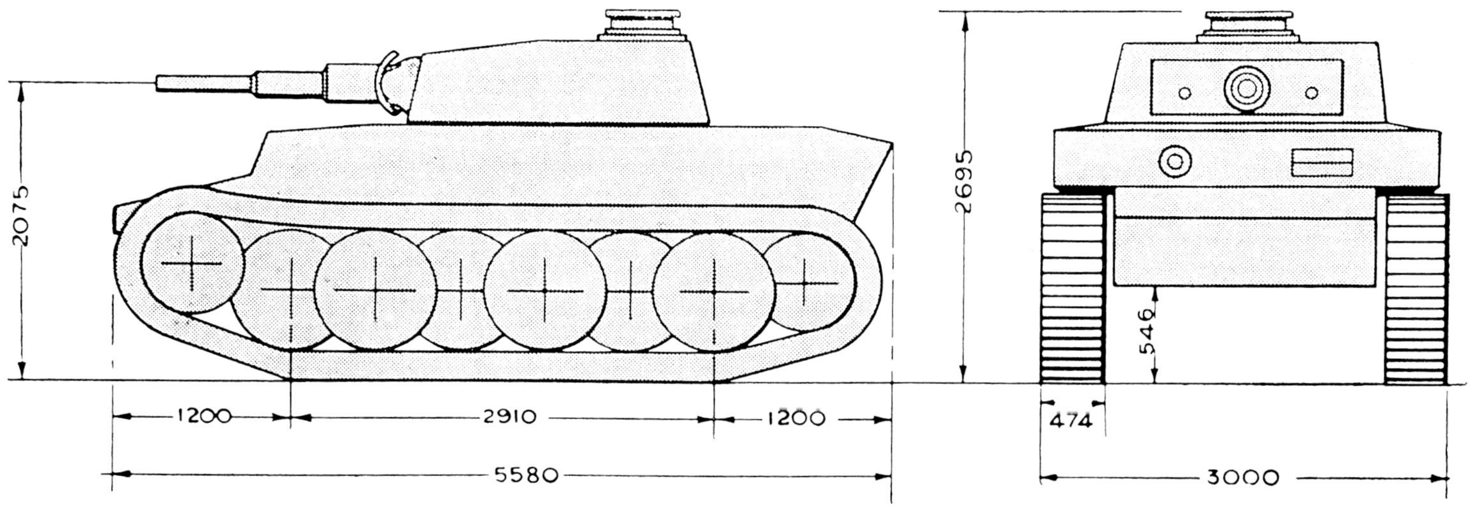

V.K. 2002 (MAN), Design Sketch

used, what with the weight of the existing vehicle. Starting in the manner of the Argus air brake might be unsuitable for the case at hand, since the Buna seals would be heated too much by the nearby steering brakes. Since only now a final clarification of the whole design is possible and the complete design can in part only begin now, the delivery of the first test vehicle may only be possible at the end of 1940. This schedule could be shortened if only the most important improvements are carried out.

On May 16, 1940, Wa Prüf 6 announced that the B.W.40 was postponed in view of the war conditions.

15.4 VK 2001 SERIES – PANZERKAMPFWAGEN III AND IV NEW TYPE

Further developments of Pz.Kpfw. in the V.K. 2001 series were carried out in competition by Daimler-Benz (VK 2001 (DB)), M.A.N. (VK 2001 (M)) and Krupp (VK 2001 (K)) from 1940 on. The VK 2001s of the firms differed mainly in the designs of running and steering gear, as well as external dimensions and armor. The VK 2001 and VK 2002 were to be built by M.A.N. and Daimler-Benz with 7.5 turrets and by Krupp with a 5 cm turret.

In this period the main specifications were not submitted through In 6 as before, but by reports from higher Army and Wehrmacht staffs to Hitler for his decision.

With such methods, the technical rules of Wa Prüf 6 (including limiting the vehicle weight for crossing bridges) were upset. The control of the designing of the VK 2001 series was taken out of Wa Prüf 6's hands and, through the following notice for uniformity of Pz.Kpfw.III and IV of September 6, 1941, taken by the Chief of Army Equipment and Commander of the Replacement Army, Stab Ruest IIa, and submitted to Hitler:

1) *Pz.Kpfw.III n.A. and KIV n.A. are armored: (identified by large road wheels like Tractor)*

	Front	Side	Stern
Pz.Kpfw.III n.A.	*50 mm*	*50 mm*	*50 mm*
Pz.Kpfw.IV n.A.	*50 mm*	*40 mm*	*40 mm*

To be able to keep the maximum weight of 23.5 tons for both tanks, the Pz.Kpfw.III n.A. has only a light hull width of 1.65 meters, as opposed to 1.8 meters of Pz.Kpfw.IV n.A.

2) *The worthwhile goal is the full uniformity of one type. The only remaining differences are the equippings of required variations. This thoroughgoing uniformity can only be attained when, in using the hull (1.8 m wide) of Pz.Kpfw.IV in the side armor of Pz.Kpfw.III n.A. to keep the weight of 23.5 tons, something has been given up, and in the sense that those surfaces of the hull sides that are covered by the running gear, and the stern, instead of 50 mm, are now being made 40 mm thick. This appears bearable in view of the extraordinarily attainable advantages, since the running-gear parts form an extra protection before the side walls. (Practice has shown that the overwhelming number of shots strike above the running gear.)*
Advantages of thoroughgoing uniformity:

- *Less work for design bureau*
- *ess work of testing*
- *Less work in terms of production of test series*
- *Complete structural identity in all parts (except equipment); thus increased production, simplified supplying and simplified and uniform training.*
- *Variability in production, that is, if necessary comparatively short-term Pz.Kpfw.III can be prepared instead of Pz.Kpfw. IV, and vice versa, thus greater mobility in organization.*
- *Essentially increased ammunition supplying for Pz.Kpfw.III (some 40% more than at present, although the future shell is larger than the present one).*
- *Possibility in terms of turret development of installing an even stronger armor-piercing weapon than the 5 cm L/60 presently being developed, such as the 0725 conical-barrel weapon.*
- *Hydraulic turning of the turret also in Pz.Kpfw.III, thus less burdensome for the crew, and higher rate of fire.*
- *Agreement is requested.*

The answer came on September 14, 1941, from the chief of

To Chief h.Ruest. u.B.d.E. Stab Ruest IIa No. 2944/41 g/ v/ 9/6/1941 re uniformity of Pz.Kpfw.III and IV, the Führer has expressed himself as follows:

1) *Uniformity of the single type, the Pz.Kpfw.IV, is purposeful and to be instituted (except only equipment).*
2) *Weakening the armor of those surfaces of the hull sides that are already protected by the running gear should nevertheless be avoided.*
3) *It is not justified to avoid any exceeding of the weight of 23.5 tons at the cost of the armor so urgently. By his order to attach fore-armor under all conditions for stronger protection of the bow sides, the former total weight must already be exceeded; the weight of 23.5 tons thus cannot be retained anyway.*

The designations of Pz.Kpfw.III n.A. and Pz.Kpfw.IV n.A. were not used by Wa Prüf 6, but rather VK 2001, VK 2002, VK 2301, etc.

The development of VK 2002 (K) is an example of how these and following orders to the building firms were presented by Wa Prüf 6:

At first a 5 cm turret with a 1350 mm servicing circle without a turntable was promoted, then (AK No. 73898 of 10/10/41) a 5 cm turret with a servicing circle of 1400 mm, to allow the optional installation of a 7.5 cm turret; finally the 7.5 cm uniform turret with KwK 44, turntable and at first 1560 mm and later 1600 mm servicing circle was called for. As a result of constant enlarging of turrets, the width (1600 to 1650 mm) and above all the length (originally 400 mm shorter) of the hull repeatedly had to be enlarged. The ammunition storage also changed basically. In addition, the study of captured Russian vehicles compelled an angled structure of the body walls, especially the driver's front wall.

A shortened history of the fate of the VK 2001 series was presented in the Fried.Krupp A.G. Technical Report for the 1940-41 business year, as follows:

The Pz.Kpfw.IV (B.W.) has proved itself extraordinarily in various campaigns and is being built in large series with the already noted stronger armament, also by branch firms. But as a result of the stronger armament and stronger armor it has meanwhile grown to a weight of 23 tons. The conditions for the Kampfwagen III made by Daimler-Benz are similar. It is therefore intended to make the two vehicles equal in their production, armor and weight and to differentiate only the armament, one with 5 cm, the other with 7.5 cm, both weapons made by Krupp. The tanks were developed in competition among Daimler-Benz, M.A.N. and Krupp. The test models are almost finished. But since, with the speedy development, the armor of these vehicles once again does not suffice, it was ordered by the fuehrer not to put this vehicle in its present form into series production, but rather to strengthen the production further, so that the totals weight will amount to 39 tons. In the interest of rational utilization of technical powers, the Führer has instructed that the development should take place only in one firm; probably M.A.N. will be entrusted with it.

15.5 BOX RUNNING GEAR

On December 14, 1940, Wa Prüf 6 gave Krupp permission to design box running gear for Pz.Kpfw.IV. Two test vehicles were released by Wa Prüf 6 on February 6, 1941, as follows:

Of the Pz.Kpfw.IV contracts for 6./B.W. (No. 210/5002/39(H) for chassis and 210-5902/39(H) for bodies), the last two vehicles are to be released after completed acceptance to be turned over to WaPrüf 6 for experimental installation of box running gear. The vehicles remain at the Gruson Works for carrying out the tests. Final delivery will be made through Wa Prüf 6.

One of these vehicles, Pz.Kpfw.IV, Type E, Chassis No. 81005, was used at Wunsdorf in December 1942 at the School for Army Motorization, 1./Instruction Battalion, for testing with the cooling-water heating device built by Fuchs.

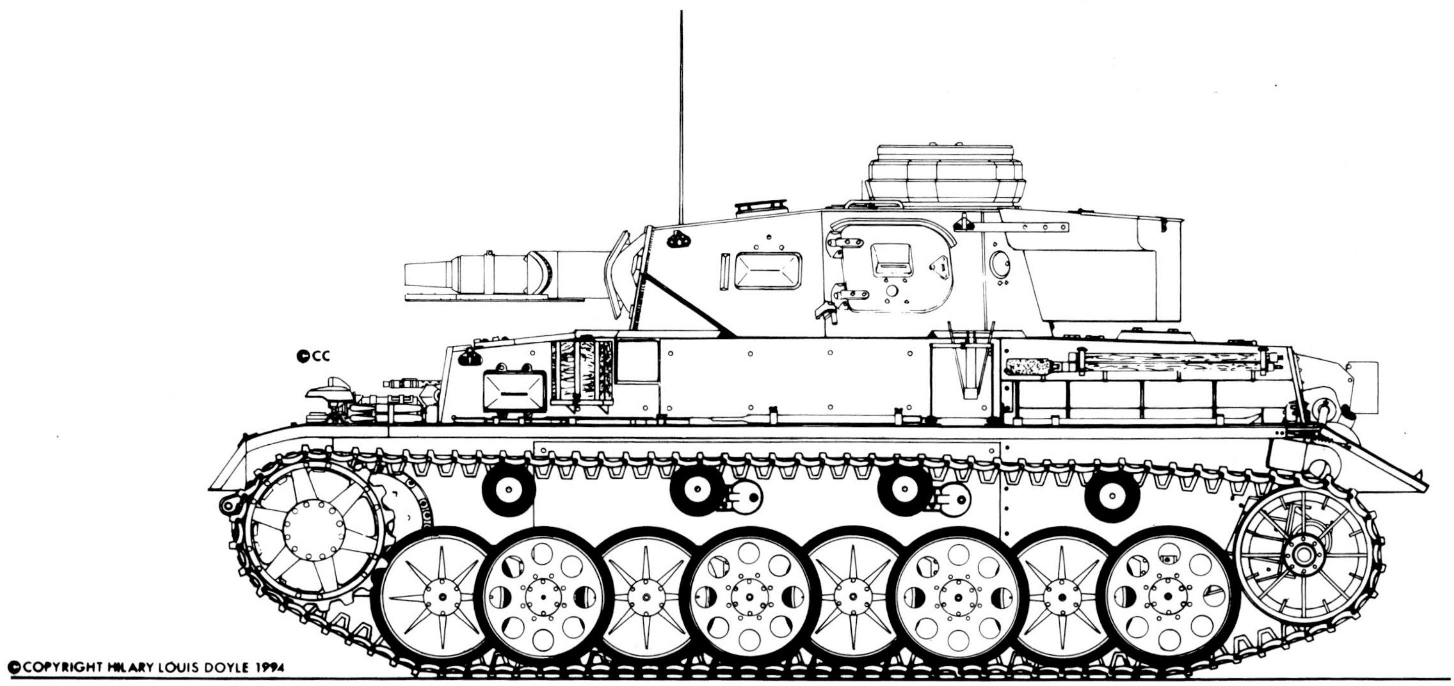

Pz.Kpfw.IV Type E with box running gear. Road wheels of the same type were used in the experimental Flak Wagon.

On August 1, 1941, it was reported that the box running gear with eight 700 x 70 road wheels per side, with a length of 3517 mm, should stay on 422 mm wide tracks.

From the notes on a discussion about developmental and experimental vehicles at the Gruson Works on November 18, 1941, and the delivery of tank parts in Essen on November 19, 1941, concerning the *...2 6./B.W. vehicles with box running gear, OKH Contract 006-6539/40 WaPrüf6, Gruson, the delivery date of the tank hubs was missing. Essen announced that four tank hubs would be delivered to the Gruson Works on November 7, 1941.*

On November 21, 1941, it was reported that the track links for the 6./B.W. box running gear should be the Kgs.62/420/120 with gliding protectors, and for the 7./B.W. box running gear should be the Kgs.62/436/120.

In the Schedule Order No. 8 of June 15, 1942, by Wa J Rue (WuG) Ataff V, it was reported as to the box running gear:

Through delays by the building firm, the tests of the two Panzerkampfwagen IV began only at the beginning of June 1942. First results for the second half of June are to be expected. Prüf 6 reported results of further tests to June 15, 1942.

The further development was reported on in a discussion on B.W. running gear at Krupp on July 17, 1942:

Box Running Gear with 70 mm wide bandages

For the two test vehicles, two sets = 4 pieces of ice-breaking leading wheels are to be completed by the works. Delivery date 10/15/1942. Box running gear with 80 mm wide bandages

For the new running gear, a track division of 120 mm is to be chosen instead of 130 mm. The drive-wheel bodies are to be set up so that the normal B.W. gear arcs can be used. It is to be investigated by A.K. whether the track width can be raised to 500 mm without major changes for the experimental box running gear of the coming series.

15.6 UK TEST VEHICLE

(UK = Underwater Combat Vehicle)

On November 11, 1941, Dipl.Ing. Blattmann (Wa Prüf 6/III H) called Mr. Woelfert (Dept. A.K.) re B.W.-UK:

Mr. Blattmann reports that the UK designs prepared by Alkett are to be worked into the set of drawings by A.K. as fast as possible. On January 8 the basis for series production must be available. I affirmed on request that drawings from Alkett had reached us.

In a discussion in Berlin on November 12, 1941, between Dipl.Ing. Blattmann (Wa Prüf 6/III H) and Mr. Lanz (Dept. A.K.) concerning Pz.Kpfw.IV UK it was reported:

Wa Prüf requests the inclusion of the drawings newly prepared by Alkett for UK capability into the set of drawings by January 8, 1942. The series shall then continue to be built UK-capable. I called attention to the fact that that it will not be possible for us, with the work force on hand, because of work in progress, and considering the especially urgent jobs such as fore-armor, box running gear, etc., to finish in the short time of about six weeks that still remain after the end of testing, such an extensive new set of drawings that includes almost all parts of the vehicles.

With a vehicle made UK-capable by the Gruson Works, tests begin at Fallersleben on November 17. The tests last about 3 weeks. On the basis of the tests, changes are to be expected. The Gruson Works also stated already that in the production of the test vehicle changes from the Alkett drawings had to be undertaken for reasons of fabrication.

On November 18, 1941, it was stated by the Gruson Works that a UK test vehicle was prepared and would go to Spandau on 20 November.

On July 22, 1942, the Army Weapons Department wrote in a letter to the Gruson Works about the installation of UK equipment in Pz.Kpfw.IV:

Contract SS 006-6911/41 WaPrüf 6/III concerning taking one Pz.Kpfw.IV from the series for the purpose of setting it up for heightened wading capability is hereby withdrawn. Thus the granting of permission for the release of this vehicle by Wa J Rue (WuG6) is eliminated.

In a discussion with Colonel Fichtner (Wa Prüf 6) on July 11, 1942, it was decided: *For the Pz.Kpfw.IV Type H (9./B.W.) the diving ability and the BV.W.40 turret (hydraulic drive and new optics) are eliminated.*

15.7 UNIFORM TURRET B.W. 40

In a discussion with Wa Prüf 6 at Krupp in Essen on November 21, 1940, a hydraulic turret drive for the BV.W. was first discussed. In a letter to Dept. A.K. of Fried.Krupp A.G. the Army High Command, Wa Prüf 6/IId explained: *Wa Prüf 6 states that as of 8./B.W. the electric turret drive will be replaced by hydraulic drive.*

In a discussion on developmental and experimental vehicles at the Gruson Works on November 18, 1941, it was reported that the hydraulic turning system would be ready for viewing as of the end of the week.

In a letter to Wa Prüf 6 dated February 24, 1942, Dept. A.K. clarified the following points about modifications to the Pz.Kpfw. IV turret drive:

It is intended by Wa Prüf that as of the autumn of 1942 the"B.W.40" uniform turret with hydraulic instead of electric drive be considered for Panzer III and IV. In terms of the chassis, we still have no official directive to change the design.

We ask that the following be considered:

1) *The electric aggregate has proved itself after initial difficulties. The one-cylinder motor was replaced by a two-cylinder motor. The dirtying of the rods for the carburetor etc. was largely prevented by covering.*
2) *The DKW motor can be used by simple means to warm the main motor in cold weather and can be used via the aggregate to produce warmth.*
3) *The main motor does not need to be is use constantly if the turret is to be turned. Save gasoline.*
4) *The electric power transmission is simpler than the mechanical transmission from the driveshaft.*
5) *The higher consumption of copper for the electric aggregate is unfavorable.*
6) *Since the electric aggregate cannot be installed in the Panzer III for reasons of space, the uniform turret for Panzer III and IV must be given different drives.*

It is seen from the data above that for vehicles of the size class of the Panzer IV and upward a turret turning with a special power source appears more favorable. We ask you to take a position on our information.

In a letter of April 8, 1942, to Dept. A.K. of Krupp in Essen, Colonel Fichtner (Wa Prüf 6) explained as to the hydraulic turret drive for the B.W.40 turret:

In reference to the discussion at the Krupp firm on April 2, 1942 between Director Dorn and Chief Engineer Woelfert (Krupp firm) and Lt. Col. Crohn (Wa Prüf 6), the following is communicated:

The hydraulic drive of the B.W. turret is judged favorably on the part of In 6 plus the troops.

On the basis of this fact and the considerably simplified possibility of manufacturing compared with the D.K.W.-Conz aggregate, the new B.W. 40 turret will be equipped with hydraulic drive.

Thus the additional carrying of 16 to 19 rounds of 7.5 cm ammunition is also possible. On the part of the Krupp firm, the necessary basic drawings for the modification of the chassis are to be made at once, and first a date for making these drawing changes is to be set.

On the basis of the date for the design drawings, a decision about them will be made along with Wa J Rue (WuG 6) as to when the B.W.40 turret will first be produced.

On April 2, 1942, it was stated in a discussion at Krupp in Essen:

The new B.W.40 turret with narrow front, organically installed weapon and UK setup can be delivered at the end of the year at the earliest. Series production hopefully at the beginning of 1943.

In the monthly discussion of the assembly firms on July 8, 1942, Oberamtmann Doesch (WuG 6) announced that the B.W.40 turret for Pz.Kpfw.IV (7./B.W.) was no longer in the picture.

In a discussion with Colonel Fichtner (Wa Prüf 6) on July 11, 1942, it was decided:

For the Pz.Kpfw.IV Type H (9./B.W.), the diving capability and the B.W.40 turret (hydraulic drive and new optics) are dropped. Strengthening the cone drive in the side shafts and the planetary drive, though, is to be carried out quickly. It was determined that the new B.W.40 turret with the cast steel front shall no longer be built, because the production of B.W. armored bodies will presumably cease at the end of 1943.

15.8 B.W. MODERNIZING

On December 31, 1942, it was decided by Mr. Woelfert (Dept. A.K.) and Lt. Col. Holzhaeuer (Wa Prüf 6) as to the Krupp design for 9./B.W. (Pz.Kpfw.IV Type H) as follows:

According to the decision of the General Staff, the B.W. housing shall be modernized as per the A.K. suggestion and receive an angled bow like the so-called "heavy" version (Drawing W1462). Added weight of 880 kg with 50 mm angled desk wall and 80 mm front wall plus from 10 to 20 mm on the front end of the strengthened bottom. Simultaneously wider tracks and 3 instead of 2 road wheels side by side shall be installed, as have already run in tests. Later, running gear with Russian rollers (rubber=sprung steel-band road wheels) will come into the picture.

The armor hull consisted of armor plates with the following thicknesses and inclinations (from the horizontal):

Plate	Thickness	Angle
Upper front wall	50 mm	35 °
Forward front wall	80 mm	76 °
Lower front wall	30 mm	29 °
Sidewalls	30 mm	90 °
Upper stern wall	20 mm	80 °
Lower stern wall	14.5 mm	16 °
Bottom, front	20 mm	0 °
Bottom, rear	10 mm	0 °
Roof	16 mm	0 °

The stern armor consisted of welded-together armor plates with the following thicknesses and inclinations (from the horizontal):

Plate	Thickness	Angle
Roof	10 mm	2 °
Sidewall (angled)	20 mm	55 °
Sidewall	20 mm	90 °
Back wall	20 mm	80 °

The turret armor consisted of welded-together armor plates with the following thicknesses and inclinations (from the horizontal):

Plate	Thickness	Angle
Front wall	50 mm	80 °
Mount cover	50 mm	curved
Sidewalls	30 mm	65 °
Back wall	30 mm	75 °
Roof, front	16 mm	7 °
Roof, rear	25 mm	0 °

Further development points were discussed by Krupp and Wa Prüf 6/II on February 5, 1942:

In case the suggestion of organic strengthening of sidewalls, and recently also back wall including turret (walls under 63 to 65 degrees on inclination) to 45-50 mm is accepted, the vehicle's weight rises with 45 mm thickness from some 25 tons with angled bow to some 27.2 tons if old tracks are retained.

In the process, the ground pressure would amount to 1 kg/cm2., or considering the lesser carrying surface in this older track version, some 1.25 kg/cm2., thus too high. With installation of intended winter tracks of 560 mm width, the ground pressure would amount to 0.72 kg/cm2. and the vehicle's weight involving a third rubber bandage to 28.2 tons. With introduction of the third rubber bandage for the wide tracks, though, the danger of melting of the inner bandage exists in particular, since with the constantly increased vehicle weight, the running-gear levers may give somewhat. Therefore we suggest switching to rubber-saving wheels as soon as possible.

Krupp design (Drawing W1466 of January 22, 1943) for the 9./B.W. (Pz.Kpfw. IV Type H). This version with inclined armor plates was officially called Type H.

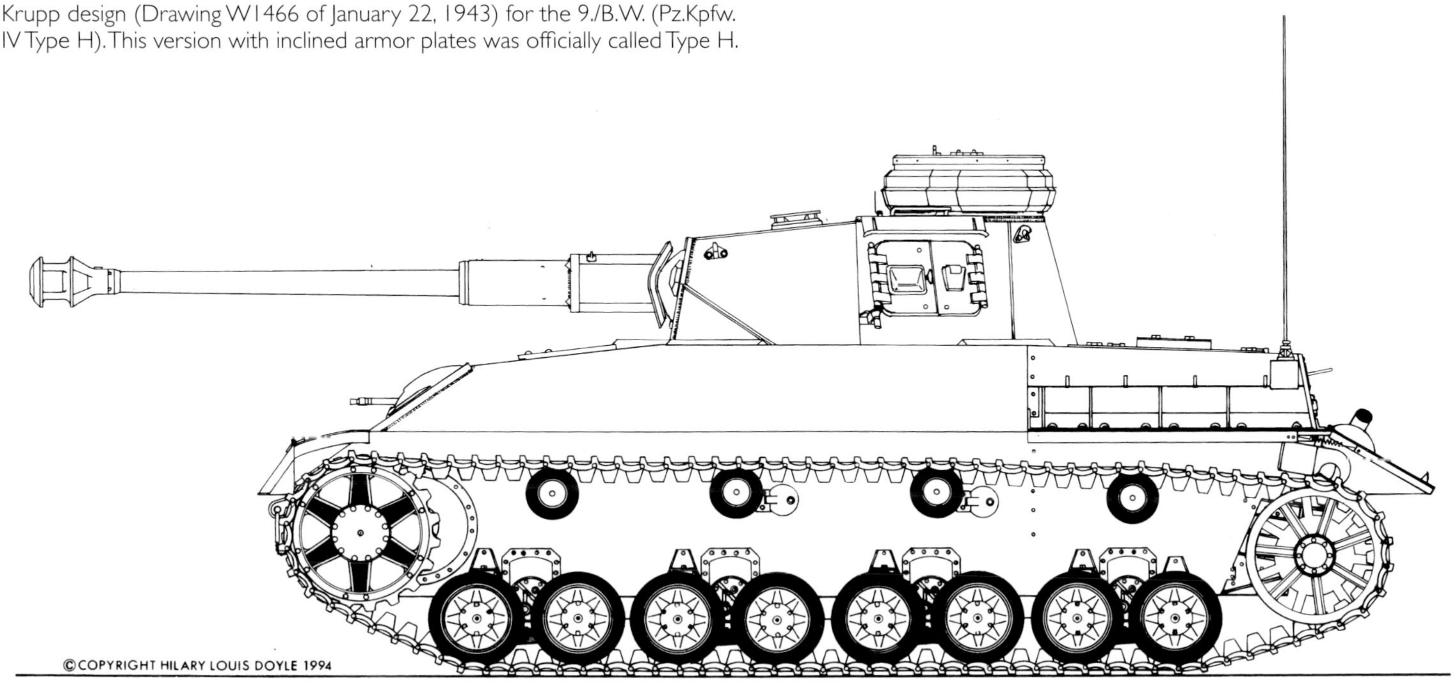

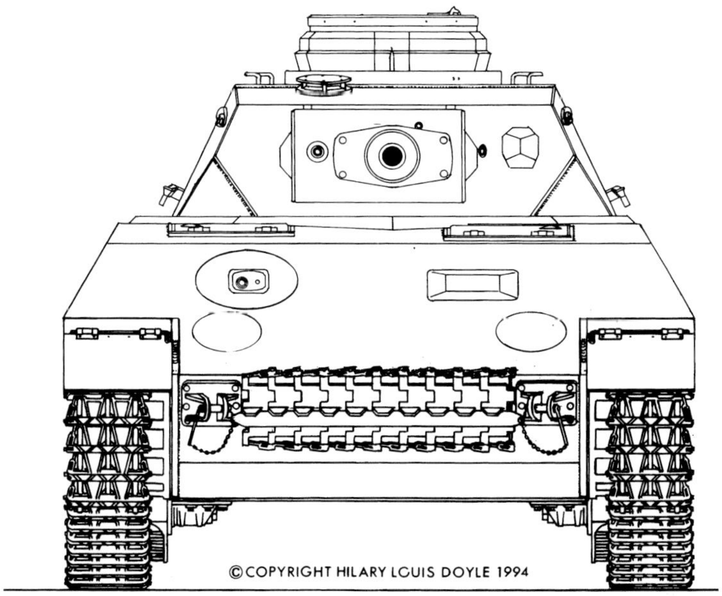

With a vehicle weight of 28.2 tons, naturally there is the greatest concern about the much greater burden on the running gear and steering drive. In our opinion, though, no other solution is possible to meet the request of the troops. A decision must be made at once.

At the same time, a driving test with 28 tons of total weight must begin at once to test the steering capability, since the activation of the steering drive, originally designed for 18 tons, becomes ever more difficult. Through the driving tests, the allowable top speed as well as the handling when braking, taking slopes and obstacles, etc., must be determined.

To save time, we immediately plan 45 mm thick side and back walls in the drawings for Type H with angled bow and upgraded performance.

We also note that through extensive rebuilding and new creation, as well as the tempering and erecting of stronger plates, a schedule delay cannot be avoided. Thus it is understood that we will use all means to shorten the schedule.

In the discussion on February 17, 1943, at the Ministry of Armament and Ammunition, the fate of the further development of the Pz.Kpfw.IV series was decided. Present were:

Reich Ministry	HDL Saur & Maj.Gen. Hartmann
Armor Commission	ol. Thomas & Prof. Dr. Porsche
Main Branch Armor	Dr. Rohland
Wa Prüf 6	Col. Von Wilcke, Lt. Cols. Holzhaeuer & Crohn
WuG 6	General Philipps
Alkett	Director Freyberg
Krupp, Dept. A.K.	Directors Dorn & Woelfert

An extensive discussion took place on rebuilding the Pz.Kpfw. IV. Basically the emphasis is to be placed on the Panther, meaning lowering the B.W. in favor of the Panther asd soon as its production is sure. It was stressed, though, from various sides that the proven Pz.Kpfw.IV cannot be dropped at all.

In the course of the Adolf Hitler Program, the Gruson Works are striving to increase the production of the Pz.Kpfw.IV in October from 80 to 140 units, the Nibelungen Works from 90 to 250. The total production of Pz.Kpfw.IV shall be increased by the spring of 1944 by 1200 units.

The following reasons not to introduce the angled bow were stated:

The more recent building firms, Nibelungen Works and Vomag, and the armor firms, Böhler and Oberdonau, make it a condition for increased production that the current design be retained unchanged. At Oberdonau difficulties will arise in changing the armor of the Pz.Kpfw.IV with the simultaneous start on the Panther.

The representative for Pz.Kpfw.IV in the Reich Ministry for Armament and Ammunition further stressed the standpoint that through additional burdens on the preparatory work for modifying the facilities, creation of new forged pieces and cast armor-steel pieces in the tense situation a decrease rather than an increase in performance is to be expected.

For the introduction of the angled bow, the following reasons were expressed:

Essential performance increase through dropping the dividing wall of the bow and stern armor and introduction of separation cutting instead of milling. Plus the introduction of an improved driver's optic with a new-type glass block and a new ball mantlet.

The troops would also gain even greater confidence in the B.W. from the introduction of the angled bow proved in the Russian T34 and presumably take a temporary decrease in production in the bargain if they received higher-performance vehicles.

The A.K. representative also indicated that for the decision on the modification of the B.W. the chassis weight must also be considered. A weight of 24 tons has been regarded to date as the outermost limit, since the steering gear and running gear built for 18 tons will no longer suffice. With KwK 40, added armor, 200 liters of added fuel, etc., the fighting weight for Type G will already add up to 25 tons.

Along with that, two essential new moments were reached: With consideration for the Russian antitank guns, a protective armor or an organic strengthening of the side and back walls. The use of wider winter tracks with rubber-sprung road wheels.

The following weight increases thus result, compared with a vehicle weight of 25 tons with KwK 40 and fore-armor:

	Approx. kg	Added weight
Strengthened turret roof	200	25.2 tons
Protective armor	600	25.8 tons
560 mm wide winter tracks and rubber-sprung wheels	1300	27.1 tons
Angled bow instead of protective armor	360	27.6 tons
Strengthening the bottom	230	27.69 tons
Strengthening the roof	230	27.92 tons

In-house road tests at the Gruson Works with a vehicle weighing 28.2 tons resulted in an essential worsening of the driving characteristics. The steering was very difficult and required constant downshifting. A running-gear lever also bent. Krupp Department A.K. considers improving the running gear and improving the steering ability by an added transmission in the side gearings to be absolutely necessary. The installation of a new single-radius steering gear built by Krupp would be favorable by allowing no-loss, easier steering. To be sure, the Maybach HL 120 motor producing only 265 HP would have to be retained, since the (new) HL 100 motor can be delivered only in about the spring of 1944.

The Armor Commission is of the opinion that a greater rebuilding must be avoided, since otherwise there will be serious doubts that the anticipated greater production of Pz.Kpfw.IV will be achieved. Besides, any weight increase is to be avoided if possible, and the weight of the winter tracks and rubber-sprung wheels limited most extremely, even at the cost of life expectancy.

The Armor Commission decided to suggest to the Führer to drop the building of the angled bow and the strengthening of the bottom and roof, and to consider protective armor at the sides and rear. The test version from Daimler-Benz with 5 mm plates before the housing and 10 mm plates before the turret (total weight some 600 kg) is also in question).

The test version, the protective armor, will be fired at the Kummersdorf Test Center along with short road tests on 2/20/1943 and then shown to the Führer.

15.9 PANZERKAMPFWAGEN III/IV

15.9.1 Basic Decisions

Wa Prüf 6 Pz III, notes on the session of the Tank Work Group on January 4, 1944.

Present:

Dir. Stieler von Heydekampf	*Tank Commission*
Dir. Dr. Blaicher	*Work Group*
Lt.Col. Schaede	*Ammunition ministry*
Colonel Mildebrath	*A H A/In 6*
Colonel Holzhaeuer	*Wa Prüf 6*
Sr. Engineer Woelfert	*Krupp firm*
Sr. Engineer Lanz	*Krupp firm*
Dr. Holbeck	*Special Branch*
Director Wiebicke	*M.A.N.*
Sr. Engineer Maier	*Zahnradfabrik firm*
Sr. Engineer Breuer	*MIAG firm*
Lt.Col. Stollberg	*Wa Prüf 6*
Dipl.-Ing. Jaeger	*Wa Prüf 6*
Dipl.Ing. Koehler	*Wa Prüf 6*
Chief Works Master Giesche	*Tank Commission*

On the advice of Chief Service Leader Saur, the possibility of making the medium and heavy tanks was discussed.

1. le.Pz.Jg. and Pz.Kpfw.IV

The uniformization can be carried out either on the basis of the Pz.Kpfw.IV or on the basis of a combination Pz.Kpfw.IV-Pz. Kpfw.III. The weakest point in both cases is in the side gearings and steering brakes.

Side gearings: *In the chosen combination of Pz.Kpfw.IV-Pz. Kpfw.III, strengthening the side gearings must be done; on the basis of the gear correction suggested by Daimler-Benz, a strengthening of about 29% results. In case it is not sufficient, a new side gearing is being designed, which is not exchangeable with the present one and requires different attachments.*

Steering Brakes: *The steering brakes are at the end of their performance capability, because of the braking power as well as the activation power. The attempts, which are not yet finished, have resulted to date in the highest steering power is produced by the steering brakes of the Pz.Kpfw.IV, less strongly by the Argus brakes, and the least from the Pz.Kpfw.III brakes. The Argus brakes, which are essentially equal to those installed in the Panther, have the advantage of being free from servicing and the disadvantage of not* *surely controllable braking power which can lead to the blocking and destruction of the side gearings. This danger is considerably less with the hand-operation in the Light Tank Destroyer than in the Panther with its oil-pressure activation.*

Running Gear: *The running gear of Pz.Kpfw.IV in its present version is too weak and cannot be chosen. There are three possibilities:*

a) Staff running gear with parts similar to those of the Pz.Kpfw. IV with 8 rubber-saving 700 mm road wheels—suspended on four wheel trucks. Armored hull unchanged here.

b) In-line running gear with parts similar to the Pz.Kpfw.IV with 6 rubber-saving double 660 mm road wheels—suspended on three wheel trucks. Armored hull must be modified.

c) In-line running gear as suggested by the Alkett firm, with 8 rubber-saving double 470 mm road wheels—suspended on four wheel trucks. Armored hull unchanged here.

The Type a) and b) running gear are based on the tested components of the Pz.Kpfw.IV and give the least risk. The staff running gear a), which offers possible interchanges, has the disadvantage of the not yet sufficiently tested behavior of staff running gear as opposed to the danger of tracks coming off. Running gear b) probably offers the best running characteristics. Running gear c), which offers the best production possibility as long as the replacement of roller bearings by sliding bearings succeeds, is not yet sufficiently tested in its components.

Cooling System: *The cooling system of the Pz.Kpfw.IV requires much power from large radiators and two V-belt drives for the two ventilators. Wa Prüf 6 is investigating the possibility of replacing this cooling system with a simplified one being developed by the Porsche firm.*

Tracks: *The presently used tracks with a width of 400 mm have too-high ground pressure. There is the possibility of installing tracks with a width of 540 mm, which provide favorable ground pressure. Because of the large number of parts, the use of tracks with a cast and a rolled link is practical.*

Conclusion: The chassis of the Light Tank Destroyer and the Pz.Kpfw. are formed the same. The following components are to be chosen:

Motor:	*Maybach HL 120 TRKM*
Cooling:	*As Pz.Kpfw.IV (if no simpler system is developed in the meantime)*
Gearbox:	*SSG 77*
Steering Gear:	*Pz.Kpfw.III*

Side Gearing: *Pz.Kpfw.III, strengthened in case it is not strong enough. (Developed by Daimler-Benz) Steering brakes: Argus brakes; the attempts to test these brakes are to be put on a broader basis at once and carried out with at least five vehicles.*

Running Gear: *Six-wheel running gear suggested by Krupp, in in-line form, with 660 mm rubber-saving road wheels. In the suspension, the following work simplifications are to be installed if possible:*

a) Dropping the rounding of the crank arm in the mount block until on a working edge for sealing.

b) Optional version of pressed-in road-wheel pins. Tests with this running gear are to be carried out on a broad basis with five vehicles.

The Alkett firm was later given a contract to develop another running gear with six rubber-saving 600 mm road wheels suspended on three wheel trucks.

Aprons: *to be installed so that they can be set vertically or angled.*

Tracks: *Tracks 540 mm wide symmetrical with single-tooth guiding, if possible each with one cast and one rolled piece.*

Leading-Wheel Suspension: *With leading wheel as on Pz.Kpfw.IV and expanded adjusting for tracks and roller links. Strengthen and simplify leading-wheel suspension.*

Hull: *Pz.Kpfw.IV and Light Tank Destroyer both with pointed bow. Fuel tanks, if necessary, not removable from below. Fire safety for fighting compartment must be retained. Turning carrier should be dropped if possible and be replaced by cable that allows limited turret turning of 270 degrees to either side, or 540 degrees in all.*

As Sketch W1773 of January 17, 1944, shows a Panzerkampfwagen IV with Panzer III drive, the front armor protection, similarly to that of the Jagdpanzer IV, was improved with a pointed front wall and angled driver's plate. The stern armor over the motor compartment was modified. Cooling air for the motor was sucked in and blown out through grids on the roof.

The armored hull consisted of armor plates with the following thicknesses and inclinations (from the horizontal).

Plate	Thickness	Inclination
Front wall, upper	60 mm	30 °
Front wall, lower	60 mm	45 °
Sidewalls	30 mm	90 °
Back wall	20 mm	80 °
Bottom	16 mm	0 °

The body consisted of armor plates with the following thicknesses and inclinations (from the horizontal):

Plate	Thickness	Inclination
Front wall	80 mm	40 °
Sidewalls	30 mm	54 °
Roof	20 mm	90 °
Back wall	20 mm	80 °

The ground clearance was raised to 520 mm with six 660 mm rubber-sprung road wheels with steel bandages. The ground pressure was lowered with 540 mm wide tracks. The range was extended significantly with 700 liters of fuel carried (Tank I with 300 liters, Tank II with 150 and Tank III with 250 liters under the turret turntable).

15.9.2 Running-Gear Testing

From May 5, 1944 on, five Pz.Kpfw.IV (chassis numbers 84715, 84557, 84779, 80625, 83945) and four Hornissen (Hornets, No. I to V) were available for testing with experimental running gear. For the five Hornets, all with six-wheel running gear, various bearing types were installed and tested. Their building state on 5 May was as follows:

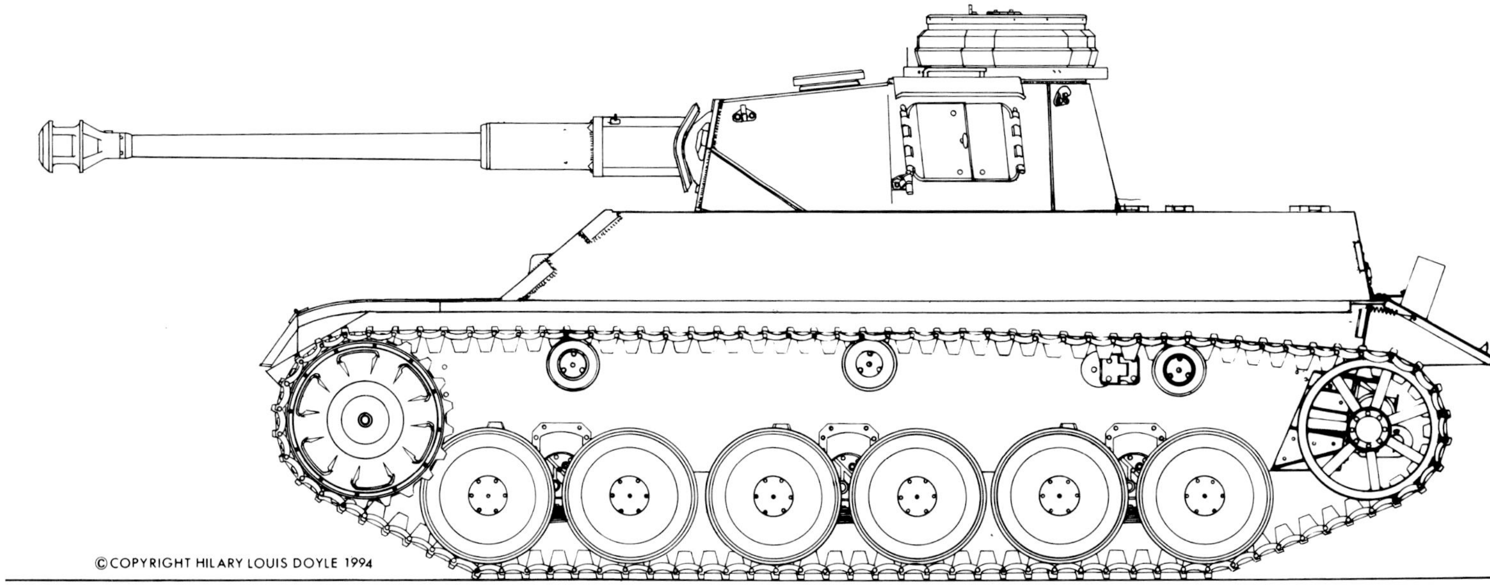
©COPYRIGHT HILARY LOUIS DOYLE 1994

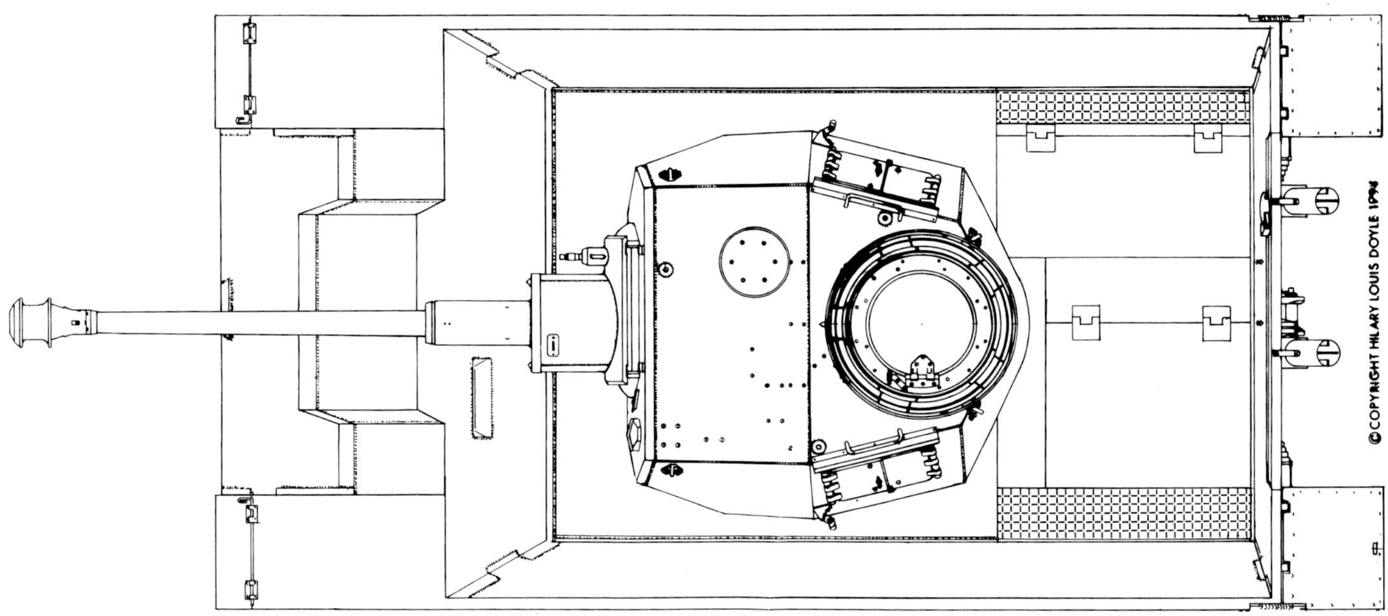
©COPYRIGHT HILARY LOUIS DOYLE 1994

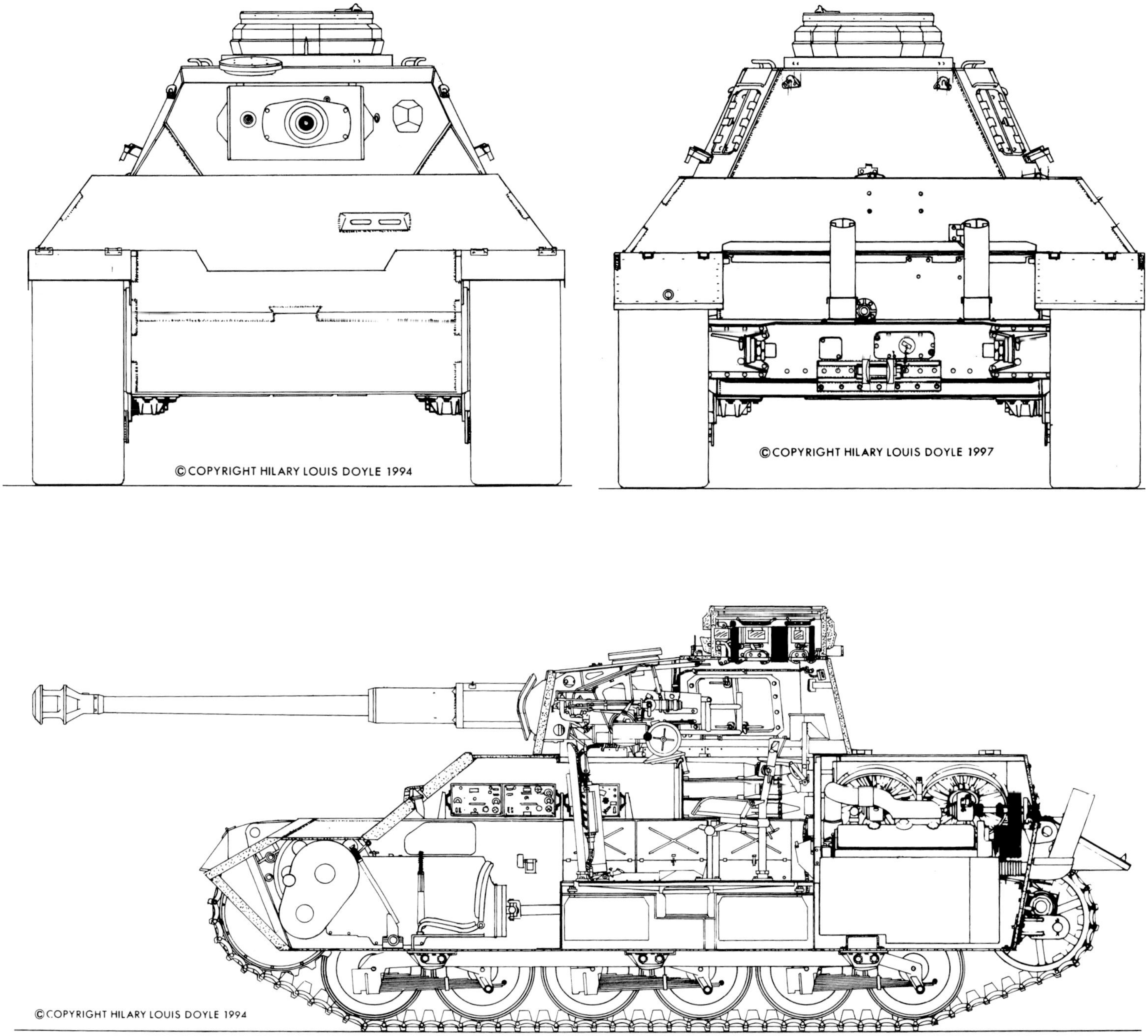

Panzerkampfwagen III/IV with uniform chassis. New angled armor. SSG 77 gearbox, six steel road wheels, increased ground clearance.

Hornisse I: To secure the road-wheel and powerplant testing, roller bearings as in SKC 6119 were to be installed.

Hornisse II: Since production had already gone too far, running-gear levers with pressed-in, hardened bearing pegs without wear boxes and with plain bearings had to be used. The total weight of this Hornet was to be brought to 30 tons because of the assault tank.

Hornisse III: Running-gear levers with pressed-in pegs, wear boxes and united cast plain bearings were installed.

Hornisse IV: Running-gear levers with pressed-in pegs, wear boxes and united cast plain bearings with varying consistency were installed.

Hornisse V: Bearing pegs pressed into the running gear, floating box. Pegs without wear boxes, hardened. This version could later be used with the floating boxes omitted.

In a discussion at Wa Prüf 6 on June 30 and July 1, 1944, in Kummersdorf about the Pz.Kpfw.III/IV rubber-suspended running gear, it was announced:

It was determined that the former shape of the rims and discs pressed too strongly on the rubber ring, so that damage to it occurs, as has been shown on the Hornet vehicles. The new shape of the profile at the tightening places for the rubber ring was determined. As for the axial pre-tension in the rubber rings, a steady 8 tons was found to be sufficient. As because of the tests at Verskraft and Krupp a strong flowing of the SMZ 50 rubber rings was found, Wa Prüf has ordered rings of a harder bandage quality for a test.

Thus the following conclusions have arisen for the test vehicles now being tested or being built: It is pointless to go on making tests with the former discs and rims. Insp. Gunkel is driving to the Deutsche Eisenwerke on July 4 with the modified drawings, as well as the Lemmertz and Kloeckner firms, and discussing the changes to be made there. The continuing production is to be maintained and, for the new parts, delivery to the Krupp Gruson Works is to begin as quickly as possible.

For the individual vehicles it was determined:

Hornisse I remains temporarily at Verskraft for further road tests. Krupp is delivering two spare wheels of the earlier type. New rims, discs and rubber rings are to be made ready.

Hornisse II is being sent back to the Gruson Works for repairs. A precise account of the damage is to be had. The hull of Hornisse IV can be used for repairing. The road wheels are to be fitted with new discs, rims and rubber rings. The welded running-gear levers are to be installed.

Hornisse III is at the Deutsche Eisenwerke at this time.

Hornisse IV running gear is at Alkett at this time. For both Hornisse running gear, new rims, discs and rubber rings are to be made ready.

Hornisse V (still being built at the Gruson Works) is to be built with road wheels of the newest type, with roller bearings. Wa Prüf hereby gives a contract for six sets of discs, rims and rubber rings.

The Pz.Kpfw.IV vehicles, chassis no. 84779 and 80625, are likewise to be fitted with new rims and discs.

15.9.3 Decisions about Components

On May 26 and July 6, 1944, it was explained that the Maybach HL 120 TRM-E motor should be used for the uniform vehicle.

Mr. Woelfert reported on a discussion concerning Pz.Kpfw. III/IV with uniform chassis on May 31, 1944 in Kummersdorf:

Baurat Kremer stated that when the two Hornets were loaded to 28 tons for drives on the Forest Route at Kummersdorf, driving could only be done up to 4th gear, since the motor performance with the increased weight is not sufficient. I suggested that the tests at Maybach with a more powerful motor absolutely had to be hastened. I spoke with Baurat Augustin about the necessity of strengthening the B.W. motor. Augustin reports that Maybach is making tests to increase power through injection instead of carburetors.

On June 7, 1944, the following decisions were made at Magedburg as to the uniform vehicle body by Major Kaupsch (Wa Prüf 6/Pz.II) and Mr. Wenzel (Dept. AK Lanz):

Hanging the Radio Set: *Wa Prüf is basically in agreement as to the removably constructed suspension of the radio set and approves the inclusion in the series. Ammunition storage: According to SKA 6215, the installation of 100 rounds of armor-piercing or explosive shells plus 3150 rounds of MG ammunition is possible. The housing of the shells is planned as follows:*

20 rounds each lying left and right in the angled body,
4 cases of six standing behind the driver,
4 cases of six standing to the right in the fighting compartment,
2 cases of four standing in the fighting compartment,
4 rounds lying in a case on the platform.

For the standing ammunition cases, Wa Prüf is in agreement, for reasons of better operation and simpler production of the cases, to leave off the lids.

Antenna: *The antenna is sensibly to be mounted on the rear armor as on Pz.Kpfw.IV.*

Armor Aprons: *The armor aprons are to be made straight through from front to back of 5 mm sheet metal, using as equal pieces as possible.*

Firing Model: *Under Official Contract No.SS4911/0006-2828-44 a contract is underway from the office to the Gruson Works for production of a firing model of the installed vision equipment according to SKA 6231 in the driver's front plate. In order to test the emergency ejection activation at the same time as the entry hatch, the model is to be made so that the bow roof is welded to the driver's front plate. The driver's front plate receives the natural width with 80 mm thickness.*

On July 9, 1944, Krupp in Essen informed the Nibelungen Works: *As we already informed you, the turret for the III/IV uniform chassis (Group No. 021 Gr 824) remains practically the same as for the currently running 10./B.W. series (Pz.Kpfw.IV Type J).*

15.9.4 Ordering Contracts and Plans

On March 22, 1944, the Army High Command issued Contract No. SS4911/0006-3036/43 to the firm of Fried.Krupp Gruson Works A.G. of Magdeburg-Buckau for the production and assembly of three Pz.Kpfw.IV on uniform chassis. For this three complete chassis (hulls) from Order Wa Prüf 6/Pz III SS4911/0006/3374/43 were delivered by the Alkett firm of Berlin-Borsigwalde. N June 1944 it was decided that the Gruson Works should already begin the production of the finished Pz./Kpfw.III/IV in February 1945.

On June 2, 1944, Contract SS 4912-0210-8808-44 for the completion of 610 hulls and 391 complete bodies for Panzerkampfwagen III/IV was issued to Krupp of Essen. The Pz.Kpfw.III/IV was no longer built as Pz.Kpfw. after notification from Obering. Wasmuth on 7/12/1944, but only as a tank destroyer with L/70.

15.10 PZ.KPFW.IV WITH 7.5 CM KWK. L/70 IN PANTHER TURRET

In the full session of the Tank Commission on April 12, 1944, in Berlin-Wannsee the question was raised of developing a modernized Pz.Kpfw.IV with 7.5 cm L/70 Kw.K.For such an investigation the narrow 7.5 cm KwK L/70 being developed by the Daimler-Benz firm might well be considered.

On November 2, 1944, the weights of the newly suggested Pz.Kpfw.IV with 7.5 cm KwK L/70 were calculated:

Weight of chassis without bow armor, with 5 men	17,300 kg
Weight of the armor aprons	800 kg
Weight of the bow armor	1900 kg
Weight of the turret (50/30 mm armor)	4500 kg
Weight of the ammunition	850 kg
Various	50 kg
Fighting weight	25,400 kg
Same with (8 + 1) road wheels if necessary	26,000 kg

15.11 SIMPLIFIED TURRET

Department A.K. of Krupp in Essen finished a design for B.W. further development of a simplified turret according to Drawing AKF 31941 of July 19, 1944. The new turret design had a six-angled turret shape, simple gun mount, no vision flaps or MP ports, no commander's cupola, two hatch covers in the turret roof, and simplified hatch covers without vision or MP ports on the left side of the turret.

The turret housing consisted of armor plates with the following thicknesses and inclinations (from the horizontal):

Plate	Thickness	Inclination
Front wall	80 mm	80 °
Gun mount cover	80 mm	—
Sidewall	42 mm	65 °
Back wall	42 mm	65 °
Roof	25 mm	0 °

On July 19, 1944, it was decided that through the new program situation, according to which the Pz.Kpfw.IV was to be converted into the Panzerjäger with L/70 gun in the next few months, the introduction of a simplified turret design for the Pz.Kpfw.IV no longer was to be considered.

15.12 SPECIAL VERSIONS OF PZ.KPFW.IV

The Army Weapons Department saw itself as exposed to growing pressure during the war from the Waffen-SS, which demanded weapons and equipment for their units that were not foreseen in the approved production programs. How far this influence had already grown in seen in the following examples:

- After taking over the Bohemian-Moravian Machine Factory in Prague in 1939, the Waffen-SS made sure of a (full-track) tractor production capacity at Praga, which reserved part of the production until the war ended.
- In order to remain mobile in the Russian terrain conditions, the Waffen-SS created a practical and reasonably priced solution by rebuilding commercial trucks into halftrack vehicles, the well-known Maultier 9 (Mule). The rebuilding could be one by the troops if necessary.

Thus, it is not surprising that the Waffen-SS also showed an interest in improving their combat vehicles. This occurred partly by circumventing the official procedures prescribed by the Army Weapons Department.

On April 6, 1944, a design, No. TE121, for a hydrostatic tank drive was finished, followed by a sketch of the main components, made by the Nibelungen Works in St. Valentin on May 18, 1944.

On October 23, 1944, under contract from the SS Command Office, T-Amt VIII F.E.P., a cutaway drawing with explanations of important modifications for a special version of the Panzerkampfwagen IV was finished. The main points were:

1) The elevating and traversing machine was hydraulically combined with a control rod with simultaneous mechanical limitation of the elevating machine. Releasing the bar automatically stopped both turning machines mechanically at the set aiming values. Electric firing for the tank gun and turret MG were on the control bar. By means of this combined weapon aiming, the turret could be turned 360 degrees in about 20 seconds; by simultaneously using all elevation values in any short time, depending on the swing of the control bar, whereby the greatest aiming speed were attained for the tank gun and turret MG. In case of an average in this combined weapon aiming, a mechanical coupling was present on each that let the aiming gunner switch immediately to old-style hand-operation.
2) Pressure oil supply for weapon aiming for optional DKW or Maybach drive, with electric and hydraulic movement of the turret.
3) Hydrodynamic rear drive, which allowed stepless driving from G to maximum load, which eliminated the nose-heaviness of the Pz.Kpfw.IV caused by the very long 7.5 cm tank gun and also created new possibilities in terms of carrying additional weapons, ammunition and fuel in the front area.

The advantages of hydrodynamics were as follows:

a) Saving the step drive, steering and support brakes and planetary drive.
b) The building of this drive required no close-call alloys with Cr, Mn or Si, since beside three ball bearings only a cast steel housing plus drive flange and piston

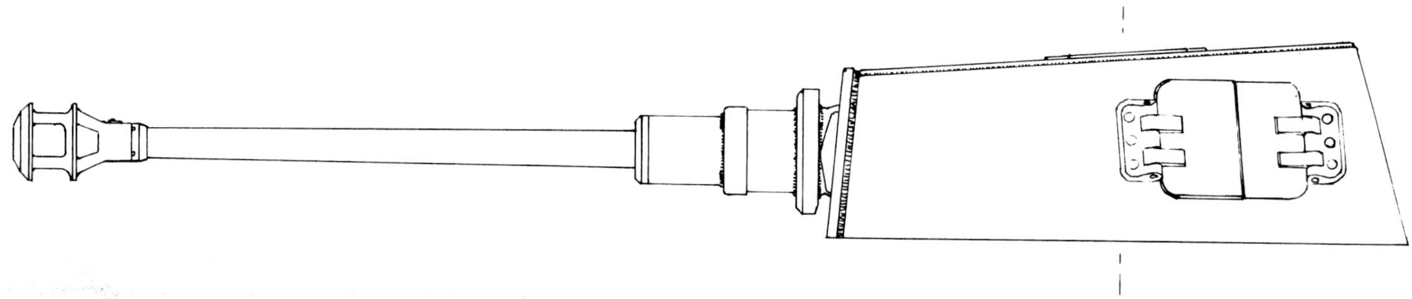

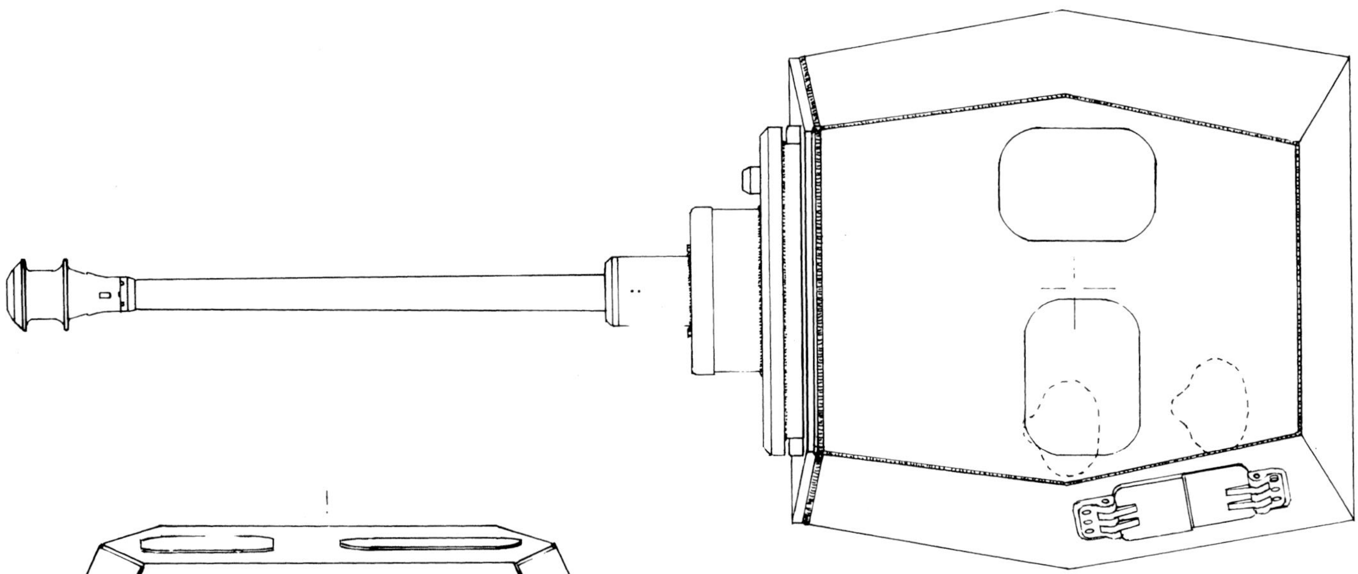

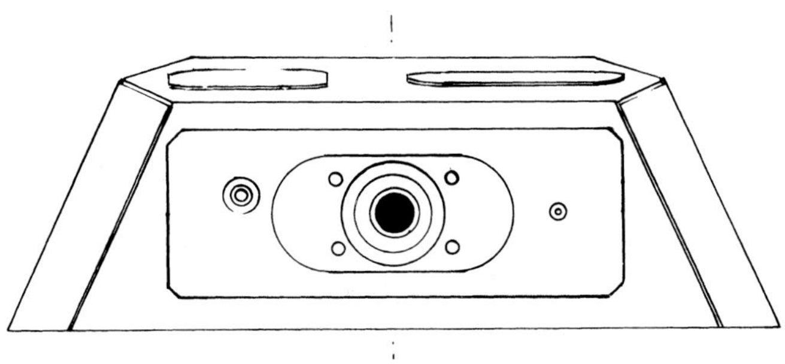

The simplified turret, from Drawing AKF 31941 of July 19, 1944. The turret widens to the rear, the commander now sits to the side of the breech. Thus it was possible to turn the gun farther to the back. The original drawing gives no information about the commander's means of vision.

rods had to be made of steel, and thus all other parts could be made of unalloyed gray castings. Of decisive significance was the fact that all parts of these drives had rounded contours, so that they could be made in quantity completely in automated machines with proper setups.

Of equally important significance was the fact that with all braking forces in a tracked vehicle (turning) were omitted from this special version, since speed correlation of the two tracks to each other would rule out any needless use of power, which considerably increased the life span of all drive organs and guaranteed a fuel saving of at least 20%.

4) Hydraulic steering, which made handling this vehicle almost like that of a normal motor vehicle. Turning in place or during forward or backward driving at any speed, depending on the pressure of a foot pedal, with simultaneous synchronizing of the engine speed. By synchronizing the engine speed with the drive pedal (forward and backward driving), any senseless racing of the motor was avoided and further sparing of the material or saving of costly fuel attained.
5) Cooling system working two-sided for the Maybach motor and hydrodynamic drive (air intake and outgo on either side of the motor).
6) The new design of the stern easily allowed any repairs to the motor by opening the rear hoods.
7) All vital parts of the motor could be reached from the fighting compartment; cranking the motor was also possible from the fighting compartment be means of a flowing-power starter, which meant that the crew no longer had to expose themselves to enemy fire in dangerous situations.
8) The leading wheels were also newly designed, and included a combination of leading-wheel track-tightening device without eccentric plus inside-shoe brakes. Tightening the tracks had no influence on the brake system and thus ruled out any readjustment of them. To steer this special version of Panzerkampfwagen IV, no mechanical brakes were needed any more, since these functions were carried out exclusively by means of hydraulic steering via the hydrodynamic drives. The inside-shoe brakes in the leading-wheel combination were used only for emergencies.
9) Tilting the bow plate was now possible.

On October 25, 1944, under contract from the SS Command Department T-Amt VIII F.E.P., instructions went to the Augsburg Zahnradfabrik to take up a special version of the

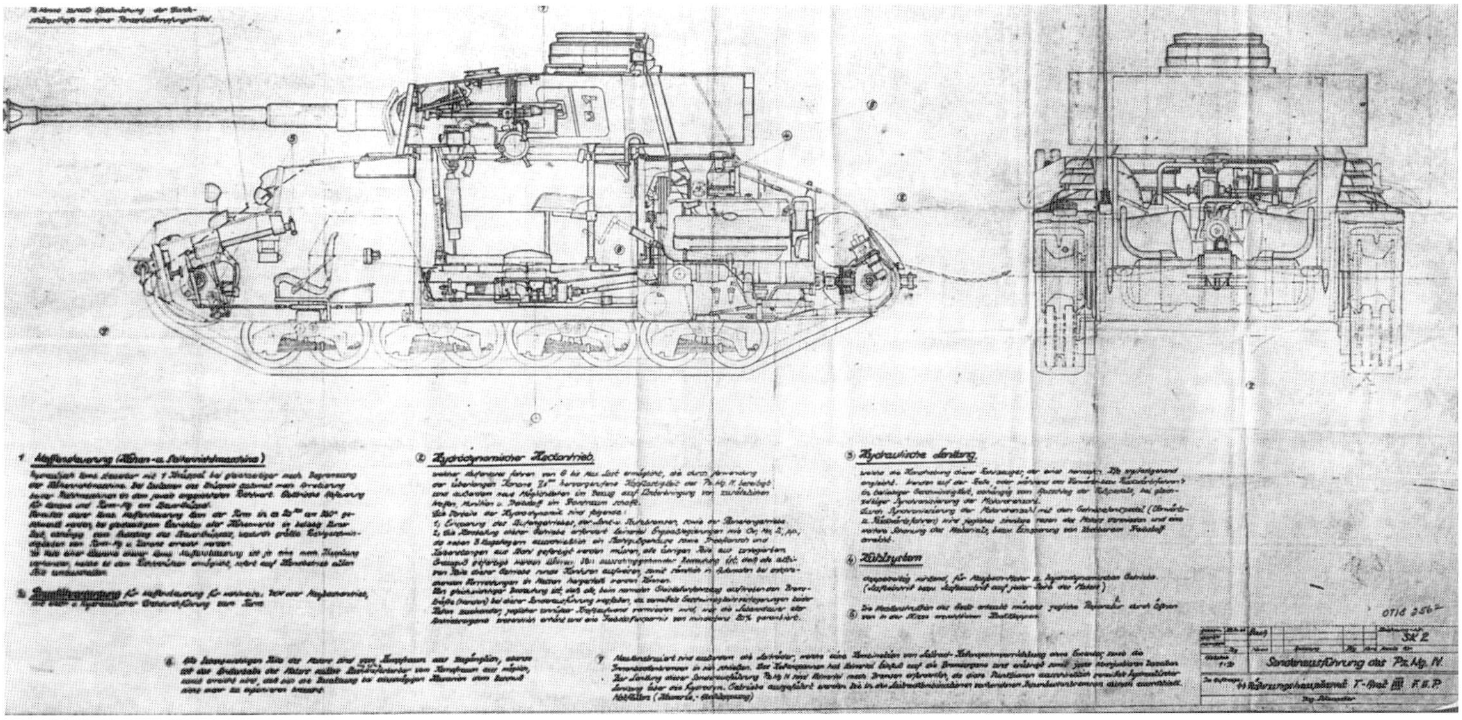

Special version of Pz.Kpfw.IV with hydraulic drive.

Panzerkampfwagen IV with hydrostatic drive in their development. It was not a fluid drive; in fact, the usual drive was completely missing from the test vehicle. The Thoma drive was laid out so that the main motor (Maybach HL 120) was linked directly with two high-performance oil pumps. These oil pumps supplied two oil motors. A whirling-disc drive under transmission drives drove the rear drive wheels. They had a diameter of some 550 mm, while the leading wheels in front measured some 780 mm. Along with the hydrostatic drive, the turret drive was also hydraulic.

The test vehicle (a rebuilt Pz.Kpfw.IV Type G) was sent by the U.S. Army in 1945, after the war ended, to the Vickers firm in Detroit, Michigan and examined thoroughly there. A final report by the Vickers firm is dated 4/12/1946. The test vehicle still stands in a tank museum in the United States today.

A look into the driver's space with all the operating gauges.

The motor compartment with built-in Maybach HL 120 motor. The drive to the oil motor is formed as a flange.

The entire oil-drive aggregate with main flange, including the two oil motors.

Special version of Pz.Kpfw.IV with hydraulic drive.

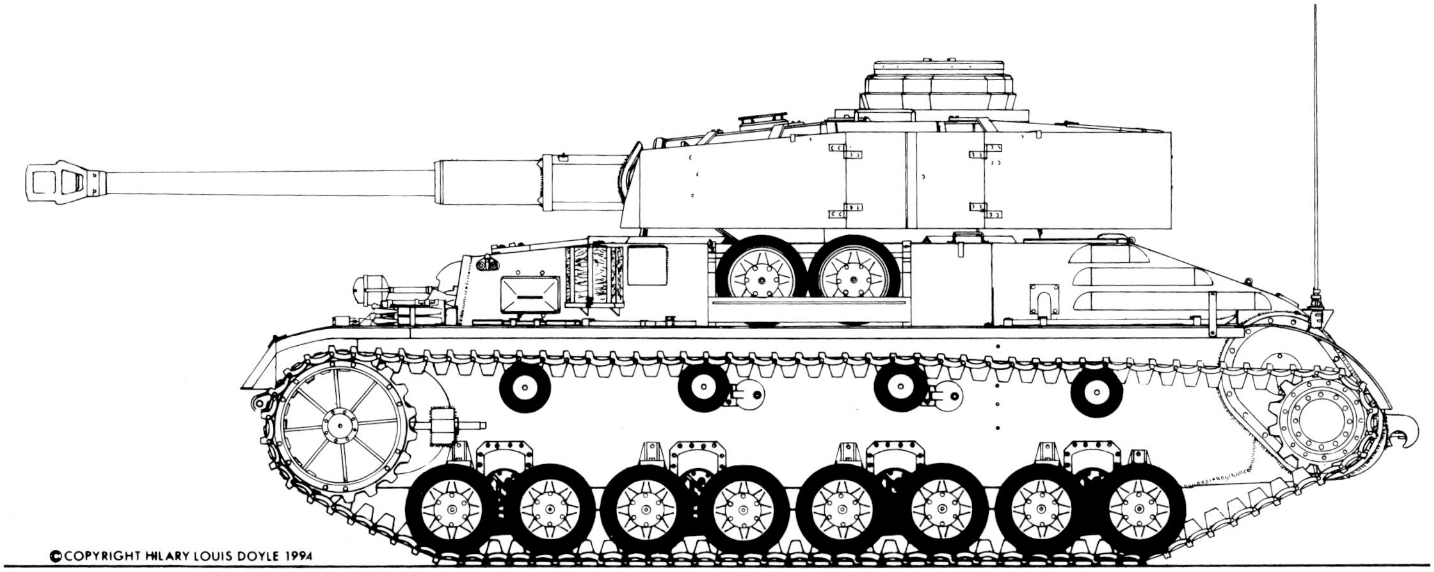

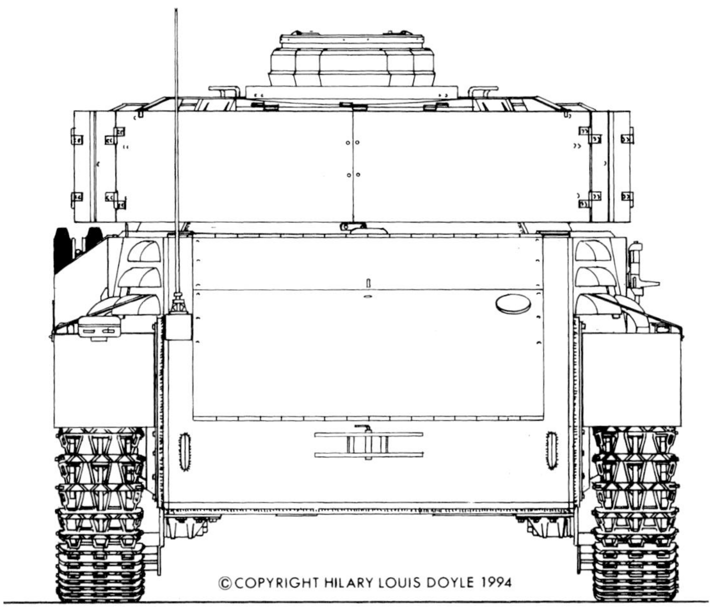

Special version of Pz.Kpfw.IV with hydraulic drive. The hydraulic drive was flanged onto the back of the Maybach HL 120 motor. Thus the drive motor could be moved 100 mm to the rear. The space that the SSG 76 gearbox formerly took up could now be used as extra storage space.

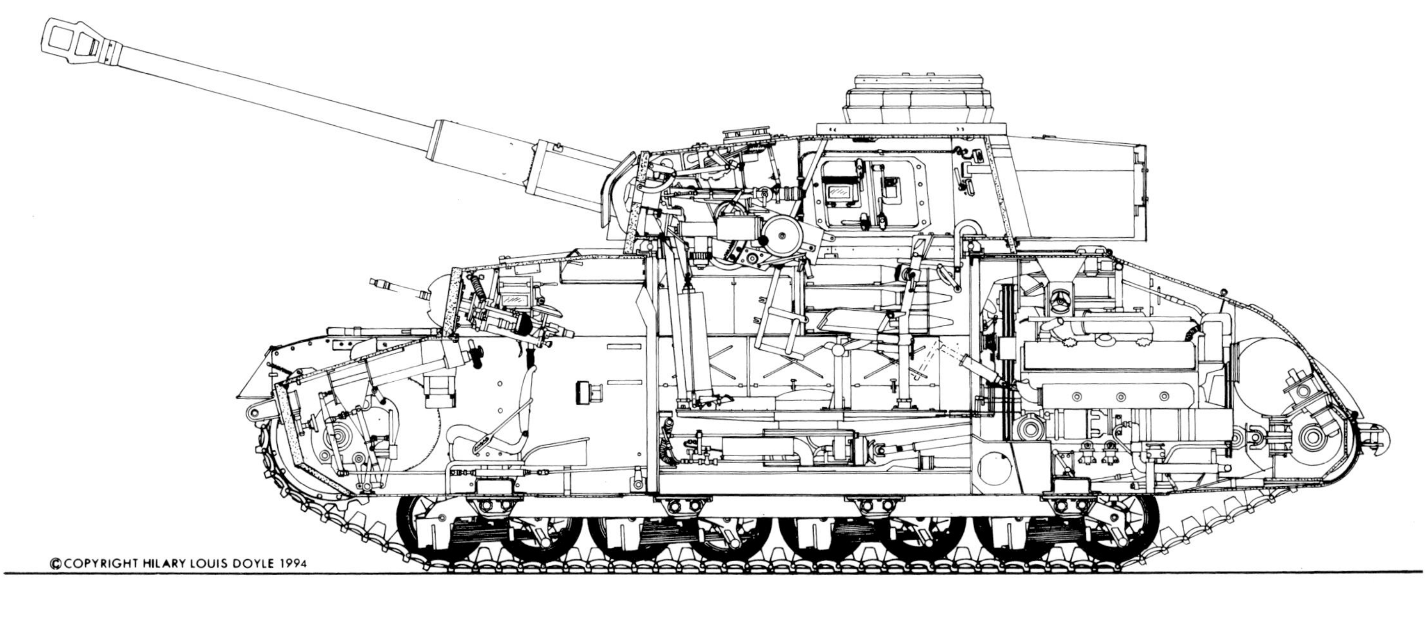

16. THE PANZERKAMPFWAGEN IV IN ACTION

16.1 The Polish Campaign as of September 1, 1939, on

At the start of the Polish campaign, Pz.Kpfw.IV were included in the structure of the tank battalions either in the Light Tank Company A (K.St.N. 1175 of 3/1/1939) or the Medium Tank Company (K.St.N. 1175(Sd.) of 9/1/1939).

At this point the medium tank companies of Tank Regiments 1, 2 and 11, plus the 1st Battalion of Armored Training Regiment and Tank Battalion 65, each equipped with 14 Pz.Kpfw.IV. Because of a lack of Pz.Kpfw.IV the other tank battalions had only three to five Pz.Kpfw.IV in their Light Tank Company A.

Pz.Kpfw.IV Type A, tactical no. 434, during the 1939 Polish campaign. The white cross as national symbol was used only during the "19-day campaign" as a vehicle marking and was replaced by the bar cross soon afterward.

Pz.Kpfw.IV Type C, tactical number 443, in Poland in 1939. The crew cleans their vehicle with the help of a fire hose.

Of the 211 Pz.Kpfw.IV that had been finished by the end of August 1939, 198 were used in the Polish campaign. The rest were scattered in the Replacement Army and at the schools, or waited at the Army Armory to be assigned. Pz.Kpfw.IV were not assigned to the armored units as substitutes during the campaign. 19 Pz.Kpfw. IV were damaged in Poland to the extent that they had to be sent to the factory for general overhauling.

The fighting value of the Pz.Kpfw.IV in Poland was clearly confirmed in an experience report of the XIX. Army Corps (General Guderian, Commander):

Primary needs hastened obtaining the Pz.Kpfw.IV and establishing medium companies. They have very strong actual and moral effects. The troops have particular trust in them.

16.2 Western Campaign as of May 10, 1940, on

Before the western campaign, all the tank battalions with the exception of Tank Battalion 40 were combined into two armored divisions. Although in the time from October 1939 to April 1940 103 Pz.Kpfw.IV were completed, these were still not enough to bring the medium tank companies up to the authorized number of 14 Pz.Kpfw.IV. Almost all tank battalions had reorganized their medium tank companies by order of February 1940, according to K.St.N. 1175(Sd.) of 9/1/1939. Yet through this order the authorization of a structure of only eight Pz.Kpfw.IV per medium tank company was given. The II. Battalion of Tank Regiment 33 and Tank Regiments 5, 6, 35 and 36 were still structured with a Light Tank Company A (K.St.N.1175 of 3/1/1939). Of the total of 290 Pz.Kpfw.IV in the entire army, 279 were ready for the western campaign with the tank units.

The following combat reports give excellent examples of the action of Pz./Kpfw.IV in this and other campaigns:

Combat Report of the 4./Tank Regiment 1 on the combat near Tourteron on 5/16/1940

Because of the possibility of an enemy tank attack from the left flank, the regiment built a hedgehog defense on the left in the night of May 15-16. The riflemen and other troop units fought on 5/15 around the crossing to Chagny. Since the enemy had moved out in the night, distance was kept from the intended surprise attack on Loubergny on the morning of May 16.

The spearhead for the advance on that day was formed by the 3./Motorcycle Riflemen Btl. 1 and the 4./Tank Regiment 1, the latter consisting of 3 Pz.Kpfw.II and 3 Pz.Kpfw.IV (Kuehne, Meier, Mattersteig, Company Chief, Lucas, Ewaldt). From Chagny the Lucas and Ewaldt tanks undertook reconnaissance on the left flank, where enemy tanks were said to have been seen. After they could see no more enemies, they were called back by radio. The Ewaldt tank broke down near Chagny. The Lucas tank reached the spearhead, which had already gone on the march again, in La Sabotterie. A platoon of Moroccan motorcycle riflemen which came into the village from the right and encountered our spearhead was captured. The advance went on. Before the spearhead reached the intersection between La

Pz.Kpfw.IV Type C, tactical no. 423, firing its 7.5 cm KwK L/24.

Pz.Kpfw.IV Type D after being hit by a French 2.5 cm antitank shell. The hit was in the middle of the front plate. France, 1940.

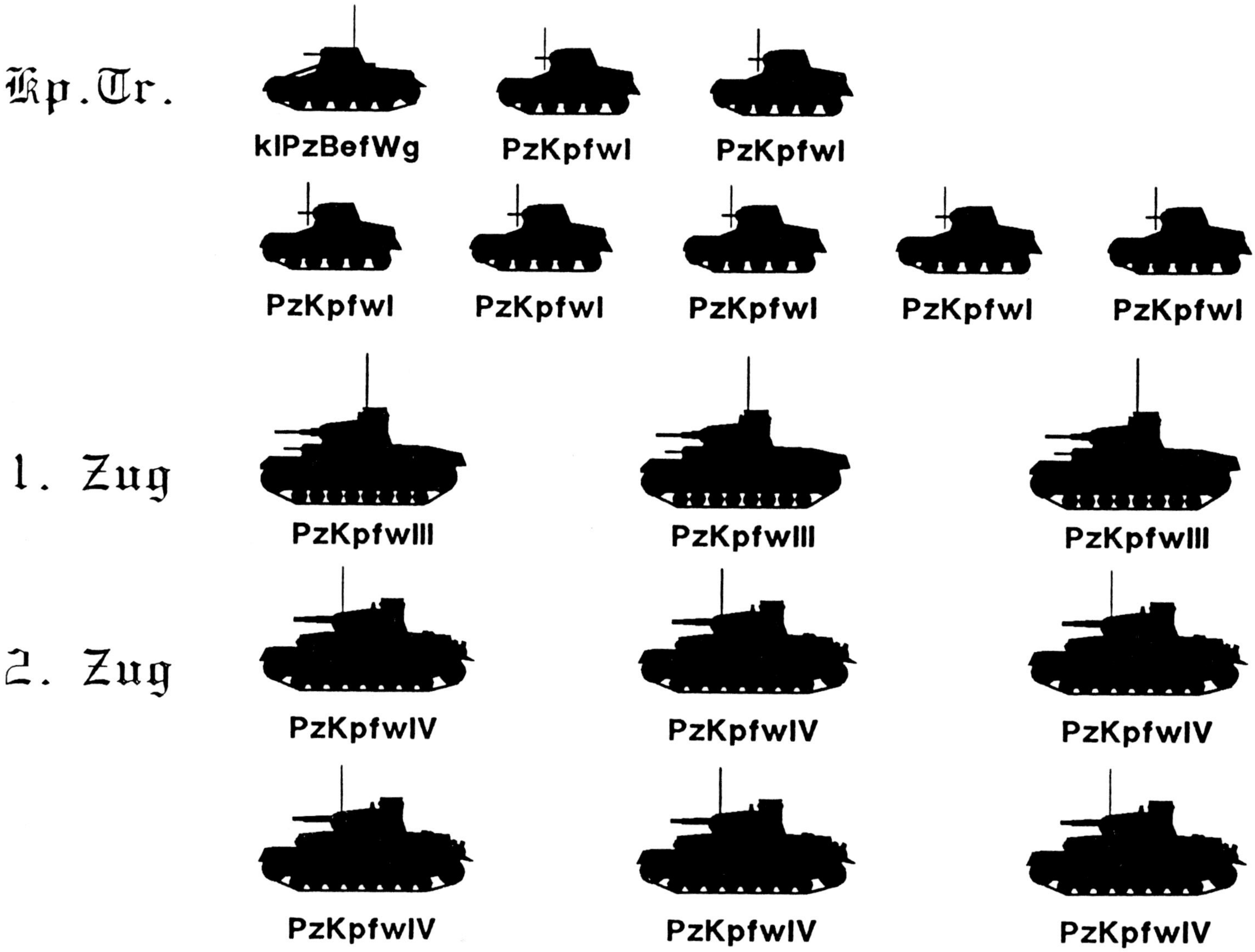

Structure of Light Tank Company A, according to K.St.N. 1175 of 3/1/1939.

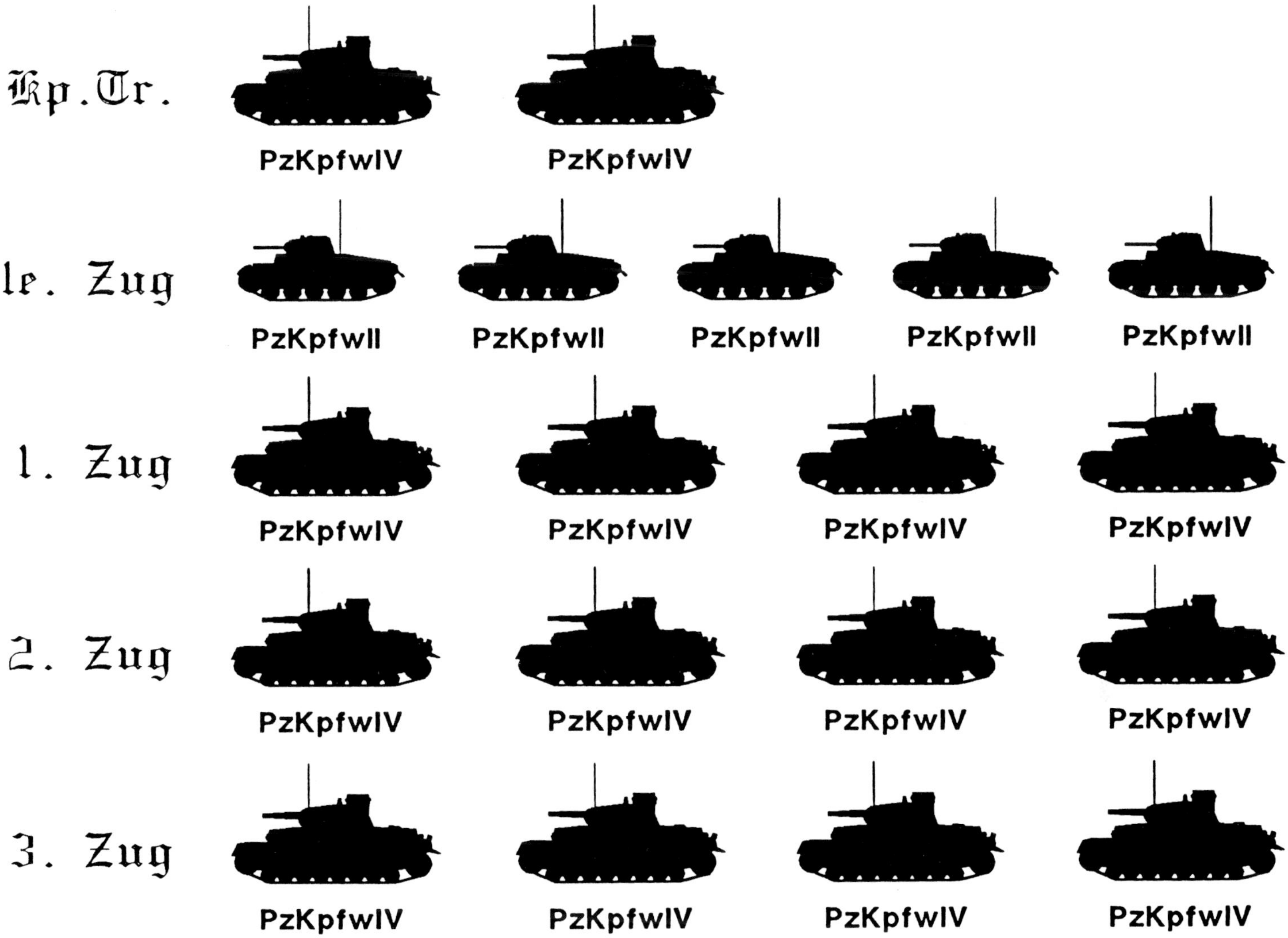

Structure of the Medium Tank Company, according to K.St.N. 1175 (Sd.) of 9/1/1939.

Sabotterie and Tourteron, three French armored scout cars and two cycles came onto the advance route at top speed and fled in the direction of Tourteron. They were fired on by the foremost vehicles, apparently without success.

When the spearhead (2 Pz.Kpfw.II, 1 Pz.Kpfw.IV) approached Tourteron, it received antitank and machine-gun fire from Tourteron and from the heights beyond Tourteron. A further advance on the road was impossible. The tanks begin to fire on known targets, but then slowly drop back. With the chief of the spearhead company of motorcycle rifles it was agreed that the Pz.Kpfw.IV should take up positions on the heights of Tourteron and the motorcycle riflemen should penetrate into the town under their covering fire. The Pz.Kpfw.II secure the road in the direction of Tourteron and La Sabotterie. The Pz.Kpfw. IV was able to drive off the enemy scout cars. One light tank and one armored ammunition tractor were destroyed. After mines in the place were removed by the motorcycle riflemen, the spearhead continued its advance again. The chief's vehicle drops out, as it received two hits, one of which damaged the tank gun. The spearhead now consists only of 3 Pz.Kpfw.II (1st Platoon) and one Pz.Kpfw.IV (Lt. Lucas). The advance went on in the old order of march, scouting troop of motorcycle rifles, tanks, motorcycle rifle company. Near Sorcy-Bauthemont the spearhead met strong resistance. One Pz.Kpfw.II of the 1st platoon took several antitank hits. The crew had to disembark. The vehicle then took several additional antitank hits, 19 in all. The only Pz.Kpfw.IV in the spearhead opened fire and put down several antitank guns. After all the high explosive and antitank shells had been fired, fog cartridges were laid on the edge ovf town to conceal the withdrawal. The crew of the Pz.Kpfw.IV also had to disembark, as the vehicle had become immobile through technical problems and the gun had also gone to pieces when it fired the last shells. Motorcycle riflemen took over the protection of the tank. Through the battalion, vehicles of the 2nd Company were summoned to get the 4th Company out. After the enemy resistance was broken, it turned out that one antitank company with motorcycle riflemen was in position in the town.

Report of the 8./Tank Regiment 1 on the combat near St. Folouin on May 24, 1940

Task: One platoon of the 4th Company (2nd Platoon), subordinated to the 8th Company, was to help the riflemen move forward to expand the bridgehead. The reconnaissance for this undertaking was carried out in the morning by Lt. von Villebois and Lt. Lucas.

Carried out: The 1st canal, which had been taken by us already the day before, is overcome without difficulty. The tanks gave the riflemen fire cover to help them get to the second canal. Then the tanks moved up to the canal and gave the riflemen fire cover from this side and to cross and extend the bridgehead. The attack went forward very smoothly at first, but then got stuck, as the riflemen had to cross the canal in float bags. The Pz.Kpfw.IV meanwhile shot down with explosive shells one house after another in which the French had nested with their machine guns. In this undertaking the Pz.Kpfw.IV carried out an absolute artillery task in closest cooperation with the riflemen. The task was fulfilled to its full extent, for the riflemen were able to extend the bridgehead without losses.

16.3 Balkan Campaign as of April 6, 1941 on

In April 1941 every armored division had only one tank regiment. The medium companies consisted at that time of one light platoon with 5 Pz.Kpfw.II and either eight or ten Pz.Kpfw.IV with the group leader and two platoons.

Only a few Pz.Kpfw.II were lost to mines or enemy antitank guns during the Balkan campaign. The main problems were mechanical breakdowns (brakes and steering gear) in almost all Pz.Kpfw.IV that took part in the long march through the mountains.

16.4 Operation Barbarossa as of Jule 22, 1941, on

At the beginning of the great offensive against the Soviet Union, all medium tank companies were organized according to K.St.N.1175 of 2/1/1941. But the preparation had only managed to equip every medium tank company with 10 Pz.Kpfw.IV. Thus, the 3rd platoons of the medium tank companies were not set up. Every unit was completely supplied with the authorized number of 10 Pz.Kpfw. IV per medium company.

Instead of ten, twelve Pz.Kpfw.IV were on hand in the III. Battalion of Tank Regiment 6 and all three battalions of Tank Regiment 18. [p. 280] These 48 Pz.Kpfw.IV (diving) were originally included as 4 Pz.Kpfw.IV (7.5 cm) (T) in the 3rd platoon of each light tank company b (K.St.N.1171b of 7/16/1940). But they were reassigned to regular medium tank companies under K.St.N.1175 before the Russian campaign began.

On June 22, 1941, there were 441 Pz.Kpfw.IV reported as the starting strength of the armored divisions.

16.5 Summer Offensive in the East as of June 1942 on

After the winter of 1941-42 had been survived, armored divisions and motorized infantry divisions were refreshed in the summer offensive under the command of Army Group South. The medium tank companies, structured according to K.St.N.1175 of 11/1/1941, were only supposed to receive 10 Pz.Kpfw.IV (still without the 3rd platoon).

In the tank units in the east there were still numerous Pz.Kpfw. II with the 7.5 cm Kw.K. L/24 (circa 222 in June 1942). The units that were to lead the offensive were assigned 8 to 12 Pz.Kpfw. IV with the 7.5 cm Kw.K.40 L/43. At the beginning of the great offensive on June 28, 1942 there were 135 Pz.Kpfw.IV with 7.5 cm Kw.K.40 L/43 on hand in the tank nits on the eastern front.

The "Grossdeutschland: and "Leibstandarte Adolf Hitler" SS armored units were the first units to be set up with three medium tank companies. The normal structure of the tank battalions in 1942 consisted of two or three light tank companies and one medium tank company. Each of these medium tank companies received 10 Pz.Kpfw.IV and 3 Pz.Kpfw.II.

The next unit that was to be set up with three medium tank companies was Tank Battalion 138, in December 1942. It still received only 8 Pz.Kpfw.II for the staff and light platoon, plus 30 Pz.Kpfw.IV with 7.5 cm Kw.K.40 (10 for each medium company).

The Pz.Kpfw.IV that proved so superior to the approximately 14,000 Russian BT and T 26 series tanks was not able to penetrate the armor of the T 34 and KW I with normal antitank shells. In February 1942 Hitler finally released the hollow-core shell, the 7.5 cm Gr. Patrone 38. This

Structure of the medium tank company, according to K.St.N.1175(Sd.) of 2/1/1941.

A shot-down Pz.Kpfw.IV Type A in the east. A typical example of maintaining fighting value through extra armor.

type of ammunition, now called the HEAT Shell, was already ready for use before April 1, 1941, but was withdrawn from the front so as not to make it known to the enemy too soon. Its immediate imitation was feared.

Tank Regiment 203 reported that the 7.5 cm Gr.Patr.38 could pierce the hull side of a T 34 between the 5th and 6th road wheels at a range of 150 to 800 meters and the MG mount in the bow at 150 to 400 meters. The 75 mm armor of the KW 1, though, withstood the HL/A version of the hollow shell (penetration performance of 45 mm at a 60-degree striking angle).

The Ziervogel Tank Battalion of the 3rd Armored Division reported shooting down enemy tanks in the time from May 12 to 22, 1942: Six Pz.Kpfw.IV (7.5 cm Kw.K. L/24) scored the following results with 7.5 cm HL shells (improved type of 7.5 cm Gr.Patr.38 HL/B): 5 KW 1 made incapable of fighting (no penetration), then blown up, a total of 24 T 34 and 2 Infantry Tank Mark II shot down, their own total losses 3 Pz.Kpfw.IV.

The following portrayal comes from the combat report of NCO Heinz Weller, commander of a Pz.Kpfw.IV with 7.5 cm Kw.K. L/24 in the 8th Company of Tank Regiment 3, of August 13, 1942:

I received the order to advance in a corridor that led to a position ahead of Point 208 as far as the edge of the woods. At the same time two Pz.Kpfw.III drove farther on the wood road leading to the north and fired into the woods where enemy tanks were suspected on the basis of engine noise. An armored grenadier reported that the T 34 firing on Point 208 turned its gun in that direction from which he received fire. This was for me the signal to advance to the free area and put the T34, now spotted by me with no problem, out of action with hollow shells at a range of about 25 meters. After [p. 283] ending the battle I viewed the now burned-out tank and spotted a lighter tank right next to it, which had also taken a hit from one of my hollow shells and was also burned out.

The 5th Armored Division, which ad only four Pz.Kpfw. IV with 7.5 cm Kw.K.40 L/43 (one of which was lost in action) reported:

In the time from 2/22 to 3/20/1943 we shot down enemy tanks in the area north of Shisdra: Pz.Kpfw.IV with 7.5 cm Kw.K.40: 17 KW 1, 26 T 34, 1 T 26, 1 Mark II, 3 Mark III and 1 General Lee. With Pz.Gr.39 at ranges of 1200 to 1600, two or three shots per tank, destructive fire caused by every hit. With Gr.HL/B ammunition seldom a fire, here 1 to 5 shots per tank.

Pz.Kpfw.IV Type C in the east on June 29, 1941, with the 8th Company of Tank Regiment 10 of the 8th Armored Division.

16.6 Campaign in North Africa, 1941 to 1943

In February 1941 Tank Regiment 5 transferred to Tripoli in Libya and arrived there with 17 of its 20 Pz.Kpfw.IV (3 were lost in a fire on board the ship). It was followed in April 1941 by 20 Pz.Kpfw. IV of Tank Regiment 8 of the 15th Armored Division. The Abt.1a of the German Africa Corps reported on August 11, 1942:

Tactical Experience and Suggested Modifications to the Pz.Kpfw.IV with 7.5 cm Kw.K. L/43

1) *From the first action on, the gun with its high penetrating power and targeting precision showed itself to be superior to all weapons previously installed in tanks. The Pzgr. penetrates at 1500 meters all American and British tank types used in the African Theater of War—even the "Pilot"—frontally.* **(Pilot = American M3 Medium Tank).** *The targeting precision decreases (poor target observation because of shimmering). With clear sight destructive effect is achieved even up to 2000 meters against the light tank types.*
2) *The vehicle was very soon recognized by the enemy as especially dangerous and, because of its striking shape, drew concentrated fire from aircraft, artillery and antitank guns. Thus it is necessary to cover this vehicle with 5 cm wagons. In general, only when approaching inviting targets, such as "Pilots." Should it enter into a firefight. The flank protection then becomes especially important. Thus it does not appear to be purposeful always to deploy the vehicles in formation.*
3) *Apportioning to scouting troops and use for flank protection should be dropped. Only focal-point tasks should be considered.*
4) *It is usually wrong to fire more than a few shots from a position. As the muzzle flash and the especially great development of dust very quickly attracts the combined fire of the enemy artillery.*
5) *The Pz.Kpfw.IV Special is not to be assigned as a command car, but equipping it with receivers and senders is to be striven for.*
6) *As long as the vehicles are available in these small numbers, it is necessary to take ammunition to them in armored vehicles during combat. The divisions test the possibility of ammunition supplying of these tanks with Carrettes, M.T.V. or other armored vehicles.*
7) *Muzzle flash and dust make observing shots more difficult, especially on desert ground.*

Pz.Kpfw.IV Type D (diving) in the winter of 1940-41 with the 5th and 6th companies of Tank Regiment 18 of the 18th Armored Division.

Pz./Kpfw.IV Type G in the summer of 1942 with Tank Regiment 4 of the 13th Armored Division.

Pz.Kpfw.IV Type D, Chassis No. 80512, completed by Krupp in October 1939. The picture was taken n Libya in the summer of 1941.

8) *To date the gun barrel extending far over the bow has not proved to be disadvantageous, even in rough terrain.*
9) *For the running gear a softer suspension is desirable to absorb the rough shocks in stony terrain.*
10) *The armor plates on top of the turret and on the armored body should be strengthened. Shot deflectors on the turret roof to protect the commander's cupola.*
11) *Attachment of frontal support for the gun barrel in a lowered position. If possible, this barrel lashing should be releasable from inside, so that driving with a lashed barrel can be done until the last moment.*
Reason: From shots to the rear, the vibration of the barrel is so strong that the guide rings in the barrel break out.
12) *Attachment of another ventilator for better and faster removal of powder gas.*
Reason: Because of the powder gas, observing from the vision slit is almost impossible.
In addition, the sharp odor of the departing powder gas bothers the crew considerably. Sensitive people tend toward nausea (CO danger). For the same reason, quick dispersal of the cartridge cases from the fighting compartment is advisable.

Other Armored Units that were transferred to North Africa were:

Tank Battalion 190 with 4 Pz.Kpfw.IV (two others were sunk) in November 1942,

Tank Regiment 7 *of* the 10th Armored Division with 8 Pz.Kpfw.IV (12 others were sunk) in November and December 1942, and

3rd Company, Tank Regiment "Hermann Göring" with 8 Pz.Kpfw.IV.

Pz.Kpfw.IV Type F2 (G), Chassis No. 82440, in action in June 1942 with the German Africa Corps.
The captain is surprised to see the water canisters attached to the turret. The vehicle was completed by Krupp in April 1942.

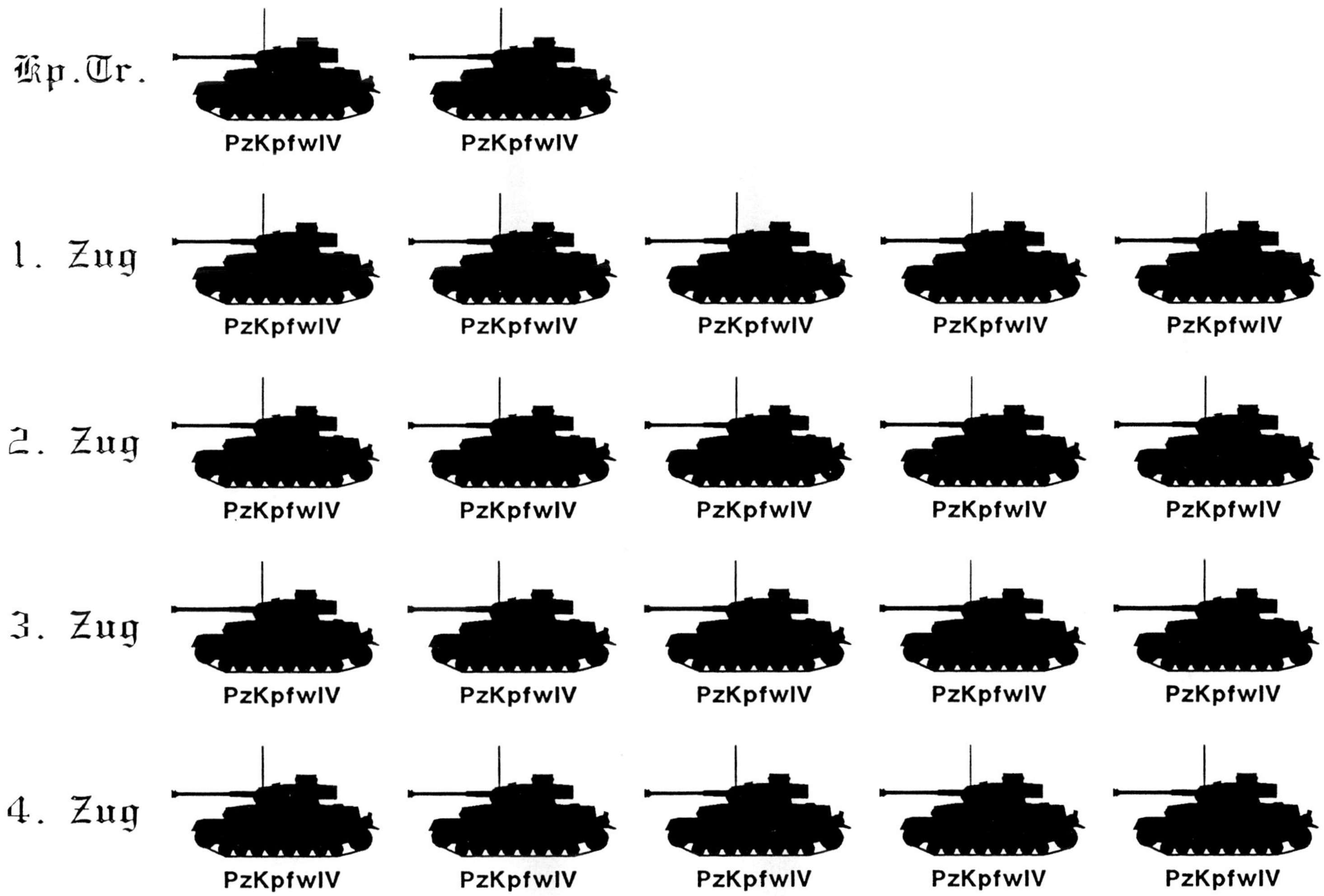

Structure of as Medium Tank Company a, according to K.St.N.1175a of 1/25/1943.

Pz/Kpfw.IV Type H in the 2nd SS Armored Division "Das Reich" in the winter of 1943-1944.

16.7 In the East as of July 1943, on in the time of "Operation Citadel"

Before July 5, 1943, all tank regiments that were to see action in the battle for the Kursk Bend had been strengthened by assigning of Pz.Kpfw.IV. The structure of the medium tank companies was still supposed to follow K.St.N.1175 of 11/1/1942, with 14 Pz.Kpfw. IV. But in reality, in almost all tank battalions the Pz.Kpfw.IV were distributed evenly to every company (light or medium) along with Pz.Kpfw.III with 5 cm Kw.K. L/60 and 7.5 cm Kw.K. L/24.

By July 5, 1943, only three tank battalions (the 4th, 5th and 13th Armored Divisions) had received authorization to restructure according to the new Medium Tank Company a (K.St.N1175a of 1/25/1943) with 22 Pz.Kpfw.IV per company. In the strength reports of July 1, 1943, 56 Pz.Kpfw.IV with 7.5 cm Kw.K. L/24, 583 Pz.Kpfw.IV with Kw.K. 40 L/43 and 302 Pz.Kpfw.IV with Kw.K.40 L/48 were on hand in the tank units in the east.

The OKH had ordered in OKH Gen.St.d.H./Org.Abt. No.1/2614/43 g.Kdos. of 6/14/1943 "Restructuring of Tank Regiments and Battalions of the Eastern Army":

By December 1943 it will presumably be possible to fill one tank battalion in each armored division of the Eastern Army with 96 Pz.Kpfw. (mainly Pz.Kpfw.IV, some long Pz.Kpfw.III).

Along with that it is intended to set up a Panther battalion for the mass of the armored divisions in the homeland war area. Restructuring the tank battalions of the Eastern Army is necessary. All tank regiments or tank battalions of the Eastern Army are to be set up or restructured according to the following structure:

Regimental Staff (if still present)	*K.St.N.1103 of 11/1/41*
Battalion Staff	*K.St.N.1107 of 4/1/43*
Staff Company	*St.N.1150 of 1/25/43*
Armored Flame Platoon (of Staff Company, subordinated, set up only on special command)	*K.St.N.1190 of 1/25/43*
4 medium tank companies	*K.St.N.1175a of 1/25/43*
Workshop Company	*K.St.N.1187b of 4/25/42*

Pz.Kpfw.IV Type H in the east, summer 1944.

By December 1943 most tank units of the Eastern Army were restructured to the new organization, which now had only four medium tank companies (some of them only with authorization for 17 Pz.Kpfw.IV per company). Exceptions were units like the tank battalion of "Feldherrnhalle," with three companies of 14 Pz.Kpfw.14 each, the III. Battalion of Tank Regiment 24, the III. Battalion of Tank Regiment 36 with two companies of 22 Pz.Kpfw. IV, and two companies with 22 assault guns.

16.8 From June 1944 on in East and West

On June 1, 1944, the 16 armored divisions and two armored grenadier divisions that awaited the start of the great Russian offensive (which began on June 22, 1944, the third anniversary of Operation "Barbarossa") were equipped with 605 Pz.Kpfw.IV (full strength was 1502 Pz.Kpfw.IV).

Awaiting the Allied landing in the west on June 6, 1944 were eleven armored divisions equipped with 863 Pz.Kpfw.IV (full strength was 965 Pz.Kpfw.IV). Additional replacement tanks were assigned to the 9th and 11th armored divisions, plus the 1st and 2nd SS armored divisions, before they were transferred to the front.

On 6/26/1944 the Inspector-General of the Armored Troops, General Guderian, advocated the continued manufacture of Pz.Kpfw.IV as follows:

Note on the Führer's Speech on 6/26/44
Converting Pz.Kpfw.IV to Assault Gun L/48 or L/70

1) *All available troop experiences state unanimously and steadily that the tank with the turnable turret is preferred to the assault gun in the armored units for action within the tank regiment. There is no known report that contradicts this.*
2) *All hitherto appearing enemy tank and assault gun types can be shot down with our current armor-piercing weapons, including 7.5 cm L/48/*
3) *Conversion of Pz.Kpfw.IV production to assault guns would result in the supplying of the existing Pz.IV battalions would have to consist of assault guns. This would result in our again having two different types of vehicles within the Pz.IV battalions, their action bases being inherently different. The Inspector-General of the Armored Troops therefore suggests that the production of Pz.Kpfw.IV, which is just sufficient at this time to cover the losses of the field battalions, be allowed to continue until the supply of Pz.Kpfw.V (Panther) tanks is so great the equipping of the 2nd Panther Battalion within the tank regiment can be started.*
   In the end, then, the Panther production must be increased to about 900 Panthers per month.

.Kpfw.IV Type H with the 6th Company of SS Tank Regiment 12 of the 12th SS Armored Division "Hitler Youth" on the invasion front, summer 1944.

Beyond this, it is suggested that, beginning as of August, according to the production of the 7.5 cm Cannon L/70, the Light Tank Destroyer IV be supplied, and as of November all Tank Destroyer III/IV (uniform chassis) be equipped with the 7.5 cm Pak L/70.

S/Guderian

The Pz.Kpfw.IV with the 7.5 cm Kw.K.40 was superior at all times to the US M 4 "Sherman" with the 7.5 cm M3 gun and the Russian T 34/76. But as is shown in the following comparison by Wa Prüf 1 of 10/5/1944, after the appearance of the newest enemy T 34/85, KW 85 and JS 122 tanks, plus the Sherman with the 76 mm M1A1 gun or 17 pounder, it was inferior.

16.9 December 16, 1944, in the West

For the last great offensive in the west as well, the Pz.Kpfw. IV again had to be depended on, though only to a limited extent. In the new organization, the numbers of Pz.Kpfw. IV were limited. The "Panther" or "IV" Tank Company (K.St.N.1177(fG) of 11/1/44), Version A, showed 17 tanks per company (two for group leaders and five per platoon), Version B had 14 tanks (two for group leaders and four per platoon). Most armored divisions were equipped with only two tank companies with Pz.Kpfw.IV. In all, 349 Pz.Kpfw.IV were reported in the original strength of the 11th Armored Division and two tank battalions.

Further losses in production and increasing losses on all fronts brought a further decrease. At the end, the organization of the "Panther" or "IV" Tank Company (K.St.N.1177a(fG) of 4/1/45) showed a structure of ten tanks (one for company leader and three per platoon).

16.10 Pz.Kpfw.IV Exports

In trade for raw materials and to support the morale of the German Reich's allies' troops, Pz.Kpfw.IV were delivered to the following countries as shown:

Delivery	Country	Pz.Kpfw.IV	Delivered by Army Armory
February 1943	Bulgaria	3	
March 1943	Bulgaria	3	
April 1943	Bulgaria	6	
May 1943	Bulgaria	4	5/6/43
June 1943	Bulgaria	15	6/30/43
July 1943	Bulgaria	15	8/15/43
August 1943	Bulgaria	15	9/8/43
September 1943	Bulgaria	15	10/28/43
December 1943	Bulgaria	3 sent	8/3/44
	Total:	**91**	

Delivery	Country	Pz.Kpfw.IV	Delivered by Army Armory
May 1943	Italy	12	5/10/43
September 1942	Romania	11	
November 1943	Romania	4	11/20/43
December 1943	Romania	24	12/28 & 30/43
January 1944	Romania	19	2/4 & 6/44
February 1944	Romania	2	2/26/44
March 1944	Romania	15	4/6/44
April 1944	Romania	15	5/11/44
May 1944	Romania	15	6/2/44
June 1944	Romania	15	7/4/44
July 1944	Romania	3 sent	7/26/44
August 1944	Romania	15	8/14/44
	Total:	**138**	

Delivery	Country	Pz.Kpfw.IV	Delivered by Army Armory
August 1944	Finland	10	8/16/44
August 1944	Finland	10	8/19/44
	Total:	**15***	

* Only 15 were reported as having arrived.

Delivery	Country	Pz.Kpfw.IV	Delivered by Army Armory
August 1943	Spain	12	11/19/43
October 1943	Spain	8	11/21/43
	Total:	**20**	

Delivery	Country	Pz.Kpfw.IV	Delivered by Army Armory
March 1943	Turkey	7	
May 1943	Turkey	8	
	Total:	**15**	

Delivery	Country	Pz.Kpfw.IV	Delivered by Army Armory
May 1942	Hungary	22 L/24	
September 1942	Hungary	10 L/43	
August 1944	Hungary	20	8/27/44
September 1944	Hungary	10	9/16/44
October 1944	Hungary	10*	10/10/44
October 1944	Hungary	10	12/17/44
November 1944	Hungary	10**	12/8/44
December 1944	Hungary	12	1/6/45
	Total:	**84**	

* to 1st Armored Division
** to 24th Armored Division

Pz.Kpfw.IV Type G turned over to Turkey in March and May 1943.

Comparison

Pz.Kpfw.IV with 7.5 cm Kw.K.40 L/48 against Enemy Tanks

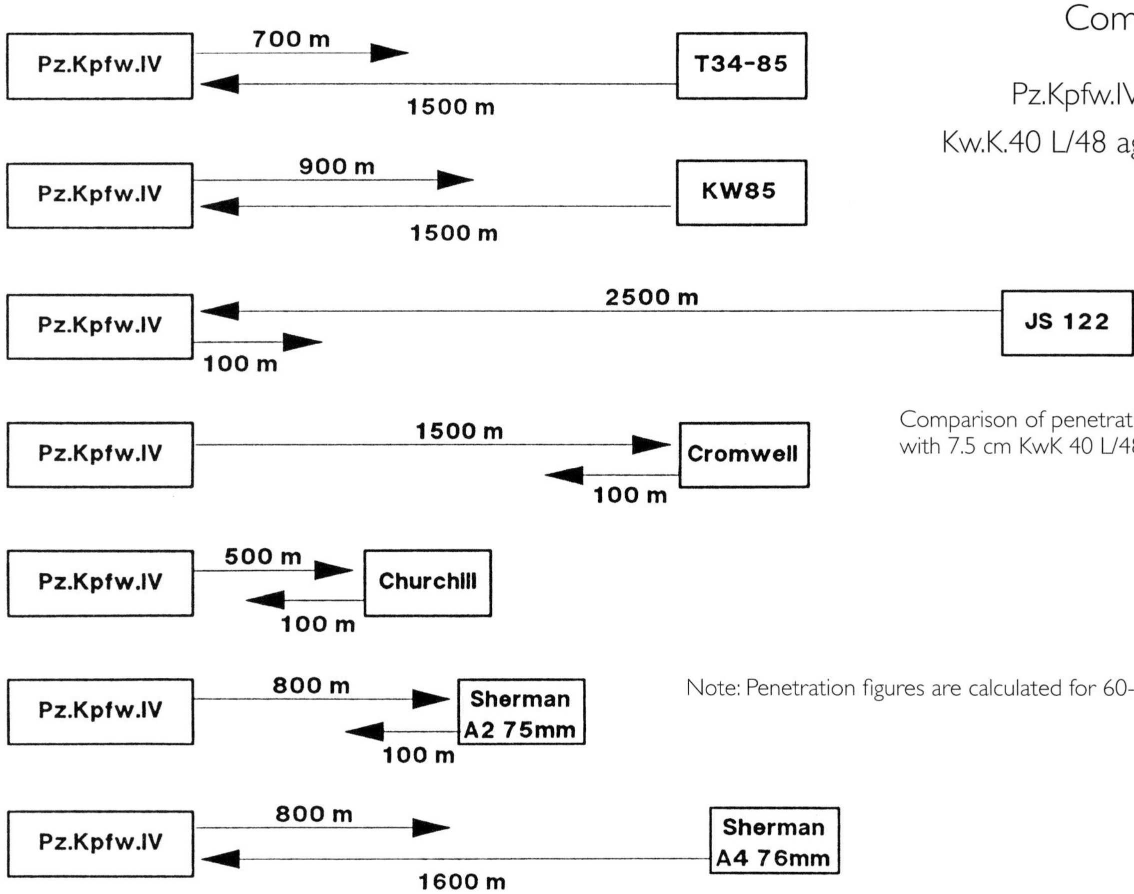

Comparison of penetrating power of Pz.Kpfw.IV with 7.5 cm KwK 40 L/48 and enemy tanks.

Note: Penetration figures are calculated for 60-degree side angle, front shot.

Pz.Kpfw.IV Type H meets a group of grenadiers in the east in 1944.

Pz.Kpfw.IV Type F, turned over to the Hungarian Army in May 1942.

Pz.Kpfw.IV Type G, turned over to the Hungarian Army in September 1942.

Pz.Kpfw.IV Type H, Chassis No. 84611, completed by Krupp in September 1943, turned over to Spain in November 1943. Back from Spain, the vehicle is now in the Military Technical Study collection in Koblenz.

Pz.Kpfw.IV Type H, turned over to the Bulgarian rmy in 1943.

Pz.Bef.Wg.IV Type J, turned over to the Bulgarian Army in August 1944.

Pz.Kpfw.IV in action with the Romanian Army.

Pz.Kpfw.IV Type G, turned over to the allied Croatian Army for training purposes.

Pz.Kpfw.IV Type J with the Czechoslovakian forces after the war ended. Other such equipment is also known.

Pz.Kpfw.IV Type J with the Syrian forces after 1945.

Pz.Kpfw.IV Type J, Chassis No. 91810, completed by the Nibelungen Works in July 1944, turned over to Finland in August 1944. The picture was taken at a museum in Parola, Finland.

Appendix

Modifications to the Road Wheels.

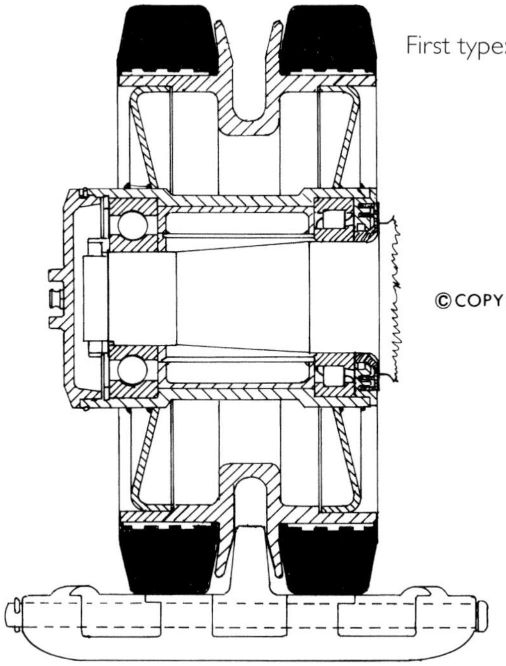

First type: Kgs 6110/380/120 tracks

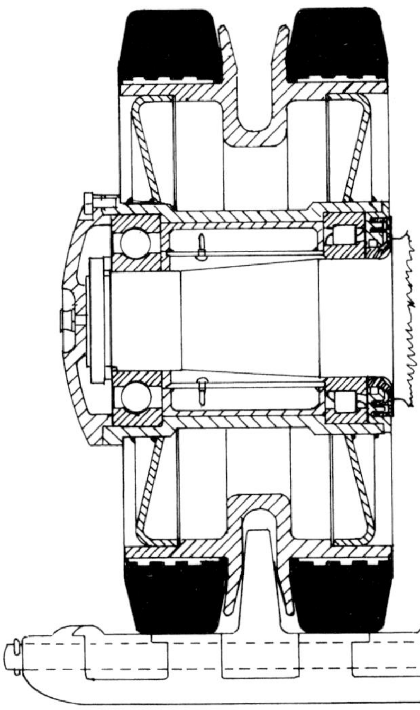

Improved version of road wheels as of Pz.Kpfw.IV Type E.
This track type: Kgs 6111/380/120, had been introduced previously.

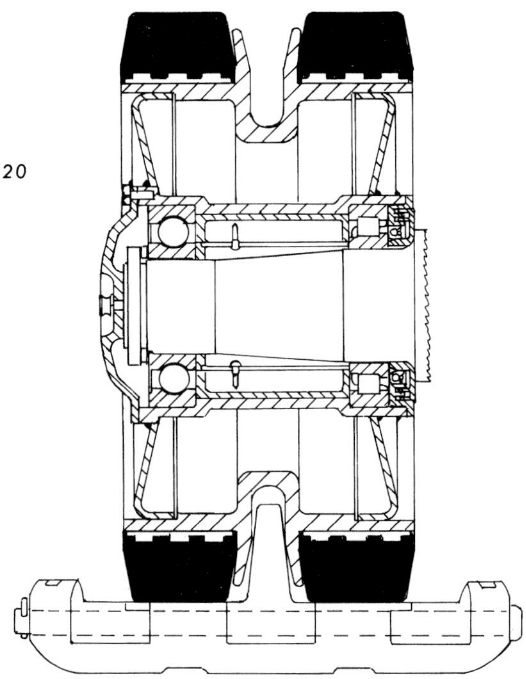

The wider type of road wheels along with the Kgs 61/400/120 tracks (Pz.Kpfw.IV Type F) had wheel hubcaps of pressed sheet steel instead of the earlier armored type.

TECHNICAL DATA FOR PZ.KPFW. IV TYPES A TO E

Armor Thickness (mm)	Degrees	Type A	Type B	Type C	Type D	Type E
Hull						
Upper front wall	(18°)	10	20	=	=	=
Forward front wall	(76°)	14.5	30	=	30 (+30)	50
Lower front wall	(29°)	14.5	20	=	20(+20)	30
Sidewall	(90°)	14.5	=	=	20(+20)	20+20
Upper back wall	(80°)	14.5	=	=	20	=
Lower back wall	(16°)	10	=	=	14.5	=
Bottom	(0°)	8	=	=	10	=
Roof	(0-5°)	11	=	=	=	=
Bow Armor						
Front wall	(81°)	14.5	30	=	30(+30)	30+30
Sidewall	(90°)	14.5	=	=	20(+20)	20=20
Roof	(0-5°)	11	=	=	=	=
Back wall	(90°)	14.5	=	=	20	=
Bottom plate	(0°)	8	=	=	=	=
Stern Armor						
Roof	(2°)	10	=	=	=	=
Back wall	(80°)	14.5	=	=	20	=
Sidewall	(90°)	14.5	=	=	20	=
Turret						
Front wall	(80°)	14.5	30	=	=	=
Mount cover	(round)	14.5	30	=	35	=
Sidewall	(65°)	14.5		=	=	20=
Back wall	(63°)	14.5		=	=	20=
Roof	(0°)	10	=	=	=	=
Main Dimensions (mm)						
Length overall		5920	=	=	=	=
Width overall		2830	=	=	=	=
Height with turret		2680	=	=	=	=
Ground clearance		400	=	=	=	=
Wading capability		800	=	=	=	=
Ditch-spanning capability		2300	=	=	=	=
Track width		380	=	=	=	=
Track division		120	=	=	=	=
Tracks touching ground		3515	=	=	=	=
Track of vehicle		2390	=	=	=	=
Track/length on ground		1 : 1.48	=	=	=	=
Weight (kg)						
Vehicle		18,000	18,500	=	20,000	22,000
Power Components						
Motor		HL 108	HL 120	=	=	=
Gearbox		SSG75	SSG76	=	=	=
Steering gear		Planetary	=	=	=	=
Running gear		8 road wh.	=	=	=	=
Springs		Leaf	=	=	=	=

TECHNICAL DATA FOR PZ.KPFW. IV TYPES F TO J

Armor Thickness (mm)	Degrees	Type F	Type G	Type H	Type J
HULL					
Upper front wall	(18°)	20	=	=	=
Forward front wall	(76°)	50	50(+30)	80	=
Lower front wall	(29°)	30	=	=	=
Sidewall	(90°)	30	=	=	=
Upper back wall	(80°)	20	=	=	=
Lower back wall	(16°)	14.5	=	=	=
Bottom	(0°)	10	=	=	=
Roof	(0-5°)	11	=	=	=
BOW ARMOR					
Front wall	(81°)	50	50(+30)	80	=
Sidewall	(90°)	30	=	=	=
Roof	(0-5°)	11	=	=	11(16)
Back wall	(90°)	20	=	=	=
Bottom plate	(0°)	8	=	=	=
STERN ARMOR					
Roof	(2°)	10	=	=	=
Back wall	(80°)	20	=	=	=
Sidewall	(90°)	20	=	=	=
TURRET					
Front wall	(80°)	50	=	=	=
Mount cover	(round)	50	=	=	=
Sidewall	(65°)	30	=	=	=
Back wall	(63°)	30	=	=	=
Roof	(0°)	10	10	16-25	=
MAJOR DIMENSIONS (mm)					
Length overall		5920	6630	7015	=
Overall width		2880	=	=	=
Height with turret		2680	=	=	=
Ground clearance		400	=	=	=
Wading capability		800	=	=	=
Ditch-spanning capability		2200	=	=	=
Track width		400	=	=	=
Track division		120	=	=	=
Tracks on ground		3515	=	=	=
Track of vehicle		2456	=	=	=
Track/length on ground		1 : 1.43	=	=	=
WEIGHT (Kg)					
Vehicle		22,300	2,600	25,000	=
POWER COMPONENTS					
Motor		HL 120	=	=	=
Gearbox		SSG76	=	=	=
Steering gear		Planetary	=	=	=
Running gear		8 road wh.	=	=	=
Springs		Leaf	=	=	=